COMPUTER SCIENCE, TECHNOLOGY AND APPLICATIONS

FOCUS ON BIOMIMETICS RESEARCH

Computer Science, Technology and Applications

Additional books in this series can be found on Nova's website under the Series tab.

Additional e-books in this series can be found on Nova's website under the e-book tab.

COMPUTER SCIENCE, TECHNOLOGY AND APPLICATIONS

FOCUS ON BIOMIMETICS RESEARCH

ABELE SCIPIONE
EDITOR

nova publishers
New York

For permission to use material from this book please contact us:
Telephone 631-231-7269; Fax 631-231-8175
Web Site: http://www.novapublishers.com

Library of Congress Cataloging-in-Publication Data

ISBN: 978-1-62417-332-5

Published by Nova Science Publishers, Inc. † New York

CONTENTS

PREFACE

In this book, the authors present current research in the study of biomimetics. Topics include the recent advances of microsystems and their application in developing new tools and techniques for investigating bone and bone-related phenomena; adhesives inspired in biological systems; biomimetic photonic crystal from butterfly wings; biomimetic molecular switches; a voltammetric sensor based on a hemin modified multiwalled carbon nanotube for hydroquinome determination in pharmaceutical samples; and trajectory formation of biomimetic robot arms with redundancy.

Chapter 1 – In the human body, bone is of tremendous importance. Studies have demonstrated the many roles of bone in providing structural support for the body, allowing for locomotion and movement, protecting vital organs and storing minerals. While the roles for bone have been established many bone phenomena are not well understood such as how bone heals itself, the mechanisms by which it adapts to altered loading environments and the metabolic diseases that result from breakdown in normal bone function. Significant research effort has been focused on understanding the interactions of the three primary bone cell types (osteoblasts, osteoclasts, and osteocytes) and their ability to coordinate activity in bone maintenance, function, repair and remodeling. However, at the cellular level bone has remained a difficult organ to model in part due to a high degree of spatial and temporal variation among these three bone cell types *in vivo* and the difficulty in isolating the impact of specific signals on these cells. These and other limitations have contributed to the delayed development of highly biomimetic models for studying bone cell phenomena. However, the use of microsystems, small devices with fluid channels ranging on the scale of nano and micro meters, is increasing for many biological systems and may permit opportunities for the study of bone-related phenomena. The precision with which microsystems can be fabricated allows for very specific geometries within the small channels, providing a means of performing a variety of experiments and analyses quickly with high resolution and sensitivity. For biological experiments these properties offer a number of unique advantages such as increasing feasibility of single cell experiments and manipulation of small volume analytes, allowing isolation of individual factors on cell behavior, and enabling precise control over the spatial and temporal organization of cells. These features suggest that microsystems have the potential to serve as powerful tools in the development of biomimetic models for bone. These models could potentially help to elucidate the mechanisms by which bone cells sense and respond to mechanical signals in their environment. Harnessing these unique mechanisms could translate to real world applications in both the public and private technology sectors.

This chapter will discuss the recent advances of microsystems and introduce their application in developing new tools and techniques for investigating bone and bone-related phenomena.

Chapter 2 – Over the years, the conception of new materials has been inspired by several processes developed by Nature over millions of years, that presenting remarkable properties. For example Velcro, a kind of fabric made of nylon and mainly used as a fastener for clothing, bags and shoes, was developed from the interlocking mechanism of burdock seeds. Similarly, self-cleaning materials have been developed based on "Lotus effect" which describes the way in which water drops roll off the super hydrophobic leaves of lotus plants, taking dirt particles away with them. A concise review of adhesives inspired by biological systems is presented, with a detailed comparison of the different components of the adhesion, as well as the potential final applications.The protein-based underwater adhesives produced by aquatic organisms, such as blue mussels, sea stars, acorn barnacles, sea urchins, freshwater caddis fly larvas and sandcastle worms (*phragmatopoma californica*) will also be reviewed, along with high-performance adhesives for aqueous milieu, based on the glue secreted by sandcastle worms and used for hold neighbouring particles which then assemble into hollow macroscopic housing colonies to provide protection from predators. The development of surgical adhesives inspired by marine mussels for different applications (prosthetic mesh fixation, islet transplantation, etc.) will also be addressed. Several approaches for creating a responsive fibrillar adhesive system inspired by the gecko, using different materials, have been shown to possess the capacity to generate strong adhesions in dry conditions. Based on that, several researchers have synthesized gecko foot-hairs as dry adhesives for wall-climbing robots. The ability of leaf beetles to climb smooth and rough surfaces using an array of micron-sized adhesive hairs (setae), of varying morphology, which help male beetles achieve strong adhesion to the elytra of females during copulation, can be used as inspiration to develop new adhesives. A successfully developed, spider-inspired hydrophobic hair structure for forming air shield underwater suggests promising applications in underwater protection and adhesives integration. The study of the pads of different insect species leads to the conclusion that smooth pads are specialized for smoother substrates, while rough pads generate stronger forces on a broader range of surfaces and have also inspired scientists in a number of ways. Tree frogs secrete a wetting liquid into the pad substrate contact area, which generates a relative long-range attractive interaction due to the formation of capillary bridges. This system is relevant for some technological applications, for instance, tires for passenger cars have drainage channels to speed up the removal of water from the tyre-road footprint area during wet road conditions. Noteworthy is the fact that it is also possible to successfully develop a biomimetic structure that adheres equally well in wet and dry conditions, for example by combining the microstructure of gecko pads with a thin layer of synthetic polymer that mimics the protein glue of mussels. The bibliographic search now presented individualizes the scope of actuation and prevalence of different adhesion mechanisms showing how they can be used for the development of new adhesion systems.

Chapter 3 – A series of approaches have been developed for the replication of photonic structures from *Morpho* butterfly wings. By selecting appropriate precursors, we can achieve exact replications of photonic structures in a variety of metal oxides, such as titania, tin oxide, silica, WO_3, etc. Interestingly, polymers such as polypyrrole (PPy) with photonic crystal structures were reported for the first time to be duplicated from *Morpho* butterfly wings by using a two-step templating process. The exact replications at the macro, micro- and even nanoscale levels were characterized by a combination of FE-SEM, TEM, EDX and Raman measurements. The optical properties of the replicas were investigated by using reflectance spectroscopy, and it was found that the interesting chromaticity of the reflected light could be adjusted simply by tuning the replica composition. The primary chemical sensor or biosensor properties of biomimetic materials thus prepared were investigated, which showed a much high performance owing to the combination of large surface area with photonic crystal structures. These reported biomimetic techniques are effective to produce photonic crystals from nature with three dimensional networks, which may open up an avenue for many passive and active infrared devices, especially high performance optical components, sensors and magneto-optical devices.

Chapter 4 – Science has always looked for inspiration in the highly effective function of biological systems. At the same time, the aspiration to mimic their functioning has been a driving force for the advancement of chemistry and material science. Light represents a usual stimulus found in very different organisms, triggering and driving diverse biological processes. In higher organisms, the process of visual perception is extremely important. But also other processes such as photosynthesis in plants and photomovement responses imply the interaction between light and biological media to provide complex responses. The basic concept behind such different behaviors is centered on photoresponsible proteins capable of regulating the biological activity by light. Thus, the use of light to control the activity of biomolecules, as well as to mimic their function, is an attractive strategy for both studying and controlling interactions in complex living systems. Photoactive biomolecules can be introduced into a cell and manipulated noninvasively using light. This allows for a high degree of spatial and temporal control. Among the different types of molecules that can be used, photoswitches are a class of chemical compounds that undergo reversible photochemical behavior so that many rounds of active/inactive states may be produced. In addition to this, it is possible to find a wide range of molecules that experience a photoswitchable behavior. In this contribution, we will describe the synthesis and photochemical properties of different families of photoswitches whose structure is based on that of natural chromophores. Also, the properties that make them attractive for the photocontrol of biomolecules will be outlined.

Chapter 5 – A simple, low cost and sensitive electrochemical method for the detection of hydroquinone (HQ) was developed based on a carbon nanotube paste (70%) electrode modified with hemin (20%) and mineral oil (10%). The aim of the present work was to evaluate the performance of a hemin (iron(II) protoporphyrin IX) functionalized carbon nanotube sensor for the voltammetric determination of HQ. The modified electrode demonstrated an electrocatalytic activity for the HQ oxidation in a 0.05 M Tris-HCl buffer solution (pH 7) using cyclic voltammetry. The anodic peak current of the sensor was 4 and 2.8 times higher when compared to glassy carbon and non-modified carbon nanotube paste electrodes, respectively. The best paste composition was found to be 3.5:1 (MWCNT:Hemin), considering the smaller separation between the cathodic and anodic peaks (ΔE) as well as the larger peak currents. The linear plot of anodic peak current *vs.* square root of scan rate demonstrated that the HQ oxidation was a diffusion-controlled process. The square wave parameters were optimized, and the best results obtained were as follows: amplitude = 70 mV, frequency = 70 Hz, and step of potential = 0.01 V. 0.15 mol L^{-1} Tris-HCl was found to be the best buffer solution condition. The analytical curve was linear in the HQ concentration range of 2.5 - 120 μmol L^{-1} (r = 0.9962, sensitivity = 0.14507 μA μmol L^{-1}, and precision = 2 %). The limits of detection and quantification for HQ were estimated to be 0.78 μmol L^{-1} and

2.60 μmol L^{-1}, respectively. The evaluation of HQ as electron mediator had insignificant effects on the sensing mechanism, showing that the proposed sensor did not act like a second-generation electrode. Therefore, peroxide was not used to measure HQ amounts. The HQ diffusion coefficient was found to be $6.34x10^{-5}$ cm^2 s^{-1}. The proposed method presented no significant differences at the 95 % confidence level compared to the chromatographic method. It can be employed for detecting other phenolic compounds, such as guaiacol, catechol, N-acetyl-p-aminophenol, ascorbic and uric acids; besides, it can be applied to pharmaceutical formulations.

Chapter 6 – It is an important issue for the design of biomimetic robots to consider not only that their appearance should resemble a human arm, but also that the arm should move along human-like trajectories. The human arm trajectories in reaching movements have been widely investigated, and it is known that the hand paths lie along either a roughly straight or a simple curved line, and the tangential velocity profile of the hand is always bell-shaped. One of the most effective methods for controlling biomimetic robots along such trajectories is to apply human arm trajectory formations. However, most of trajectory formations proposed to date are limited to either planar motion or the arm model within three degrees of freedom. Accordingly, it seems to be difficult to apply existing trajectory formations to the human arm model with redundancy since the human arm consists of a shoulder with three degrees of freedom and an elbow with one degree of freedom even when the motion of a wrist is frozen. This study investigates the trajectory formation of biomimetic robot arms with redundancy. Here, a redundant arm model with three joints at the shoulder and one joint at the elbow movable in a three-dimensional space is considered. The basic trajectories of the shoulder and elbow joints are first determined by the minimum jerk criterion in joint angle space. Then, an idea for a time adjustment of the joint motion of the elbow relative to the shoulder has been brought. This time adjustment supposed at the beginning or the end of movement can be numerically determined so as to demonstrate the property of human arm trajectories. Also, a systematical scheme for formulating the required trajectory without complex analytical calculation has been developed. Consequently, the hand trajectory derived from the joint angle trajectories can be uniquely produced once the initial and target states of the arm and the movement duration are given. The trajectories produced under the proposed trajectory formation are evaluated by the trajectories observed by experiments. The results show that the proposed trajectory formation is an advantageous scheme for biomimetic robots to control along human-like trajectories.

In: Focus on Biomimetics Research
Editor: Abele Scipione

ISBN: 978-1-62417-332-5

Chapter 1

BIOMIMICRY, MICROSYSTEMS, AND BONE

Spencer L. York*, Ahmad R. Arida, Karan S. Shah, and Marnie M. Saunders

The University of Akron, Akron, Ohio, US

ABSTRACT

In the human body, bone is of tremendous importance. Studies have demonstrated the many roles of bone in providing structural support for the body, allowing for locomotion and movement, protecting vital organs and storing minerals. While the roles for bone have been established many bone phenomena are not well understood such as how bone heals itself, the mechanisms by which it adapts to altered loading environments and the metabolic diseases that result from breakdown in normal bone function. Significant research effort has been focused on understanding the interactions of the three primary bone cell types (osteoblasts, osteoclasts, and osteocytes) and their ability to coordinate activity in bone maintenance, function, repair and remodeling. However, at the cellular level bone has remained a difficult organ to model in part due to a high degree of spatial and temporal variation among these three bone cell types *in vivo* and the difficulty in isolating the impact of specific signals on these cells.

These and other limitations have contributed to the delayed development of highly biomimetic models for studying bone cell phenomena. However, the use of microsystems, small devices with fluid channels ranging on the scale of nano and micro meters, is increasing for many biological systems and may permit opportunities for the study of bone-related phenomena. The precision with which microsystems can be fabricated allows for very specific geometries within the small channels, providing a means of performing a variety of experiments and analyses quickly with high resolution and sensitivity. For biological experiments these properties offer a number of unique advantages such as increasing feasibility of single cell experiments and manipulation of small volume analytes, allowing isolation of individual factors on cell behavior, and enabling precise control over the spatial and temporal organization of cells. These features suggest that microsystems have the potential to serve as powerful tools in the development of biomimetic models for bone.

* Corresponding Author: Department of Biomedical Engineering, The University of Akron, 260 S. Forge St., Akron, O. H., 44325, US. E-mail: *sly15@zips.uakron.edu.*

These models could potentially help to elucidate the mechanisms by which bone cells sense and respond to mechanical signals in their environment. Harnessing these unique mechanisms could translate to real world applications in both the public and private technology sectors. This chapter will discuss the recent advances of microsystems and introduce their application in developing new tools and techniques for investigating bone and bone-related phenomena.

Biomimicry

Harnessing the behavior of nature to create increasingly more accurate models and a greater repertoire for solving problems remains a central goal of biomimicry. One example where this may be extremely beneficial is bone. Mimicking and exploiting the powerful renewable and adaptable properties of bone has vast potential in many fields, including aeronautical, civil, mechanical and nautical engineering. Bone exists as a dynamic and complex composite structure undergoing a constant remodeling process which maintains the structure while allowing for adaptation over time to changing mechanical load and repair of localized damage.

Through this process, old bone can be removed and replaced with new bone. This has been illustrated in tennis players where the radius and ulna of the dominant forearm have qualitative and quantitative increases in bone in comparison to the bones in the non-dominant forearm (Huddleston et al., 1980; Krahl et al., 1994). Bone's ability to remodel itself is of tremendous importance, as the structure adapts to have sufficient strength to meet the loading demands placed upon it (Wolff, 1892). Activity at the microscopic (cellular) level guides the remodeling process resulting in observable changes in the mechanical response at the macroscopic (organ) level.

As we work to develop accurate biomimetic model systems we move closer toward developing self-renewing engineering structures. Materials that change strength relative to alterations in mechanical loading may be highly beneficial for variable stress environments. Loaded structures may be impacted with the development of materials that become stronger with time or with increases in in-service loads. The microscopic response with macroscopic changes in properties may also be beneficial in other industries that require high strength with limited space or weight. Before any potential biomimetic applications of bone remodeling are realized, the process needs to be thoroughly understood. We believe microsystems have tremendous potential to provide new insights into bone remodeling and other bone cell-related phenomena.

Bone

Given the functional diversity of bone its complex structure is not surprising, Figure 1. Bone is a natural composite largely composed of an organic protein (type I collagen (CTI)) and an inorganic mineral (hydroxyapatite (HA)). Together these components largely account for the mechanical properties of bone. However, due to the different structural and mechanical properties of CTI and HA, each imparts to bone unique mechanical characteristics.

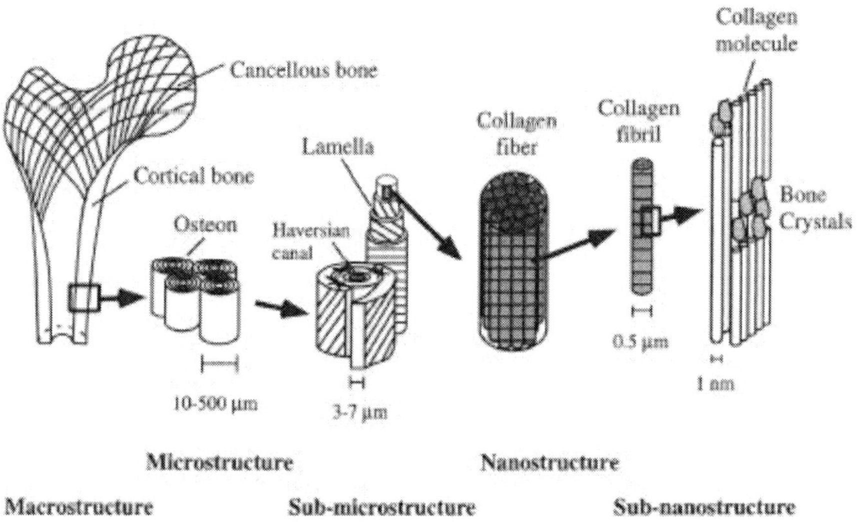

Figure 1. A diagram illustrating the length scales of bone from the gross structure to the sub-nanoscale where CTI and HA form matrix. Reprinted from Rho et al., 1998, with permission from Elsevier.

Type I collagen is an organic protein found in many tissues including bone and tendon. In bone, CTI makes up almost the entirety of the organic matrix (90%), which corresponds to approximately 30% of the dry weight of bone (Eastoe and Eastoe, 1954). As an extracellular matrix molecule CTI is laid down as part of the bone formation process. Structurally, CTI is a triple helix of polyproline peptide chains, two identical α1(I) chains and one α2(I) chain. CTI self-assembles in a quarter-staggered array (Hodge and Petruska, 1963). The banding together of the contiguous collagen fibrils results in the development of fibers which contribute to the elasticity, tensile strength and toughness of bone. However, the fibril array does not provide adequate mechanical support under compressive loading. As bone is frequently exposed to compressive loads, the mechanical properties CTI alone imparts to bone is not sufficient to support the bone.

Hydroxyapatite ($Ca_{10}(PO_4)_6(OH)_2$) is an inorganic mineral comprising approximately 70% of the weight of bone (Neuman and Neuman, 1953). The assembly of collagen in the quarter-staggered array results in a discontinuous pattern (holes and overlapping regions) that allow for nucleation sites where bone mineralization occurs (Landis, 1995). Through the process of mineralization, the organic matrix is converted to an inorganic substance. The mineralization is not immediate and osteoid is the term given to the matrix before it has completely mineralized (Baron, 1996). Mineralization is regulated by the extracellular matrix proteins. Proteins such as CTI, bone sialoprotein (BSP), phosphoproteins and proteolipids promote the nucleation and growth of apatite (Termine and Robey, 1996), while proteins such as pyrophosphates, adenosine triphosphate (ATP) and citrate inhibit nucleation and growth (Termine and Robey, 1996). In addition mineral proliferation is regulated by the extracellular matrix proteins including osteocalcin (OCN), the phosphoproteins and the proteoglycans (PG) (Termine and Robey, 1996). Metabolically, with calcium as a major component of HA, bone functions as a mineral reservoir in calcium homeostasis (Parfitt, 1987). Mechanically, the hydroxyapatite contributes to the compressive strength of bone.

The synthesis of the CTI and HA properties of the bone matrix imparts to bone a balance of flexibility and strength which is essential for locomotion and mechanical support. This level of biomimicry can be seen in current engineering structures, as buildings and bridges are rarely built with only one type of material. For example, reinforced concrete contains metal bar to provide tensile strength. In addition, towards the development of earthquake-resistant buildings, new materials and structural designs are being pursued to augment the normally vertically loaded materials with the ability to handle the wave propagation and lateral movements imparted to structures in earthquakes and tsunamis. However, unlike these man-made structures bone is capable of self-renewal through bone remodeling.

Bone Remodeling

Bone remodeling is an intricate and tightly regulated process in which mature bone is removed and replaced with new bone. This process that occurs throughout the lifetime of an individual enables localized bone damage, such as damage-induced microcracks (microdamage) to repair (Launey et al., 2010). In addition this process contributes to fracture healing and is responsible for alterations in bone geometry and density by coupling bone resorption and formation. Geometric and density changes occur in response to subphysiologic and supraphysiologic alterations in the mechanical loading environment.

Credited to Julius Wolff, the observation of bone adaptation suggests that the form and function of bone are intricately linked to internal changes in its structure. These changes are assumed necessary to accommodate the loading demands and alterations in the loading demands of the environment (Wolff, 1892). Increases in mechanical loads stimulate bone formation while inhibiting resorption. Reduction in mechanical loads has the opposite effect in stimulating bone resorption while inhibiting bone formation (Ehrlich and Lanyon, 2002; Rodan, 1997). Bone quality and quantity are affected with subsequent changes that given the extent (magnitude and duration) of the loading alterations may include size, shape and density. Normally, the processes of resorption and formation work in concert to maintain a net bone equilibrium with respect to both bone quality and quantity. A number of metabolic bone diseases, most recognizably osteoporosis, are the result of an imbalance in this equilibrium.

For osteoporosis the imbalance results in a net bone loss with time (Grubber et al., 1986). However, osteoporosis is not simply an age-related disease. Whereas primary osteoporosis is associated with senility (men and women) and a post-menopaUSl state (women), secondary osteoporosis is subsequent to other medical conditions including hyperthyroidism, diabetes and anorexia. In addition, idiopathic (juvenile) osteoporosis may be secondary or associated with a genetic condition, such as osteogenesis imperfecta (Gilsanz, 2006). Regardles of etiology, osteoporosis results in the loss of both bone mass and strength (Peck, 1993). Currently there is no cure for osteoporosis. Given that our understanding of the cellular and organ level aspects of normal bone function are incomplete, our ability to address pathologies related to breakdown in normal function is greatly compromised.

The ability of bone to respond to load and remodel is highly dynamic and has been under study over the past several decades (Bakker and Klein Nulend, 2010). During this time the process of bone remodeling has been found to consist of a period of bone resorption followed by a period of bone formation and regulated by the bone cells. More specifically, the sequential, cyclical process of remodeling may be classified by an Activation-Resorption-Formation (ARF) cycle (Baron, 1996). In the (A) resorption activation stage, localized microdamage to osteocytes results in osteoclast recruitment and bone resorption. Following resorption by osteoclasts (R), apoptosis in these cells is initiated (Roodman, 1999). Following osteoblast precursor recruitment, the (F) formation stage occurs in which osteoblasts form bone (Henriksen et al., 2009). While much is known about the biology of the cells, very little is known about how these cells interact with each other and coordinate their activity to collectively alter bone at the organ level. In addition, as we move toward understanding this interplay among the cells it is becoming increasingly clear that key discoveries will require novel approaches that enable us to study the cells not in isolation but in environments that are as physiologically relevant as possible. Ideally these biomimetic systems will enable fine control and study of individual factors of interest while maintaining the multicellular, three-dimensional (3D) bone environment complete with native elements including the bone matrix, hormones, cytokines. With further investigation into the regulation of bone remodeling via biomimetic models, insights into the mechanisms of self-renewal may enable load-induced material alterations in aeronautical, civil, mechanical and nautical structures to become possible in the distant future. Of interest to us, in the immediate future these insights may contribute to the development of targeted treatments and therapies for bone disorders.

THE BONE CELLS

Osteoblasts are the principal bone forming cells. Originating from mesenchymal (marrow stromal) tissue, they are found on free surfaces of bone. These surfaces include the periosteal and endosteal surfaces, osteonal (Haversian) surfaces of cortical bone and the trabecular surfaces. Osteoblasts can exist in both an active (bone forming), and an inactive (quiescent) state (Clarke, 2008; Everts et al., 2002). Quiescent osteoblasts residing on the bone surface are more accurately referred to as bone lining cells; the cells possess a thin, elongated appearance (Clarke, 2008). While these cells do not actively form bone, lining cells have been suggested to participate in mineral exchange and ion balance between the circulation and bone (Rubinacci et al., 1998). In addition, bone lining cells may also play a role in bone formation as it has been proposed that they degrade the residual bone matrix following osteoclast resorption, thereby contributing to an essential step in the bone forming process (Everts et al., 2002). Upon activation, bone lining cells become osteoblasts. Osteoblasts possess a cuboidal appearance owing to their role in protein production and the large amount of rough endoplasmic reticulum and a prominent Golgi apparatus. Osteoblasts are responsible for laying down the bone matrix and regulating its subsequent mineralization (Bailey et al., 2002; Clarke, 2008; Eastoe and Eastoe, 1954). Biomarkers common to osteoblast activity and bone formation include CTI, bone sialoprotein (BSP), (OC) and alkaline phosphatase (AP) (Kamiya and Takagi, 2001; Liao et al., 2003; Rochefort et al., 2010).

Osteoblasts also possess hemichannels. These proteins channels are comprised of a connexon, which in turn is comprised of six protein subunits, or connexins. Connexins are identified by molecular weight (for example, connexin43 (cx43)). When hemichannels in juxtaposed plasma membranes dock, they form a gap junction channel. This channel physically unites the internal compartments of the cells enabling the passage of small molecules such as calcium ions, phosphates, and cyclic nucleotides, to pass between neighboring cells (Donahue, 2000). Connexin 43 and 45 are abundant in osteoblasts (Civitelli, 2008; Donahue, 2000; Vander Molen et al., 1996). In addition, bone lining cells also possess the ability to link to each other and osteoblasts via gap junction channels (Jones et al., 1993; Miller et al., 1989).

Osteocytes are the most abundant bone cell accounting for more than 90% of the bone cells in the body, roughly ten times the number of osteoblasts (Dallas and Bonewald 2010). Osteocytes are senescent, fully differentiated osteoblasts that have become entrapped within the osteoid during bone formation. Whereas all osteocytes were previously osteoblasts, the converse is not true; between 10-20% of osteoblasts further differentiate into osteocytes (Puzas, 1996). As these cells become encased within the bone matrix they undergo morphologic changes that are concurrent with the decreased cellular demands for matrix synthesis (Neve et al., 2012). In their new role as osteocytes, these cells experience a reduction in volume correlating with a significant decrease in the size and number of organelles (Dudley and Spiro, 1961). Following cytoskeletal reorganization, the shape of the osteocyte changes from the cuboidal shape of the osteoblast to a significantly smaller cell body with numerous (~ 50) dendritic processes (Neve et al., 2012; Sugawara et al., 2005). In addition, a rigid primary cilium originating at the basal body of the cell extends into the extracellular space (Temiyasathit and Jacobs, 2010). While the cell body of the osteocyte occupies a lenticular shaped cavity (lacunae) in the mineralized matrix, the dendritic processes extending from the cell body reside in little canals (canaliculi) establishing the lacunocanalicular network in bone. This fluid-filled network provides a unique and critical syncytium throughout the bone matrix for cellular nutrition and communication (Knothe Tate, 2003). While the osteocyte body is not physically attached to the walls of the lacunae the dendritic processes are ‘spot-welded’ to the canalicular walls at discrete locations (McNamara et al., 2006). In addition, the tips of the dendritic processes are in physical contact with neighboring osteocytes such that hemichannels in contiguous cell membranes dock forming gap junctions. Furthermore, this communication network extends beyond the embedded osteocytes to include the osteoblasts and bone lining cells. In this state, it has been proposed that osteocytes may have a half-life as long as 25 years (Franz-Odendaal et al., 2006).

Biomarkers common to osteocyte activity include BSP, CD44, SOST/sclerostin and Dickkopfs-1 (Dkk-1) (Rochefort et al., 2010). SOST/sclerostin is emerging as a key target for study in bone remodeling given its role as an inhibitor of bone formation. As a delayed secreted product following osteocyte encasement, sclerostin appears to halt further bone formation by osteoblasts contributing to the regulation of bone size, shape and density (Poole et al. 2005).

Given their longevity, volume, communication network and physical location within the bone matrix the osteocytes possess a physically unique environment. Our understanding of the significance of this environment is still in its infancy. However, fluid flow shifts through the lacunocanaliular network with changes in mechanical load (Burger and Klein-Nulend, 1999; Piekarski, 1977); the physical dendritic tip connections via gap junctions enabling intercellular trafficking of molecules and early messengers (Nakahama, 2010); and, the primary cilia extending from the osteocyte cell surface (Malone et al., 2007; Whitfield, 2003) unite to form an incredibly intricate system enabling the cells to sense their physical surroundings.

Osteoclasts are large, multinucleated cells responsible for bone resorption (Geddes, 1913). Osteoclasts are derived from the fusion and differentiation of monocyte/macrophage precursors of hematopoietic origin. Bone resorption by the osteoclast is a multi-step process consisting of osteoclast polarization with the subsequent formation of the ruffled border; adherence to the bone surface with the subsequent formation of a sealing ring; and, the secretion of acids and proteases for mineral and matrix destruction, respectively (Zaidi, et al., 2003).

Integrin binding is central to the ability of osteoclasts to selectively isolate and enclose an area between their basal surface and the bone matrix. A principal integrin in osteoclasts is vitronectin receptor ($\alpha_v\beta_3$). The osteoclasts' capacity to polarize on the bone surface results in the creation of a ruffled border on the basal cell surface (Teitelbaum, 2000). This infolded structure has been proposed to represent acidifying vesicle transport along microtubules and their polarized insertion into the plasma membrane (Teitelbaum, 2000). Further contributing to an isolated microenvironment, the sealing zone is an actin rich ring surrounding the ruffled border (Vaananen and Horton, 1995) that localizes into podosomal protrusions containing matrix-recognizing integrins including talin and vinculin (Marchisio et al., 1984). The ring is present during active bone resorption (Vaananen and Horton, 1995). Active resorption occurs first with the mineral dissolution and then proceeds to the organic matrix destruction. The mineral destruction is accomplished with the creation of an acidic (pH ~4.5) microenvironment (Silver et al., 1988) which is mediated by a vacuolar H^+ - adenosine triphosphatase (H^+ -ATPase) (Blair et al., 1989). Intracellularly the pH is maintained by an energy-independent Cl^-/HCO_3^- exchange system (Teti et al., 1989). Through this system HCl is pumped into the cell and the mineral is mobilized. Subsequently the exposed organic material is removed by lysosomal proteases such as cathepsin K (Li et al., 1999). Endocytosed degradative products are released through the apical surface of the osteoclast (Nesbitt and Horton, 1997).

As part of this process, the mineral deposits in the bone are released. This allows bone, and osteoclasts, to serve an important role in maintaining serum calcium concentrations (Glorieux et al., 2000). Given the role of the osteoclast in bone resorption, biomarkers of osteoclast activity correlate to clinical markers of resorption and include the amino acids hydroxyproline, pyridinolines and the peptide fragments N- and C-telopeptides (Singer and Eyre, 2008).

The regulation of osteoclast differentiation and function relies on receptor activator of nuclear factor kappa-B (RANK), receptor activator of nuclear factor kappa-B ligand (RANKL), and osteoprotegrin (OPG), Figure 2. It has been known for some time that osteoblasts are critical for osteoclastogenesis. That is, *in vitro* osteoclasts would not form in culture in the absence of osteoblasts.

With the discovery of the RANKL/RANK/OPG system it became clear that macrophage colony-stimulating factor (M-CSF) and RANKL from the osteoblast are both necessary and sufficient for osteoclastogenesis (Boyce and Xing, 2007). M-CSF promotes osteoclast progenitor proliferation and survival (Zaidi et al., 2003). With the binding of the RANK receptor of the osteoclast to the RANK ligand (soluble and membrane-bound) of the osteoblast a functional osteoclast capable of bone resorption evolves. In addition, the osteoblast produces OPG, a soluble decoy receptor to RANK. If OPG and RANKL bind, the binding site is blocked to RANK and osteoclastogenesis cannot occur (Simonet et al., 1997; Yasuda et al., 1998).

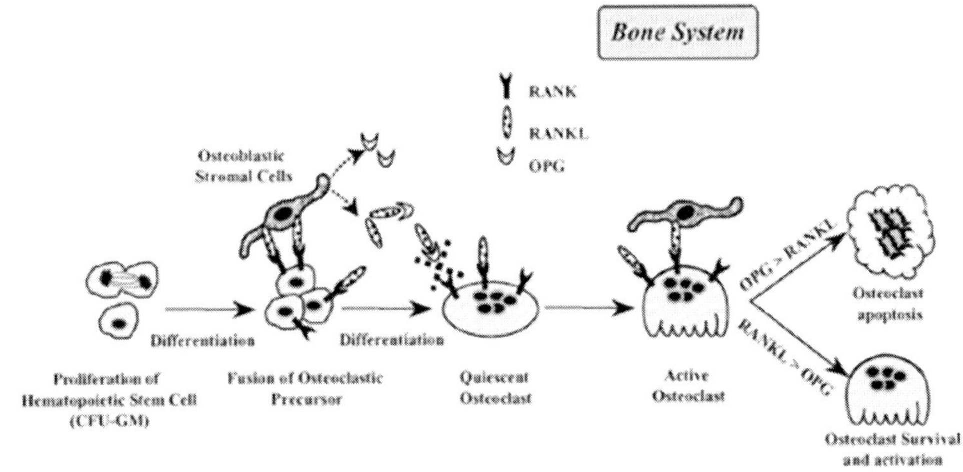

Figure 2. The RANKL/RANK/OPG system. In the presence of RANKL, osteoclastic precursors are able to develop into functional osteoclasts capable of resorption. OPG is a soluble decoy receptor that binds to RANKL and in so doing competes with RANKL/RANK binding. Adapted with permission from Theoleyre et al., 2004, with permission from Elsevier.

Unlike cells of the osteoblast lineage that likely communicate via gap junctions, osteoclast-osteoblast communication via the RANKL/RANK/OPG system is at least in part a soluble event that has the potential to regulate and coordinate osteoblast/osteoclast activity. While this has been shown to have implications for mechanical loading *in vitro* (Saunders and Donahue, 2004), it is also likely to have future implications for bone remodeling (Boyce and Xing, 2007).

Via physical structures and soluble interactions osteoblasts, osteoclasts, and osteocytes are able to communicate and impact the function of each other. Signaling pathways are also critical to this process. For example, while nonspecific to bone with roles in embryogenesis and cell differentiation (Logan and Nusse, 2004), the Wnt/β-catenin signaling pathway is important to bone regulation.

Wnt Signaling Pathway and Bone

Basic communication between the bone cells as well as the translation of mechanical into biochemical signals likely involved in bone remodeling is the result of a complex network of cellular signaling (Bonewald and Johnson, 2008; Li et al., 2011; Lin et al., 2009; Robling et al., 2006). While a number of signaling pathways have been implicated in the regulation of bone remodeling, one of the best characterized to date is the Wnt/β-catenin (canonical Wnt) signaling pathway (Armstrong et al., 2007; Case et al., 2008; Mantila Roosa et al., 2011).

The canonical Wnt signaling pathway functions by controlling the cytoplasmic concentration of β-catenin, which affects downstream β-catenin-dependent targets of gene transcription. (Bonewald and Johnson, 2008; Westendorf et al., 2004).

When the appropriate Wnts are present, β-catenin is stabilized in the cytosol, translocates to the nucleus, and interacts with Tcf/Lef family of transcription factors to activate transcription of genes needed for bone formation.

This process occurs via a co-receptor complex which is formed with the appropriate Wnt, a member of the Frizzled (Fz) family of seven-pass transmembrane receptors, and low-density lipoprotein receptor-related proteins 5 or 6 (LRP5, LRP6) on the outer surface of plasma membrane. Within the cytoplasm, Dishevelled (Dsh) becomes activated leading to the phosphorylation and subsequent inactivation of Glycogen Synthase Kinase-3β (GSK-3β), which is responsible for phosphorylating β-catenin.

Non-phosphorylated β-catenin levels increase in the cytoplasm causing translocation into the nucleus where gene expression is altered to promote bone formation in bone cells.

Antagonism of Wnt signaling occurs by competitive inhibition of the co-receptor complex formation via several methods. Proteins such as SOST/sclerostin and Dkk-1 act as Wnt mimetics and bind LRP5/6, but prevent formation of the co-receptor complex (Lin et al., 2009; Mason and Williams, 2010; Semenov et al., 2001).

Secreted frizzled-related proteins (sfrp) bind Wnts and prevent their binding of LRP5/6. In both cases, β-catenin becomes phosphorylated by GSK-3β. Following phosphorylation β-catenin becomes ubiquitinated and later degraded by the 26S proteasome.

Mechanotransduction of Bone Cells

While biomechanics is a field that incorporates the application of mechanical load to biological tissues to assess mechanical performance, mechanobiology is a field that recognizes that mechanical loading contributes to the development, growth, maintenance, function and pathologies of biological tissues. Within this field, mechanotransduction aims to understand the mechanisms and pathways by which organs, such as bone, can sense and respond to a mechanical signal. It is therefore at the heart of biomimicry.

Mechanotransduction as it applies to bone has been designated as a series of events that begin with mechanical load inducing a local signal that may be sensed by the osteocyte. This signal must then be capable of being transduced by the osteocytes into a biochemical signal that is capable of being sent to and received by osteoblasts and osteoclasts to bring about a functional response (bone formation/resorption) (Burger et al., 1995; Temiyasathit and Jacobs, 2010). A variety of key receptors, physical structures and signaling pathways have been implicated in bone cell mechanotransduction. In addition, a variety of mechanical signals (fluid shear, substrate deformation, pressure) have been shown to affect mechano-responsive cells.

To date *in vitro* studies have begun to provide us with an understanding of how bone cells respond to mechanical load. While this is by no means an exhaustive review, we will touch upon the role some of the key elements we have introduced earlier in this chapter are thought to play in mechanotransduction. Specifically we will address physical structures such as cilia, gap junctions and the lacunocanalicular network and soluble factors/signaling pathways such as RANKL/RANK/OPG and Wnt/β-catenin. As you will note, these elements work in concert in affecting mechanically-induced alterations in bone.

Flow-induced mechanotransduction centers on the seminal work of Piekarski who proposed that fluid flow through the lacunocanalicular bone network resulted from fluid shifts generated by mechanical loading (Piekarski and Munro, 1977). Parallel-plate flow chambers have been used to subject cells to presumably physiologic levels of fluid shear. These systems have been used to demonstrate flow-induced increases in intracellular calcium concentration ((Ca2+)) in osteoblasts (Chen et al., 2000; Hung et al., 1995). This response was later shown to be dependent upon waveform (steady vs oscillating) (Jacobs et al., 1998). in that an oscillatory waveform was demonstrated to be more stimulatory to bone cells than a comparable substrate deformation (Owan et al., 1997; You et al., 2000). Given the 3D nature of the bone environment, more recent efforts have been made to subject bone cells in 3D matrices to fluid flow (Tanaka et al., 2005). Microprinting techniques are being realized for their potential to create a more biomimetic environment for bone cell mechanotransduction studies (You et al., 2008) (Guo et al., 2006).

Mechanical loading may also play a role in the mineralization of bone via a pathway involving SOST/sclerostin and Wnt/β-catenin (Bonewald, 2006). As previously noted, osteocytes are mechanosensitive cells that produce sclerostin, an inhibitor of bone formation during the laying down of osteoid (Poole et al., 2005). Given the highly ordered structure of bone, as osteoid is laid down, signals possibly transmitted between osteocytes and osteoblasts via gap junctions could allow for selectivity of specific osteoblasts (for osteocyte differentiation) in the local environment that are ideally located to maintain the ordered structure of bone (Bonewald, 2006). As an antagonist to Wnt/β-catenin signaling and given a role for SOST in regulation of mineralization, it is possible that mechanical loading regulates formation and mineralization (Bonewald, 2006).

Mechanotransduction in bone has been shown to use the canonical Wnt signaling pathway in both osteoblasts and osteocytes (Armstrong et al., 2007; Case et al., 2008; Mantila Roosa et al., 2011). Up-regulation of genes involved in canonical Wnt signaling is essential for bone formation as a result of mechanical stimulation. In mouse primary osteoblasts exposed to fluid shear stresses, expression of Wnt1 and Wnt3a were shown to increase (Lau et al., 2006). In addition, overexpression of β-catenin (downstream LRP5 mediator) in osteoblasts, results in a decrease in the RANKL/OPG ratio and a subsequent decline in osteoclastogenesis (Spencer et al., 2006). These studies are part of a growing body of evidence which strongly supports the essential role of Wnt signaling in the mechano-transduction response of bone to load.

There is growing evidence to support the role of osteocyte primary cilium in mechanotransduction. This rigid structure projects from the basal body of the osteocyte and extends into the extracellular space (Temiyasathit and Jacobs, 2010). These microtubule structures have been proposed to act as mechanosensors in bone, a role previously proposed in kidney and liver cells. When exposed to fluid flow, the cilia deform resulting in cilia-dependent increases in Cox-2 gene expression and prostaglandin (PGE_2) release. Unlike kidney and liver cells, the response in osteocytes appears to be independent of Ca^{2+} flux (Malone et al., 2007). With respect to the RANKL/RANK/OPG system, flow-induced OPG/RANKL ratio increases were negated with cilia removal (Malone et al., 2007). Interestingly, similar findings were demonstrated in osteoprogenitors with gap junction pharmacological blockage (Jekir and Donahue, 2009). More recent evidence suggests fluid flow to bone cells is accompanied by cAMP decrease that is cilia-dependent (Kwon et al., 2008). Cellular communication between the bone cells is critical to mechanotransduction and bone remodeling (Henriksen, 2009). Throughout the ARF cycle direct and indirect cell communication between the osteocytes, osteoblasts and osteoclasts guides cellular activity. With respect to a role for gap junctions in bone cell mechanotransduction, osteocytes are thought to communicate with osteoblasts to control the bone formation and destruction process (Duncan and Turner, 1995). In *in vitro* studies, osteocytes exposed to fluid flow demonstrated the ability to regulate osteoblast expression through gap junctions (Taylor et al., 2006). We have previous acknowledged the role of RANKL/RANK/OPG in osteoblast regulation of osteoclastogenesis and the role of LRP5 in the Wnt/β catenin signaling, these systems also interact.

Potential Limitations of *In Vitro* Model Systems

While *in vitro* systems are effective in their ability to isolate and study key factors in mechanotransduction, they have largely relied upon single cell type stimulation models, artificial substrates and environment isolation. Given the complexity and intricacy of the microstructure of bone, increasing the physiologic relevance of these *in vitro* approaches may provide new avenues for investigation and discovery. Research in mechanobiology parallels that in any field of science in that for progress and discovery to advance the field, tools and technologies need to advance such that the necessary experiments may be conducted and not hampered by the lack of these tools and technologies. In an effort to provide the necessary tools and technologies, researchers must constantly be aware of advancements in other fields and how those advancements may be used to further discovery in one's own field. We believe an active area of research that has tremendous potential to offer new insights into research in the bone field is the area of microfluidics and microsystems.

Microfluidics

Microfluidic devices are a new wave of technology platforms that deal with precise control and manipulation of fluids and particles at small-scale, frequently in the sub-millimeter range. These systems and devices run a varied amount of experiments and perform analyses by controlling fluid flow through various miniscule channels, known as microchannels. Using modern microfabrication techniques, channel sizes have been developed in the nano and micrometer scale (Linder et al., 2003). The small geometry of these devices allow them to perform vast and time consuming experiments and analyses very quickly with high resolution and sensitivity, saving not only time but the amounts of reagents and other materials needed in the experiments (Whitesides, 2006). Additionally, due to the nano and micro scale of these devices, they are often used for biological studies and are capable of single cell analysis (Gao et al., 2004). Microfluidic devices offer a new method for biomimetics and are helping to revolutionize biology and biochemistry by improving studies focusing on the impact of individual factors on the behavior of a small group of cells. This may prove particularly effective for developing improved mechanotransduction studies of bone cells, as they can be exposed to isolated types of loads prior to single cell analysis. The individual components of microfluidic devices, such as micromixers, microvalves, microchannels, detectors, etc., can be combined to form a device, micro total analysis systems (μTAS), capable of running an entire experiment on a lab-chip. These μTAS devices are capable of running biochemical assays such as enzyme-linked immunsorbent assay (ELISA), capillary electrophoresis (CE), and flow cytometry through fluorescence activated cell sorting

(FACS) (Effenhauser et al., 1997; Fu et al., 1999; Rossier et al., 2001). Here we discuss some of the key microfluidic components, devices and methods that can be combined to develop a need based μTAS system.

Fabrication Techniques

The advancements in the field of microelectronics have aided in the development of various fabrication techniques for microfluidic device construction. These techniques are inexpensive and allow for the development of both simple and highly complex devices that can be scaled for mass production. Microfluidic device fabrication techniques such as photolithography have led to the development of new devices, recently reaching precision down to the nanometer scale (Sundaramuurty, et al., 2006). Photolithography is performed by placing a layer of photoresist on a silicon wafer. When the photoresist is exposed to light it undergoes a color change. By placing a mask on top of the photoresist layer, a pattern can be transferred onto the photoresist and then onto the silicon wafer. The immediate impact of photolithography was the rapid development of microchips for computers and micro-electromechanical machines, composed primarily of silicon. These devices excelled at carrying electronic signals and vastly improving the computer chips, in large part due to the nano and micro level scale and precision of the fabrication process (Nguyen and Werely, 2006). For use in creation of microfluidic devices, photolithography was adapted to develop soft-lithography. With soft-lithography the precision of photolithography can be applied towards the fabrication of softer materials such as poly-dimethylsiloxane (PDMS). PDMS is a common material for microfluidic devices, and preferred due to the relative ease of processing, its permeability to gas for cells, and its optical transparency, permitting the use of microscopy to observe the inside of the channels (Whitesides, 2006). Figure 3 breaks down the process of soft-lithography. Briefly, the soft lithography process consists of pouring a mixture of PDMS cross linker and curing agent onto a mold or mask. After baking overnight, the PDMS can be removed from the mold. Holes are then punched into the PDMS for the necessary inlets and outlets for fluid flow, and the PDMS can be treated with plasma oxygen before adhering to a glass base (Duffy et al., 1998). As discussed earlier, soft lithography is most commonly used to process PDMS, due to the relative inexpense ease of processing and fast turnaround times. Other alternate materials can also be used to fabricate microfluidic devices via this method, such as: glass, silicon, paper, and other polymers. These materials are more commonly found for specialized purposes, or in commercial settings for disposable chips (Fiorini and Chiu, 2005; Martinez et al., 2008).

Laminar Flow

Fluid flow through a microchannel is characterized by the Reynolds number, $Re = \frac{\rho v D}{\mu}$, where: ρ corresponds to the fluid density, ν corresponds to the velocity of the fluid, μ corresponds to fluid viscosity, and D corresponds to the width of the channel. The fluid flow is characterized as laminar when the Reynolds number remains low, such as below 1800 – 2000 (Nguyen and Werely, 2006). Due to the micron level dimensions of microfluidic devices, the Reynolds number is usually much less than 1000, often less than 100, and therefore the flow in these devices is almost always laminar (Nguyen and Werely, 2006; Stone et al., 2004). Laminar flow is referred to when a fluid flows in parallel layers/streams without any disruption or cross-current between its streams (See Figure 4) This well characterized flow is important for many microfluidic applications as the fluid can be precisely manipulated allowing for controlled and predictable particle gradients within a channel, due to the steady streamlines (Dertinger et al., 2001; Jeon et al., 2000).

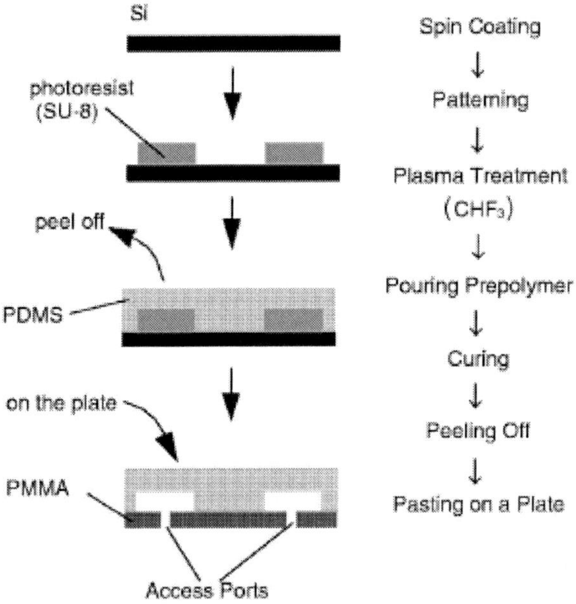

Figure 3. Fabrication process for microfluidic devices, using soft lithography. The photoresist is a component of the mask, and considered the pattern. Glass plates are used as well as PMMA plates as a base for microfluidic devices. Reprinted from Teruo Fujii et al., 2002, with permission from Elsevier.

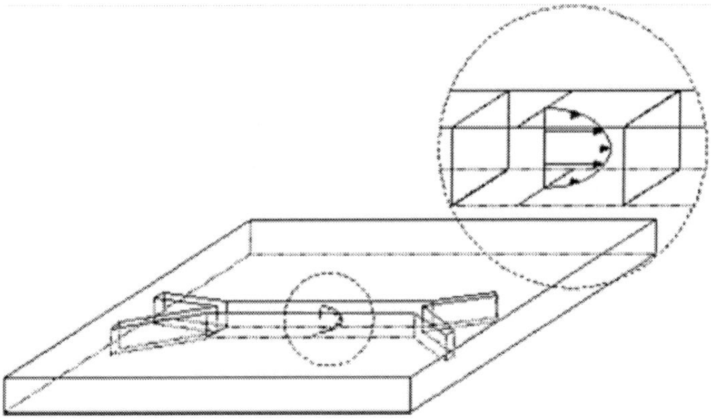

Figure 4. Laminar fluid flow profile in a microfluidic device. As the close-up of the flow profile shows, the streamlines in the microchannel are stacked parallel to each other. In this flow, the maximum fluid velocity is seen in the middle of the microchannel and the lowest velocities are experienced near the walls of the microchannel, giving the flow a parabolic shape.

MICROFILTERS

Laminar flow in microfluidic devices offer a number of advantages such as the ability to create precise gradients, and separating molecules through the use of microchannels (Dertinger et al., 2001; Hatch et al., 2001; Jeon et al., 2000; Sethu et al., 2006; Yang et al., 2006). Various microfluidic filters (shown in Figures 5 and 6), a precise arrangement of microchannels, use the diffusion and convection properties of the fluid as well as the geometrical characteristics of the channels, for the purposes of molecule/particle transport.

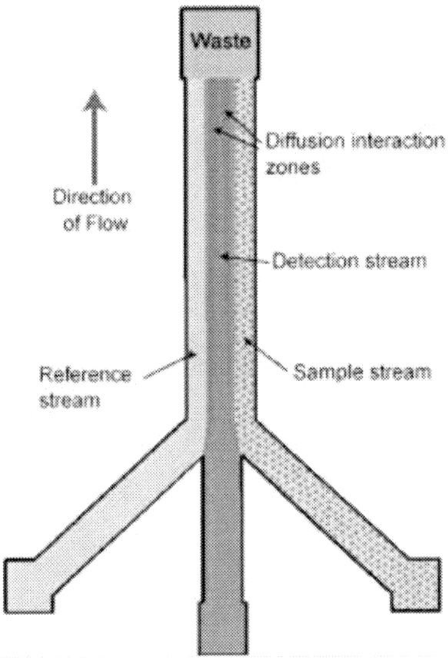

Figure 5. A t sensor, used to measure the diffusivity of a fluorescently labeled particle. Reprinted from Schulte et al., 2002, with permission from Elsevier.

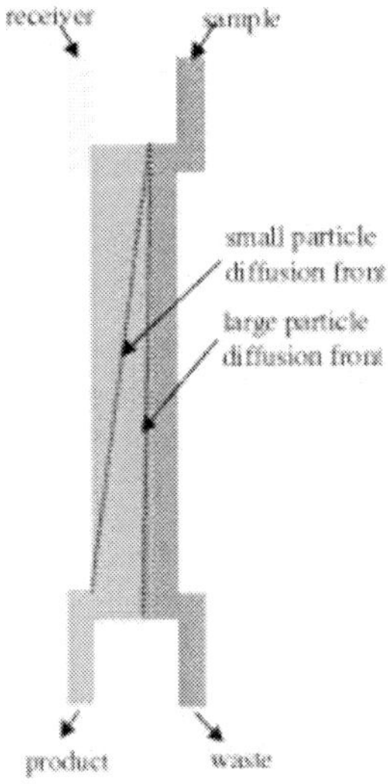

Figure 6. A H filter, used to sort molecules based on size. Reprinted from Jandik et al., 2002, with permission from Elsevier.

The precise control over this transportation phenomenon is what helps in the generation of different gradients within the microchannels (Nguyen and Werely, 2006). These precise gradients can be used to study the impact of a molecule on cell viability, such as with drug screening studies (Sung and Shuler, 2009). Microfluidic systems are also effective at separating molecules. Through the use of an H filter, shown in figure 6, molecules can be separated by diffusivities.

By running a fluid with particles to sort through the sample inlet, and changing the length of the device, fluids can separate based on diffusivity. Molecules with higher diffusivity diffuse faster, thus moving further away from the input stream. By selecting the output based on what cell or molecule is of interest, H filters can be used to filter molecules and cells as needed. This is particularly helpful for separating and analyzing various components from whole blood such as leukocytes, albumin, or plasma (Hatch et al., 2001; Sethu et al., 2006; Yang et al., 2006).

Micromixers

When mixing is necessary, such as the combining of reagents for a polymerase chain reaction (PCR), there are many different methods and mixers that are available (Lee et al., 2011; Nguyen and Werely, 2006; Suh and Kang, 2010). Mixing in microfluidic devices have a tricky and complicated path due to the small scale of the microchannels in these devices that force the Reynolds number to stay well below the turbulence flow regime. Because of the inability to achieve turbulent flow, mixing at micro scale is achieved mainly via two methods: passive and active mixing. Passive mixing relies mainly on the concepts of molecular diffusion and convection, whereas active mixing relies on external inputs to induce chaos in the flow of fluid through the microchannels (Nguyen and Werely, 2006).

This section focuses on the passive mixers currently being used by researchers in the microfluidic field. In order to enhance the diffusive mixing effect in microchannels, it is necessary to increase the contact area between the fluids, molecules and, particles to be mixed (Lee et al., 2011). This is generally achieved via increasing the length of the microchannels, which increases the time the two fluids are in contact with each other, providing sufficient time for the diffusion and convection properties of the fluids to take over while achieving full mixing at the end of the microchannel. Currently there are many types of micromixers that play on the geometrical lengths and arrangements of their microchannels to achieve passive mixing.

Parallel lamination micromixers and serial lamination micromixers are two of the most common used by researchers in this field (Lee et al., 2011; Nguyen and Werely, 2006). Parallel lamination mixers rely completely on the diffusive/convective transport equations, increasing the length of the microchannels to achieve mixing. A long T-mixer is an example of a parallel lamination micromixer that introduces the solute and solvent through two different inlets and merges these two microchannels for appropriate lengths to achieve full mixing.

Serial lamination mixers work in a similar manner, except they reduce the length of the microchannels, by introducing several splitters and joiners to enhance the mixing process (Nguyen and Werely, 2006).

MICROVALVES

Control and directed flow of fluids through a microchannel is performed with the help of valves. Primarily, there are two types of techniques used for the control of valves in microfluidic applications: utilizing direct mechanical means, and environmentally sensitive polymers. A direct control mechanism for the operation of valves in microchannels is through the application of pressure. This system, developed by Quake et al., consists of running a microchannel perpendicular to, and above the channel that must be blocked. By applying pressure to the upper channel, the PDMS membrane is deflected downward. This can cause the lower channel to be blocked, therefore cutting off the fluid flow through the channel. Thus the upper channel serves as a pneumatically activated valve for the lower channel (Quake and Scherer, 2000; Unger et al., 2000). Using a similar principle, 4 stacked channels can be used, with the top and bottom channels used for applying pressure. When pressure is applied, the direction of flow can rotate 90 degrees, leading to a set of pneumatically activated switches, as shown in figure 7 (Ismagilov et al., 2001). These valves have been used for a variety of applications, such as trapping sample solutions in a portion of the channel, providing time for a chemical reaction to occur or for cell sorting (Fu et al., 2002; Hisamoto et al., 2005). However, the pneumatic valves require input from the user. An alternative type of valve, based on polymers, can respond to the environment of the microfluidic device (Beebe et al., 2000). These polymer systems can alternate between blocking flow, and allowing the flow of fluid, based on the environment. When the concentration of a target molecule remains low, the channel can be blocked. Over time, as more of the target molecule nears the valve, the hydrogel valve opens and allows fluid flow to continue. Previous research has demonstrated valves sensitive to temperature, electric fields, visible light, and the concentration of specific molecules (Beebe et al., 2000; Hu et al., 1995; Kataoka et al., 1998; Miyata et al., 1999; Suzuki et al., 1990; Tanaka et al., 1980; Tanaka et al., 1982). While these valves do not require user input for function, allowing for automated response to chemical reactions, the hydrogel reaction takes more time than other valve systems. Timing for these valves ranges from less than 10 to over 100 seconds depending on the design (Beebe et al., 2000).

DATA ANALYSIS

Data collection in microfluidic devices is commonly accomplished by the use of various detectors that analyze the contents of the microchannels as the fluid travels from the inlet to the outlet. Through the use of fluorescent dyes, a variety of properties can be measured; for example they can be used to measure the concentration of fluorescently labeled molecules (Brody and Yager, 1996; Hatch et al., 2001). While these studies measured the rate of diffusion, microfluidic devices could also be used to measure the rate of a chemical reaction that increases or reduces the level of fluorescence. Alternatively, temperature dependent fluorescent dyes can be used to monitor the temperature inside the microfluidic device (Ross et al., 2001; Samy et al., 2008).

These fluorescence methods allow for the detection of exothermic or endothermic reactions. Mechanical systems such as cantilevers may also be used as a method of data collection in microfluidics.

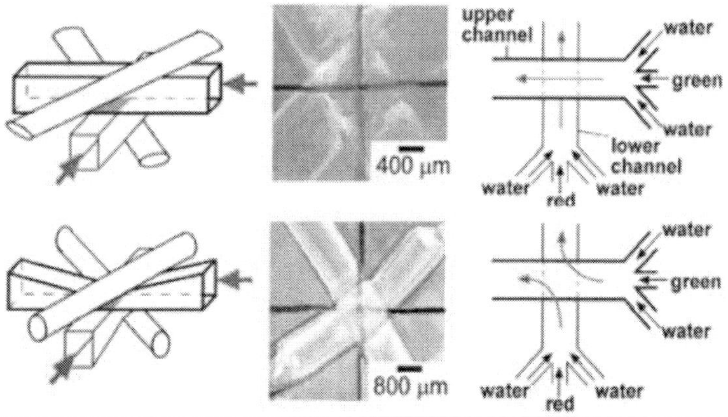

Figure 7. A system of 4 channels, used to generate a switch. When the top and bottom channel are filled with air, the switch is activated, and direction of fluid flow can turn 90 degrees. Reprinted with permission from Ismagilov et al., 2001. Copyright 2001, American Chemical Society.

Through the use of binding agents, cantilevers can be targeted for specific analytes, and can be used to study the rate of cell growth through repeated measurements (Anderson et al., 2011; Godin et al., 2010). Alternatively, piezoelectric detectors can be utilized to measure the quantity of nanoparticles that flow through the microchannel (Fraikin et al., 2011). While a number of different designs exist for measuring passing particles, the specific detector chosen varies based on the needs of the device.

MICROFLUIDICS AND BONE

As discussed in this chapter, fluid flow has been shown to be a key regulator of bone cell activity and a means of nutrient exchange and waste removal. As such, fluid flow has been used as a source of mechanical stimulation in mechanotransduction research. Given the importance of fluid flow to bone and the ability to generate similar flow streams in microfluidic devices, it would appear to be a logical approach to use microfluidic techniques and tools for bone research, specifically bone mechanotransduction research. Benefits of using microfluidics to subject bone cells to fluid flow include the ability to alter flow rate and viscosity of the fluid (Wang et al., 2011). The advantage of this has been demonstrated. Estrada et al. utilized microfluidic devices to successfully replicate the fluid shear stresses that endothelial cells are exposed to, in both physiologic and pathologic conditions (Estrada et al., 2011). Wang et al. were capable of observing the production of angiogenic factors and the drug response of endothelial cells at different fluid shear stresses (Wang et al., 2011). Rahimzadeh et al. used fluid shear stress to study changes in the cytoskeleton of epithelial cells (Rahimzadeh et al., 2011). Leclerc et al. completed an early bone microfluidic study, and demonstrated that osteoblasts grow and proliferate in microfluidic devices when exposed to fluid shear stresses (Leclerc et al., 2006).

Microfluidic devices are powerful largely because of their modular nature. By combining several components such as microchannels, micromixers, microvalves, and detectors, devices can be designed for a wide variety of applications. The modular nature of these devices assists in the development of new microfluidic devices through various combinations and organizations of the basic components, for a variety of experiments. Along with the modular nature of these devices, their ability to run and analyze experiments with high-throughput and increased sensitivity make them a great tool for use in the construction of biomimetic devices. For example, a μTAS system with the correct components can be used to develop a bone model to study the various events occurring in the bone-remodeling process. As will be discussed further in the chapter, the various key microfluidic components have been used in devices used for PCR, ELISA, mass spectrometry, and drug research experiments (Chan et al., 1999; Eteshola and Leckbend, 2001; Jiang et al., 2001; Sung et al., 2009; White et al., 2011; Zeng et al., 2010).

Applications of Microfluidic Devices

The modular nature of microfluidic devices provides for a vast number of possible adaptations to common techniques used in biomedical research laboratories. As the technology develops and fabrication processec are refined, new miniaturization technologies will become feasible and new applications for microfluidics will be found. While some microfluidic devices are small-scale replicas of conventional laboratory procedures, such as DNA sequencing and analysis, protein/proteomic analysis, enzyme kinetics, immunoassays, cytometry, and cell culture, others exploit the unique physical properties inherent to the microscale (Beebe et al., 2002; Cho et al., 2011; Sia and Whitesides, 2003).

Cell Culture

Recent advances in microfluidic techniques have led to the development of fully automated cell culture systems, which use less reagents and end-user labor, while more closely approximating *in vivo* conditions (Skafte-Pedersen et al., 2012; Warrick et al., 2010). Using a 10 x 10 microfluidic cell culture array, human cervical cancer (HeLa) cells were loaded into the chambers and allowed to grow (Hung et al., 2005). When cells were approaching confluence, trypsin was flowed through the channel and cells were detached, washed and diluted with PBS, then reseeded, allowing for fully automated cell culture combined with real-time data monitoring. In addition to normal cell culture techniques, some microfluidic devices can be used for more individualized culture conditions. Using an automated microfluidic cell culture chip, human primary mesenchymal stem cells were grown in 96 independent chambers (Figure 8), each capable of independent culture conditions (Gomez-Sjoberg et al., 2007). While such an experiment can be performed using 96-well tissue culture plates, the microfluidic device was less than seven centimeters wide and fully capable of maintaining cell viability on the order of weeks. This combination of consistent, uniform microenvironments with multi-chamber cell culture arrays delivered by microfluidic devices allows for a vast range of biological applications (Hung 2005), such as long term cell culture and cellular differentiation (Tourovskaia et al., 2005; and Chung et al., 2005). In addition to use in mammalian cell culture systems, microfluidic devices have been developed for long-term monitoring and large-scale integration of complex networks for bacteria (Balagadde et al., 2005; Thorsen et al., 2002).

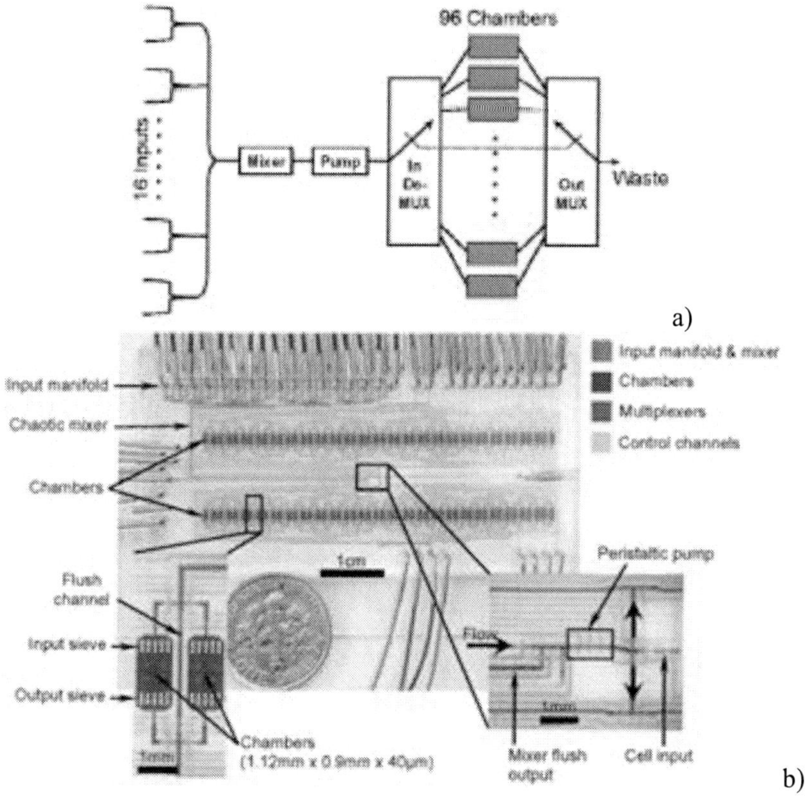

Figure 8. A microfluidic system designed with 96 independent chambers, each of which are capable of independent culture conditions. Reprinted with permission from Gomez-Sjoberg et al., 2007. Copyright 2007 American Chemical Society.

Progressing from studies of individual populations of cells, microfluidic devices have been developed for co-culture studies. Human umbilical vein endothelial cells (HUVECs) and HeLa cells have been co-cultured onto an integrated microfluidic platform to study the interactions between tumor and endothelial cells (Zheng et al., 2012). Outside of the tumor metastasis systems, co-culture microfluidics have been used to study nephrotoxicity. To simulate *in vitro* the metabolism of ifosfamide by the liver and study the role of downstream metabolites on nephrotoxicity, a hepatocellular carcinoma cell line (HepG2/C3a) or a terminally differentiated human liver progenitor cell line (HepaRG) were cultured with a dog kidney tubular cell line (Choucha-Snouber et al., 2012). These co-culture systems have also been used for "infection on a chip" assays to study the interactions of bacteria and spread of viruses with respect to mammalian cells (Lee et al., 2010; Zhu et al., 2009).

As tissue culture systems using three-dimensional (3D) scaffolds are widely accepted to more closely mimic the *in vivo* environment than cell culture on plastic dishes, microfluidic devices have been adapted for 3D culture environments (Zhang and van Noort, 2011). In some cases, the 3D structure is given by a gel matrix. For example, human breast cancer (MCF-7) cells were grown within a collagen-gel matrix on a multichannel microfluidic chip (Torisawa et al., 2005). In other cases, the 3D structure is imparted by physical properties of the microfluidic polymer. For example, polymer chips composed of non-degradable polycarbonate, polymethylmethacrylate, and cycloolefin were used to culture either rat primary hepatocytes or hepatocellular carcinoma cells (HepG2/C3a) (Gottwald et al., 2007). Co-culture of cells within a 3D matrix has also been performed. Salivary gland adenoid cystic carcinoma (ACC-M) cells were grown in co-culture with either human fetal lung fibroblasts (HFL1) or primary carcinoma-associated fibroblasts (CAF) in Cultrex basement membrane extract (BME) as a substitute for extracellular matrix (Liu et al., 2010). In a similar manner to the automation of normal culture conditions, recently automation of 3D cell culture in microfluidic devices has been described (Montanez-Sauri et al., 2011). Human mammary fibroblasts (HMF) and human breast carcinoma (T47D) cells were suspended in liquid CTI and loaded into the microfluidic device, where the collagen is then polymerized to form a 3D matrix. Cell viability, image acquisition (phase contrast and fluorescence images), and cell growth analysis were all automated for the microfluidic device.

Whole Cell Analysis

The use of microfluidic devices as cell culture systems are just a small part of their overall capabilities. As part of their repertoire, microfluidic devices have been adapted to utilize one of the most commonly used methods of single-cell analysis, flow cytometry (Lindstrom and Andersson-Svahn, 2002; Meldrum and Holl, 2002). Flow cytometry rapidly analyzes individual cells suspended within a flowing stream of fluid that passes a detector recording various parameters, such as cell number, size, shape, DNA content, viability or fluorescence intensity (Beebe et al., 2002; Huh et al., 2005; Sia and Whitesides, 2003). Fluid control in microfluidic flow cytometers is generated and controlled by pumps (syringe, vacuum, or both) or electroosmotic flow (Huh et al., 2005). The use of fluorescence allows for a semi-quantitative measurement of the desired levels of an analyte (Lindstrom and Andersson-Svahn, 2010). Using a human promyelocytic leukemia (HL-60) cell line, intracellular levels of heat shock protein 72 (Hsp72) were measured at various times following exposure to heat shock (Gottwald et al., 2003). In contrast to conventional flow cytometers, this microscale application used only one-tenth the number of cells. In addition, adaptation of automated cell culture on microfluidic chips to include sample preparation and flow cytometry have been developed. Partial profiling of the toll-like receptor 4 (TLR4) signaling pathway in a mouse macrophage (RAW264.7) cell line in response to lipopolysaccharide (LPS) was performed using microfluidic flow cytometry by monitoring TLR4 receptor complex formation and downstream phosphorylation of ERK (Wu et al., 2012).

With additional equipment, fluorescence activated cell sorting (FACS) can be used to sort cells based on desired properties for later analysis. For microfluidic devices, the microfabricated FACS (μFACS) system has been used to sort mixed populations of cells or enrich for specific types of cells within a population (Huh et al., 2005). The separation of cells can be achieved using a variety of methods. The most common in conventional FACS is achieved using mechanical means to divide the liquid stream into individual droplets that are charged after passing a detector and attracted or repulsed based on predetermined experimental conditions (Huh et al., 2005); however, the microscale devices more commonly use precise fluid control (Figure 9) via a number of methods, some of which include hydrodynamic and electrokinetic flow switching (Fu et al., 1999; Fu et al., 2002; Wolff et al.,

2003). Recently, μFACS systems have been developed that are capable of high-throughput and high sensitivity (Cho et al., 2010). With this high performance μFACS system, a human myelogenous leukemia (K562) cell line was sorted to enrich for labeled cells with a 230 fold enrichment.

Using a method similar to FACS, magnetic-activated cell sorting (MACS) enriches for a subpopulation of cells based on attachment of a magnetic bead (Sia and Whitesides, 2003). As the cell suspension passes a magnetic field, the labeled cells are deflected from the main flow path (Figure 10). For example, HeLa cells have been sorted based upon uptake of magnetic nanoparticles (Pamme and Wilhelm, 2006) and hematopoietic stem cells have been sorted based upon expression of the specific cell surface antigen CD34 (Wu et al., 2010). In addition to sorting based on one parameter, a multitarget MACS (MT-MACS) system has been developed to sort multiple magnetic tags at the same time (Adams et al., 2008). In terms of cellular assays, MACS has been used to selectively separate cells that expose phosphatidylserine on the outer leaflet of the plasma membrane, which is one marker of a cell undergoing apoptosis (Kim, H. S. et al., 2007).

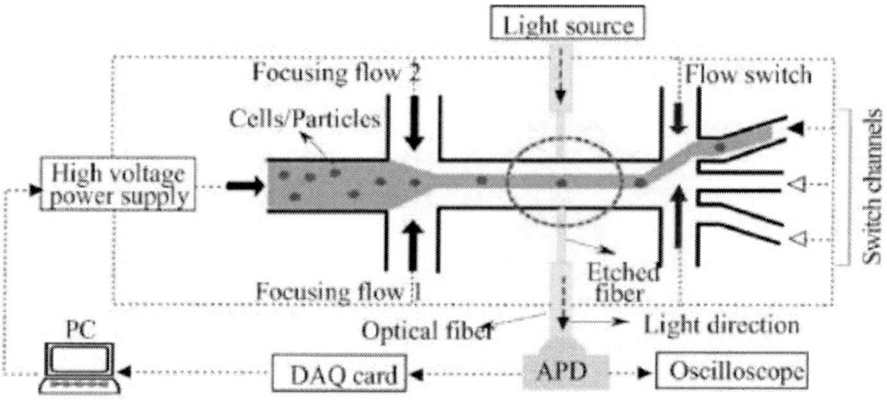

Figure 9. A diagram displaying fluorescence-based cell sorting in a microfluidic FACS device. The light source is able to detect fluorescently labeled cells and directs cells to desired channels at the end of the device. Reused from Fu et al., 2004, with permission from Elsevier.

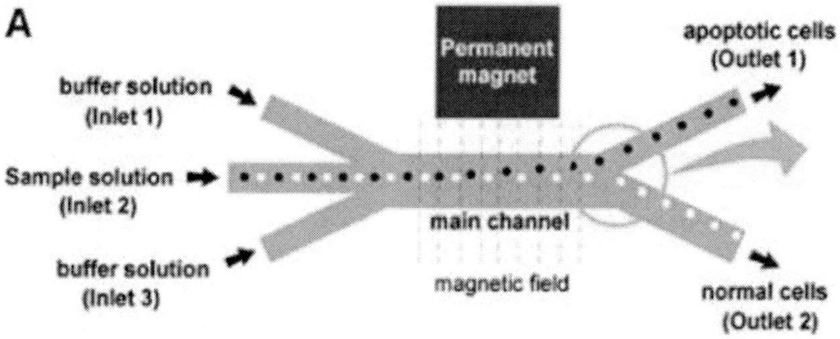

Figure 10. Drawing of the microfluidic channel. Black dots and white dots represent a magnetically labeled apoptotic cell and a normal cell, respectively. Reprinted with kind permission from Springer Science and Business Media. Kim, H. S. et al., 2007.

As microfabrication technology has progressed, the study of culture conditions more closely approximating *in vivo* conditions can be performed. Preliminary studies using osteocytes grown within microchambers resembling artificial lacunae with engineered canaliculi developed extended processes, and future studies will determine the effects of fluid flow within these artificial canaliculi (You et al., 2008). In addition, multishear microfluidic devices have been used for determining effects of bone cells exposed to varying levels of fluid shear stress.

Precise response profiles were generated of osteoblast cytosolic calcium dynamics as a result of multiple fluid shear stresses (Kou et al., 2011). In contrast to growth under normal cell culture conditions, exposing an osteoblast cell line to fluid shear stress within a microfluidic device results in elongation and alignment of cells in the direction of fluid flow (Leclerc et al., 2006). These osteoblasts also had marked increases in alkaline phosphatase activity (Leclerc et al., 2006), which has been associated with matrix producing osteoblasts (Bonewald, 2011).

Sample Preparation and Techniques

To analyze cellular contents, lysis is generally needed and is usually performed using either physical or chemical methods to disrupt the plasma membrane of cells (Yan et al., 2008). Some benefits of combining cellular lysis with microfluidic devices is the ability to lyse and analyze single cells or small populations of cells at the same time (Yan et al., 2008). Some conventional methods of lysing cells that have been adapted for microfluidic devices include chemical agents, such as solvents and detergents (Figure 11), enzymatic digestion, osmotic shock, or mechanical and electrical lysis (Lee and Tai, 1999; Yan et al., 2008; Zare and Kim, 2010). The method for disrupting the lipid bilayer is dependent on the downstream application. For example, if analysis of protein-DNA complexes is desired and lysis occurs through the use of a detergent, the effects of common surfactants on these complexes can be referenced (Pang et al., 2006).

In addition to lysis through detergents, microfluidic systems utilizing electroosmotic systems, which are capable of lysing cells through generation of a strong electric field (Figure 12), have been used to lyse mammalian and bacterial cells (Jen et al., 2012; Lee and Tai, 1999; Nevill et al., 2007; Yan et al., 2008). While capable of electrically lysing cells, electroosmotic systems are also able to precisely control the flow of droplets, and therefore well suited to the specific mixing of a chemical agent with samples for lysis, such as a detergent (Huang et al., 2007; Li and Harrison, 1997). Mechanical means of lysis have also been developed for microfluidic devices (Figure 13).

Applying compressive stress by deflection of a PDMS membrane, mammary gland epithelium (MCF7) cells were lysed (Kim, Y. C. et al., 2007). Also, integration of multiple technologies on a single microfluidic chip (FACS and sonication) has allowed for lysis of HL-60 cells and bacterial spores as they passed through a microchannel (Marentis, 2005).

While adaptation of macro-scale cell lysis techniques for micro-scale samples is feasible, this conversion to microfluidic devices can be challenging due to analyte dilution. As a result, microfluidic devices capable of cell lysis on the pico-scale have been designed for single cell chemical cytometry analysis (He et al., 2005; Irimia et al., 2004).

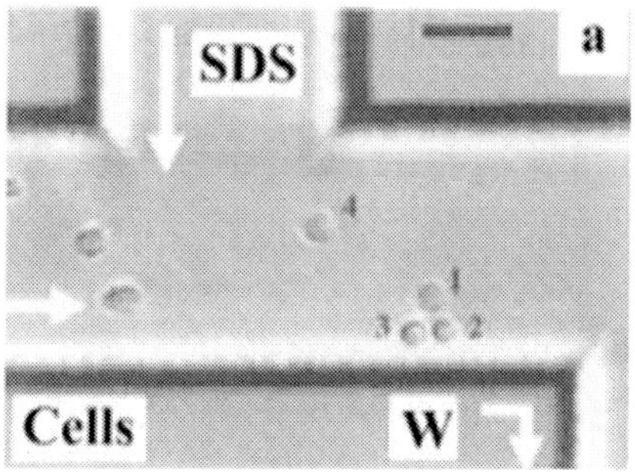

Figure 11. Example of chemical lysis of cells in a microfluidic device. As cells flow through the microchannel, they are exposed to a lysing agent (SDS) to release cellular contents. Reprinted with permission from Irimia et al., 2004. Copyright 2004 American Chemical Society.

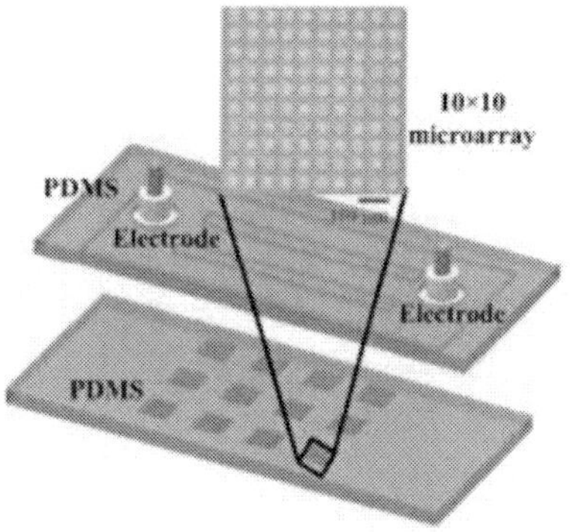

Figure 12. Example of an electrical lysis microfluidic device. Cells in each microwell can be lysed with the application of voltage between the two electrodes. Reprinted with permission from Jen et al., 2012.

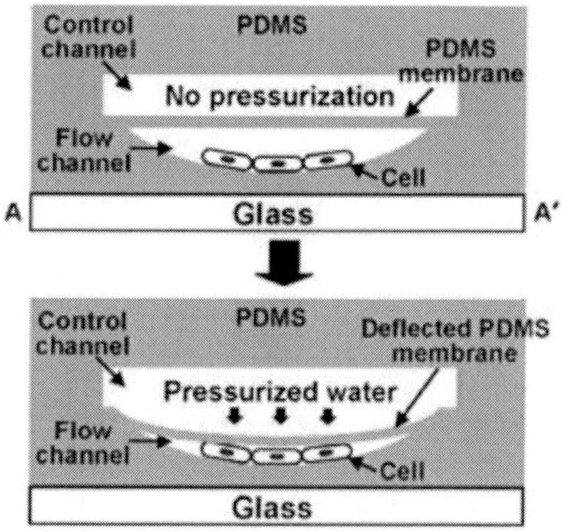

Figure 13. Example of a mechanical lysis microfluidic device. Cells are seeded into a flow channel. This flow channel can be compressed by a valve, leading to mechanical lysis of the cells. Reprinted from Kim, Y. C. et al., 2007, with permission from Elsevier.

Depending upon the downstream application, further sample preparation may be needed before chemical cytometry can be employed. For example, isoelectric focusing (IEF) can be used to separate proteins and peptides by application of an electric field in a pH gradient.

These molecules migrate through the gradient until their isoelectric point (pI) is reached and the net charge is zero (Kenyon et al., 2011).

Microfluidic devices using isoelectric focusing have separated individual organelles such as mitochondria and nuclei for further analysis. As each organelle contains different membrane proteins, each organelle mobilizes to a different position within the channel (Lu et al., 2004).

Capillary electrophoresis (CE) is another common method of separating contents based on the charge of a molecule and its mobility in an electric field. On microfluidic devices, CE offers a number of advantages, such as the ability to run multiple samples in parallel with faster separation than large scale capillary electrophoresis (Effenhauser et al., 1994; Manz et al., 1992).

Based on the characteristic of microfluidic devices, separation can occur in under two minutes with shorter sample times occurring with higher voltages or a shorter distance between the voltage application and sample (Effenhauser et al., 1994; Wooley et al., 1994). A variety of CE devices have been developed to analyze and separate the DNA, amino acid, and protein contents of cells (Effenhauser et al., 1997; Kenyon et al., 2011; Woolley et al., 1994; Woolley et al., 1996; Ye et al., 2010).

POLYMERASE CHAIN REACTION ANALYSIS IN MICROFLUIDIC SYSTEMS

As the polymerase chain reaction (PCR) is an essential tool in many fields, the ability of microfluidic devices to incorporate this technique is extremely useful. For PCR, these devices require significantly less reagents and time than conventional PCR and are capable of high-throughput analysis (Cho et al., 2011; Bebee et al., 2002).

While a large number of variations on PCR exist, the basic process includes amplification of a target region of double-stranded DNA (dsDNA). The incorporation of microheaters throughout the device allows compliance with the precise temperature requirements for DNA amplification and microfluidic devices accomplish this through the utilization of microheaters within specific channels (Chang et al., 2006; White et al., 2011; Zeng et al., 2010).

These microheaters are also able to lyse cells prior to the PCR reaction, allowing the amplification of DNA (Waters et al., 1998). For stationary samples (Figure 14), the reaction containing the DNA to be amplified, buffer, polymerase, nucleotides, and primers are all heated and cooled within the same chamber (Giordano et al., 2001; Yuen et al., 2001).

More recently, continuous flow PCR systems (Figure 15) have been developed that pump the reaction mixture through zones of heating or cooling to achieve amplification (Obeid, 2003). Compared to most conventional applications, microfluidic devices are also capable of performing PCR experiments with a high-through-put capacity with less reagent volume (White et al., 2011).

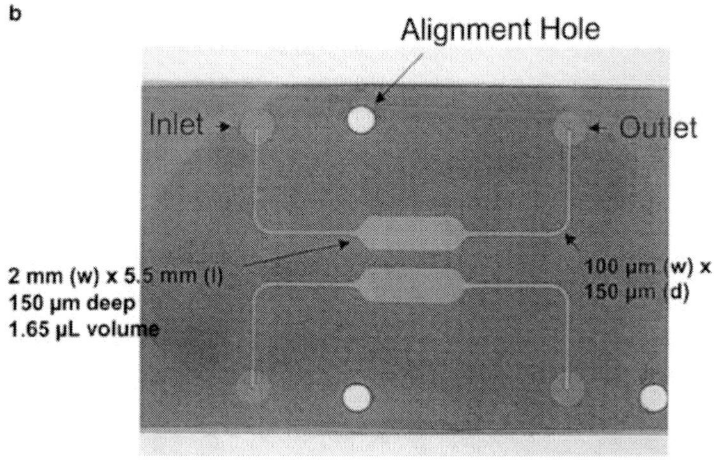

Figure 14. Example of a stationary PCR microfluidic device. The two chambers in the middle of the device are used as reaction chambers. Reprinted from Giordano et al., 2007, with permission from Elsevier.

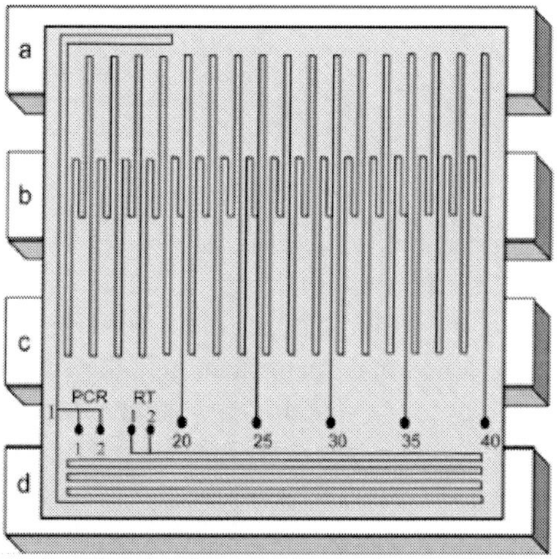

Figure 15. Example of a continuous-flow PCR microfluidic device. The reaction mixture passes over different temperature zones (a-d) during the PCR reaction. Reused with permission from Obeid et al., 2003. Copyright 2003 American Chemical Society.

PROTEIN ANALYSIS

While a number of tools for proteomic analysis have been developed for microfluidic devices, this section will only superficially discuss a few selected methods. Some of the proteomic methods that have been used in microfluidic devices include protein crystallization, protein-protein interactions, and immunoassays. In order to develop inhibitors and novel ligands, create more stable molecules, and extract valuable information concerning structure-function relationships, protein crystallization is one of the many methods used to determine the 3D structures of proteins. Briefly, the observation that proteins will form crystals in solution when a point of supersaturation is reached is exploited. When high quality crystals composed of repeating structures called unit cells are formed, a method called X-ray crystallography can be used to determine the 3D structure of a protein. Screening the sensitive conditions needed to form crystals of a protein can be extremely time intensive, laborious, and require large amounts of reagents, especially important protein samples. However, the ability of microfluidic devices to rapidly screen these conditions in droplet-based microfluidic systems requires only four nanoliters of a protein solution (Zheng et al., 2003).

In addition, microfluidic devices can be coupled with downstream mass spectrometry for the detection of peptides, proteins, and other small molecules. While these require an external electrospray ionization mass spectrometry, they are capable of processing the samples to be analyzed, and some advantages to this coupling include reduced dead volume, interference, sample loss and consumption (Chan et al., 1999; Jiang et al., 2001). Briefly, peptides are ionized using a number of common methods to form charged particles, separated based on mass-to-charge (m/z) ratio, then detected and processed into a collision-induced dissociation spectra (CID) (Chan et al., 1999). This profile is unique to the protein or peptide amino acid sequence and the identity and relative amount of protein can be determined. In combination with other methods discussed above, myoglobin has been digested into fragments which were separated based on their isoelectric points, then directly transferred to a mass spectrometer for spectral analysis (Nesbitt et al., 2008).

Immunoassays

Incorporation of immunoassays into microfluidic devices overcomes the long incubation times needed for conventional methods by limiting the antigen-antibody diffusion distance to provide rapid analyses (Cho et al., 2011; Dodge et al., 2001; Sia and Whitesides, 2003). A common method of antigen detection and quantification is the enzyme-linked immunosorbent assay (ELISA). Briefly, antibodies to a specific antigen are immobilized on a solid surface, exposed to a sample containing the antigen, and then linked to an enzyme that will cause a colorimetric change. A number of permutations on this theme have been developed and adapted to microfluidic devices for rapid analysis (Cho et al., 2011). For example, ELISAs on microfluidic devices have been developed to detect a number of disease causing viruses and bacteria, such as the foot-and-mouth disease virus (FMDV), dengue virus (DV), and *E. coli* O157:H7 (Dong et al., 2011; Weng et al., 2011; Sen et al., 2011). In addition, newer techniques are continually being developed as needs arise for more specialized microfluidic ELISAs, such as detection of bone remodeling markers (Numthuam et al., 2011).

Pharmacological Treatment of Cells

While there has been considerable research with respect to microfluidic devices and treatment of cells with pharmacological agents, this chapter will only briefly highlight a few studies. The chemical gradients generated by microfluidic devices (Figure 16) are extremely precise and capable of exposing cells to a variety of treatments and factors (Chung et al., 2005; Dertinger et al., 2001; Hung et al., 2005; Jeon et al., 2000). By varying the concentration and flow rate of the inputs of a microfluidic device, a precise gradient can be established at the output of the device, and the effects of various chemicals and compounds on cells can be monitored in real time. By using many individual chambers, the effects of different cytokine gradients to enrich neural progenitor cells from human embryonic stem cells have been analyzed (Park et al., 2009). In addition, the precise gradients allow for the determination of specific conditions for optimizing growth and differentiation of select cells, such as neural stem cells (Chung et al., 2005). These treatments are not limited to monolayer cultures. Human breast cancer (MCF-7) cells grown within a collagen-gel matrix on a multichannel microfluidic chip have been treated to determine drug sensitivity to various anticancer agents (Torisawa et al., 2005). In order to test the effectiveness of various osteoporosis therapies, a microfluidic platform was designed using a human osteosarcoma cell line (SaOS-2) (Nason et al., 2011). While the study has optimized various paramaters of the device, such as material, chamber size, future studies may test the effects of osteoporosis drugs. Biomimetic bone models using microfluidics have also been used for developing models of wound healing and infection resistance of novel biomaterials for orthopedic implants (Lee et al., 2010; Lee et al., 2011).

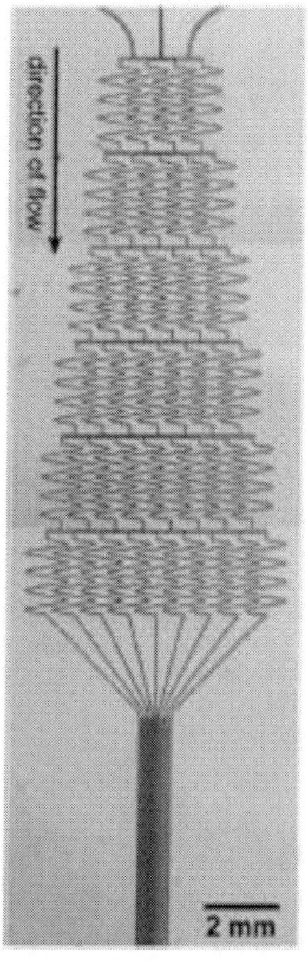

Figure 16. Method of generating a chemical gradient in microfluidic devices. As more branching occurs vertically down the device, slight mixing occurs between the above inputs. By varying the concentration and flow rate of each input, a precise gradient can be established at the output of the device. Reprinted with permission from Dertinger et al., 2001. Copyright 2001, American Chemical Society.

Communication Studies Using Microfluidic Devices

Microfluidic devices have also been utilized as a method of examining intercellular communication (Wei et al., 2006). In order to study the effects of one cell type on another, a platform was developed (Figure 17) with multiple upstream wells that were connected to downstream wells by microchannels (Wei et al., 2006). The study was designed to determine the effects that macrophages responding to debris at the bone implant interface have on local osteoblasts. This was modeled by seeding and stimulating macrophages in wells upstream of osteoblasts, and then determining the response of macrophage secreted factors on osteoblasts. This stimulation caused the release of bone resorption markers from osteoblasts interleukin-6 and prostaglandin E2 (Wei et al., 2006). Recently, intercellular communication studies have been completed by seeding one type of cell on the surface of a microchannel and then flowing a second cell type through the device, allowing studies into the role of cellular contact (Lawrence et al., 2012). These types of studies may have unique implications for studying novel bone modeling interactions. While initial studies utilizing bone cells and microfluidic devices have been completed, this foundation presents tremendous potential as a model for bone studies.

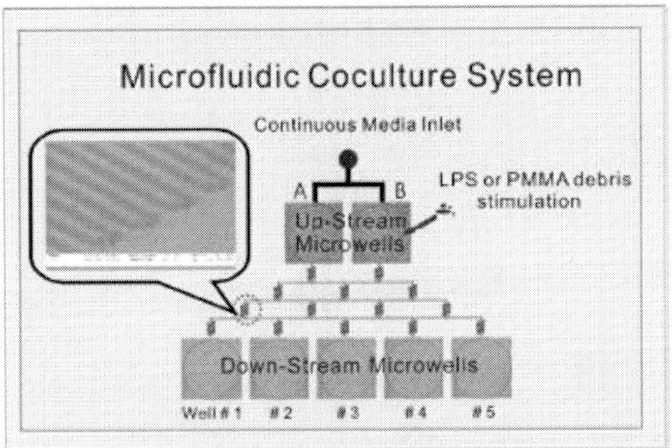

Figure 17. Schematic of the microfluidic coculture system. The macrophages were cultured in up-stream microwells and osteoblasts were cultured in down-stream microwells. Wei et al., 2006. Reprinted with kind permission from Springer Science and Business Media.

Conclusion

Many naturally occurring materials have unique characteristics that if harnessed and exploited could be used to benefit society. As we have tried to show, bone is one such material. Given its ability to adapt to its loading environment and its capacity for self-renewal, bone possesses innate properties that are enviable in engineering and architectural fields. Moreover, unlocking the mechanisms and pathways by which bone adapts and self-renews is critical to understanding the normal function of bone and a necessary step to understanding the breakdown in normal function that results in pathologies. Expanding the tools, techniques and modeling approaches we use in the research setting is key to advancing discovery. Here we provided a brief introduction of the function of bone, the tools and techniques of microfluidics and applications of the current capabilities of microfluidic devices in research. We believe the use of microfluidics has tremendous potential to mimic the complexity of the bone environment in *in vitro* studies and will lead to new and exciting discovery in the bone field.

REFERENCES

Adams, J. D., Kim, U., and Soh, H. T. (2008). Multitarget magnetic activated cell sorter. *Proc. Natl. Acad. Sci. US*, 105(47), 18165-70.

Anderson, J. R., Chiu, D. T., Jackman, R. J., Cherniavskaya, O., McDonald, C. J., Wu, H., Whitesides, S. H., and Whitesides, G. M. (2000). Fabrication of topologically complex three-dimensional microfluidic systems in P. D. M. S. by rapid prototyping. *Anal. Chem.*, 72(14), 3158-64.

Armstrong, V. J., Muzylak, M., Sunters, A., Zaman, G., Saxon, L. K., Price, J. S., and Lanyon, L. E. (2007). Wnt/β-catenin signaling is a component of osteoblastic bone cell early responses to load bearing and requires estrogen receptor α. *J. Biol. Chem.*, 282(28), 20715-27.

Bailey, A. J., Sims, T. J., and Knott, L. (2002). Phenotypic expression of osteoblast collagen in osteoarthiritic bone: production of type I homotrimer. *Int. J. Biochem. Cell Biol.*, 34(2), 176-82.

Bakker, A. D. and Klein-Nulend, J. (2010). Mechanisms of osteocyte mechanotransduction. *Clinic. Rev. Bone Miner. Metab.*, 8(4),163-9.

Balagadde, F. K., You, L., Hansen, C. L., Arnold, F. H., and Quake, S. R. (2005). Long-term monitoring of bacteria undergoing programmed population control in a microchemostat. *Science,* 309(5731), 137-49.

Baron, R. E. (1996). Anatomy and ultrastructure of bone. In Favus, M. J. editor. *Primer on the Metabolic Bone Diseases and Disorders of Mineral Metabolism.* (3rd edition, 3-10). Philadelphia, S. T.: Lippincott-Raven Publishers.

Beebe, D. J., Mensing, G. A., and Walker, G. M. (2002). Physics and applications of microfluidics in biology. *Annu. Rev. Biomed. Eng.*, 4, 261-86.

Beebe, D. J., Moore, J. S., Bauer, J. M., Yu, Q., Liu, R. H., Devadoss, C., and Jo, B. H. (2000). Functional hydrogel structures for autonomous flow control inside microfluidic channels. *Nature*, 404(6778), 588-90.

Blair, H. C., Teitelbaum, S. L., Ghiselli, R., and Gluck, S. (1989). Osteoclastic bone resorption by a polarized vacuolar proton pump. *Science*, 245(4920), 855-7.

Bonewald, L. F. (2006). Mechanosensation and transduction in osteocytes. *Bonekey Osteovision*, 3(10), 7-15.

Bonewald, L. F. (2011). The amazing osteocyte. *J. Bone Miner. Res.*, 26(2), 229-38.

Bonewald, L. F., and Johnson, M. L. (2008). Osteocytes, mechanosensing, and Wnt signaling. *Bone*, 42(4), 606-15.

Boyce, B. F., and Xing, L. (2007). Biology of R. A. N. K., R. A. N. K. L., and osteoprotegerin. *Arthiritis Res. Ther.,* 9(Suppl. 1), S1-7.

Brody, J. P., and Yager, P. (1997). Diffusion-based extraction in a microfabricated device. *Sens. Actuators*, A, 58(1), 13-18.

Burger, E. H., and Klein-Nulend, J. (1999). Mechanotransduction in bone – role of the lacuna-canalicular network. *FASEB J*, 13(Suppl.), S101-12.

Burger, E. H., Klein-Nulend, J., van der Plas, A., and Nijweide, P. J. (1995). Function of osteocytes in bone-their role in mechanotransduction. *J. Nutr.*, 125(Suppl. 7), 2020S-3.

Case, N., Ma, M., Sen, B., Xie, Z., Gross, T. S., and Rubin, J. (2008). B-catenin levels influence rapid mechanical responses in osteoblasts. *J. Biol. Chem.*, 283(43), 29196-205.

Chan, J. H., Timperman, A. T., Qin, D., and Aebersold, R. (1999). Microfabricated polymer devices for automated sample delivery of peptides for analysis by electrospray ionization tandem mass spectrometry. *Anal. Chem.*, 71(20), 4437-44.

Chang, Y. H., Lee, G. B., Huang, F. C., Chen, Y. Y., and Lin, J. L. (2006). Integrated polymerase chain reaction chips utilizing digital microfluidics. *Biomed. Microdevices*, 8(3), 215-25.

Chen, N. X., Ryder, K. D., Pavalko, F. M., Turner, C. H., Burr, D. B., Qui, J., and Duncan, R. L. (2000). Ca^{2+} regulates fluid shear-induced cytoskeletal reorganization and gene expression in osteoblasts. *Am. J. Physiol. Cell Physiol.*, 278(5), C989-97.

Cho, S. H., Chen, C. H., Tsai, F. S., Godin, J. M., and Lo, Y. H. (2010). Human mammalian cell sorting using a highly integrated micro-fabricated fluoresence-activated cell sorter (μFACS). *Lab. Chip.*, 10(12), 1567-73.

Cho, S., Kang, D. K., Choo, J., de Mello, A. J., and Chang, S. . (2011). Recent advances in microfluidic technologies for biochemistry and molecular biology. *B. M. B. Rep.*, 44(11), 705-12.

Choucha-Snouber, L., Aninat, C., Grsicom, L., Madalinski, G., Brochot, C., Poleni, P. E., Razan, F., Guillouzo, C. G., Legallais, C., Corlu, A., and Leclerc, E. (2012). Investigation of ifosfamide nephrotoxicity induced in a liver kidney coculture biochip. *Biotechnol. Bioeng.*, In press.

Chung, B. G., Flanagan, L. A., Rhee, S. W., Schwartz, P. H., Lee, A. P., Monuki, E. S., and Jeon, N. L. (2005). Human neural stem cell growth and differentiation in a gradient-generating microfluidic device. *Lab. Chip.*, 5(4), 401-6.

Civitelli, R. (2008). Cell-cell communication in the osteoblast/osteocyte lineage. *Arch. Biochem. Biophys.*, 473(2), 188-92.

Clarke, B. (2008). Normal bone anatomy and physiology. *Clin. J. Am. Soc. Nephrol.*, 3(Suppl. 3), S131-9.

Cui, H., Horiuchi, K., Dutta, P., and Ivory, C. F. (2005). Multistage isoelectric focusing in a polymeric microfluidic chip. *Anal. Chem.*, 77(24), 7878-86.

Dallas, S. L., and Bonewald, L. F. (2010). Dynamics of the transition from osteoblast to osteocyte. *Ann. N. Y. Acad. Sci.*, 1192, 437-43.

Dertinger, S. K. W., Chiu, D. T., Jeon, N. L., and Whitesides, G. M. (2001). Generation of gradients having complex shapes using microfluidic networks. *Anal. Chem.*, 73(6), 1240-6.

Dodge, A., Fluri, K., Verpoorte, E., and de Rooij, N. F. (2001). Electrokinetically driven microfluidic chips with surface-modified chambers for heterogeneous immunoassays. *Anal. Chem.*, 73(14), 3400-9.

Donahue, H. J. (2000). Gap junctions and biophysical regulation of bone cell differentiation. *Bone*, 26(5),417-22.

Dong, Y., Xu, Y., Liu, Z., Fu, Y., Ohashi, T., Tanaka, Y., Mawatari, K., and Kitamori, T. (2011). Rapid screening swine foot-and-mouth disease virus using micro-ELISA system. *Lab. Chip.*, 11(13), 2153-5.

Dudley, H. R., and Spiro, D. (1961). The fine structure of bone cells. *J. Biophys. Biochem. Cytol.*, 11(3), 627-49.

Duffy, D. C., McDonald, J. C., Schueller, O. J. A., and Whitesides, G. M. (1998). Rapid prototyping of microfluidic systems in poly(dimethylsiloxane). *Anal. Chem.*, 70(23), 4974-84.

Duncan, R. L., and Turner, C. H. (1995). Mechanotransduction and the functional response of bone to mechanical strain. *Calcif. tissue Int.*, 57(5), 344-58.

Eastoe, J. E., and Eastoe, B. (1954). The organic constituents of mammalian compact bone. *Biochem. J.*,57(3), 453-9.

Effenhauser, C. S., Bruin, G. J., Paulus, A., and Ehrat M. (1997). Integrated capillary electrophoresis on flexible silicone microdevices: analysis of D. N. A. restriction fragments and detection of single D. N. A. molecules on microchips. *Anal. Chem.*, 69(17), 3451-7.

Effenhauser, C. S., Paulus, A., Manz, A., and Widmer, M. H. (1994). High-speed separation of antisense oligonucleotides on a micromachined capillary electrophoresis device. *Anal. Chem.*, 66(18), 2949-53.

Ehrlich, P. J., and Lanyon, L. E. (2002). Mechanical strain and bone cell function: a review. *Osteoporos. Int.*, 13(9), 688-700.

Estrada, R., Giridharan, G. A., Nguyen, M. D., Roussel, T. J., Shakeri, M., Parichehreh, V., Prabhu, S. D., and Sethu, P. (2011). Endothelial cell culture model for replication of physiological profiles of pressure, flow, stretch, and shear tress in vitro. *Anal. Chem.*, 83(8), 3170-7.

Eteshola, E., and Leckband, D. (2001). Development and characterization of an E. L. I. S. A. assay in P. D. M. S. microfluidic channels. *Sens. Actuators B*, 72(2), 129-33.

Everts, V., Delaisse, J. M., Korper, W., Jansen, D. C., Tigchelaar-Gutter, W., Saftig, P., and Beertsen, W. (2002). The bone lining cell: its role in cleaning Howship's lacunae and initiating bone formation. *J. Bone Miner. Res.,* 17(1), 77-90.

Fiorini, G. S., and Chiu, D. T. (2005). Disposable microfluidic devices: fabrication, function, and application. *Biotechniques*, 38(3), 429-46.

Fraikin, J. L., Teesalu, T., McKenney, C. M., Ruoslahti, E., and Cleland, A. N. (2011). A high-throughput label-free nanoparticle analyser. *Nat. Nanotechnol.*, 6(5), 308-13.

Franz-Odendaal, T. A., Hall, B. K., and Witten, P. E. (2006). Buried alive: how osteoblasts become osteocytes. *Dev. Dyn.*, 235(1), 176-90.

Fu, A. Y., Chou, H. P., Spence, C., Arnold, F. H., and Quake, S. R. (2002). An integrated microfabricated cell sorter. *Anal. Chem.*, 74(11), 2451-7.

Fu, A. Y., Spence, C., Scherer, A., Arnold, F. H., and Quake, S. R. (1999). A microfabricated fluorescence-activated cell sorter. *Nat. Biotechnol.*, 17(11), 1109-11.

Fu. L. M., Yang, R. J., Lin, C. H., Pan, Y. J., and Lee, G. B. (2004). Electrokinetically driven micro flow cytometers with integrated fiber optics for on-line cell/particle detection. *Anal. Chim. Acta.*, 507(1), 163-9.

Fujii, T. (2002). PDMS-based microfluidic devices for biomedical applications. *Microelectron. Eng.*, 61-62, 907-14.

Gao, J., Yin, X. F., and Fang, Z. L. (2004). Integration of single cell injection, cell lysis, separation and detection of intracellular constituents on a microfluidic chip. *Lab. Chip.*, 4(1), 47-52.

Geddes, A. C. (1913). The origin of the osteoblast and of the osteoclast. *J. Anat. Physiol.*, 47(pt 2), 159-76.Glorieux, F. H., Travers, R., Taylor, A., Bowen, J. R., Rauch, F., Norman, M., and Parfitt, A. M. (2000). Normative data for iliac bone histomorphometry in growing children. *Bone*, 26(2), 103-9.

Gilsanz, V. (2006). Imaging in children and adults. In Langman, C. B., and Jan de Beur, S. M., editors. *Primer on the Metabolic Bone Diseases and Disorders of Mineral Metabolism*. (6th edition, 133-150). Washington, D. C., S. T.: American Society for Bone and Mineral Research.

Giordano, B. C., Ferrance, J., Swedberg, S., Huhmer, A. F. R., and Landers, J. P. (2001). Polymerase chain reaction in polymeric microchips: D. N. A. amplification in less than 240 seconds. *Anal. Biochem.*, 291(1), 124-32.

Glorieux, F. H., Travers, R., Taylor, A., Bowen, J. R., Rauch, F., Norman, M., and Parfitt, A. M. (2000). Normative data for iliac bone histomorphometry in growing children. *Bone,* 26(2), 103-9.

Godin, M., Delgado, F. F., Son, S., Grover, W. H., Bryan, A. K., Tzur, A., Jorgensen, P., Payer, K., Grossman, A. D., Kirschner, M. W., and Manalis, S. R. (2010). Using buoyant mass to measure the growth of single cells. *Nat. Methods*, 7(5), 387-90.

Gomez-Sjoberg, R., Leyrat, A. A., Pirone, D. M., Chen, C. S., and Quake, S. R. (2007). Versatile, fully automated, microfluidic cell culture system. *Anal. Chem.*, 79(22), 8557-63.

Gottwald, E., Giselbrecht, S., Augspurger, C., Lahni, B., Dambrowsky, N., Truckenmuller, R., Piotter, V., Gietzelt, T., Wendt, O., Pfleging, W., Welle, A., Rolletschek, A., Wobus, A. M., and Weibezahn, K. F. (2007). A chip-based platform for the in vitro generation of tissues in three-dimensional organization. *Lab. Chip.*, 7(6), 777-85.

Gottwald, E., Lahni, B., Ludke, G., Preckel, T., and Buhlmann, C. (2003). Intracellular HSP72 detection in HL60 cells using a flow cytometry system based on microfluidic analysis. *Biotechniques*, 35(2), 358-62, 364, 366-7.

Gruber, H. E., Ivey, J. L., Thompson, E. R., Chesnut, C. H. 3rd, and Baylink, D. J. (1986). Osteoblast and osteoclast cell number and cell activity in postmenopaUSl osteoporosis. *Miner. Electrolyte. Metab.*, 12(4), 246-54.

Guo, X. E., Takai, E., Jiang, X., Xu, Q., Whitesides, G. M., Yardley, G. T., Hung, C. T., Chow, E. M., Hantschel, T., and Costa, K. D. (2006). Intracellular calcium waves in bone cell networks under single cell nanoindentation. *Mol. Cell Biomech.*, 3(3), 95-107.

Hatch, A., Kamholz, A. E., Hawkins, K. R., Munson, M. S., Schilling, E. A., Weigl, B. H., and Yager, P. (2001). A rapid diffusion immunoassay in a T-sensor. *Nat. Biotechnol.*, 19(5), 461-5.

He, M., Edgar, J. S., Jeffries, G. D. M., Lorenz, R. M., Shelby, P. J., and Chiu, D. T. (2005). Selective encapsulation of single cells and subcellular organelles into picoliter- and femtoliter-volume droplets. *Anal. Chem.*, 77(6), 1539-44.

Henriksen, K., Neutzsky-Wulff, A. V., Bonewald, L. F., and Karsdal, M. A. (2009). Local communication on and within bone controls bone remodeling. *Bone*, 44(6), 1026-33.

Hisamoto, H., Funano, S. I., and Terabe, S. (2005). Integration of valving and sensing on a capillary-assembled microchip. *Anal. Chem.*, 77(7), 2266-71.

Hodge, A. J., and Petruska, J. A. (1963). Recent studies with the electron microscope on ordered aggregates of the tropocollagen molecule. In Ramachandran, G. N., editor. *Aspects Protein Structure*. (1st edition, 289-300). New York, S. T.: Academic Press.

Hu, Z., Zhang, X., and Li, Y. (1995). Synthesis and application of modulated polymer gels. *Science*, 269(5223), 525-7.

Huang, B., Kim, S., Wu, H., and Zare, R. N. (2007). Use of a mixture of n-dodecyl-β-D-maltoside and sodium dodecyl sulfate in poly(dimethylsiloxane) microchips to suppress adhesion and promote separation of proteins. *Anal. Chem.*, 79(23), 9145-9.

Huddleston, A. L., Rockwell, D., Kulund, D. N., and Harrison, R. B. (1980). Bone mass in lifetime tennis athletes. *J. Am. Med. Assoc.*, 244(10), 1107-9.

Huh, D., Gu, W., Kamotani, Y., Grotberg, J. B., and Takayama, S. (2005). Microfluidics for flow cytometric analysis of cells and particles. *Physiol. Meas.*, 26(3), R73-98.

Hung, C. T., Pollack, S. R., Reilly, T. M., and Brighton, C. T. (1995). Real-time calcium response of cultured bone cells to fluid flow. *Clin. Orthop. Relat. Res.*, 313, 256-69.

Hung, P. J., Lee, P. J., Sabounchi, P., and Lee, L. P. (2005). Continuous perfusion microfluidic cell culture array for high-throughput cell-based assays. *Biotechnol. Bioeng.*, 89(1), 1-8.

Irimia, D., Tompkins, R. G., and Toner, M. (2004). Single-cell chemical lysis in picoliter-scale closed volumes using a microfabricated device. *Anal. Chem.*, 76(20), 6137-43.

Ismagilov, R. F., Rosmarin, T. D., Kenis, J. A., Chiu, D. T., Zhang, W., Stone, H. A., and Whitesides, G. M. (2001). Pressure-driven laminar flow in tangential microchannels: an elastomeric microfluidic switch. *Anal. Chem.*, 73(19), 4862-7.

Jacobs, C. R., Yellowley, C. E., Davis, B. R., Zhou, Z., Cimbala, J. M., and Donahue, H. J. (1998). Differential effect of steady versus oscillating flow on bone cells. *J. Biomech.*, 31(11), 969-76.

Jandik, P., Weigl, B. H., Kessler, N., Cheng, J., Morris, C. J., Schulte, T., and Avdalovic, N. (2002). Initial study of using a laminar fluid diffusion interface for sample preparation in high-performance liquid chromatography. *J. Chromatogr. A*, 954(1-2), 33-40.

Jekir, M. G., and Donahue, H. J. (2009). Gap junctions and osteoblast-like cells gene expression in response to fluid flow. *J. Biomech. Eng.*, 131(1), 011005-1-6.

Jen, C. P., Amstislavskaya, T. G., Liu, Y. H., Hsiao, J. H., and Chen, Y. H. (2012). Single-cell electric lysis on an electroosmotic-driven microfluidic chip with arrays of microwells. *Sensors*, 12(6), 6967-77.

Jeon, N. L., Dertinger, S. K. W., Chiu, D. T., Choi, I. S., Stroock, A. D., and Whitesides, G. M. (2000). Generation of solution and surface gradients using microfluidic systems. *Langmuir.*, 16(22), 8311-16.

Jiang, Y., Wang, P. C., Locascio, L. E., and Lee, C. S. (2001). Integrated plastic microfluidic devices with ESI-MS for drug screening and residue analysis. *Anal. Chem.*, 73(9), 2048-53.

Jones, S. J., Gray, C., Sakamaki, H., Arora, M., Boyde, A., Gourdie, R., and Green, C. (1993). The incidence and size of gap junctions between the bone cells in rat calvaria. *Anat. Embryol.* (Berl.), 187(4), 343-52.

Kamiya, N., and Takagi, M. (2001). Differential expression of dentin matrix protein 1, type I collagen and osteocalcin genes in rat developing mandibular bone. *Histochem. J.*, 33(9-10), 545-52.

Kataoka, K., Miyazaki, H., Bunya, M., Okano, T., and Sakurai, Y. (1998). Totally synthetic polymer gels responding to external glucose concentration: their preparation and application to on-off regulation of insulin release. *J. Am. Chem. Soc.*, 120(1998), 12694-5.

Kenyon, S. M., Meighan, M. M., and Hayes, M. A. (2011). Recent developments in electrophoretic separations on microfluidic devices. *Electrophoresis*, 32(5), 482-93.

Kim, H. S., Son, O. T., Kim, K. H., Kim, S. H., Maeng, S., and Jung, H. I. (2007). Separation of apoptotic cells using a microfluidic device. *Biotechnol. Lett.*, 29(11), 1659-63.

Kim, Y. C., Kang, J. H., Park, S. J., Yoon, E. S., and Park, J. K. (2007). Microfluidic biomechanical device for compressive cell stimulation and lysis. *Sens. Actuators, B*, 128(1), 108-16.

Knothe Tate, M. L. (2003). "Whither flows the fluid in bone?" An osteocyte's perspective. *J. Biomech.*, 36(10), 1409-24.

Kou, S., Pan, L., van Noort, D., Meng, G., Wu, X., Sun, H., Xu, J., and Lee, I. (2011). A multishear microfluidic device for quantitative analysis of calcium dynamics in osteoblasts. *Biochem. Biophys. Res. Commun.*, 408(2), 350-5.

Krahl, H., Michaelis, U., Pieper, H. G., Quack, G., and Montag, M. (1994). Stimulation of bone growth through sports: a radiologic investigation of the upper extremities in professional tennis players. *Am. J. Sports Med.*, 22(6), 751-7.

Kwon, R. Y., Temiyasathit, S., Tummala, P., Quah, C. C., and Jacobs, C. R. (2008). Primary cilia mediate intracellular cAMP responses in bone cells exposed to dynamic fluid flow. A. S. M. E. Summer Bioengineering Conference 2008.

Landis, W. J. (1995). The strength of a calcified tissue depends in part on the molecular structure and organization of its constituent mineral crystals in their organic matrix. *Bone*, 16(5), 533-44.

Lau, K. H., Kapur, S., Kesavan, C., and Baylink, D. J. (2006). Up-regulation of the Wnt, estrogen receptor, insulin-like growth factor-I, and bone morphogenetic protein pathways in C57BL/6J osteoblasts as opposed to C3H/HeJ osteoblasts in part contributes to the differential anabolic response to fluid shear. *J. Biol. Chem.*, 281(14), 9576-88.

Launey, M. E., Buehler, M. J., and Ritchie, R. O. (2010). On the mechanistic origins of toughness of bone. *Annu. Rev. Mater. Res.*, 40(1), 25-53.

Lawrence, M. B., McIntire, L. V., and Eskin, S. G. (2012). Effect of flow on polymorphonuclear leukocyte/endothelial cell adhesion. *Blood*, 70(5), 1284-90.

Leclerc, E., David, B., Griscom, L., Lepioufle, B., Fujii, T., Layrolle, P., and Legallasia, C. (2006). Study of osteoblastic cells in a microfluidic environment. *Biomaterials*, 27(4), 586-95.

Lee, C. Y., Chang, C. L., Wang, Y. N., and Fu. L. M. (2011). Microfluidic mixing: a review. *Int. J. Mol. Sci.*, 12(5), 3263-87.

Lee, J. H., Wang, H., Kaplan, J. B., and Lee, W. Y. (2010). Effects of Staphylococcus epidermidis on osteoblast cell adhesion and viability on a Ti alloy surface in a microfluidic co-culture environment. *Acta Biomater.*, 6(11), 4422-9.

Lee, S. W., and Tai, Y. C. (1999). A micro cell lysis device. *Sens. Actuators, A*, 73(1-2), 74-9.

Li, J., Zhang, F., and Chen, J. Y. (2011). An integrated proteomics analysis of bone tissues in response to mechanical stimulation. *B. M. C. Syst. Biol.*, 5(Suppl. 3), S7.

Li, P. C. H., and Harrison, D. J. (1997). Transport, manipulation, and reaction of biological cells on-chip using electrokinetic effects. *Anal. Chem.*, 69(8), 1564-8.

Li, Y. P., Chen, W., Liang, Y., Li, E., and Stashenko, P. (1999). Atp6i-deficient mice exhibit severe osteopetrosis due to loss of osteoclast-mediated extracellular acidification. *Nat. Genet.*, 23(4), 447-51.

Liao, H., Andersson, A. S., Sutherland, D., Petronis, S., Kaesmo, B., and Thomsen, P. (2003). Response of rat osteoblast to microstrucutred model surfaces in vitro. *Biomaterials*, 24(4), 649-54.

Lin, C., Jiang, X., Dai, Z., Guo, X., Weng, T., Wang, J., Li, Y., Feng, G., Gao, X., and He, L. (2009). Sclerostin mediates bone response to mechanical unloading through antagonizing Wnt/β-catenin signaling. *J. Bone Miner. Res.*, 24(10), 1651-61.

Linder, V., Wu, H., Jiang, X., and Whitesides, G. M. (2003). Rapid prototyping of 2D structures with feature sizes larger than 8 μm. *Anal. Chem.*, 75(10), 2522-27.

Lindstrom, S., and Andersson-Svahn, H. (2010). Overview of single-cell analyses: microdevices and applications. *Lab. Chip.*, 10(24), 3363-72.

Liu, T., Lin, B., and Qin, J. (2010). Carcinoma-associated fibroblasts promoted tumor spheroid invasion on a microfluidic 3D co-culture device. *Lab. Chip.*, 10(13), 1671-7.

Logan, C. Y., and Nusse, R. (2004). The wnt signaling pathway in development and disease. *Annu. Rev. Cell Dev. Biol.*, 20, 781-810.

Lu, H., Gaudet, S., Schmidt, M. A., and Jensen, K. F. (2004). A microfabricated device for subcellular organelle sorting. *Anal. Chem.*, 76(19), 5705-12.

Malone, A. M., Anderson, C. T., Tummala, P., Kwon, R. Y., Johnston, T. R., Stearns, T., and Jacobs, C. R. (2007). Primary cilia mediate mechanosensing in bone cells by a calcium-independent pathway. *Proc. Natl. Acad. Sci. US*, 104(33), 13325-30.

Mantila Roosa, S. M., Liu, Y., and Turner, C. H. (2011). Gene expression patterns in bone following mechanical loading. *J. Bone Miner. Res.*, 26(1), 100-12.

Manz, A., Harrison, J. D., Verpoorte, E. M. J., Fettinger, J. C., Paulus, A., Ludi, H., and Widmer, M. H. (1992). Planar chip technology for miniaturization and integration of separation techniques into monitoring systems: capillary electrophoresis on a chip. *J. Chromatog.*, 593(1-2), 253-8.

Marchisio, P. C., Cirillo, D., Naldini, L., Primavera, M. V., Teti, A., and Zambonin-Zallone, A. (1984). Cell-substratum interaction of cultured avian osteoclasts is mediated by specific adhesion structures. *J. Cell Biol.*, 99(5), 1696-1705.

Marentis, T. C., Kusler, B., Yaralioglu, G. G., Liu, S., Haeggstrom, E. O., and Khuri-Yakub, B. T. (2005). Microfluidic sonicator for real-time disruption of eukaryotic cells and bacterial spores for D. N. A. analysis. *Ultrasound in Med. and Biol.*, 31(9), 1265-77.

Martinez, A. W., Philips, S. T., and Whitesides, G. M. (2008). Three-dimensional microfluidic devices fabricated in layered paper and tape. *Proc. Natl. Acad. Sci. US*, 105(50), 19606-11.

Mason, J. J., and Williams, B. O. (2010). S. O. S. T. and D. K. K.: antagonists of L. R. P. family signaling as targets for treating bone disease. *J. Osteoporos.*, 2010, 460120-1-9.

McNamara, L. M., Van der Linden, J. C., Weinans, H., Prendergast, P. J. (2006). Stress-concentrating effect of resorption lacunae in trabecular bone. *J. Biomech.*, 39(4), 734-41.

Meldrum, D. R., and Holl, M. R. (2002). Microscale bioanalytical systems. *Science*, 297(5584), 1197-8.

Miller, S. C., de Saint Georges, L., and Jee, W. S. (1989). Bone lining cells: structure and function. *Scanning Microsc.*, 3(3), 953-60.

Miyata, T., Asami, N., and Uragami, T. (1999). A reversibly antigen-responsive hydrogel. *Nature*, 399(6738), 766-9.

Montanez-Sauri, S. I., Sung, K. E., Puccinelli, J. P., Pehlke, C., and Beebe, D. J. (2011). Automation of three-dimensional cell culture in arrayed microfluidic devices. *J. Lab. Autom.*, 16(3), 171-85.

Nakahama, K. (2010). Cellular communications in bone homeostasis and repair. *Cell Mol. Life Sci.*, 67(23), 4001-9.

Nason, F., Morganti, E., Collini, C., Ress, C., Bersini, S., Pennati, G., Boschetti, F., Colombini, A., Lombardi, G., Banfi, G., Lorenzelli, L., and Dubini, G. (2011). Design of microfluidic devices for drug screening on *in-vitro* cells for osteoporosis therapies. *Microelectron Eng.*, 88(8), 1801-6.

Nesbitt, C. A., Jurcic, K., and Yeung, K. K. (2008). Nanoliter-volume protein enrichment, tryptic digestion, and partial separation based on isoelectric points by C. E. for M. A. L. D. I. mass spectral analysis. *Electrophoresis*, 29(2), 466-74.

Nesbitt, S. A., and Horton, M. A. (1997). Trafficking of matrix collagens through bone-resorbing osteoclasts. *Science*, 276(5310), 266-9.

Neuman, W. F., and Neuman, M. W. (1953). The nature of the mineral phase of bone. *Chem. Rev.*, 53(1), 1-45.

Neve, A., Corrado, A., and Cantatore, F. P. (2012). Osteocytes: central conductors of bone biology in normal and pathological conditions. *Acta Physiol.* (Oxf.), 204(3), 317-30.

Nevill, J. T., Cooper, R., Dueck, M., Breslauer, D. N., and Lee, L. P. (2007). Integrated microfluidic cell culture and lysis on a chip. *Lab. Chip.*, 7(12), 1689-95.

Nguyen, N. T., and Wereley, S. T. *Fundamentals and applications of microfluidics*. 2nd Edition. Norwood, M. A.: Artech House Inc., 2006.

Numthuam, S., Ginoza, T., Zhu, M., Suzuki, H., and Fukuda, J. (2011). Gold-black micropillar electrodes for microfluidic E. L. I. S. A. of bone metabolic markers. *Analyst*, 136(3), 456-8.

Obeid, P. J., Christopoulos, T. K., Crabtree, H. J., and Backhouse. C. J. (2003). Microfabricated device for D. N. A. and R. N. A. amplification by continuous-flow polymerase chain reaction and reverse transcription-polymerase chain reaction with cycle number selection. *Anal. Chem.*, 75(2), 288-95.

Owan, I., Burr, D. B., Turner, C. H., Qiu, J., Tu, Y., Onyia, J. E., and Duncan R. L. (1997). Mechanotransduction in bone: osteoblasts are more responsive to fluid forces than mechanical strain. *Am. J. Physiol.*, 273(3 pt 1), C810-5.

Pamme, N., and Wilhelm, C. (2006). Continuous sorting of magnetic cells via on-chip free-flow magnetophoresis. *Lab. Chip.*, 6(8), 974-80.

Pang, Z., Al-Mahrouki, A., Berezovski, M., and Krylov, S. N. (2006). Selection of surfactants for cell lysis in chemical cytometry to study protein-DNA interactions. *Electrophoresis*, 27(8), 1489-94.

Parfitt, A. M. (1987). Bone and plasma calcium homeostasis. *Bone*, 8(Suppl. 1), 1-8.

Park. J. Y., Kim, S. K., Woo, D. H., Lee, E. J., Kim, J. H., and Lee, S. H. (2009). Differentiation of neural progenitor cells in a microfluidic chip-generated cytokine gradient. *Stem. Cells*, 27(11), 2646-54.

Peck, W. A. (1993). Consensus development conference: diagnosis, prophylaxis, and treatment of osteoporosis. *Am. J. Med.*, 94(6), 646-50.

Piekarski, K., and Munro, M. (1977). Transport mechanism operating between blood supply and osteocytes in long bones. *Nature*, 269(5623), 80-2.

Poole, K. E. S., van Bezooijen, R. L., Loveridge, N., Hamersma, H., Papapoulos, S. E., Lowik, C. W., and Reeve, J. (2005). Sclerostin is a delayed secreted product of osteocytes that inhibits bone formation. *FASEB J.*, 19(13), 1842-4.

Puzas, J. E. (1996). Osteoblast cell biology. In Favus, M. J, editor. *Primer on the Metabolic Bone Diseases and Disorders of Mineral Metabolism.* (3rd edition, 11-16). Philadelphia, S. T.: Lippincott-Raven Publishers.

Quake, S. R., and Scherer, A. (2000). From micro- to nanofabrication with soft materials. *Science*, 290(5496), 1536-40.

Rahimzadeh, J., Meng, F., Sachs, F., Wang, J., Verma, D., and Hua, S. Z. (2011). Real-time observation of flow-induced cytoskeletal stress in living cells. *Am. J. Physiol., Cell Physiol.,* 301(3), C646-52.

Rho, J. Y., Kuhn-Spearing, L., and Zioupos, P. (1998). Mechanical properties and the hierarchical structure of bone. *Med. Eng. Phys.*, 20(2), 92-102.

Robling, A. G., Bellido, T., and Turner, C. H. (2006). Mechanical stimulation in vivo reduces osteocyte expression of sclerostin. *J. Musculoskelet. Neuronal. Interact.*, 6(4), 354.

Rochefort, G. Y., Pallu, S., and Benhamou, C. L. (2010). Osteocyte: the unrecognized side of bone tissue. *Osteoporosis Int.*, 21(9), 1457-69.

Rodan, G. A. (1997). Bone mass homeostasis and bisphosphonate action, *Bone*, 20(1), 1-4.

Roodman, G. D. (1999). Cell biology of the osteoclast. *Exp. Hematol.*, 27(8), 1229-41.

Ross, D., Gaitan, M., and Locascio, L. (2001). Temperature measurement in microfluidic systems using a temperature-dependent fluorescent dye. *Anal. Chem.* 73(17), 4117-23.

Rossier, J. S., and Girault, H. H. (2001). Enzyme linked immunosorbent assay on a microchip with electrochemical detection. *Lab. Chip.*, 1(2), 153-7.

Rubinacci, A., Villa, I., Dondi Benelli, F., Borgo, E., Ferretti, M., Palumbo, C., and Marotti, G. (1998).Osteocyte-bone lining cell system at the origin of steady ionic current in damaged amphibian bone. *Calcif. Tissue Int.*, 63(4), 331-9.

Samy, R., Glawdel, T., and Ren, C. L. (2008). Method for microfluidic whole-chip temperature measurement using thin-film poly(dimethylsiloxane)/rhodamine b. *Anal. Chem.* 80(2), 369-75.

Saunders, M. M., and Donahue, H. J. (2004). Development of a cost-effective loading machine for biomechanical evaluation of mouse transgenic models. *Med. Eng. Phys.*, 26(7), 595-603.

Schulte, T. H., Bardell, R. L., and Weigl, B. H. (2002). Microfluidic technologies in clinical diagnostics. *Clin. Chim. Acta.*, 321(1-2), 1-10.

Semenov, M. V., Tamai, K., Brott, B. K., Kuhl, M., Sokol, S., and He, X. (2001). Head inducer Dickkopf-1 is a ligand for wnt coreceptor LRP6. *Curr. Biol.*, 11(12), 951-61.

Sen, A., Harvey, T., and Clausen, J. (2011). A microsystem for extraction, capture and detection of E-Coli O157:H7. *Biomed. Microdevices*, 13(4), 705-15.

Sethu, P., Sin, A., and Toner, M. (2006). Microfluidic diffusive filter for apheresis (leukapheresis). *Lab. Chip.*, 6(1), 83-9.

Sia, S. K., and Whitesides, G. M. (2003). Microfluidic devices fabricated in poly(dimethylsiloxane) for biological studies. *Electrophoresis*, 24(21), 3563-76.

Silver, I. A., Murrills, R. J., and Etherington, D. J. (1988). Microelectrode studies on the acid microenvironment beneath adherent macrophages and osteoclasts. *Exp. Cell Res.*, 175(2), 266-76.

Simonet, W. S., Lacey, D. L., Dunstan, C. R., Kelley, M., Chang, M. S., Luthy, R., Nguyen, H. Q., Wooden, S., Bennett, L., Boone, T., Shimamoto, G., DeRose, M., Elliott, R., Colombero, A., Tan, H. L., Trail, G., Sullivan, J., Davy, E., Bucay, N., Renshaw-Gegg, L., Hughes, T. M., Hill, D., Pattison, W., Campbell, P., Sander, S., Van, G., Tarpley, J., Derby, P., Lee, R., and Boyle, W. J. (1997). Osteoprotegerin: a novel secreted protein involved in the regulation of bone density. *Cell*, 89(2), 309-19.

Singer, F. R., and Eyre, D. R. (2008). Using biochemical markers of bone turnover in clinical practice. *Cleve. Clin. J. Med.*, 75(10), 739-50.

Skafte-Pedersen, P., Hemmingsen, M., Sabourin, D., Blaga, F. S., Bruus, H., and Dufva, M. (2012). A self-contained, programmable microfluidic cell culture system with real-time microscopy access. *Biomed. Microdevices*, 14(2), 385-99.

Spencer, G. J., Utting, J. C., Etheridge, S. L., Arnett, T. R., and Genever, P. G. (2006). Wnt signaling in osteoblasts regulates expression of the receptor activator of NFkB ligand and inhibits osteoclastogenesis in vitro. *J. Cell Sci.*, 119(Pt 7), 1283-96.

Stone, H. A., Stroock, A. D., and Ajdari, A. (2004). Engineering flows in small devices: microfluidics toward a lab-on-a-chip. *Annu. Rev. Fluid. Mech.*, 36(1), 381-411.

Sugawara, Y., Kamioka, H., Honjo, T., Tezuka, K., and Takano-Yamamoto, T. (2005). Three-dimensional reconstruction of chick calvarial osteocytes and their cell processes using confocal microscopy. *Bone*, 36(5), 877-83.

Suh, Y. K., and Kang, S. (2010). A review on mixing in microfluidics. *Micromachines*, 1(3), 82-111.

Sundaramurthy, A., Schuck, P. J., Conley, N. R., Fromm, D. P., Kino, G. S., and Moerner, W. E. (2006). Toward nanometer-scale optical photolithography: utilizing the near-field of bowtie optical nanoantennas. *Nano. Lett.*, 6(3), 355-70.

Sung, J. H., and Shuler, M. L. (2009). A micro cell culture analog (μCCA) with 3-D hydrogel culture of multiple cell lines to assess metabolism-dependent cytotoxicity of anti-cancer drugs. *Lab. Chip.* 9(10), 1385-94.

Suzuki, A., and Tanaka, T. (1990). Phase transition in polymer gels induced by visible light. *Nature*, 346(6282), 345-7.

Tanaka, T., Fillmore, D., Sun, S. T., Nishio, I., Swislow, G., and Shah A. (1980). Phase transitions in ionic gels. *Phys. Rev. Lett.*, 45(20), 1636-9.

Tanaka, T., Nishio, I., Sun, S. T., and Ueno-Nishio, S. (1982). Collapse of gels in an electric field. *Science,* 218(4571), 467-9.

Tanaka, S. M., Sun, H. B., Roeder, R. K., Burr, D. B., Turner, C. H., and Yokota, H. (2005). Osteoblast responses one hour after load-induced fluid flow in a three-dimensional porous matrix. *Calci Tissue Int.*, 76(4), 261-71.

Taylor, A. F., Saunders, M. M., Shingle, D. L., Cimbala, J. M., Zhou , Z., and Donahue, H. J. (2006). Mechanically stimulated osteocytes regulate osteoblastic activity via gap junctions. *Am. J. Physiol. Cell Physiol.*, 292(1), C545-52.

Teitelbaum, S. L. (2000). Bone resorption by osteoclasts. *Science*, 289(5484), 1504-8.

Temiyasathit, S., and Jacobs, C. R. (2010). Osteocyte primary cilium and its role in bone mechanotransduction. *Ann. N. Y. Acad. Sci.*, 1192, 422-8.

Termine, J. D., and Robey, P. G. (1996). Bone matrix proteins and the mineralization process. In Favus, M. J., editor. *Primer on the Metabolic Bone Diseases and Disorders of Mineral Metabolism*. (3rd edition, 11-16). Philadelphia, S. T.: Lippincott-Raven Publishers.

Teti, A., Blair, H. C., Teitelbaum, S. L., Kahn, A. J., Koziol, C., Konsek, J., Zambonin Zallone, A., and Schlesinger, P. H. (1989). Cytoplasmic pH regulation and chloride/bicarbonate exchange in avian osteoclasts. *J. Clin. Invest.*, 83(1), 227-33.

Theoleyre, S., Wittrant, Y., Tat, S. K., Fortun, Y., Redini, F., and Heymann, D. (2004). The molecular triad: OPG/RANK/RANKL: involvement in the orchestration of pathophysiological bone remodeling. *Cytokine Growth Factor Rev.*, 15(6), 457-75.

Thorsen, T., Maerkl, S. J., and Quake, S. R. (2002). Microfluidic large-scale integration. *Science*, 298(5593), 580-4.

Torisawa, Y. S., Shiku, H., Yasukawa, T., Nishizawa, M., and Matsue, T. (2005). Three-dimensional micro-culture system with a silicon–based cell array device for multi-channel drug sensitivity test. *Sens. Actuators, B*, 108(1-2), 654-9.

Tourovskaia, A., Figueroa-Masot, X., and Folch, A. (2005). Differentiation-on-a-chip: a microfluidic platform for long-term cell culture studies. *Lab. Chip.*, 5(1), 14-9.

Unger, M. A., Chou, H. P., Thorsen, T., Scherer, A., and Quake, S. R. (2000). Monolithic microfabricated valves and pumps by multilayer soft lithography. *Science*, 288(5463), 113-6.

Vaananen, H. K., and Horton, M. (1995). The osteoclast clear zone is a specialized cell-extracellular matrix adhesion structure. *J. Cell Sci.*, 108(Pt 8), 2729-32.

Vander Molen, M. A., Rubin, C. T., McLeod, K. J., McCauley, L. K., and Donahue, H. J. (1996). Gap junctional intercellular communication contributes to hormonal responsiveness in osteoblastic networks. *J. Biol. Chem.*, 271(21), 12165-71.

Wang, L., Zhang, Z. L., Wdieczak-Bakala, J., Pang, D. W., Liu, J., and Chen, Y. (2011). Patterning cells and shear flow conditions: convenient observation of endothelial cell remoulding, enhanced production of angiogenesis factors and drug response. *Lab. Chip.*, 11(24), 4235-40.

Warrick, J., Casavant, B., Frisk, M., and Beebe, D. (2010). A microfluidic cell concentrator. *Anal. Chem.*, 82(19), 8320-6.

Waters, L. C., Jacobson, S. C., Kroutchinina, N., Khandurina, J., Foote, R. S., and Ramsey, M. J. (1998). Microchip device for cell lysis, multiplex P. C. R. amplification, and electrophoretic sizing. *Anal. Chem.*, 70(1), 158-62.

Wei, C. W., Cheng, J. Y., and Young, T. H. (2006). Elucidating in vitro cell-cell interaction using a microfluidic coculture system. *Biomed. Microdevices*, 8(1), 65-71.

Weng, C. H., Huang, T. B., Huang, C. C., Yeh, C. S., Lei, H. Y., and Lee, G. B. (2011). A suction-type microfluidic immunosensing chip for rapid detection of the dengue virus. *Biomed. Microdevices*, 13(3), 585-95.

Westendorf, J. J., Kahler, R. A., and Schroeder, T. M. (2004). Wnt signaling in osteoblasts and bone diseases. *Gene*, 341, 19-39.

White, A. K., VanInsberghe, M., Petriv, O. I., Hamidi, M., Sikorski, D., Marra, M. A., Piret, J., Aparicio, S., and Hansen, C. L. (2011). High-throughput microfluidic single-cell RT-qPCR. *Proc. Natl. Acad. Sci. US*, 108(34), 13999-14004.

Whitesides, G. M. (2006). The origins and the future of microfluidics. *Nature*, 442(7101), 368-73.

Whitfield, J. F. (2003). Primary cilium – is it an osteocyte's strain-sensing flowmeter? *J. Cell Biochem.*, 89(2), 233-7.

Wolff, A., Perch-Nielsen, I. R., Larsen, U. D., Friis, P., Goranovic, G., Pulsen, C. R., Kutter, J. P., and Telleman, P. (2003). Integrating advanced functionality in a microfabricated high-throughput fluorescent-activated cell sorter. *Lab. Chip.*, 3(1), 22-7.

Wolff, J. *Das Gesetz der Transformation der Knochen*. Berlin: Verlag von August Hirschwald, 1892.

Woolley, A. T., Hadley, D., Landre, P., deMello, A. J., Mathies, R. A., and Northrup, M. A. (1996). Functional integration of PCR amplification and capillary electrophoresis in a microfabricated D. N. A. analysis device. *Anal. Chem.*, 68(23), 4081-6.

Woolley, A. T., and Mathies, R. A. (1994). Ultra-high-speed D. N. A. fragment separations using microfabricated capillary array electrophoresis chips. *Proc. Natl. Acad. Sci. US*, 91(24), 11348-52.

Wu, H. W., Hsu, R. C., Lin, C. C., Hwang, S. M., and Lee, G. B. (2010). An integrated microfluidic system for isolation, counting, and sorting of hematopoietic stem cells. *Biomicrofluidics*, 4(2), 024112-1-12.

Wu, M., Perroud, T. D., Srivastava, N., Branda, C. S., Sale, K. L., Carson, B. D., Patel, K. D., Branda, S. S., and Singh, A. K. (2012). Microfluidically-unified cell culture, sample preparation, imaging and flow cytometry for measurement of cell signaling pathways with single cell resolution. *Lab. chip.*, 12(6), 2823-31.

Yan, H., Zhang, B., and Wu, H. (2008). Chemical cytometry on microfluidic chips. *Electrophoresis*, 29(9), 1775-86.

Yang, S., Undar, A., and Zahn, J. D. (2006). A microfluidic device for continuous, real time blood plasma separation. *Lab. Chip.*, 6(7), 871-80.

Yasuda, H., Shima, N., Nakagawa, N., Mochizuki, S. I., Yano, K., Fujise, N., Sato, Y., Goto, M., Yamaguchi,K., Kuriyama, M., Kanno, T., Murakami, A., Tsuda, E., Morinaga, T., and Higashio, K. (1998). Identityof osteoclastogenesis inhibitory factory (O. C. I. F.) and osteoprotegrin
(O. P. G.): a mechanism by which O. P. G. / O. C. I. F. inhibits osteoclastogenesis in vitro. *Endocrinology*, 139(3), 1329-37.

Ye, F., Huang, Y., Xu, Q,, Shi, M., Zhao, S. (2010). Quantification of taurine and amino acids in mice single fibrosarcoma cell by microchip electrophoresis coupled with chemiluminesence detection. *Electrophoresis*, 31(10), 1630-6.

You, J., Yellowley, C. E., Donahue, H. J., Zhang, Y., Chen, Q., and Jacobs, C. R. (2000). Substrate deformation levels associated with routine physical activity are less stimulatory to bone cells relative to loading-induced oscillatory fluid flow. *J. Biomech. Eng.*, 122(4), 387-93.

You, L., Temiyasathit, S., Tao, E., Prinz, F., and Jacobs, C. R. (2008). 3D microfluidic approach to mechanical stimulation of osteocyte processes. *Cell Mol. Bioeng.*, 1(1), 103-7.

Yuen, P. K., Kricka, L. J., Fortina, P., Panaro, N. J., Sakazume, T., and Wilding, P. (2001). Microchip module for blood sample preparation and nucleic acid amplification reactions. *Genome. Res.*, 11(3), 405-12.

Zaidi, M., Blair, H. C., Moonga, B. S., Abe, E., and Huang, C. L. (2003). Osteoclastogenesis, bone resorption, and osteoclast-based therapeutics. *J. Bone Miner. Res.*, 18(4), 599-609.

Zare, R. N., and Kim, S. (2010). Microfluidic platforms for single-cell analysis. *Annu. Rev. Biomed. Eng.*, 12, 187-201.

Zeng, Y., Novak, R., Shuga, J., Smith, M. T., and Mathies, R. (2010). High-performance single cell genetic analysis using microfluidic emulsion generator arrays. *Anal. Chem.*, 82(8), 3183-90.

Zhang, C., and van Noort, D. (2011). Cells in microfluidics. *Top Curr. Chem.*, 304, 295-321.

Zheng, B., Roach, L. S., and Ismagilov, R. F. (2003). Screening of protein crystallization conditions on a microfluidic chip using nanoliter-size droplets. *J. Am. Chem. Soc.*, 125(37), 11170-1.

Zheng, C., Zhao, L., Chen, G., Zhou, Y., Pang, Y., and Huang, Y. (2012). Quantitative study of the dynamic tumor-endothelial cell interactions through an integrated microfluidic coculture system. *Anal. Chem.*, 84(4), 2088-93.

Zhu, Y., Warrick, J. W., Haubert, K., Beebe, D. J., and Yin, J. (2009). Infection on a chip: a microscale platform for simple and sensitive cell-based virus assays. *Biomed. Microdevices*, 11(3), 565-70.

In: Focus on Biomimetics Research
Editor: Abele Scipione

ISBN: 978-1-62417-332-5

Chapter 2

BIOMIMETIC ADHESIVES

A. P. Duarte [1,2,3]***, J. F. J. Coelho*** [1]***,***
M. T. Cidade [2]***, and J. C. Bordado*** [3]
[1]CIEPQPF, FCT/UC, Pólo II, Rua Silvio Lima, Coimbra, Portugal
[2]DCM e CENIMAT, FCT/UNL, Campus Caparica, Caparica, Portugal
[3] IBB, DEQ, IST, Lisboa, Portugal

ABSTRACT

Over the years, the conception of new materials has been inspired by several processes developed by Nature over millions of years, that presenting remarkable properties. For example Velcro, a kind of fabric made of nylon and mainly used as a fastener for clothing, bags and shoes, was developed from the interlocking mechanism of burdock seeds. Similarly, self-cleaning materials have been developed based on "Lotus effect" which describes the way in which water drops roll off the super hydrophobic leaves of lotus plants, taking dirt particles away with them.

A concise review of adhesives inspired by biological systems is presented, with a detailed comparison of the different components of the adhesion, as well as the potential final applications. The protein-based underwater adhesives produced by aquatic organisms, such as blue mussels, sea stars, acorn barnacles, sea urchins, freshwater caddis fly larvas and sandcastle worms (*phragmatopoma californica*) will also be reviewed, along with high-performance adhesives for aqueous milieu, based on the glue secreted by sandcastle worms and used for hold neighbouring particles which then assemble into hollow macroscopic housing colonies to provide protection from predators. The development of surgical adhesives inspired by marine mussels for different applications (prosthetic mesh fixation, islet transplantation, etc.) will also be addressed.

Several approaches for creating a responsive fibrillar adhesive system inspired by the gecko, using different materials, have been shown to possess the capacity to generate strong adhesions in dry conditions.

Based on that, several researchers have synthesized gecko foot-hairs as dry adhesives for wall-climbing robots. The ability of leaf beetles to climb smooth and rough surfaces using an array of micron-sized adhesive hairs (setae), of varying morphology, which help male beetles achieve strong adhesion to the elytra of females during copulation, can be used as inspiration to develop new adhesives.

A successfully developed, spider-inspired hydrophobic hair structure for forming air shield underwater suggests promising applications in underwater protection and adhesives integration.

The study of the pads of different insect species leads to the conclusion that smooth pads are specialized for smoother substrates, while rough pads generate stronger forces on a broader range of surfaces and have also inspired scientists in a number of ways. Tree frogs secrete a wetting liquid into the pad substrate contact area, which generates a relative long-range attractive interaction due to the formation of capillary bridges. This system is relevant for some technological applications, for instance, tires for passenger cars have drainage channels to speed up the removal of water from the tyre-road footprint area during wet road conditions.

Noteworthy is the fact that it is also possible to successfully develop a biomimetic structure that adheres equally well in wet and dry conditions, for example by combining the microstructure of gecko pads with a thin layer of synthetic polymer that mimics the protein glue of mussels.

The bibliographic search now presented individualizes the scope of actuation and prevalence of different adhesion mechanisms showing how they can be used for the development of new adhesion systems.

INTRODUCTION

Definition of Biomimetics

Over the years, Man has been aware that the evolution of Nature has helped create high-performance objects using commonly existing materials.

The understanding of the functions provided by organisms and processes found in Nature can help us to imitate and produce nanomaterials, nanodevices and processes.

The word "biomimetics", derived from the Greek word "biomimesis", was coined by Otto Schmitt in 1957 when, during his doctoral research, he developed a physical device that mimicked the electrical action of a nerve.

In 1974 the word "biomimetics" first appeared in Webster's dictionary and was defined as "the study of the formation, structure or function of biologically produced substances and materials (such as enzymes or silk) and biological mechanisms and processes (such as protein synthesis or photosynthesis) especially for the purpose of synthesizing similar products by artificial mechanisms which mimic natural ones". Other words used to describe it include bionics (coined in 1960 by Jack Steele of Wright-Patterson Air Force Base in Dayton, OH), biomimicry and biognosis [1]. Even if the word "biomimetic" is relatively new to us, many centuries ago our ancestors found inspiration from Nature to develop several devices and materials in different areas [1-9]. For example, Leonardo da Vinci proposed designs of flying machines based on studies of birds' flight, and 3000 years ago the Chinese tried to make artificial silk. Gustavo Eiffel gained inspiration from human skeleton and the fine structure of bone, bearing considerable body weight with great stability to design the Eiffel Tower in Paris [7]. The Crain's design, which is one of the great examples of modern progress in architecture [3], was also inspired by the cross section of bone presenting the porous architecture of trabecule, enduring fine balance of tension and compression [8, 9].

The first step in learning from Nature is to select unique properties of biological systems and use them as bioinspiration [10]. Nature's intelligent responses to external factors can inspire us to answer many scientific questions. Materials and fibres with great mechanical strength, molecular scale devices, biological self-assembly, antireflection, self-cleaning, structural coloration, thermal insulation, self-healing and sensory aid mechanisms are some of the examples found in Nature that are of commercial interest. Biologists, physicists, chemists and material scientists have not only started to better understand biological functions and principles of various structures found in Nature but have also applied their knowledge in the design and fabrication of various materials and devices of commercial interest to engineers, material scientists, chemists and others. Biomimetics is thus considered highly interdisciplinary [1].

As previously stated, over the years, Materials Science has been inspired by many biological surfaces with remarkable properties. For example, Velcro, a kind of fabric made of nylon and mainly used as a fastener for clothing, bags and shoes, was developed from the interlocking mechanism of burdock seeds. Similarly, self-cleaning materials have been developed based on the "Lotus effect" which describes the way in which water drops roll off the super hydrophobic leaves of lotus plants, taking dirt particles away with them [11]. Inspired by polar bear fur, artificial furs and textiles have also been developed (1).

Other special features of biologic systems are suitable for bio-inspiration (Figure 1), such as: the locomotion of geckos and octopi via highly adhesive feet and suckers [12,13], the superhydrophobic compound found in the eyes of mosquitoes [14], the non-wetting phenomenon of water striders walking on water [15], the color of peacock feathers, butterfly wings, and beetle shells which is caused by a periodic microstructure [16-18], the special nanostructures causing anti-reflectivity in the cicada's wings, the moth's compound eyes [19, 20] and lastly, the special photonic reflectivity of sponge spurs due to their unique microstructure [21]. Nowadays, scientists are applying advanced scientific methods of research and nanomicroscale analysis to understand finer details of Nature's designs, and to bridge the gap between synthetic and natural materials.

To achieve a daily improvement in the basic needs and comforts of human life, devices that result from mimicking Nature are gradually integrating new solar cells, smart sensors and advanced electronics, thus defining a new era of scientific progress [9, 22]. The major emphasis on nanotechnology and nanoscience observed in the 1990s has provided a significant impetus in this process, using nanofabrication techniques for commercial applications [23]. It is estimated that the 100 largest biomimetic products generated approximately US $1.5 billion between 2005-2008 and it is expected that annual sales will continue to increase dramatically [1]. Although the biological phenomena investigated have been of various kinds, adhesion in Nature is one of the most promising, because recent advances in surface engineering and nanotechnology have made available a wide range of applicable research tools and fabrication techniques [24]. For this reason biomimetic adhesives will be analysed in detail in this chapter.

Biomimetic Adhesives

A variety of adhesives exists in Nature, which meet the individual demands of many organisms, and in their natural context usually offer impressive performance, thus inspiring novel, superior, industrial adhesives for an increasing variety of high-technology applications [24, 25].

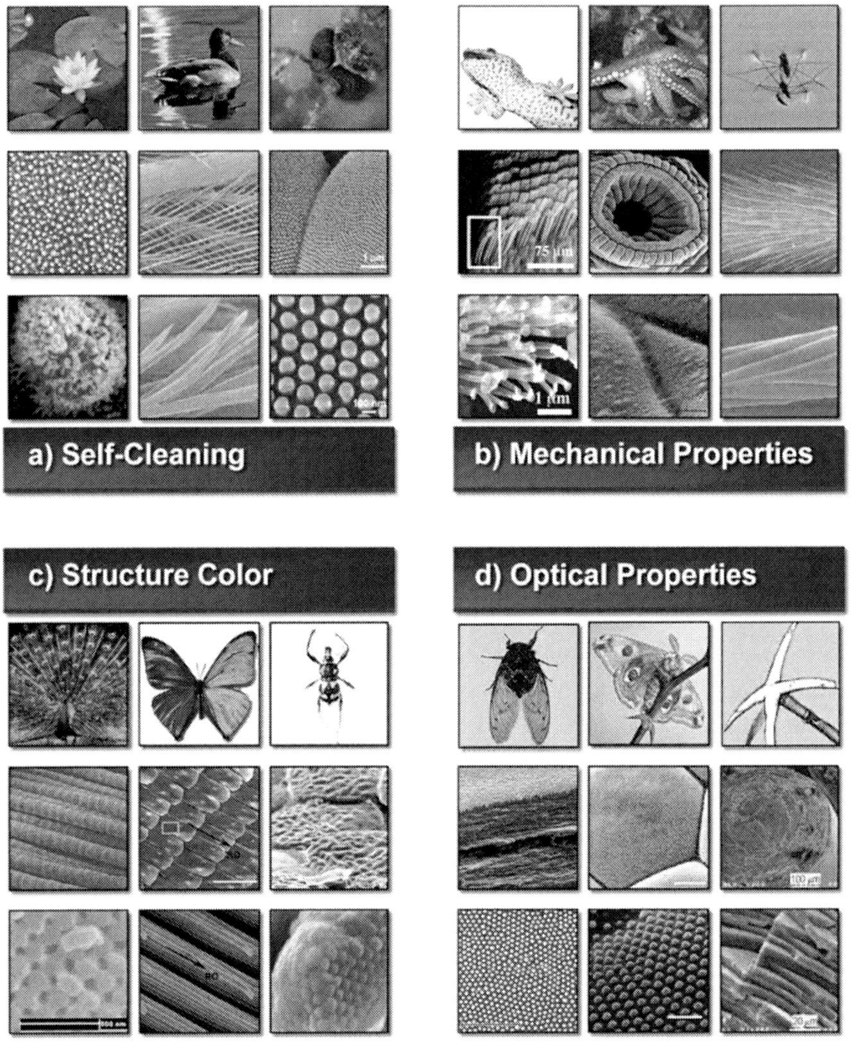

Figure 1. Four types of interesting biological properties that can be found in Nature. In each case the first row shows a photo of the organism, while the second and third rows show SEM images of the corresponding micro- and nanometer-scale structures. a) self-cleaning properties: lotus leaf, duck feather and mosquitos eye (from left to right); b) mechanical properties: gecko feet, octopus suckers and water strider; c) color through structure: peacock feather, butterfly wings and beetle shells and d) optical properties: cicada wings, moth compound eyes and sponge spur [10].

Depending on the organisms involved, the attachment of biomimetic adhesives can occur in the air, underwater or with fixation or locomotion, with the animal's body part attached to foreign materials or by joining different foreign materials together. Similarly, the adhesion process can present differences in the size of the attachment area, surface cleanliness, differing times required to attain a full strength of attachment and differences in the direction (tensile or shear) and weight to be loaded [25].

The biological adhesives are excellent models to develop general theories about interface science and, mainly, to learn how to attach, either temporary or permanently, materials to surfaces in order to develop innovative bio-inspired adhesives and put them on the market [24, 25].

The development of a new generation of adhesives with various applications (e.g. reversible pressure sensitive adhesives or adhesives working in an aqueous environment) is an important scientific and technological challenge which can find applications in several areas such as medicine, the automotive industry, robotics, micro and nanotechnology, etc., while providing environmentally-friendly features. Furthermore, biological adhesives are safe to produce because they do not contain volatile organic or toxic compounds, and present the potential to be reversible and recyclable.

Despite the progress made in recent years in this research area, the knowledge of natural adhesive systems still does not fulfil the needs of innovative adhesives for specific industrial, medical and surgical applications. This may be due to the complexity of some biological adhesives systems, involving a wide range of interactions and components with different functions. To design improved biomimetic adhesives it is essential to understand and simplify Nature while retaining the desired efficacy. Thus, one of the foundation stones of biomimetics is a broad, intensive knowledge exchange and a multidisciplinary collaboration between scientists from different areas (biology, chemistry and materials science) on all aspects of bioadhesion [24].

Adhesion Mechanisms

In Nature, despite the huge diversity of organism attachment devices and its dependence on the biology of the species and on the environmental constraints, some functional solutions have evolved independently in different biological lineages [11, 26]. For example, terrestrial animals which climb on rocks, plants or other types of substrata possess adhesive organs on feet that come in two basic designs: pads densely covered with specialized micro or nano sized setae or pads with a relatively smooth surface profile [26, 27].

Other examples are beetles, flies, large spiders and gekkonid lizards which have feet possessing adhesive organs of the hairy type, while bees, tree frogs and grasshoppers have adhesive organs of the smooth type [27].

The adhesion systems can be classified according to a number of factors, such as the presence or absence of lubricant liquids. If the attached pads are not supplemented with fluids, the phenomenon is called dry adhesion [28] but if these pads are supplemented with various kinds of fluids it is a case of wet adhesion [29].

To enable animals to climb on different surfaces, their adhesive systems must meet a considerable number of requirements: 1. to be reusable allowing the organism to quickly and repeatedly attach and detach the feet from the surface; 2. to be wear-resistant and self-cleaning (30); 3. to be reversible, 4. to be substrate tolerant, 5. to have a non-sticky default state allowing adhesion to take place only when required [31].

Contrary to what happens to terrestrial organisms, marine organisms are subjected to unpredictable and considerable strong shear forces, requiring robust adhesion systems in all directions. Thus, in the marine environment, the devices responsible for the organisms' adhesion are usually highly viscous or based on solid secretions containing specialized adhesive proteins [32]. Depending on the type and habitat of marine organisms, adhesion is either permanent or temporary [33]. The latter type occurs in organisms such as sea stars and limpets which attach themselves strongly and temporarily to the substratum through viscoelastic secretions but retain the ability to move [33, 34].

The adhesive systems in Nature can be dynamic or reversible depending on the mobile characteristics of animals, and on the physical and chemical processes to provide adhesion. So, in biologic systems adhesion can result from different mechanisms [35]:

1) Chemical adhesion
2) Wet adhesion or capillarity
3) Dry adhesion (Van der Waals adhesion)

1. Chemical Adhesion

All adhesive systems have a chemical component, either from the formation of ionic, covalent or hydrogen bonding to promote the adhesion, or from the adhesive system [35].

Several marine organisms such as barnacles, mussels, tubeworms (*Phragmatopoma californica*), sea cucumbers and echinoderms show chemical adhesion [36] and have the ability to attach to materials with both hydrophobic and hydrophilic surfaces. The adhesion presented by marine species in many cases results from proteins and amino acid 3,4-dihydroxy-L-phenylalanine (DOPA). This group is able to coordinate metal ions and form hydrogen bonds with inorganic surfaces; additionally, upon oxidation to the quinine form, it reacts with thiol or amine groups forming covalent bonds (Figure 2).

Carbohydrates, proteins and polysaccharides can also form bonds to promote the adhesion of marine species [35].

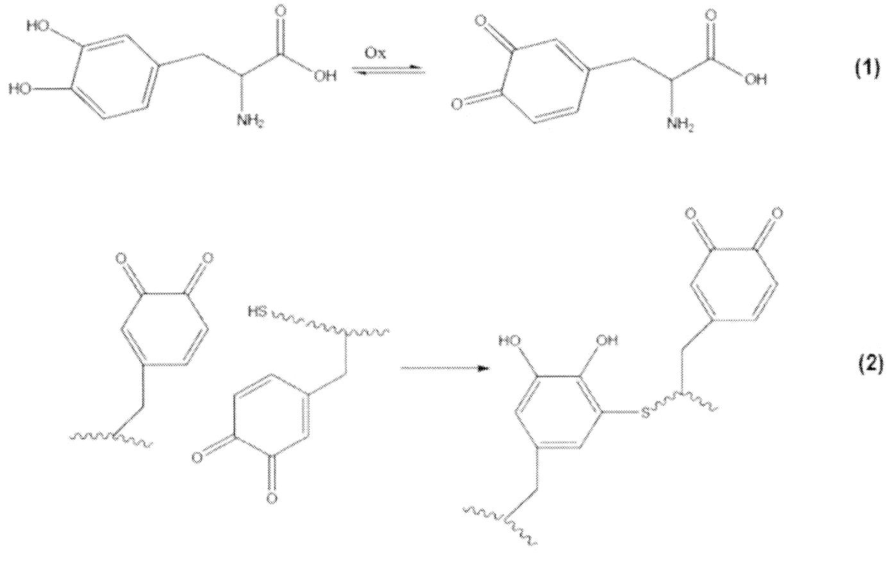

Figure 2. (1) Oxidation of DOPA to quinine; (2) Quinone reaction with thiol forming cross-linking.

2. Wet Adhesion or Capillarity

The wet adhesion mechanism is found in organisms with smooth adhesive pads such as salamanders, arboreal frogs [37-39], some river frogs and some mammals. The adhesion results from the attraction between the pad surface and the substratum surface, due to the capillarity and viscosity of the liquid film secreted onto the pad. Some hairy insect species also possess a capillarity adhesion mechanism [37-38, 40-41].

3. Dry (van der Waals) Adhesion

The dry adhesion is a dynamic (reversible) mechanism observed in species with hair-like structures on their toes or legs, called setae, which allow organisms to climb vertical surfaces, and even walk upside down. This form of adhesion is common in some insects (beetles, flies, spiders [42]), skinks, geckos and anoline lizards [43].

The attachment of the animals to the surfaces results from the application of a slight loading of the setae, normal to the surface, and then from a small drag back towards the body and parallel to the surface. The detachment from the surface results from increasing the setae shaft angle (rolling the toes) with the surface to above 30 degrees [44].

Some studies of different species show that setae have fine structures as the animal mass increases, and that the adhesion increases with the setae density [45].

The adhesion and the release from the substrates depends on several aspects: the hierarchical structure, the specific dimensions of the fibre arrays, the aspect ratios, the slope between the fibre and substrate, the size and shapes of the contact regions and the material properties [35].

The different types of adhesion mechanisms in Nature suggest that evolution lead to the development of an optimal design for each system. This justifies the great interest of scientists in these adhesion mechanisms, despite understanding the complexity and difficulty in reproducing them artificially [24].

A comprehensive review of adhesives inspired by biological systems is presented here, with a detailed comparison of the different components of the adhesion, as well as the potential final applications.

Adhesive Types and Applications

Chemical Adhesion

Sandcastle Worm

The sandcastle worm (*Phragmatopoma californica*), also called honeycomb worm, [46] lives in colonies on the rocky beaches of the Californian coast, building tube reefs several inches long, using sand grains and shell fragments [47]. Sandcastle worms remain in sandcastle-like tube-shaped colonies, and are almost never seen [48, 49] (Figure 3). They are usually found in sheltered areas, such as concave shorelines or rock faces covering an area up to 2 square meters [48, 49].

Figure 3. a) sandcastle worm in captivity, building its tube out of sand and white ceramic beads. b) a sandcastle worm colony on California's coast (adapted from [47]).

The sandcastle worms have the ability to attach themselves to surfaces, and form a protective coating (tubes) with a solid foam like structure that withstands waves and tidal action [50].

This structure possesses properties between elastic and hard solid, which increases the toughness and elasticity, and reduces the crack propagation of the material. It is chemically composed by polyphenolic proteins which contain side chains of phosphate and amine groups, such as O-phosphoserine and DOPA [35, 51].

Laboratory experiments have shown that the sandcastle worm has the ability to build tubes by gluing together tiny pieces of egg shell, glass beads, red sand, bone, zirconium oxide, and even pieces of a silicon chip, using their natural adhesives [47].

Potential Uses of Sandcastle Worms Inspired Adhesives

The proteins in the glue used by sandcastle worms to build their protective coating were replicated and used as bioadhesive [52]. Potential medical applications for this glue include the fixation of small bone fragments in fractured knees, wrists, elbows, ankles, face, skull, etc. It can also be used in sealing skin cuts and corneal incisions [53, 54].

The scientists consider this glue appropriate for the aforementioned applications because, contrary to most adhesives, it works even when the surfaces are wet and bloody, as can happen with bones. It can also be applied in anti-inflammatory medication, antibiotics and stem cells, to glue bone fragments together, or even to attach tissue scaffolds to promote the re-growth of the missing bone [47].

Echinoderms

The echinoderms are found at every ocean depth and include well-known animals such as sea stars, sea cucumbers, sand dollars and sea urchins. All the species of this phylum use adhesive secretions for various vital functions, which means that adhesion is a way of life for echinoderms. The adhesive systems presented by echinoderms depend on the species or on the development stage under consideration, and differ in structure, way of operating and adhesive composition [35, 55].

1. Tube Feet

Tube feet or podia are organs which enable the attachment to the surface, locomotion, feeding or burrowing of the animal [55]. Due to sea stars, sea cucumbers and sea urchins mobility these organisms possess a reversible adhesion mechanism. The temporary adhesion relies on a duo-gland adhesive system resorting to both adhesive and de-adhesive secretions. The adhesive produced is comprised of 20% protein, 8% carbohydrates, 40% inorganic residue and 2.5% lipids. Due to their soft feet, sea urchins and sea stars present considerable adhesion to the surface because they allow a conformation to rough surfaces, and consequently an increased contact area [56].

2. Larval Adhesion

The echinoderms undergo a larval to adult metamorphosis. In this phenomenon, the larvae attachment during settlement and strong fixation is ensured by the larval adhesive organs [55].

3. Sea Cucumber

When threatened, sea cucumbers, via their cuvierian tubules, instantaneously discharge a quick-setting adhesive to bond to, and ensnare the threat in a matter of seconds. It is a mixture of protein and carbohydrate (3:2), rich in small side-chain amino acids, especially glycine, and in a charged and polar amino acids, which may assist in ionic and hydrogen bonding to surfaces [57,58].

Potential Uses of Echinoderms Inspired Adhesives

The knowledge of echinoderm adhesives enables scientists to study their potential applications in two broad fields; the development of new antifouling systems (e.g:. to be incorporated into antifouling paints), and the design of water-resistant adhesives for several applications, such as tissue adhesives and dentistry [35, 55, 59].

Barnacles

The barnacle adhesive, also called cement, presents such a high adhesive and cohesive strength that barnacles stay attached to ships that travel at high speed [60].

Like the mussel adhesive, the barnacle adhesive is a proteinaceous material, consisting of five proteins that account for approximately 90 % of the content. The remaining 10% is formed of carbohydrate ash and trace amounts of lipid [61].

The largest portion of proteins in these adhesives is a protein with a molecular weight of 36 kD that presents hydrophobic and hydrophilic alternating sequences, which influences the adhesive performance in seawater [35].

Abbott [62] reported that the most durable and toughest connection in the marine world is barnacle cement, whose adhesion results from a complex of proteins, cross-linked through cysteine residues [63].

There is also evidence of the presence in the cement of a 3kD protein, which is thought to be the smallest unit polymerizing to larger molecules. This finding, besides being unique in biological adhesives, increases the probability of producing such an adhesive because the small monomers are firstly cross-linked to obtain larger chains that later interact during the curing process [60, 63, 64].

Potential Uses of Barnacles Inspired Adhesives

The barnacle proteins resulted from the base plate or the hemolymph of barnacle can be used as high-strength medical adhesive to act as an interface between implants and tissue or bone, the scaffolding for tissue engineering, as well as to avoid biofouling on surfaces underwater [65, 66].

The design of materials based on a protein unit with a molecular weight of 20 kD is also being investigated [66].

Marine Mussels

At any given beach, taking a closer look at how blue mussels (*Mytilus edulis*) adhere, it is possible to see the byssal (or 'byssus' or 'beard') adhesive assembly, much like that shown in Figure 4. Mussels affix themselves to surfaces by depositing a series (~10-40) of small (~2 mm diameter, ~0.1 mm thick) adhesive plaques, each connected to the animal using a long thread. An impressive amount of force is required to break the adhesive bonding. Mussels can stick to nearly all surfaces, including Teflon and PTFE [67-69].

Mussels produce and secrete specialized adhesives enabling them to attach to surfaces in marine environments, which are typically characterized by salinity, humidity, tides, turbulence, and waves [71]. These organisms adhere tightly to surfaces underwater using the byssus secreted from their feet, which consist of a bundle of threads. At the end of each thread there is an adhesive plaque containing water-resistant glue which enables the plaque to anchor to wet solid surfaces [72].

Mussel adhesive proteins are known as the most powerful adhesive in the world. Even though mussel adhesion is much stronger than other polymer-based adhesives, such as epoxy and phenolic resins, it is flexible and elastic. Most important is the fact that the mussel adhesive protein maintains its adhesion in wet environments. All adhesive events in living organisms occur in the presence of water, but adhesions using chemical-based adhesives are for the most part broken down by water.

Figure 4. Marine mussel clinging to a glass sheet (left) and Teflon (right). Note the attachment comprising many adhesive plaques and threads [70].

The mussel adhesive protein can also attach to various substrates, including plastic (e.g Teflon), glass, metals, and living bodies. Adhesion of mussel adhesive proteins (extracted and recombinant) to various solid materials has already been verified [73-78].

Burkett *et al.* [79] describe a method to quantify the adhesion strength of mussels. On high energy surfaces such as aluminium, mussels stick with strengths of almost 300 kilopascal (kPa), or a little over 40 pounds per square inch (psi) [79]. By comparison, common synthetic adhesives such as poly (vinyl acetate) (PVA, 'white glue') with ~900 kPa and cyanoacrylate 'super' glue with ~8000 kPa are stronger (80) but the mussel has enough sticking power for its life in the turbulent waters.

Three different types of collagenous proteins from byssus thread [81-83] and six distinct foot proteins type from adhesive plaque [72,74,84-88] have been identified. However studies reveal that mussel adhesive proteins closer to the adhesion interface have a higher proportion of 3,4-dihydroxy-L-phenylalanine (DOPA) residues [84-87]. A huge reduction in the ability for adhesion was verified in similar mussel adhesive proteins without DOPA [77, 89, 90]. Its residues also enable mussel adhesive protein molecules to cross-link by oxidative conversion to DOPA-quinone. The reactive quinine is thought to provide the water-resistance characteristics of mussel adhesion [89,90].

Ninan *et al.* [91] describe the process of mussel adhesive protein (''mussel extract'') from *M. edulis* feet as well as its adhesive properties. The results of the tests performed show that (1) mussel extract is capable of forming strong tissue joints (~1MPa) in the end-to-end configuration, when given adequate curing time; (2) the time required for mussel extract to reach the maximum adhesive strength on porcine connective tissue substrate is between 12 and 24 h; and (3) mussel extract joints are similar in strength to fibrin joints, although they cure more slowly than fibrin. In addition, the results show that the type of adhesive had no significant effect on joint strength. The study revealed that fibrin and mussel adhesive were similarly influenced by the curing conditions.

Both types of joints were stronger for ''dry'' curing conditions than for corresponding ''humid'' curing conditions at cure times up to 12 h. However the authors note that the moisture had more influence on the strength of the mussel extract joint than on the fibrin joint. The effect of curing time was similar for both types of joints. These results show that mussel extract curing kinetics is very sensitive to moisture. After 48 h of ''dry'' curing, a reduction in average strengths, not only in mussel extract but also in fibrin joints, was observed, probably due to the dehydration of the skin tissue.

Wolowacz and Chivers [92] reported a very weak bond strength due to the delay in the curing of adhesives, which probably results from the difficulty in removing moisture from the mussel extract when the interface was not exposed to air. In order to avoid this problem of very weak bond strength, some authors [93, 94] have studied the effect of the addition of various chemical cross-linkers on the strength and on the biocompatibility of mussel extract.

Despite having been identified as the major protein responsible for the mussel adhesive properties and the thorough examination of its adhesive properties and biocompatibility, fp-1 (foot protein from the mussel specium *M. Edulis*, which was the first identified and the most studied mussel) practical applications have been severely limited by the uneconomical extraction process and unsuccessful large-scale production. The natural extraction used to isolate mussel adhesive proteins is a labour-intensive and inefficient process, requiring around 10000 mussels for 1g of the fp-1 adhesive protein [85, 95-97]. The chemical extraction process used does not yield pure or individual adhesive proteins [98].

Therefore, and to our knowledge, bioadhesives developed from the marine environment are not marketed yet. For the stated reasons, recombinant DNA technology has also been used to obtain large amounts of their components for further conventional adhesion tests and practical applications. Genetic production of fp-1, which is considered a key protein for adhesion of mussels in wet environments, has been attempted in several systems [74, 97, 99], but they have failed for several reasons [75, 97, 99]. Researchers [74, 75, 99-102] have attempted to use other technologies for the mass production of functional recombinant mussel adhesive but without success.

The development of a novel hybrid type of mussel adhesive protein fp-151, which is easy to purify and presents significantly greater production yields, is described in the literature [77, 78]. This purified recombinant demonstrated comparable adhesive characteristics to recombinant Mefp-5, which is a plaque specific protein and contains the highest DOPA level (~30 mol%) among mussel adhesive proteins [78, 87]. The results show efficient adhesion and good biocompatibility for various cell types including both anchorage-independent and anchorage-dependent cells, as well as good adhesion to laboratory plastic consumables, such as polypropylene (PP).

The macro-scale adhesion strength tests of recombinant fp-151 in comparison with the commercially available tissue adhesive fibrin glue, as a positive control, showed that the shear strength of fp-151 was about four times greater (~0.8 MPa) than that of fibrin glue (~0.2 MPa), when cowhide square pillars (10 mm $\times$10 mm) and 10 mg samples were used [98].

Potential Uses of Marine Mussels Inspired Adhesives

One of the most thoroughly explored biomimetic approaches, which encompasses artificial materials that mimic natural forms, is the development of adhesives based on the marine mussels [103]. These mussel adhesive proteins have been considered as potential underwater bioadhesives due to their many fascinating features [71, 104-106]. Mefp-1, for example, has been marketed (Cell-Tak™) as an adhesive to immobilize cells and tissues to inert substrates for further manipulation [55].

The marine mussel inspired adhesives may be suitable as surgical adhesives, thus avoiding the need of sutures, staples and even anaesthetics. The surgical adhesives used nowadays are based on fibrin and cyanoacrylates. However these do not work in the areas of the body where there is high fluid concentration [35].

Recently, studies have been performed examining the attachment of mussel adhesive proteins to living body substances, especially porcine skin [91], but also several types of mammalian cells [78,107,108].

Strong and water-insoluble mussel adhesives have provoked interest for potential uses in diverse biotechnological applications because these systems could be used as cell and tissue adhesives, and have the advantages of being environmentally-friendly as they are biodegradable [109], apparently harmless to the human body, and do not impose immunogenicity [104,105,109,110]. In addition, they are efficient for use under aqueous conditions.

The development of a hybrid type of mussel adhesive protein for such a specific application as cell-adhesion biomaterial is reported by Hwang *et al.* [107]. These novel hybrid type mussel adhesive proteins, fp-151 and fp-151-RGD, if available in large quantities, have the potential to become the raw material for bioadhesives for diverse applications, such as surgical sealants, medical coatings, and gene and drug delivery systems [98].

The antifouling coatings can also be inspired by marine organisms. One example of this is the design of coatings using the concept of Mefp-1 peptides or DOPA units, grafted onto polyethylene glycol (PEG) chains with the aim of bonding to the surface and preventing cell attachment, respectively [55, 67, 69, 111].

Wet Adhesion or Capillarity

Tree Frogs

Tree frogs have the ability to cling on to smooth surfaces using large toe pads, whose skin is characterized by peg-studded hexagonal cells separated by deep channels into which mucus glands open (Figure 5) [41, 112].

This microstructure provides a mechanism to remove water or mucus from the toe pads and increase the development of surface contacts and frictional forces and may also help toe pads conform to surface irregularities [35].

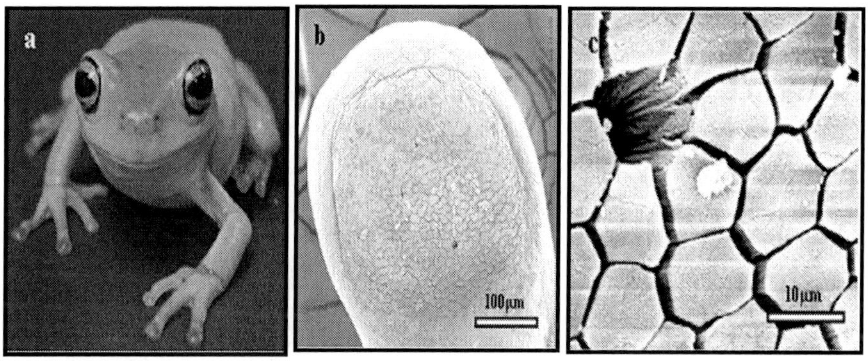

Figure 5. a) White's tree frog (*Litoria caerulea*); b) SEM of toe pad; c) SEM of epidermis with hexagonal epithelial cells separated from each other at their distal ends by channels (adapted from (112)).

Despite the microstructure on the toe pads, cells may be the responsible for the first substrate-substrate contact. The work described elsewhere [37, 39] shows that tree frog adhesion forces increase linearly with the area of the toe pad, indicating that capillarity is the predominant adhesion mechanism, even though frictional forces are also involved, contributing to the adhesion in a wet environment [41]. The tree frogs detachment from a surface occurs by toe pad peeling. In order to reduce peel adhesion, frogs tend to maintain a head's up orientation on vertical surfaces [35].

Potential Uses of Tree Frogs Inspired Adhesives

The remarkable wet adhesion properties presented by tree frogs have lead scientists to consider that these animals' toe pads have biomimetic relevance.

Since the tree frog toe pads have three systems of grooves at different scales and an effective elastic modulus akin to rubber that surrounds a much softer material, potential applications are the development of improved boots and tyres to wet weather [113-116].

Although the similarities between the tree frog toe pads and wet weather tyres seems uncanny, their performance (easy detachment combined with high coefficient of friction) for this application is impressive. The effectiveness of applying these mechanisms to wet weather tyres remains to be tested. Another potential use for adhesives inspired by the tree frog is in biomedical applications, such as bandages. [112].

Insects

Many insects, such as bees, ants, cockroaches and grasshoppers cling to vertical and inverted surfaces with adhesive pads, or have hairy flexible setae, such as flies and earwigs. Both groups secrete a fluid, which is a hydrophilic-hydrophobic emulsion of watery droplets in an oil continuous phase, into the contact zone [35].

In the case of insects with adhesive pads, the aforementioned fluid may be only responsible for providing adhesion to very rough surfaces by filling in the non-contacting areas, because the adhesion and the friction forces are greater when less fluid is present.

Dirks *et al.* [117] noted that the adhesive emulsions from insect adhesive pads present non-newtonian properties, combining the benefits of wet adhesion and resistance against shear forces.

Similarly, in hairy insects the hairs tend to conform to a rough surface and the contact area is increased by the adhesive fluid [40].

Potential Uses of Insects Inspired Adhesives

Although scientists have not yet been inspired by the mechanisms of adhesion used by insects, their secretion-aided attachment system also provides a source of inspiration for new biomimetic adhesives.

Understanding the adhesion mechanisms may certainly lead to the development of new applications, such as non-toxic, wear-resistant insect-repellent coatings [117].

Dry Adhesion (van der Waals Adhesion)

Geckos

Geckos' ability to adhere to rough, smooth, vertical and inverted surfaces [118] is known since ancient time. However, little was understood about this phenomenon until the hair covering the toes of gecko were viewed by scanning electron microscopy (SEM) in the late of 20th century [119]. With the aim to develop synthetic adhesives inspired by gecko, over the past 50 years, scientific studies have been conducted to analyse gecko-like morphology and its impact on the adhesive properties. The surface structure of the skin on the toes of geckos is responsible for their adhesive properties. It is a complex fibrillar structure of lamellae (scansors), setae, branches and spatulae. It consists of an intricate hierarchy of structures beginning with lamellae, soft ridges that are 1-2mm in length located on the attachment pads (toes) that compress easily, so that contact can be made with rough surfaces [120].

Tiny curved hairs, known as setae, typically 30-130 μm in length and 5-10 μm in diameter [119-122], are extend from the lamellae. At the end of each seta, 100-1000 spatulae (because of their shape) with a diameter of 0.1-0.2 μm [120] branch out and form the points of contact with the surface, thus maximizing the interfacial adhesion [123, 124]. The tips of the spatulae are approximately 0.2-0.3 μm in width [62], 0.5 μm in length, and 0.01 μm in thickness [125] (Figure 6).

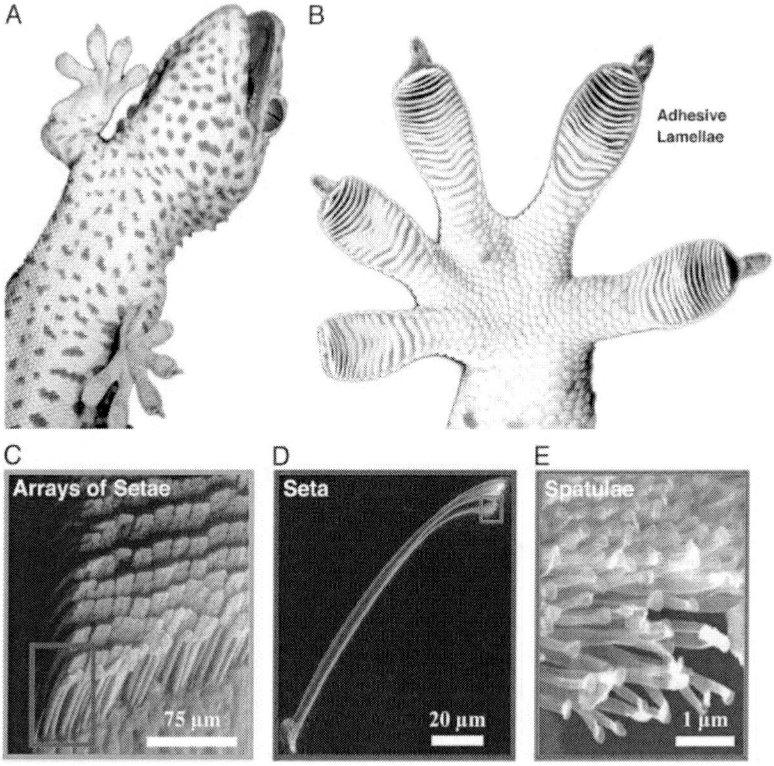

Figure 6. Hierarchical structures of a Tokay gecko (*G. gecko*): (***A***) Macrostructure: ventral view of a tokay gecko climbing vertical glass. (***B***) Mesostructure: ventral view of the foot with adhesive lamellae (scansors) visible as overlapping pads. (***C***) Microstructure: proximal portion of a single lamella, with individual setae in an array visible. (***D* and *E***) Nanostructure: single seta with branched structure at upper right, terminating in hundreds of spatular tips [30].

The adhesion of these spatulae to the surfaces is controlled by the combination of Van der Waals [124, 126] and capillary forces [127].

Potential Uses of Gecko Inspired Adhesives

Unlike commercial adhesives, such as tape or glue, which involve the use of wet adhesives that permanently attach two surfaces, the replication of gecko feet characteristics would enable the development of a super-adhesive polymer capable of clean and dry adhesion [119, 128-130].

Several scientists worldwide have made progress in developing biologically-inspired adhesives, which mimic the geometry of biological hairy systems at different scales [128, 131-133]. Nosonovsky and Bhushan [134] show that patterned surfaces usually present significantly higher adhesion on a smooth surface than a smooth sample made out of the same material. Futhermore, the patterned surfaces contact on various surfaces, which increase tolerance to defects of individual contacts. To mimic these surfaces materials, have been used polydimethylsiloxane elastomers (PDMS) or carbon nanotubes because they present self-cleaning properties resultant from their extreme hydrophobicity [135].

The developed patterned surfaces have been used as adhesives in various applications. For example, in recent years a gecko mimetic adhesive has been optimized for a wet tissue-like environment, which proved effective underwater with reversible noncovalent bonding to inorganic surfaces [68].

In the study described in [136] an elastomeric, biocompatible, and biodegradable gecko-inspired tissue adhesive was developed from a biocompatible and biodegradable elastomer, combined with a thin tissue-reactive biocompatible surface coating. This adhesive is based on poly(glycerol sebacate acrylate) (PGSA), a tough biodegradable elastomer with elastic and biodegradable properties, which can be used for specific tissue applications and can easily incorporate growth factors or drugs. Tissue adhesion was optimized by varying dimensions of the nanoscale pillars, including the ratio of tip diameter to pitch and the ratio of tip diameter to base diameter. Coating these nanomolded pillars of biodegradable elastomers with a thin layer of oxidized dextran significantly increased the interfacial adhesion strength on porcine intestine tissue in vitro and in the rat abdominal subfascial in vivo environment. The authors developed a biocompatible tissue adhesive with promising covalent cross-linking to wet tissue, obtained through combined morphology and chemistry effects, which may be have applied in medical therapies ranging from suture/staple replacements/supplements, waterproof sealants for hollow organ anatomoses, mesh grafts to treat hernias, ulcers, and burns and haemostatic wound dressings [137]. However, before such adhesive can be widely used, some obstacles need to be overcome, mainly its adhesion performance which is not currently strong enough for clinical applications, such as bandages (Figure 7) [138].

Karp [138] describes a tape-based tissue adhesive that seems to prevent air leaks in lung resection procedures, haemostatic wound dressing and mesh grafts to treat ulcers, hernias and burns, among other applications.

The Gecko's ability to stick and climb on various surfaces was also studied to develop wall climbing robots which are capable of climbing vertical walls, without contaminating the environment. These robots are of great interest for the development of activities such as cleaning, inspection, repair and exploitation. Menon *et al.* [139] describes a new technique for producing synthetic microfibers to be used as dry adhesives, as well as the successful results obtained using two robotic prototypes inspired in gecko.

The reusable adhesives inspired by geckos also have a potential use in toys [140] and in advanced technology, such as microelectronic and space applications [130, 132].

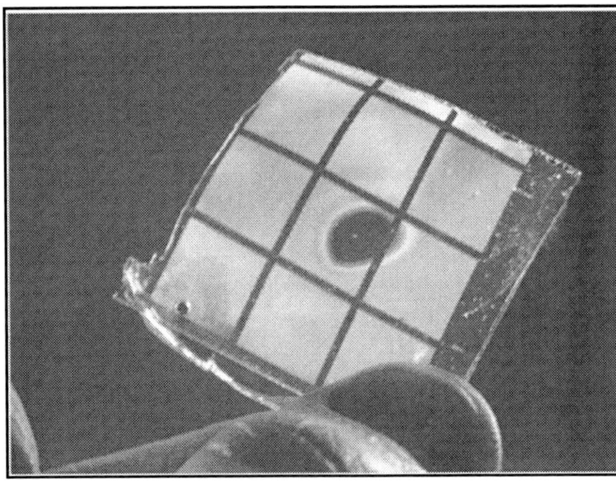

Figure 7. Gecko's feet inspired bandage [137].

CONCLUSION

Nature has evolved and optimized a considerable number of materials and surfaces with remarkable adhesive properties. The understanding of these mechanisms will help exploit these properties for commercial applications, including adhesives for medicine, robotic, microelectronic, toys, space application, etc.

So far marine mussels and geckos have been the most studied for the development of biomimetic adhesives. This review shows that many other organisms, currently under study, also demonstrate great potential.

REFERENCES

[1] Bhushan, B. Biomimetics: lessons from nature-an overview. *Phil. Trans. R. Soc. A* 2009; 367: 1445-1486.

[2] Ball, P. Natural strategies for the molecular engineer. *Nanotechnology* 2002; 13:R15-R28.

[3] Bar-Cohen, Y. (ed.). Biomimetics: biologically inspired technologies; Boca Raton, FL: Taylor & Francis, 2006.

[4] Vincent, JFV; Bogatyreva, OA; Bogatyrev, NR; Bowyer, A; Pahl, AK. Biomimetics: its practice and theory. *J. R. Soc. Interface* 2006; 3: 471-482.

[5] Anon. Biomimetics: strategies for product design inspired by nature. Bristol, UK: Department of Trade and Industry, 2007.

[6] Meyers, MA; Chen, PY; Lin, AYM; Seki, Y. Biological materials: structure and mechanical properties. *Prog. Mater. Sci.* 2008; 53: 1-206.

[7] Krausse, E. In: Engels E-M, editor. Die Rezeption von Evolutionstheorien im Jahrhundert, Nineteenth ed., Frankfurt; 1995.

[8] Thompson, DW. On growth and form. Cambridge Univ. Press, 1942.

[9] Singh, AV; Rahman, A; Sudhir Kumar, NVG; Aditi, AS; Galluzzi, M; Bovio, S; Barozzi, S; Montani, E; Parazzoli, D. Bio-inspired approaches to design smart fabrics, *Materials and Design* 2012; 36: 829-839.

[10] Xia, F; Jiang, L. Bio-Inspired, Smart, Multiscale Interfacial Materials. *Interfacial Materials Adv. Mater.* 2008; 20: 2842-2858.

[11] Jon, W; Barnes, P. Biomimetic Solutions to Sticky Problems. *Science* 2007; 318: 203-204.

[12] Atumn K, Liang YA, Hsieh ST, Zesch W, Chan WP, Kenny TW, Feating R, Full RJ. Adhesive force of a single gecko foot-hair. *Nature* 2000; 405: 681-685.

[13] Kier, WM; Smith, AM. The structure and adhesive mechanism of Octopus suckers. *Integr. Comp. Biol.* 2002; 42: 1146-1153.

[14] Gao, X; Yan, X; Yao, X; Xu, L; Zhang, K; Zhang, J; Yang, B; Jiang, L. The dry-style antifogging properties of mosquito compound eyes and artificial analogues prepared by soft lithography. *Adv. Mater.* 2007; 19(17): 2213-2217.

[15] Gao, X; Jiang, L. Biophysics: water-repellent legs of water striders. *Nature* 2004; 432: 36.

[16] Zi, J; Yu, X; Li, Y; Hu, X; Xu, C; Wang, X; Liu, X; Fu, R. Coloration strategies in peacock feathers. *Proc. Natl. Acad. Sci. USA* 2003; 100(22): 12576-12578.
[17] Zheng, Y; Gao, X; Jiang, L. Directional adhesion of superhydrophobic butterfly wings. *Soft Matter*. 2007; 3: 178-182.
[18] Parker AR, Welch VL, Driver D. Structural colour: Opal analogue discovered in a weevil. *Nature* 2003; 426**:** 786-787.
[19] Guo, C; Feng, L; Zhai, J; Wang, G; Song, Y; Jiang, L; Zhu, D. Large-Area Fabrication of a Nanostructure-Induced Hydrophobic Surface from a Hydrophillic Polymer. *ChemPhysChem*. 2004; 5: 750-753.
[20] Stavenga, DG; Foletti, S; Palasantzas, G; Arikawa, K. Light on the moth-eye corneal nipple array of butterflies. *Proc. R. Soc. B* 2006; 273: 661-667.
[21] Sarikaya, M; Fong, H; Sunderland, N; Flinn, BD; Mayer, G. Biomimetic model of a sponge-spicular optical fiber-mechanical properties and structure. *Journal Mater. Res.* 2001; 16: 1420-1428.
[22] Ball, P. Life's lessons in design. *Nature* 2001; 409(6818):413-416.
[23] Bhushan, B. Springer Handbook of Nanotechnology, 2nd edn, Germany Heidelberg; Springer-Verlag, 2007.
[24] Memorandum of Understanding for the implementation of a European Concerted Research Action designated as COST Action TD0906; Biological Adhesives: from Biology to Biomimetics, Brussels, European Cooperation in the field of Scientific and Technical Research, Dec 2009.
[25] Kamino, K. Biological adhesive, package of wonder. Abstracts book of 1st International Conference on Biological and Biomimetic Adhesives, 9th to 11th May 2012, Lisbon, Portugal.
[26] Federle, W. Why are so many adhesive pads hairy?.*The Journal of Experimental Biology* 2006; 209: 2611-2621.
[27] Creton, C; Gorb, S. Sticky Feet: From Animals to Materials, *MRS Bull.* 2007; 32:466-468.
[28] Huber, G; Mantz, H; Spolenak, R; Mecke, K; Jacobs, K; Gorb, SN; Arzt, E. Evidence for capillarity contributions to gecko adhesion from single spatula nanomechanical measurements. *PNAS* 2005; 102(45): 16293-16296.
[29] Federle W, Riehle M, CURTIS ASG, Full RJ. An Integrative Study of Insect Adhesion: Mechanics and Wet Adhesion of Pretarsal Pads in Ants. *Integr. Comp. Biol.* 2002; 42:1100-1106.
[30] Hansen, WR; Autumn, K. Evidence for self-cleaning in gecko setae. *PNAS* 2005; 102(2): 385-389.
[31] Autumn, K; Hansen, W. Ultrahydrophobicity indicates a non-adhesive default state in gecko setae. *J Comp Physiol A* 2006; 192:1205-1212.
[32] Smith, AM; Callow, JA. Editors. Biological adhesives. Berlin Heidelberg; Springer-Verlag 2006, pp.284.
[33] Flammang, P. Adhesive secretions in echinoderms: an overview. In: Smith, AM, Callow, JA (Eds.). Biological Adhesives. Berlin; Springer-Verlag, 2006, pp. 183-206.
[34] Walker, G. Marine organisms and their adhesion. In: Wake WC (ed) Synthetic adhesives and sealants. Chichester, Wiley, 1987, pp. 112-135.
[35] Saville, P. Bio-Inspired Adhesion. Technical Memorandum. Defence Research and Development Canada, June 2009.

[36] Silverman, HG; Roberto, FF. Understanding Marine Mussel Adhesion. *Marine Biotechnology* 2007; 9: 661-681.

[37] Smith, JM; Barnes, WJP; Downie, JR; Ruxton, GD. Structural correlates of increased adhesive efficiency with adult size in the toe pads of hylid tree frogs. *J Comp Physiol* A 2006; 192: 1193-1204.

[38] Barnes, WJP. Dynamic adhesion in animals: mechanisms and biomimetic implications. *J Comp Physiol A* 2006; 192: 1165-1168.

[39] Barnes, WJP; Oines, C; Smith, JM. Whole animal measurements of shear and adhesive forces in adult tree frogs: insights into underlying mechanisms of adhesion obtained from studying the effects of size and scale. *J Comp Physiol A* 2006; 192: 1179-1191.

[40] Dreschsler, P; Federle, W. Biomechanics of smooth adhesive pads in insects: influence of tarsal secretion on attachment performance. *J Comp Physiol A* 2006; 192: 1213-1222.

[41] Federle, W; Barnes, WJP; Baumgartner, W; Drechsler, P; Smith, JM. Wet but not slippery: boundary friction in tree frog adhesive toe pads. *J. R. Soc. Interface* 2006; 3: 689-697.

[42] Gasparetto, A; Vidoni, R; Zanotto, V; Brusa, E. Attaching mechanisms and strategies inspired by the spiders' leg: Final Report, University of Udine, 2008.

[43] Irschick, DJ; Herrel, A; Vanhooydonck, B. Whole-organism studies of adhesion in pad-bearing lizards: creative evolutionary solutions to functional problems. *J Comp Physiol A* 2006; 192: 1169-1177.

[44] Autumn, K. Gecko adhesion: Structure, function, and applications. *MRS Bull.* 2007; 32: 473-478.

[45] Arzt, E; Gorb, S; Spolenak, R. From micro to nano contacts in biological attachment devices. *Proc. Natl. Acad. Sci. U. S. A.* 2003; 100: 10603-10606.

[46] Hinton, S. Seashore life of southern California: an introduction to the animal life of California beaches south of Santa Barbara. California natural history guides. University of California Press. 1987; 26: 48.

[47] Superglue from the sea - Synthetic sea worm glue may mend shattered knee, face bone. University of Utah, published online in www.asknature.org, November 2008.

[48] Ricketts, EF; Calvin, J; Hedgepeth, JW; Phillips, DW. Between Pacific Tides. 5th ed. Stanford, Stanford University Press, 1985, pp.652.

[49] Hinton, S. Seashore Life of Southern California: An Introduction to the Animal Life of California Beaches South of Santa Barbara. Nidottu, Pehmeäkantinen, 1988, pp.31.

[50] Stewart, RJ; Weaver, JC; Morse, DE; Waite, JH. The tube cement of Phragmatopoma calfornica: a solid foam, *J. Exp. Biol.* 2004; 207: 4727-4734.

[51] Jensen, RA; Morse, DE. The bioadhesive of Phragmatopoma californica tubes: a silk-like cement containing L-DOPA. *Journal of Comparative Physiology B* 1988; 158 (3): 317-324.

[52] Rzepecki, LM; Chin, SS; Waite, JH; Lavin, MF. Molecular diversity of marine glues: Polyphenolic proteins from five mussel species. Molecular marine biology and biotechnology 1991; 1 (1): 78-88.

[53] Secrets of the sandcastle worm could yield a powerful medical adhesive. Published online in www.sciencedaily.com, September 2009.

[54] Fountain, H. Studying Sea Life for a Glue That Mends People. New York Times. Published online in www.nytimes.com, April 2010.

[55] Flammang, P; Santos, R; Haesaerts, D. Echinoderm Adhesive Secretions: From Experimental Characterization to Biotechnological Applications. *Progress in Molecular and Subcellular Biology*, 2005; 39: 201-220.

[56] Santos, R; Gorb, S; Jamar, V; Flammang, P. Adhesion of echinoderm tube feet to rough surfaces. The Journal of *Experimental Biology* 2005; 208: 2555-2567.

[57] Demoor, S; Waite, JH; Jangoux, M; Flammang, P. Characterization of the adhesive from cuvierian tubules of the sea cucumber Holothuria forskali (Echinodermata, Holothuroidea). *Marine Biotechnology* 2003; 5: 45.

[58] Flammang, P; Ribense, J; Jangoux, M. Biomechanics of adhesion in sea cucumber cuvuerian tubules (Echinodermata, Holothuroidea), *Integrative and Comparative Biology* 2002; 42: 1107-1115.

[59] Zardus, JD; Nedved, BT; Huang, Y; Tran, C; Hadfield, MG. Microbial Biofilms Facilitate Adhesion in Biofouling Invertebrates. *Biology Bulletin* 2008; 214: 91-98.

[60] Wiegemann, M. Adhesion in blue mussels (*Mytilus edulis*) and barnacles (genus *Balanus*): Mechanisms and technical applications, *Aquat. Sci.* 2005; 67: 166-176.

[61] Walker, G. Cirripedia. In: F. W. Harrison and A. G. Humes (eds.), Microscopic Anatomy of Invertebrates 9: Crustacea, John Wiley, New York, Chichester, 1999, pp. 249-311.

[62] Abbott, A. Bioadhesives: potential for exploitation. *Sci. Progress*. 1990; 74: 131-146.

[63] Naldrett, MJ; Kaplan, DL. Characterization of barnacle (*Balanus eburneus* and *B. crenatus*) adhesive proteins. *Marine Biology* 1997; 127: 629-635.

[64] Khandeparker, L; Anil, AC. Underwater adhesion: the barnacle way. *International Journal of Adhesion & Adhesives* 2007; 27(2): 165-172.

[65] Kaplan, DL; Gatenholm, P; Berglin, KM; Platko, JD; Pepper, LR; Ngangan, AV. Barnacle adhesion proteins, 2003, US Patent 0143673 A1.

[66] Kamino, K. Barnacle underwater adhesive - Biochemistry of self-organized multi-functional complex. Abstract, Workshop on New Perspectives in Marine Biofouling and Biofouling Control, Fiskebäckskil, Sweden, 2003.

[67] Lee, H; Dellatore, SM; Miller, WM; Messersmith, PB. Mussel-Inspired Surface Chemistry for Multifunctional Coatings, *Science* 2007; 318: 426.

[68] Lee, H; Lee, BP; Messersmith, PB. A reversible wet/dry adhesive inspired by mussels and geckos, *Nature* 2007; 448: 338-341.

[69] Lee, H; Scherer, NF; Messersmith, PB. Single-molecule mechanics of mussel adhesion, *PNAS* 2006; 103: 12999-13003.

[70] Wilker, JJ. Marine bioinorganic materials: mussels pumping iron, *Curr. Opin. Chem. Biol.* 2010; 14: 276-283.

[71] Waite, JH. Adhesion in byssally attached bivalves. *Biol. Rev*. 1983; 58(2): 209-231.

[72] Rzepecki, LM; Hansen, KM; Waite, JH. Characterization of a cysteine-rich polyphenolic protein family from the blue mussel Mytilus edulis L. *Biol. Bull.* 1992;183:123-137.

[73] Yu, M; Deming, TJ. Synthetic polypeptide mimics of marine adhesive. *Macromolecules*. 1998;31:4739-4745.

[74] Filpula, DR; Lee, S-M; Link, RP; Strausberg, SL et al. Structural and functional repetition in a marine mussel adhesive protein. *Biotechnol. Prog*. 1990;6:171-177.

[75] Kitamura, M; Kawakami, K; Nakamura, N; Tsumoto, K et al. Expression of a model peptide of a marine mussel adhesive protein in Escherichia coli and characterization of its structural and functional properties. *J. Polym. Sci.* 1999;37:729-76.
[76] Fank, BP; Belfort, G. Atomic force microscopy for low-adhesion surfaces: Thermodynamic criteria, critical surface tension, and intermolecular forces. *Langmuir* 2001;17:1905-1919.
[77] Hwang, DS; Yoo, HJ; Jun, JH; Hoon, WK et al. Expression of functional recombinant mussel adhesive protein Mgfp-5 in Escherichia coli. *Appl. Environ. Microbiol.* 2004;70(6):3352-3359.
[78] Hwang, DS; Gim, Y; Yoo, HJ; Cha, HJ. Practical recombinant hybrid mussel bioadhesive fp-151. *Biomaterials* 2007;28:3560-3568.
[79] Burkett, JR; Wojtas, JL; Cloud, JL; Wilker, JJ. A method for measuring the adhesion strength of marine mussels. *J Adhes* 2009;85:601-615.
[80] Benson, RS. Use of radiation in biomaterials science. Nuclear Instruments and Methods in Physics Research Section B: Beam Interactions with Materials and Atoms 2002;191:752-757.
[81] Qin, X-X; Coyne, KJ; Waite, JH. Tough tendons. Mussel byssus has collagen with silk-like domains. *J. Biol. Chem.* 1997;272(51):32623-32627.
[82] Corne, KJ; Qin, X-X; Waite, JH. Extensible collagen in mussel byssus: A natural block copolymer. *Science* 1997;277:1830-1832.
[83] Waite, JH; Qin, X-X; Corne, KJ. The peculiar collagens of mussel byssus. *Matrix Biol.* 1998;17:93-106.
[84] Waite, JH. Evidence for a repeating 3,4-dihydroxyphenylalanine andhydroxyproline containing decapeptide in the adhesive protein of the mussel, Mytilus edulis L. *J. Biol.Chem.* 1983;258:2911-2915.
[85] Inoue, K; Takeuchi, Y; Miki, D; Odo, S. Mussel adhesive plaque proteins is a novel member of epidermal growth factor-like gene family. *J. Biol. Chem.* 1995;270:6698-6701.
[86] Papov, VV; Diamond, TV; Biemann, K; Waite, JH. Hydroxyarginine-containing polyphenolic proteins in the adhesive plaques of the marine mussel Mytilus edulis. *J. Biol. Chem.* 1995 270:20183-20192.
[87] Waite, JH; Qin, X. Polyphosphoprotein from the adhesive pads of Mytilus edulis. *Biochemistry* 2001;40:2887-2893.
[88] Zhao, H; Waite, JH. Linking adhesive and structural proteins in the attachment plaque of Mytilus californianus. *J. Biol. Chem.* 2006;281-36:26150-26158.
[89] Yu, M; Hwang, J; Deming, TJ. Role of L-3,4-dihydroxylphenylalanine in mussel adhesive proteins. J. *Am. Chem. Soc.* 1999;121:5825-5826.
[90] Yamada, K; Chen, T; Kumar, G; Vesnovsky, O; Topoleski, LD; Payne, GF. Chitosan based water-resistant adhesive. Analogy to mussel glue. *Biomacromolecules* 2000;1:252-258.
[91] Ninan, L; Monahanh, J; Stroshine, RL; Wilker, JJ; Shi, R. Adhesive strength of marine mussel extracts on porcine skin. *Biomaterials* 2003; 24:4091-4099.
[92] Chivers, RA; Wolowacz, RG. The strength of adhesive-bonded tissue joints. *Int. J. Adhesion and Adhesives* 1997;17:127-132.

[93] Benedict, CV; Picciano, PT. Adhesives from marine mussels In: Hemingway RW, Conner AH, Branham SJ. editors. Adhesives from renewable resources; 385; Washington: *American Chemical Society*; 1989. p.465-483.
[94] Monahan, J; Wilker, JJ. Cross-Linking the Protein Precursor of Marine Mussel Adhesives: Bulk Measurements and Reagents for Curing. *Langmuir* 2004;20:3724-3729.
[95] Strausberg, RL; Link, RP. Protein-based medical adhesives. *Trends Biotechnology*. 1990; 8:53-57.
[96] Waite, JH; Tanzer, ML. Polyphenolic substance of Mytilus edulis: Novel adhesive containing L-Dopa and hydroxyproline. *Science* 1981;212:1038-1040.
[97] Morgan, D. Two firms race to derive profits from mussels' glue. *Scientist* 1990; 4:1-6.
[98] Cha, HJ; Hwang, DS; Lim, S. Development of bioadhesives from marine mussels. *Biotechnology* J. 2008; 3:631-638.
[99] Salerno, AJ; Goldberg, I. Cloning, expression, and characterization of a synthetic analog to the bioadhesive precursor protein of the sea mussel Mytilus edulis. *Appl. Microbiol. Biotechnol.* 1993;58:209-214.
[100] Wong, HH; Kim, YC; Lee, SY; Chang, HN. Effect of post-induction nutrient feeding strategies on the production of bioadhesive protein in *Escherichia coli. Biotechnol. Bioeng*. 1998;60(3):271-276.
[101] Marks, P. Mussel power. *New Scientist* 1999;164:12.
[102] Takeuchi, Y; Inoue, K; Miki, D; Odo S et al. Cultured mussel foot cells expressing byssal protein genes. *J. Exp. Zool.* 1999;283:131-136.
[103] Bitton, R; Bianco-Peled, H. Novel biomimetic adhesives based on algae glue. *Macromol Biosci*. 2008;8:393-400.
[104] Dove, J; Sheridan, P. Adhesive protein from mussels: Possibilities for dentistry, medicine, and industry. *J. Am. Dent. Assoc.* 1986;112-6:879.
[105] Grande, DA; Pitman, MI. The use of adhesives in chondrocyte transplantation surgery: preliminary studies. *Bull. Hosp. Jt. Dis. Orthop. Inst.* 1988;48:140-148.
[106] Pitman, MI; Menche, D; Song, EK; Ben-Yishay, A et al. The use of adhesive in chondrocyte transplantation surgery: In vivo studies. *Bull. Hosp. Jt. Dis. Orthop. Inst.* 1989;49:213-220.
[107] Hwang, DS; Sim, SB; Cha, HJ. Cell adhesion biomaterial based on mussel adhesive protein fused with RGD peptide. *Biomaterials* 2007;28:4039-4046.
[108] Hwang, DS; Gim, Y; Kanga, DG; Kim, YK; Cha, HJ. Recombinant mussel adhesive protein Mgfp-5 as cell adhesion biomaterial. *J. Biotechnol.* 2007;127:727-735.
[109] Waite, JH. Adhesion à la Moule. *Integr.Comp. Biol.* 2002;42:1172-1180.
[110] Sáez, C ; Pardo, J ; Gutierrez, E ; Brito, M et al. Immunological studies of the polyphenolic proteins of mussels. *Comp. Biochem. Physiol.* 1991;98B-4:569-572.
[111] Dalsin, JL; Lin, L ; Tosatti, S; Vörös, J ; Textor, M ; Messersmith, PB. Protein resistance of titanium oxide surfaces modified by biologically inspired mPEG-DOPA. Langmuir 2005; 21: 640-646.
[112] Scholz, I; Barnes, WJP; Smith, JM; Baumgartner, W. Ultrastructure and physical properties of an adhesive surface, the toe pad epithelium of the tree frog, *Litoria caerulea* White. *The Journal of Experimental Biology* 2009; 212(2): 155-162.
[113] Barnes, WJP. Functional Morphology and Design Constraints of Smooth Adhesive Pads, *MRS Bull.* 2007; 32: 479-485.

[114] Barnes, WJP. Tree frogs and tire technology. *Tire Technol. Int.* March 1999: 42-47.
[115] Barnes, WJP; Smith, J; Oines, C; Mundl, R. Bionics and wet grip. *Tire Technol. Int.* December 2002: 56 60.
[116] Persson, BNJ. Wet adhesion with application to tree frog adhesive toe pads and tires. *J. Phys. Condens. Matter.* 2007; 19. 376110.
[117] Dirks, J-H; Clemente, CJ; Federle, W. Insect tricks: two-phasic foot pad secretion prevents slipping. *J. R. Soc. Interface* 2010; 7: 587-593.
[118] Pennisi, E. Geckos climb by the hairs of their toes. *Science* 2000; 288:1717-1718.
[119] Bhushan, B; Sayer, RA. Gecko Feet: Natural Attachment Systems for Smart Adhesion, Applied Scanning Probe Methods Vol. VII - Biomimetics and Industrial Applications, Heidelberg Germany, Springer-Verlag, 2007, pp. 41-76.
[120] Ruibla, R; Ernst, V. The structure of the digital setae of lizards. *J. Morph.* 1965; 117:271-294.
[121] Russell, AP. A contribution to the functional morphology of the foot of the tokay, Gekko gecko. *J. Zool. Lond.* 1975; 176:437-476.
[122] Williams, EE; Peterson, JA. Convergent and alternative designs in the digital adhesive pads of scincid lizards. *Science* 1982; 215:1509-1511.
[123] Ghandi, M. The ughly buglies. Swagat. Media Transasia (India), Bangalore, India, 2002.
[124] Autumn, K; Liang, Y; Hsieh, ST; Zesch, W; Chan, WP; Kenny, TW; Fearing R; Full, RJ. Adhesive force of a single gecko foot-hair. *Nature* 2000;405:681-685.
[125] Persson, BNJ; Gorb, S. The effect of surface roughness on the adhesion of elastic plates with application to biological systems. *J Chem Phys* 2003; 119:11437-11444.
[126] Autumn, K; Sitti, M; Liang, YA; Peattie, AM; Hansen, WR; Sponberg, S; Kenny, TW; Fearing, R; Israelachvili, JN; Full, RJ. Evidence for van der Waals adhesion in gecko setae. *Proc Natl Acad Sci U S A* 2002;99:12252-12256.
[127] Sun, W; Neuzil, P; Kustandi, TS; Oh, S; Samper, VD. The nature of the gecko lizard adhesive force. *Biophys J Biophys Lett* 2005;89: L14-7.
[128] Geim, AK; Dubonos, SV; Grigorieva, IV; Novoselov, KS; Zhukov, AA; Shapoval, SY. Microfabricated adhesive mimicking gecko foot-hair. *Nat Mater* 2003; 2:461-463.
[129] Sitti, M; Fearing, RS. Synthetic gecko foot-hair for micro/nano structures as dry adhesives. *J Adhes Sci Technol* 2003; 18:1055-1074.
[130] Yurdumakan, B; Raravikar, NR; Ajayan, PM; Dhinojwala, A. Synthetic gecko foothairs from multiwalled carbon nanotubes. *Chem Commun* 2005; 30: 3799-3801.
[131] Peressadko, A; Gorb, SN. When less is more: experimental evidence for tenacity enhancement by division of contact area. *J Adhes* 2004; 80:247-261.
[132] Northen, MT; Turner, KL. A batch fabricated biomimetic dry adhesive. *Nanotechnology* 2005; 16:1159-1166.
[133] Parness, A; Soto, D; Esparza, N; Gravish, N; Wilkinson, M; Autumn, K; Cutkosky, M. A microfabricated wedge-shaped adhesive array displaying gecko-like dynamic adhesion, directionality and long lifetime. *J. R. Soc. Interface* 2009; 6: 1223-1232.
[134] Nosonovsky, M; Bhushan, B. Multiscale effects and capillary interactions in functional biomimetic surfaces for energy conversion and green engineering. *Phil. Trans. Royal. Soc.* 2009; A367:1511-1539.
[135] Sethi, S; Ge, L; Ci, L; Ajayan, PM; Dhinojwala, A. Gecko-inspired carbon nanotube-based self-cleaning adhesives. *Nano Letters, Am. Chem. Soc.*, 2008; 8(3):822-825.

[136] Mahdavi, A; Ferreira, L; Sundback, C; Nichol, JW; Chan, EP; Carter, DJD; Bettinger, CJ; Patanavanich, S; Chignozha, L; Joseph, EB; Galakatos, A; Pryor, H; Pomerantseva, I; Masiakos, PT; Faquin, W; Zumbuehl, A; Hong, S; Borenstein, J; Vacanti, J; Langer, R; Karp, JM. A biodegradable and biocompatible gecko-inspired tissue adhesive. *Proc Natl Acad Sci U S A* 2008;105:2307-2312.

[137] Yanik, MF. Towards gecko-feet- inspired bandages. *Trends in Biotechnology* 2009; 27:1-2.

[138] Karp, JM. Tape-based tissue adhesives: inspiration from nature. Boston, MA: Harvard Medical School; 2011, 6 pp. http://scholar. google.pt/scholar?hl=pt-PT&q=Tape-Based+Tissue+Adhesives%3 A+Inspiration+From+Nature&lr=&as ylo=&as vis=0; consulted 6 December 2011.

[139] Menon, C; Murphy, M; Sitti, M. Gecko Inspired Surface Climbing Robots. *Proceedings of the 2004 IEEE International Conference on Robotics and Biomimetics* August 22-26, 2004, Shenyang, China.

[140] Full, RJ; Fearing, RS; Kenny, TW; Autumn, K. Adhesive microstructure and method of forming same. 2004, US Patent 6,737,160.

In: Focus on Biomimetics Research
Editor: Abele Scipione

ISBN: 978-1-62417-332-5

Chapter 3

BIOMIMETIC PHOTONIC CRYSTAL FROM BUTTERFLY WINGS

Shenmin Zhu**, Di Zhang and Yao Li
State Key Laboratory of Metal Matrix Composites,
Shanghai Jiao Tong University, Shanghai, P. R. China

ABSTRACT

A series of approaches have been developed for the replication of photonic structures from *Morpho* butterfly wings. By selecting appropriate precursors, we can achieve exact replications of photonic structures in a variety of metal oxides, such as titania, tin oxide, silica, WO_3, etc. Interestingly, polymers such as polypyrrole (PPy) with photonic crystal structures were reported for the first time to be duplicated from *Morpho* butterfly wings by using a two-step templating process. The exact replications at the macro, micro- and even nanoscale levels were characterized by a combination of FE-SEM, TEM, EDX and Raman measurements.

The optical properties of the replicas were investigated by using reflectance spectroscopy, and it was found that the interesting chromaticity of the reflected light could be adjusted simply by tuning the replica composition. The primary chemical sensor or biosensor properties of biomimetic materials thus prepared were investigated, which showed a much high performance owing to the combination of large surface area with photonic crystal structures.

These reported biomimetic techniques are effective to produce photonic crystals from nature with three dimensional networks, which may open up an avenue for many passive and active infrared devices, especially high performance optical components, sensors and magneto-optical devices.

Keywords: photonic crystal; butterfly wings; metal oxide; PPY; sensor

---

* E-mail: smzhu@sjtu.edu.cn; zhangdi@sjtu.edu.cn. Tel.: +86-21-34202584. Fax: +86-21-34202749.

INTRODUCTION

The study of photonic crystals (PCs), has become an important area of research for potential applications in the photonic and optical industries, including optical communication, smart windows, memory devices, and biosensors. Generally, templatings from the sedimentation of monodispersed silica or polystyrene spheres, are used to fabricate photonic crystals with opal or inverse-opal structure, including semiconductors [1], metals [2], polymers [3] and hydrogels [4]. However, the preparation of the ordered periodic dielectric structures relies on the assembling of colloidal crystals through the vertical deposition–evaporation method, which is not easy to control. In fact, the nature offers an astonishing variety of photonic structures which cannot easily be synthesized using available nanotechnology: arm ossicles from light sensitive species of the brittle star have regular arrays of inorganic microstructures with a characteristic double lens as photonic elements, to focus light towards nerve bundles of photoreceptors; a two-dimensional (2D) hexagonal lattice of voids within the cross section of hair-like setae of polychete worm creates a natural pseudo-photonic crystal fiber, to generate strong iridescence and/or to appear metallic. Particularly, butterfly wings are very well known for the diversity of colors and patterns, all of which derives from the surface periodic scales of the wings. The biological organizations produced by self-assembly of highly ordered functional units provide us numerous opportunities in producing a wide range of photonic structural materials. Unfortunately, the exact imitation of biological structures by using a man made assembly synthesis route is an extremely difficult task, since the formation mechanisms of biological structures are tremendously complex. Therefore, research has turned towards naturally occurring photonic structures for inspiration and several nanofabrication techniques have been developed to replicate natural photonic structures, including electron beam lithography, focused ion-beam chemical vapor deposition [5], chemical vapor deposition [6], atomic layer deposition [7], the conformal-evaporated-film-by-rotation technique [8] and the soaking method [9] that rely on the formation of inorganic, organic and polymeric layers or stacks. An interesting work reported by Huang et al shows that the fabricated Al_2O_3 replicas from the *Morpho* butterfly can reflect different colors depending on the thickness of the replicas [10]. In our previous work, iridescent large-area ZrO_2 photonic crystals have been prepared by a socking and then calcination method [11].

Although these technologies are promising, they often involve fairly long or complicated processes. In 2008, we successfully developed a new technology of using ultrasonic irradiation to replicate natural hierarchical structures in manganese oxide [12]. The advantages of the sonochemical approach are not only that the replication is precise but also the replication process is simple, efficient and the whole process can be accomplished in only a few hours. Unfortunately, no optical properties can be measured in these brown/dark manganese oxide replicas.

Transparent materials with refractive indexes (RIs) ranging from 1.4 (SiO_2) to 3.8 (Sb_2S_3), including anatase phase (RI = 2.65), and SnO_2 (RI = 1.45) have long been candidate materials for use as PCs. Structural colors of PCs can be modulated by their optical coefficients and internal submicrostructures. Therefore, synthesis of butterfly wings with high dielectric constants and excellent optical properties is of great interest.

Our purpose is to develop new, efficient approaches for the fabrication of photonic crystal from biotemplates. The recent results for biomimetic replication of photonic structures from butterfly wings in TiO_2, SnO_2, SiO_2, WO_3, and even polypyrrole are presented. The optical properties of the replicas were investigated by using reflectance spectroscopy, and it was found that the interesting chromaticity of the reflected light could be adjusted simply by tuning the replica materials.

The primary chemical sensor or biosensor properties of biomimetic materials thus prepared were investigated, which showed a much high performance owing to the large surface area as well as photonic structures. The strategies developed here are expected to be extended in the design of other functional materials with 3D network photonic crystal structures.

Sonochemical Method towards Replication of Photonic Structures from Butterfly Wings

Firstly, we report a general facile technique with ultrasound irradiation for producing photonic structural metal oxides with the morphology and ordered nanoscale arrays inherited from the wings of the *Morpho* butterfly. By adjusting the kinds of precursors, we can achieve the exact replication of photonic structures in a variety of metal oxides, including TiO_2, SnO_2 and silica [13]. The replication mechanism is proposed. The sensing properties of the SnO_2 replica with photonic structures were investigated, which shows an extremely high sensitivity in ethanol, and short response and recovery time.

Figure 1a shows a photo of a male *Morpho* butterfly used in our replication experiments. The uniform blue color covering the majority of the wings was from millions of ordered scales, as shown in the optical image in Figure 1b.

It is known that the blue color of the *Morpho* wings results from arrays of vertically aligned net-like skeleton structures. A typical dimension of a scale is 150 μm in length and 60 μm in width.

These overlapping scales possess an overall rectangular shape with pointed tips as shown in Figure 1c. Thirty five to forty rows of lamellae align on the scale surface with an almost identical interspacing. Figure 1d reveals the porous architecture of one such scale.

Each scale is comprised of parallel ridges, spaced several microns apart, aligned along the scale length. The lamellae (ridges) are supported by cross-ribs, which dictate the inter-lamellae spacing of about 1.6 μm and hold the lamellae 1 μm above the bottom surface of the scales. A closer view of these lamellae is shown in Figure 1d (inset). The ridges are decorated with nanoscale ribs (approx. 50 nm in width) that are spaced approximately 150 nm apart.

The tile-like arrangement of the scales and the ridges with nanoscale ribs were retained in TiO_2 replicas after being calcined at 500 °C for 3 h (Figures 2a and b).

An extremely good replication of the fine detail of the original scale structures can clearly be seen. In particular, the excellent reproduction of the longitudinal ridges with overlapping slanted lamellae running along the top of the ridge, transverse ribs and micro-ribs running down the sides of the longitudinal ribs indicate that the TiO_2 had coated on the fine structure of the scales. Between the lamellae are the supporting cross-ribs, each of which had one or two legs standing on the bottom surface and exhibiting a hollow body (Figure 2b).

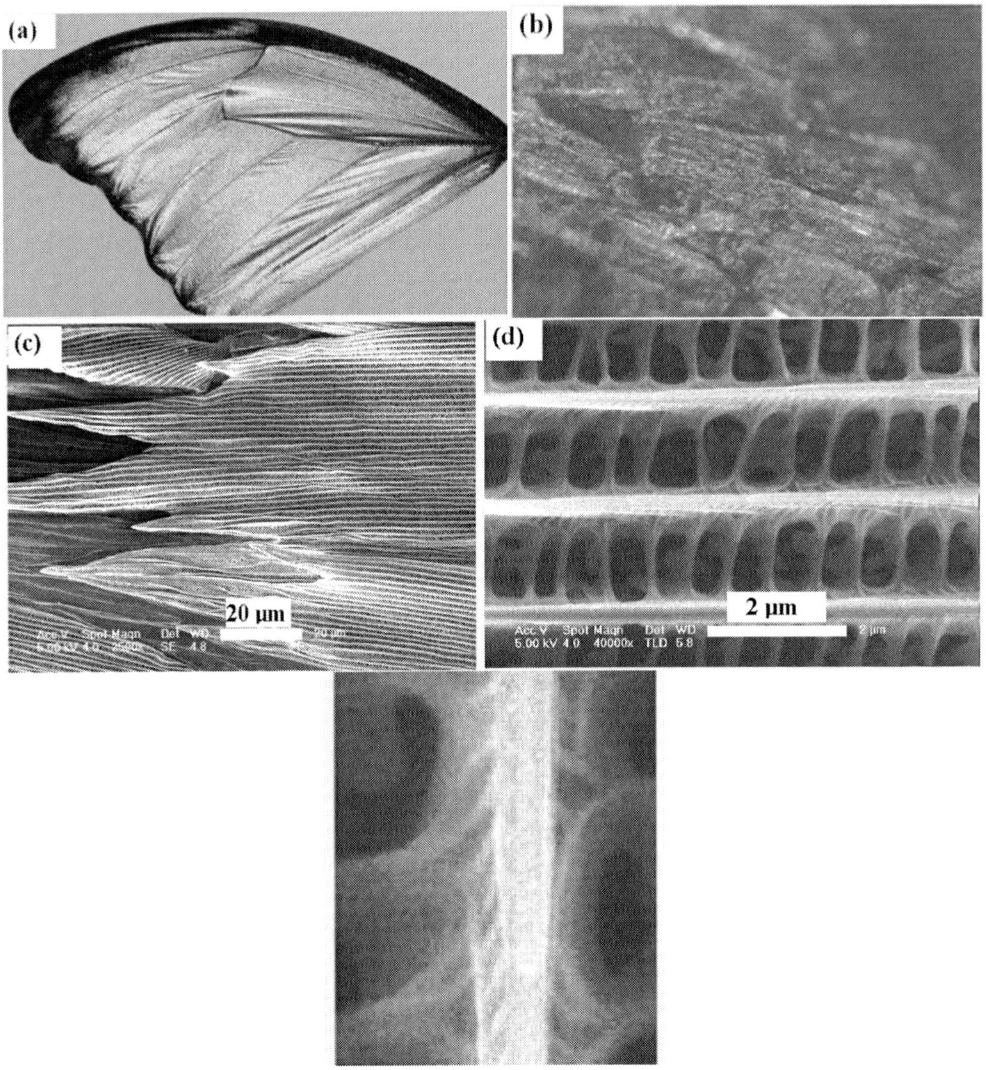

Figure 1. (a) Photograph of the original wing, (b) optical micrograph, (c) FE-SEM image of sections of butterfly wing scales, (d) a higher magnification of a wing scale. Produced with permission [13].

The seamless continuous shell at the contacting junction between the ribs and the scale bottom suggests that the TiO_2 layer grew simultaneously around both the ribs and the bottom surfaces (Figure 2c). The high quality of the cross section of a replica scale, presented in Figure 2d, and the smooth surface indicate that no additional short- or long range structural disorder was introduced. The longitudinal ridges are approximately 1.25 μm apart and the transverse ribs between the ridges have spaces between them in the region of 100 nm; a 22% shrinkage was obtained because of the elevated calcination temperature. TEM images (Figures 2e–h) reveal that the TiO_2 retained the overall scale morphology of the original butterfly (Figures 2e and f). HRTEM reveals the thin scale bottom was comprised of TiO_2 nanocrystals measuring approximately 5–10 nm (Figure 2g). The crystal structure of the nano TiO_2 was characterized by selected-area electron diffraction (SAED) (see Figure 2g (inset)) and can be consistently indexed to the crystal structure of anatase TiO_2 by the diffraction rings.

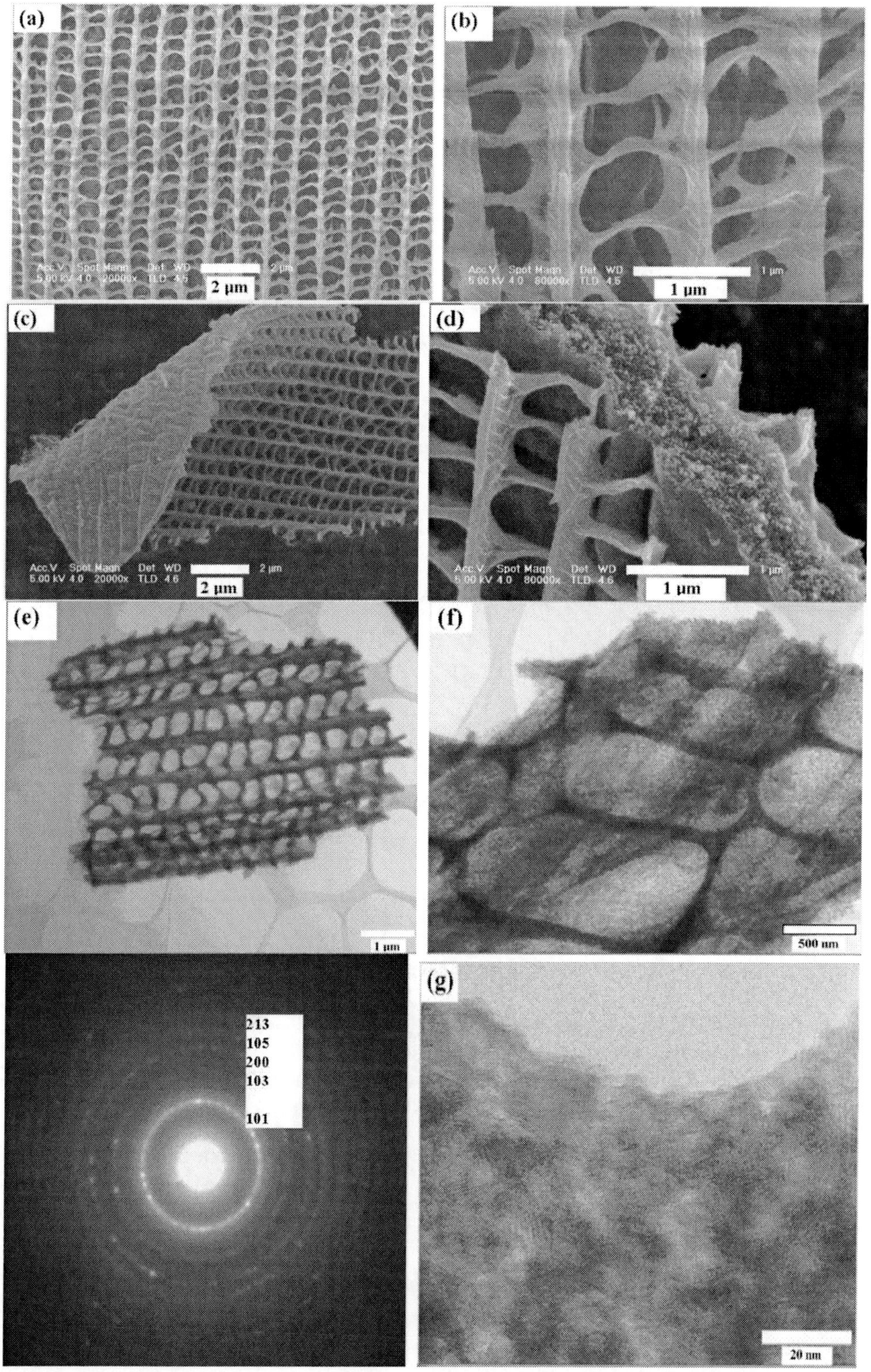

Figure 2. (Continued).

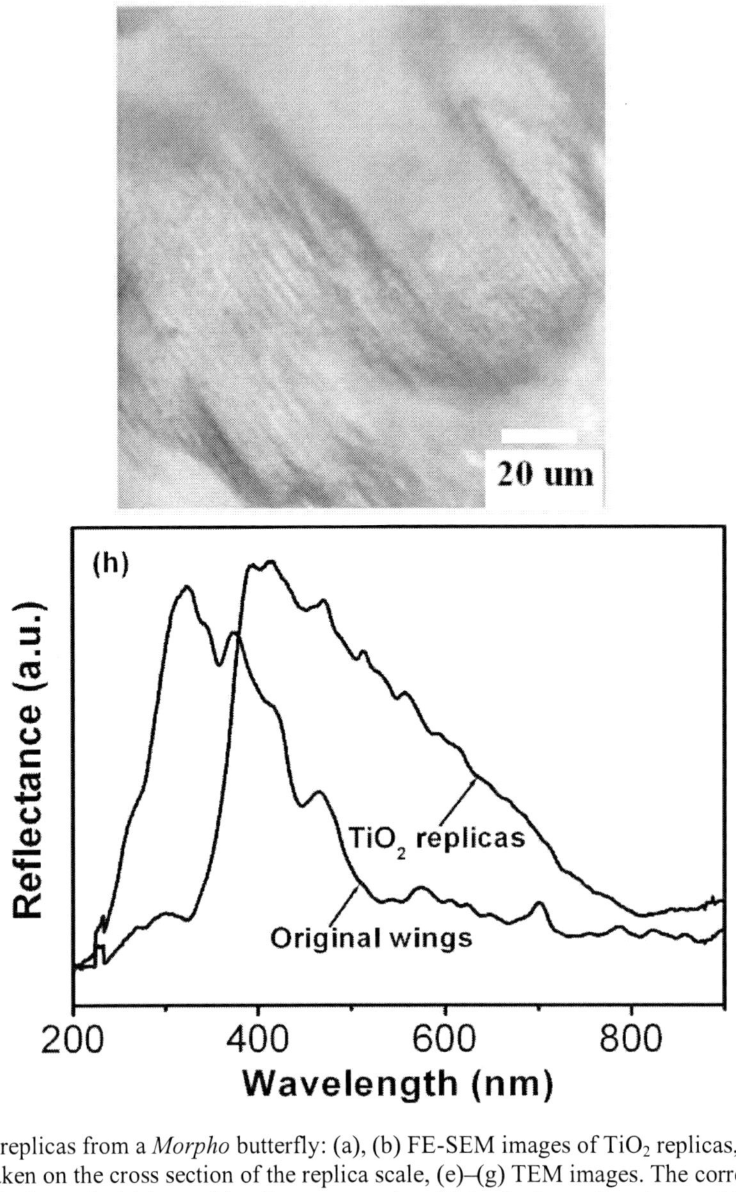

Figure 2. TiO_2 replicas from a *Morpho* butterfly: (a), (b) FE-SEM images of TiO_2 replicas, (c), (d) FE-SEM images taken on the cross section of the replica scale, (e)–(g) TEM images. The corresponding SAED pattern is shown in (g) inset, (h) reflectance results of Morpho butterfly wings and TiO_2 replica; inset is the optical image of the TiO_2 replicas. Produced with permission [13].

Raman spectroscopy analyses also indicated that those TiO_2 were comprised of anatase. Energy dispersive x-ray (EDX) analysis confirms that the replica was essentially TiO_2. The success of the replication was not only on the morphologies but also on the optical property.

Reflection measurements were performed in the visible and near-infrared wavelength regions to reveal the interactions of the replicated structures with light (Figure 2h). The original butterfly wing exhibited the highest reflection at 323 nm, which was in the UV region. The relatively large reflectance in the short visible wavelength region contributed to its natural violet/blue color. The TiO_2 fossil shows a redshifted reflection peak at 415 nm, which is consistent with the green/pink color observed under an optical microscope (Figure 2h(inset)). The results clearly show the progressive shift, broadening and increase of the main peak compared with that of the original wing. From the point of view that an obvious response to an imposed electromagnetic (EM) wave requires a large refractive index (RI) difference between two components of a PC, the appearance of this peak indicates that there must be a PBG in the replicated structure, where the original organic lamellae lattice (RI = 1.57) was replaced by air and covered by a TiO_2 matrix with higher RI (2.65). In an effort to convert individual *Morpho* wing scales into SnO_2 replicas, the sonochemistry was modified by introducing tin (II) chloride as a precursor.

The resulting oxide structure retained the overall scale shape, parallel ridges and nanoscale ribs (Figure 3a). The fracture cross section shown in Figure 3b also reveals delicate supportive trusses connecting the ridges to a wavy substructure (inherited from the scale template).

The thickness of the oxide coating was approximately 100 nm (Figure 3c). After calcination, the amorphous SnO_2 shell was gradually crystallized into a more robust polycrystalline shell structure, with crystal grain sizes smaller than 4 nm (Figure 3d).

Electron diffraction and EDX microanalysis of the replica suggest that the replica consists of nanocrystalline tin oxide and the carbon associated with the organic composition of the wing of the butterfly was almost completely removed by the calcination. Raman microspectroscopic analyses of these structures also yielded vibration bands similar to those expected for rutile polymorphs of tin dioxide forms. Owing to the excellent uniformity of the tin oxide coating both the large-scale arrangement of the wing scales and the nanometer-scale period structures was perfectly preserved after the calcinations.

Figure 3 presents the reflectance spectra both from the SnO_2 and the original butterfly wings. It shows that the SnO_2 replica has strong reflection peaks at around 300–400 and 500–650 nm, though somewhat lower in intensity than the original butterfly wings. The small variations can be attributed to wavelength-dependent absorption in the material used for replication. The SnO_2 fossil gives blue, brown or yellow colors under an optical microscope (Figure 3e(inset)). To the best of our knowledge, this is the first time that the replication of SnO_2 from a butterfly wing shows photonic properties, although hierarchical SnO_2 has been synthesized from paper, cotton and cloth by Huang et al using a multiple sol–gel process before [14]. A similar process was also extended to producing SiO_2 replicas. SiO_2 replicas have the same shape, orientation and distribution as their 'parent' scales. As the technique was implemented at room temperature in a noncorrosive environment, the spatial features of the butterfly wing were preserved during the replication. The main and sub-ribs on the lamellae were well preserved without any obvious distortion. This structure is an inverse two dimensional (2D) photonic crystal slab, where periodic air spaces with particular shapes were confined and surrounded by a dielectric outer shell.

Relative to the original wing, the replica is slightly 'chemically colored' under visible light and the reflection peak was at around 450–600 nm, with a similar intensity to the original one, indicating a red/pink color. However, both the colors of the butterfly wing and the replica are mainly physical colors and due to optical interference involving grating structures (lamellae) and a pseudo-photonic crystal (reticular network). Early in 2003, replication of butterfly wings in silica was reported by Cook and his group using a chemical vapor deposition method [15]. Unfortunately, the unique morphological features, especially down to the nanometer level, were not maintained in the final materials. Therefore the method presented in this study presents a breakthrough and offers a route for synthesizing biomorphic silica with unique optical properties.

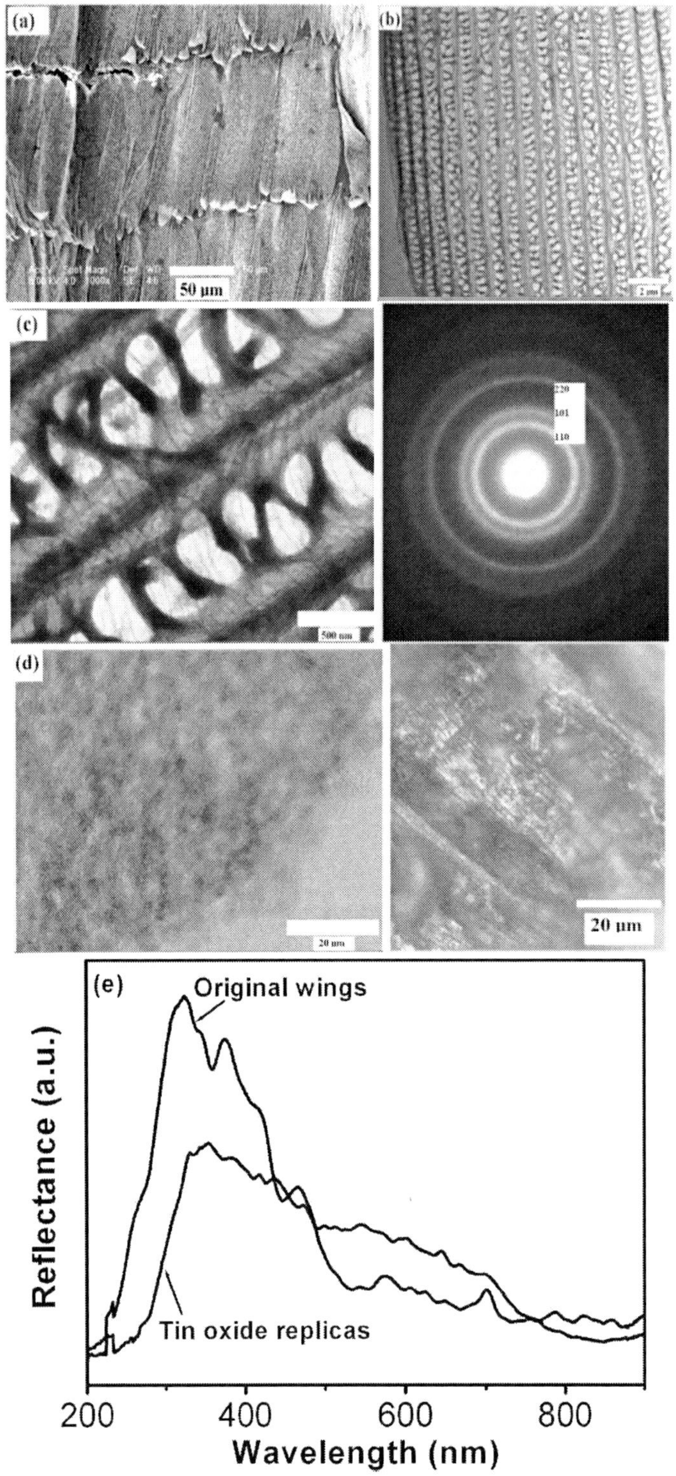

Figure 3. SnO_2 replica structures prepared by ultrasonication: (a), (b) FE-SEM images of SnO_2 replicas, (c), (d) TEM images of the SnO_2 replicas. The corresponding SAED pattern is shown in (d) inset, (e) reflectance results of Morph butterfly wings and the SnO_2 replicas, inset is the optical image of the SnO_2 replicas. Produced with permission [13].

The mechanism is proposed as below: As is well known, the butterfly wings consist of chitin. The pretreatment of *Morpho* butterfly wings is to remove the inorganic salts and to get rid of the pigment. These modifications make it easy for the precursors to access the pore channels of the biotemplates and ensure the final obtained films preserve well the microstructures of the nature butterfly wings. Moreover, sonochemical processing has proven to be a useful technique for generating novel materials with unusual properties [16]. Extreme high temperature (>5000 K), high pressure (>20 MPa) and high cooling rate (>1010 K s^{-1}) during cavitation collapse have been attributed to promote many unique reactions in the irradiated solution. Agitation by ultrasound helps to disperse solutes homogeneously in the mixture and enhances the diffusion rate of the precursor to the wing surfaces. The continuous irradiation favors the progressive departure of HCl and removes the HCl in situ and allows an extensive condensation of the inorganic framework.

Therefore, the application of ultrasound is to catalyze the hydrolysis and condensation of the titania (tin oxide or silica) precursor to TiO_2 (SnO_2 or SiO_2). The increased reactivity induced by the sonication is due to the localized high temperatures, which speeds up the condensation and polymerization reactions. Therefore the synthesis could be achieved within hours through the use of ultrasound.

Recently, Potyrailo found such a highly ordered photonic structure of scales produces remarkably diverse differential reflectance spectra, achieving a highly selective response to individual vapors [17]. This knowledge aids us in designing and producing artificial nanostructures for high performance gas sensors if we replace chitin with material having gas sensitivity, such as tin oxide, as a result of the coupling effect. The chemical gas sensing properties of the SnO_2 replica were measured using a static test system. Figure 4a shows the response of the SnO_2 replica sensor as a function of ethanol concentration at 300 °C. The response increased with the rise of ethanol concentration. There is an approximately linear response–condition relationship in the range 1-500 ppm ethanol on the SnO_2 replica. The fitted curve shows the linear relationship lg S = 0.43835 + 0.66531lgC. The correlation coefficient R of the sensor fitted curve is 0.9993.

According to the literature, high correlation coefficients are necessary for high sensitivity. As expected, the sensors exhibited an extremely high sensitivity response to ethanol gas with sensitivities of 2.9, 7.7, 12.2, 35.3, 64.8, 91.0 and 171.7 corresponding to concentrations of 1, 5, 10, 50, 100, 200 and 500 ppm, respectively. The response of the sensors can reach to 2.9 for an ethanol concentration as low as 1 ppm. From the literature, it is known that the sensitivity of SnO_2 nanofibers with a grain size of 11.5 nm is around 4.5 for 10 ppm [18], whereas the sensitivity of the SnO_2 replicas with photonic structures reaches as high as 12.2. The results are much higher than that of the sensitivity of SnO_2 obtained in other ways shown in other works, including doping materials [19]. The sensitivity of a sensor is well known to be controlled by grain size and connectivity between grains [20]. As the grain size decreases to values near the Debye thickness, the thickness of the space-charge region, approx. 10–100 nm for SnO_2, the sensitivity could be exponentially enhanced [21]. Examination of the powder x-ray diffraction results shows that the walls of the SnO_2 replicas consist of nanocrystalline SnO_2 with particle sizes of 6–7 nm estimated using the Scherrer equation, which are in approximate agreement with those observed by TEM analysis. That means the small grain size can make the replica SnO_2 more pronounced for ethanol sensing. However, the sensitivity for the SnO_2 replicas is still three times that of mesoporous ZnO–SnO_2 nanofibers reported with grain sizes of 8–14 nm (12.8 for 50 ppm). This may be partly due to the hierarchical 3D structure of the SnO_2 replica as the 3D structure provides a high surface to volume ratio and large surface accessibility for the interaction between SnO_2 and ethanol, which favors the gas response. A Brunauer–Emmett–Teller (BET) analysis showed that the surface area of the SnO_2 replica was 75 $m^2\ g^{-1}$, only half that of mesoporous ZnO–SnO_2 nanofibers. That means the ultrasensitive SnO_2-based chemical sensor resulting from SnO_2 replicas cannot be mainly attributed to the surface area of the obtained replica. Differences in behavior of the sensor should be related to the changes in geometry as described by Ozin [22].

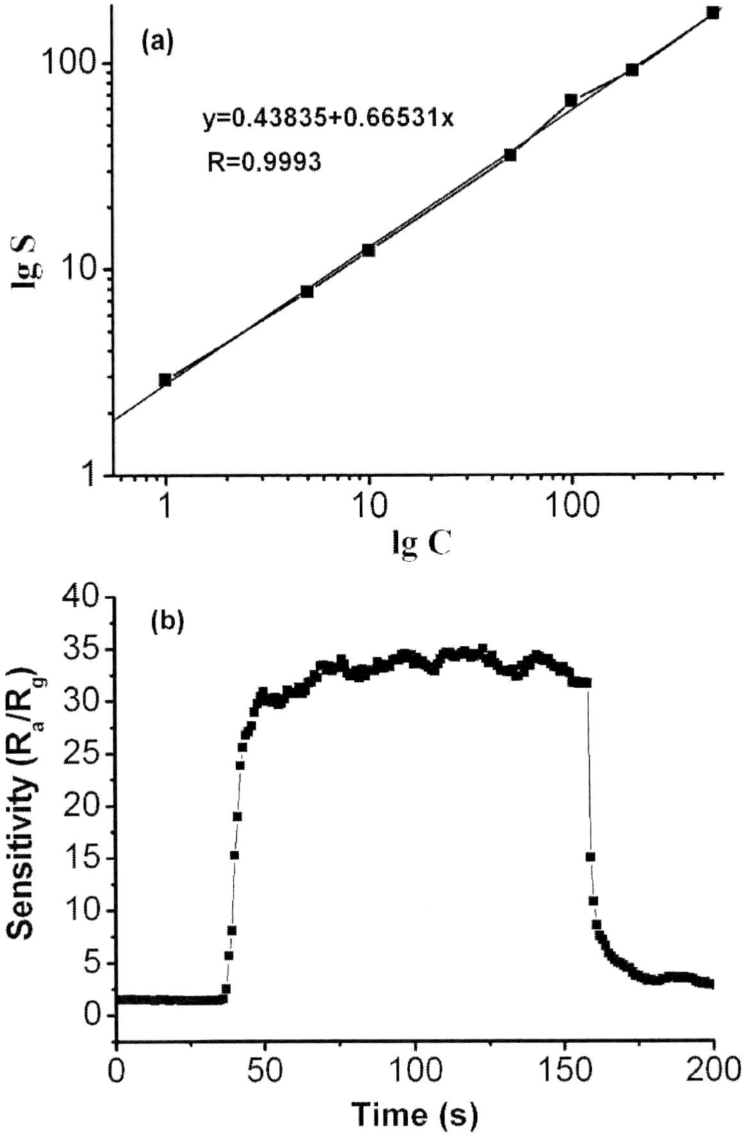

Figure 4. The sensitivity of SnO_2 replica sensors exposed to ethanol at concentrations ranging from 1 to 500 ppm measured at 300 °C and di-logarithm fitted curve of the sensitivity of the sensors with the concentration of ethanol (a), response–recovery plot of SnO_2 replica sensor to 50 ppm of ethanol (b). Produced with permission [13].

Early in 2001, Ozin found that periodic macroporous forms of SnO_2 with opal and inverse opal structures can be used as structurally well-defined gas sensors, which is close to the theoretical ideal for gas sensor behavior. Thus, a similar explanation can be used here for the chemical sensing behavior of SnO_2 replicas with photonic structures thus prepared. To further exploit the application of SnO_2 replicas for gas sensors, we analyzed their recovery upon exposure to ethanol vapor (concentration: 50 ppm) at a working temperature of 300 °C (Figure 4b). The sensor exhibited a fast response time (8 s) and short recovery time (15 s) for the ethanol gas tested. Taking into account their high sensitivities and quick response and recovery speed, our materials are very promising materials for fabricating gas sensors. In conclusion, we have demonstrated a general and efficient sonochemical process to produce photonic structures from *Morpho* butterfly wings.

A variety of photonic structural semiconductors TiO_2, SnO_2 and SiO_2 were obtained after exposing the butterfly wing templates to a suitable metal oxide precursor under ultrasonic irradiation, followed by calcination at an elevated temperature.

All the samples can reflect iridescent visible light, suggesting that the simple sono-chemical process does replicate not only the outer but also the inner and more complex surfaces of the hierarchical structure of the butterfly wings. The results confirm that the interesting chromaticity of the reflected light can be adjusted simply by tuning the replica materials. Thus this process could be developed to produce photonic structures tailored to design optical devices. The SnO_2 replica thus prepared demonstrated enhanced sensing properties, including high sensitivity, rapid response and recovery (around 8–15 s), and linearity in the range 1–500 ppm. It may provide a new means of improving gas sensing properties of SnO_2 materials besides noble metal addition and UV light irradiation.

Cu-Doped WO_3 Materials with Photonic Structures from Butterfly Wings for High Performance Sensors

Nanostructured tungsten oxide (WO_3) is of great interest due to its broad range of applications, such as photocatalysts, electrochromic devices, as well as gas sensors [23]. Several strategies have been developed to improve its performance in these potential applications. One of the easiest routes for improving the gas sensing and photocatalytic performance of tungsten oxide is doping with suitable cations, because it alters the electronic and catalytic properties of the base material [24]. Further research found that the electronic, catalytic and optical performance of WO_3 can also be enhanced by the control of its morphology, since these parameters are key elements that determine their electrical and optical properties. For example, great interest has been focused on developing semiconductors with photonic crystal (PC) structures for applications in photonic and optical industries, because their sharp spectral features allow them to be probed using low power laser probes [25]. In 2001, Ozin investigated the chemical sensing behavior of synthetic opals and inverse opals composed of SnO_2 and found that the resulting tin oxide opals showed a fairly large response to carbon monoxide, which is believed to be related to the gas-sensitive necks between adjacent spheres in geometry [22]. The effects mean that control of microstructure is necessary for command over the sensitivity to the detected gases. As a sensor material, WO_3 is widely used for the detection of amides and other pollutants from combustion or automotive emissions. The combination of WO_3 with photonic crystal structures is expected to result in an improved sensor behavior because the near-ideal microstructures of WO_3 can be formed, which can be used as structurally well-defined gas sensors. Nowadays, many shape-controlled synthesis routes have been widely employed for the preparation of semiconductors with photonic crystal (PC) structures. Among them, the sol–gel method has proved to be an effective way of obtaining a material with a desired nanostructure and morphology without altering the precursor structure or morphology. Unfortunately, these nanostructural oxides are usually obtained on specially prepared substrates with a long reaction time, which hinder its practical usage. Inspired by the nature, herein, we report the synthesis of hierarchical Cu-doped WO_3 with distinguished photonic crystal (PC) structures by using *Morpho* butterfly wings as a hard template. Replicas without photonic crystal structures were also prepared from Euploea mulciber butterfly wings as a comparison. The gas sensing properties of the photonic replicas were tested for a range of gases: $(CH3)_3N$ (TMA), NH_3, C_2H_5OH, HCHO, CH_3OH, acetone, H_2, CO and NO_2. It was found that the Cu-doped WO_3 replicas showed much higher sensitivity to trimethylamine than pure WO_3 and the replicas with PC structures possess even more enhanced sensitivity to trimethylamine. The effect of the Cu doping and PC structure on the sensor's properties will be discussed [26].

In an effort to convert every individual *Morpho* wing scale into the Cu-W-PC replicas, the sol–gel method was modified by introducing a solution of phosphotungstic acid and $CuCl_2$ in ethanol in order to retard the condensation of tungstic acid. After being calcined at 430 °C for 3 h to remove the template, the Cu-W-PC replicas were thus obtained. As shown in Figure 5, the tile-like arrangement of the scales and the ridges decorated with nanoscale ribs were retained in the Cu-W-PC replicas (Figure 5a).

An extremely good replication of the fine detail of the original scale structures can clearly be detected. In particular, the excellent reproduction of the longitudinal ridges with over lapping slanted lamellae running along the top of the ridge, transverse ribs and micro-ribs running down the sides of the longitudinal ribs indicate that the WO_3 had coated on the fine structure of the scales (Figure 5b). Between the lamellae are the supporting cross ribs. Each of the lamellae had one or two legs standing on the bottom surface and exhibited a hollow body (Figure 5c).

The high quality cross section of a replica scale is presented in Figure 5c, and the smooth surface indicates that no additional short- or long range structural disorder was introduced. The seamless continuous shell at the contacting junction between the ribs and the scale bottom suggests that the WO_3 layer grew simultaneously around both the ribs and the bottom surfaces (Figure 5d). The longitudinal ridges are approximately 1.2 mm apart and the transverse ribs between the ridges have spaces between them in the region of 100 nm; 30% shrinkage was obtained because of the elevated calcination temperature. Transmission electron microscopy (TEM) (Figure 5e) reveals that the Cu-W-PC retained the overall scale morphology of the original butterfly wing.

A typical high resolution TEM image (Figure 5f) provided further insight into the structure of the shell wall, indicating that the walls of the hollow spheres were composed of numerous small nanocrystals. The nanoparticles are randomly oriented in the framework and rather uniform in size because of the confined growth in the template. The crystal structure of the Cu-W-PC replicas was characterized by selected area electron diffraction (SAED) (Figure 5f (inset)) and the diffraction rings can be consistently indexed to the crystal structure of monoclinic WO_3, which is consistent with XRD analysis. The EDX analysis confirms that the replicas were composed of mainly W and O and a small amount of Cu (Figure 5g).

For the Cu-W replica from Euploea mulciber butterfly wings, as is expected, the original butterfly wing exhibited no reflection in the UV-vis range thus giving a natural black colored appearance due to lack of photonic crystal structures. Similarly, the Cu-W replica also showed no reflection peaks in 300–800 nm range thus giving a white color under an optical microscope (Figure 6).

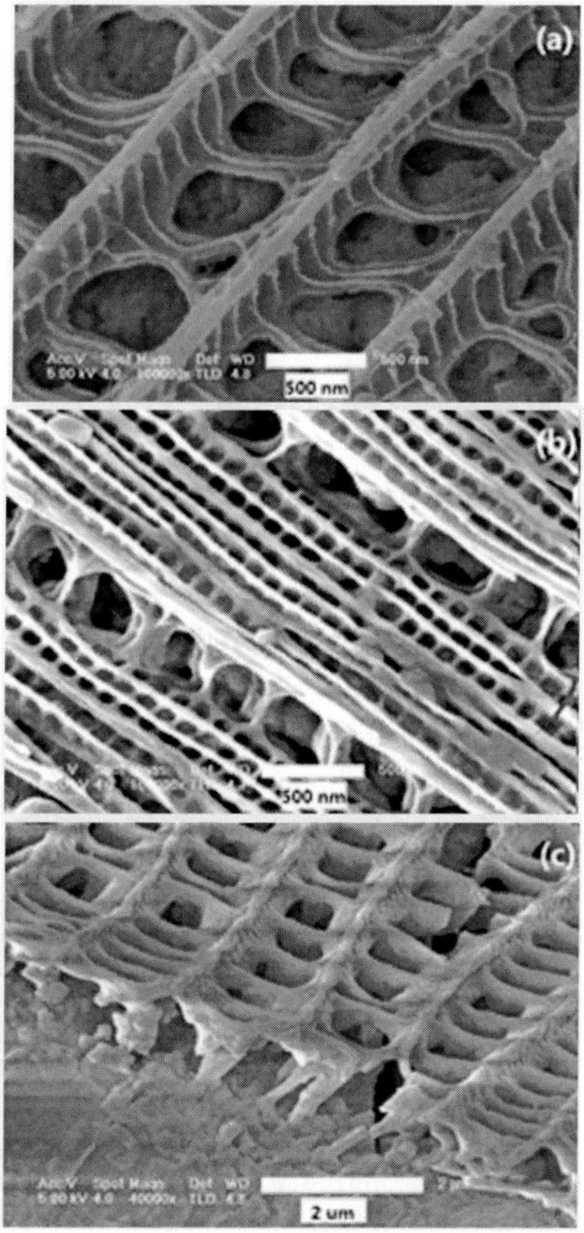

Figure 5. (Continued).

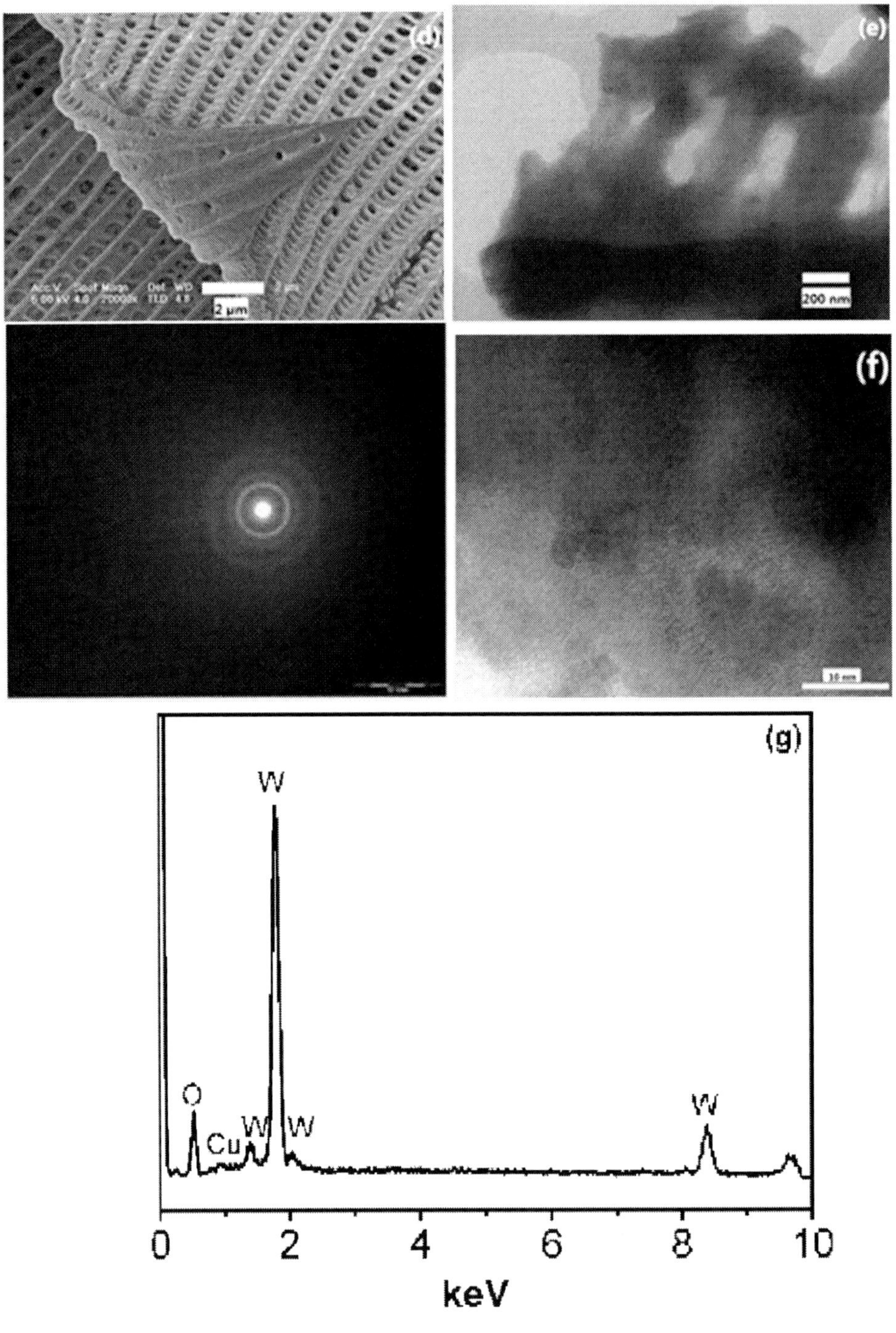

Figure 5. Cu-W-PC replicas from a *Morpho* butterfly: (a), (b) FE-SEM images of the replicas, (c), (d) FE-SEM images taken on the cross section of the replica scale, (e) TEM image, (f) a high resolution TEM image. The corresponding SAED pattern is shown in (f) inset, (g) an EDX spectrum obtained from Cu-W-PC replica, revealing the presence of W, O along with Cu in the structure. Produced with permission [26].

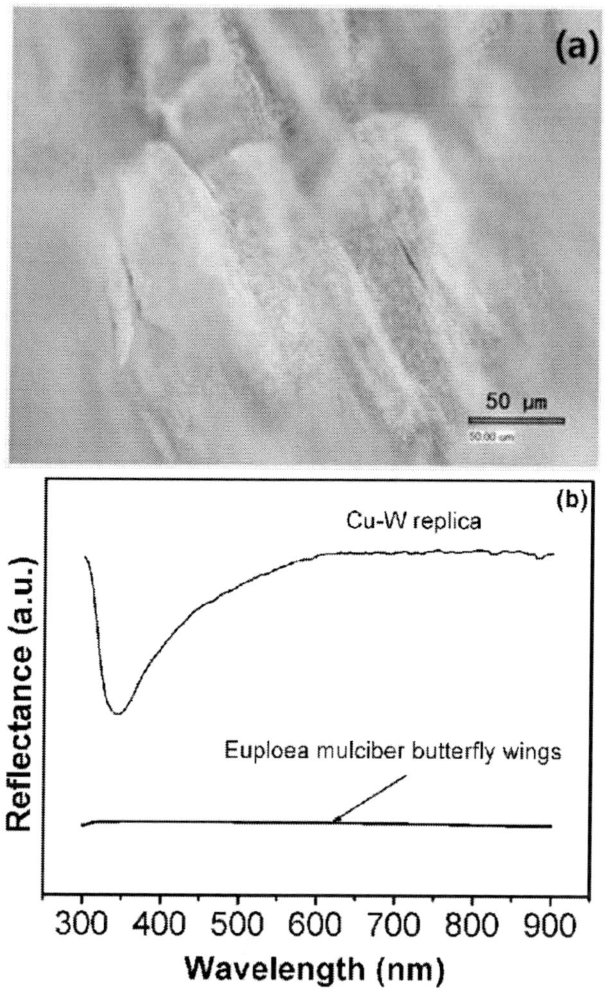

Figure 6. (a) Optical image of the Cu-W replica, (b) reflectance results of Euploea mulciber butterfly wings and Cu-W replicas. Produced with permission [26].

The XPS survey spectra of the pure and copper-doped WO_3 do not show any appreciable contribution from Cu above the detection limit. The tungsten $4p_{1/2}$, $4p_{3/2}$, $4d_{3/2}$, $4d_{5/2}$, $4f_{5/2}$, $4f_{7/2}$ peaks and oxygen 1s (O1s) peak can be detected in the spectrum. As shown in Figure 7(b), the photoelectron peak of O1s is at 531.1 eV, and is assigned to lattice oxygen in the WO_3 crystal.

As for the replicas, the O1speak shifted to 530.6 eV, probably due to the doping of Cu. The photoelectron peaks of W $4f_{5/2}$ and W $4f_{7/2}$ at 38.2 and 36.94 eV observed in the pure WO_3 crystal shifted to 35.6 (or 35.7) and 37.8 (or 37.9) in the Cu-doped WO_3 replicas. The electronic configuration of the copper centres was firstly characterized using the XPS signal in the Cu 2p3/2 region. The amount, calculated from the Cu $2p_{3/2}$ peak, is in agreement with the nominal copper concentration loaded in the synthesis route for both replicas. Gas sensors were constructed with the Cu-W-PC replica and tested with $(CH_3)_3N$ (TMA), NH_3, C_2H_5OH, HCHO, CH_3OH, acetone, H_2, CO and NO_2. The Cu-W-PC replica sensor is very sensitive to TMA at 290 °C, but not sensitive to NH_3, C_2H_5OH, HCHO, CH_3OH, acetone, H_2, CO and NO_2 (Figure 7a).

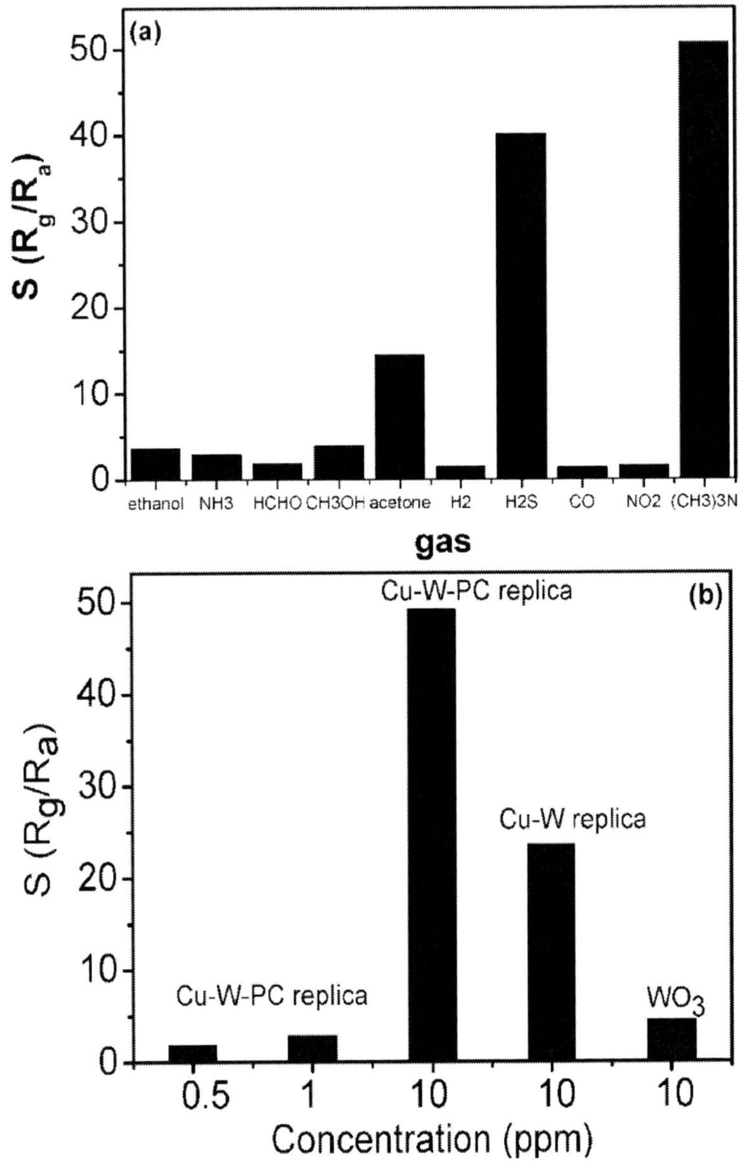

Figure 7. (a) Sensitivity of Cu-W-PC sensors to different gases (10 ppm) at 290 °C and (b) the relationship between the gas sensitivity and TMA concentration. Produced with permission [26].

The Cu-W-PC replica sensor response to TMA was evaluated in the range of 0.5–10 ppm. The response sensitivity increased with the rise of the $(CH_3)_3N$ concentration and exhibited an extremely high sensitivity to the $(CH_3)_3N$ gas and the sensitivities are 2.0, 3.3 and 49.6 corresponding to the concentrations of 0.5, 1, 10 ppm, respectively (Figure 7b).

From Figure 7b, it is known that the sensitivity of pure WO_3 is around 4.5 for 10 ppm, whereas the sensitivity of the Cu-W-PC replicas reaches as high as 49.6 for the same concentration. This much improved sensitivity of the Cu-W-PC replicas over the pure WO_3 is probably due to the doping of Cu which occupies the atomic sites instead of interstitial sites of the WO_3 lattice. The Cu doping caused negligible lattice distortion and acted as an acceptor-type impurity, thereby increasing the number of oxygen vacancies. Thus the Cu doping enhanced the interaction between the target gas and the semiconductor oxide.

In the case of reducing gases, these changes are often due to the interaction of the gas with oxygen species present on the surface of metal oxides. Sensing properties can be enhanced by the addition of suitable transition metal ions that catalyse these surface reactions which have been the key research interest in using metal oxides for gas sensing applications. As a result, the response of both the Cu-W-PC and Cu-W replicas display an enhanced sensitivity to TMA compared with the pure WO_3. It is interesting to note that the Cu-W-PC replicas sensor shows much better performance than the Cu-W replica sensor and the sensitivity of the Cu-W-PC replicas to TMA is twice that of the Cu-W replica. It is well known that increasing oxide surface area makes a great contribution to the improved sensor response. In order to get an explanation, the samples were characterized by using N_2 adsorption/desorption measurements. A Brunauer–Emmett–Teller (BET) analysis showed that the surface area of the Cu-W-PC replica was 4.9 $m^2 g^{-1}$, similar to that of the Cu-W replica (4.2 $m^2 g^{-1}$). Thus the better performance of the Cu-W-PC replica sensor over the Cu-W replica sensor cannot be explained by the different surface areas.

However it may be explained in terms of different geometries of these two replica sensors as described elsewhere. Ozin et al. found that periodic macroporous forms of SnO_2 with opal and inverse opal structures are close to the theoretical ideal structure for a gas sensor. Thus, a similar explanation can be used here for the chemical sensing behaviors of the Cu-W-PC replicas with and without photonic structure. The actual active surface area ''seen'' by the gas in the Cu-W-PC replica sensors is probably higher than for the Cu-W replica sensor due to the photonic crystal structure.

Replication of Polypyrrole with Photonic Structures from Butterfly Wings as Biosensor

As is described above, we successfully developed a new technology of using ultrasonic waves to the synthesis of photonic structures in TiO_2, SnO_2 and SiO_2 [13]. However, attempts to prepare polymer replicas using these approaches had often failed. Generally the formation of pores and tiny structural details in polymers is not probable thermodynamically. Even when they are formed, the pores tend to be closed or fractured by plastic deformation of the walls. Another general hindrance for the nanostructuration of polymers is depletion, i.e. the entropic restriction of conformations in the nanoconfinement of particles and thin walls. Therefore, to obtain an organic material with interconnected meso- or microporous structures normally requires a stiff or highly cross-linked polymer system [27]. Caruso et al. have successfully developed polypyrrole (PPy) inverse opals for biosensing [28] and fluorescence control [29]. However, the exact replication of PPy with hierarchical and photonic crystal structures from biological species is still a tough task, since biological structures are tremendously complex and consist of 3D interconnected macro, meso and nanoporous. In 2008, Saison et al. reported the replication of butterfly wings with poly(dimethylsiloxane) (PDMS), unfortunately, only the surface structures was replicated by this casting process [30]. The general restrictions of this approach still hold and mutual compatibility of the template with both the monomer and the polymer phase is one of the many problems need to be addressed. Thus, it would be very interesting to develop an approach for the fabrication of polymer with hierarchical structures from biotemplates. Inspired by our previous work on the precise replication of SiO_2 photonic structures from butterfly wings, herein, we present a two step template-directed approach for the fabrication of PPy with predetermined photonic crystal and hierarchical structures of *Morpho* butterfly wings. As a conducting polymer, PPy has been widely explored for a comprehensive range of applications in biosensors [31], gas sensors [32], microactuators [33], electrochemical supercapacitor [34] for its good environmental stability, facile synthesis, and higher conductivity. In particular, PPy has gained a distinguished popularity in biosensor applications because of its good biocompatibility, high sensitivity and fast response. It is conceivable that electrical activity of PPy combined with obvious advantages of photonic crystals in sensitive biomolecular detection, will improve its biosensor characteristics. The morphology of the PPy replicas was investigated by SEM and the results are shown in Figure 8a–d [35]. It is clear that a good detail of the original scale structures were retained in the PPy replicas.

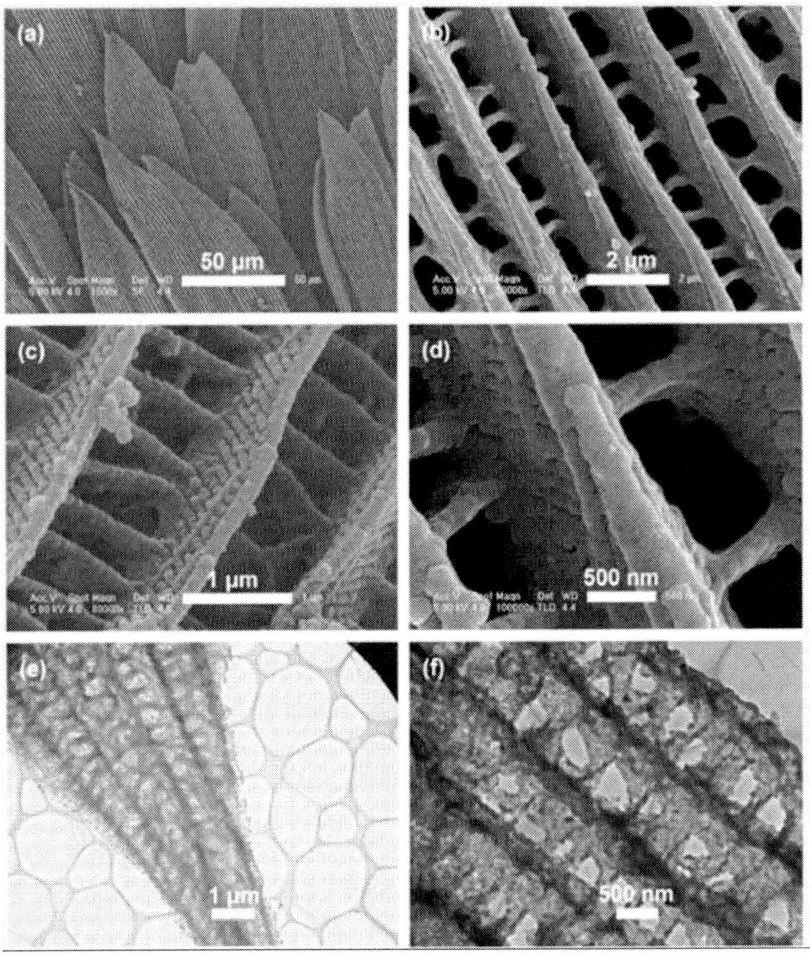

Figure 8. (a)–(d) SEM images of PPy replica and (e) and (f) TEM images of PPy replica. Produced with permission [35].

Figure 8a and b shows that the PPy replica maintained the overall scale shape, parallel ridges and nanoscale ribs of the SiO_2 template. Higher magnification SEM images (Figure 8c and d) reveal that the inter-lamellae spacing are approximately 1.5 μm for the PPy replica, which indicates that the overall dimension is reduced by 25% due to the shrinkage of PPy molecular chains during the polymerization process.

Each of the cross-ribs has one or two legs standing on the bottom surface and exhibiting a hollow body. Between the ridges, the nanoscale ribs also exhibit a periodic lattice that extends from the surface toward the scale substrate beneath. Figure 8e and f shows the morphology of the PPy replicas in TEM. Figure 8e shows the microstructure of the PPy replica at a low magnification. The dark and dense walls are the ridges orderly ranged like parallel tubes, consistent with the results of SEM. As the technique was implemented at room temperature, the spatial features of the original butterfly wing were preserved without any obvious distortion during the replication (Figure 8e).

High magnification image in Figure 8f further reveals that the well-organized parallel ridges were composed of numerous small nanoparticles down to nanometers. EDS microanalyses of the SiO_2 and PPy replicas confirm that the PPy replica is mainly composed of carbon and nitrogen, corresponding to the chemical composition of PPy, with a small amount of fluorine, silicon and gold, which means that almost the entire SiO_2 template was removed by soaking in HF solution. Fluorine was introduced when the samples were soaked in HF solution and gold was due to the gold coating to minimize charging effects. It is reasonable to conclude that the unique morphological features of the butterfly wings, especially down to the nanometer level, were maintained in the final PPy replicas. The chemical composition and the molecular structure of the PPy replicas, together with the SiO_2 template and the PPy powder fabricated without using the SiO_2 templates, were further investigated using FTIR analysis (Figure 9).

The SiO_2 template shows the characteristic absorption bands of Si–O–Si bond at 1090 cm^{-1}. The absence of this absorption band in the FTIR spectrum of the PPy replica further confirms that the SiO_2 templates were removed by the HF solution soaking. No obvious difference was observed in the FTIR spectra of the PPy replicas and PPy powders.

The peaks centered at 1175, 1457 and 1549 cm^{-1} were assigned to C–N, C–C and C=C backbone stretching vibrations, respectively. While the peaks at 1303 and 1040 cm^{-1} originated from C–H in-plane vibration. The peaks of the two PPy samples were consistent with the characteristic of pure PPy [36]. In contrast to nanostructured inorganic materials, the introduction of a nanostructure into organic materials is restricted because of the more stringent requirements on the resulting framework. In general, polymer chains pack space efficiently, because polymer macromolecules can bend and twist to maximize intermolecular interactions. Therefore, an interconnected meso- or microporosity in an organic material normally requires a stiff and highly cross-linked system. For comparison, we used poly N-isopropyl acrylamide (PNIPA), a relatively flexible polymer, to replicate the structures of *Morpho* butterfly wings in a similar process as that of the replication of the PPy. The SEM images indicate that the overall structure of a butterfly wing was obtained in the PNIPA replica. However, the hierarchical structures, especially the photonic crystal structures, were not kept during the replication and almost all the cross-ribs were torn apart due to the motion and shrink of the molecular chains and plastic deformation of the walls after the remove of the SiO_2 template.

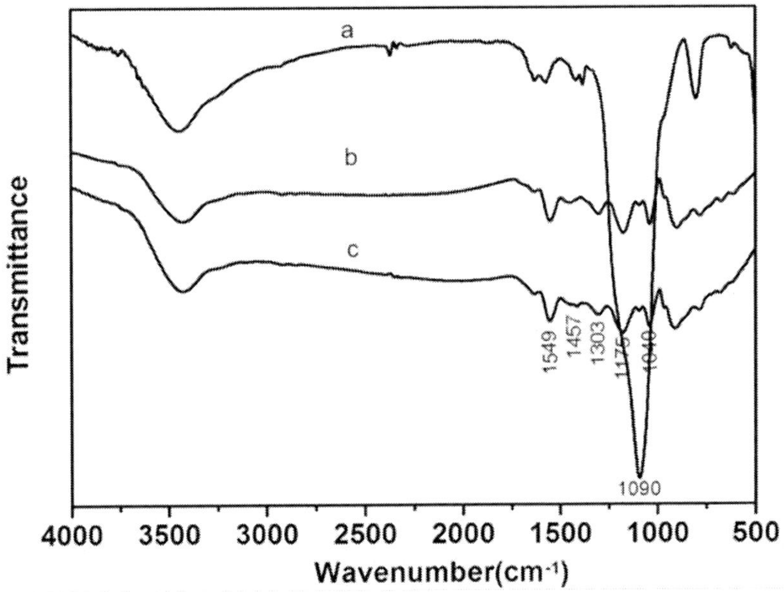

Figure 9. FT-IR spectra of SiO_2 replica and PPy with different microstructures. (a) SiO_2 replica, (b) PPy replica with hierarchical and photonic crystal structures and (c) PPy particles synthesized in a common way. Produced with permission [35].

This means that the final structures are interesting but far from the starting structure. This proved that a direct templating and in situ polymerization is usually not suitable for the soft polymers [37].

Figure 10 shows optical micrographs of the original butterfly wings, SiO_2 templates and PPy replicas. Reflection measurements were performed between near ultraviolet and near infrared wavelength regions to reveal the interactions of the replicated structures with light. In the UV region, the original butterfly wing has a strongest reflection peak at 312 nm. The relatively strong reflectance in the short visible wavelength region agrees with its natural violet color observed under an optical microscope (Figure 10a). The SiO_2 replica shows a red shifted reflection peak at around 500–700 nm, which is a progressive shift, broadening and increase of the main peak intensity compared with that of the original butterfly wings. The measured wavelength is consistent with the orange/pink color observed under an optical microscope (Figure 10b).

The reflectance spectrum from the PPy replica is presented in Figure 10e. It shows that the PPy replica has strong reflection peaks at around 490–580 and 600–650 nm, which also are consistent with the green/yellow color obtained under an optical microscope (Figure 10c and d). These reflections or colors are due to the periodicity of the photonic crystal inherited from the SiO_2 template. All the colors of the *Morpho* butterfly wings and the replicas are mainly physical colors and attribute to optical interference involving photonic crystal structures, hierarchical structures (lamellae) and ordered mesopores structures (reticular network). Therefore, this study offers a simple and effective route for synthesizing biomorphic stiff polymers with photonic crystal structures. Sensing properties: Applications of materials with photonic crystal structures to biosensors have attracted a great deal of interest. The inverse opal film that contains platinum nanoparticles showed improved electrocatalytic activity toward glucose oxidation. Polyaniline (PANI) doped with poly(styrene sulfonate) (PSS) inverse opals for use as biosensor has been reported [38]. It is reasonable to expect the PPy replicas thus prepared with the promise of high surface area as well as controlled porosity from photonic crystal structure, would be an candidate for biosensors. The response of the biosensor to DA was used to characterize the biological sensibility of PPy replica.

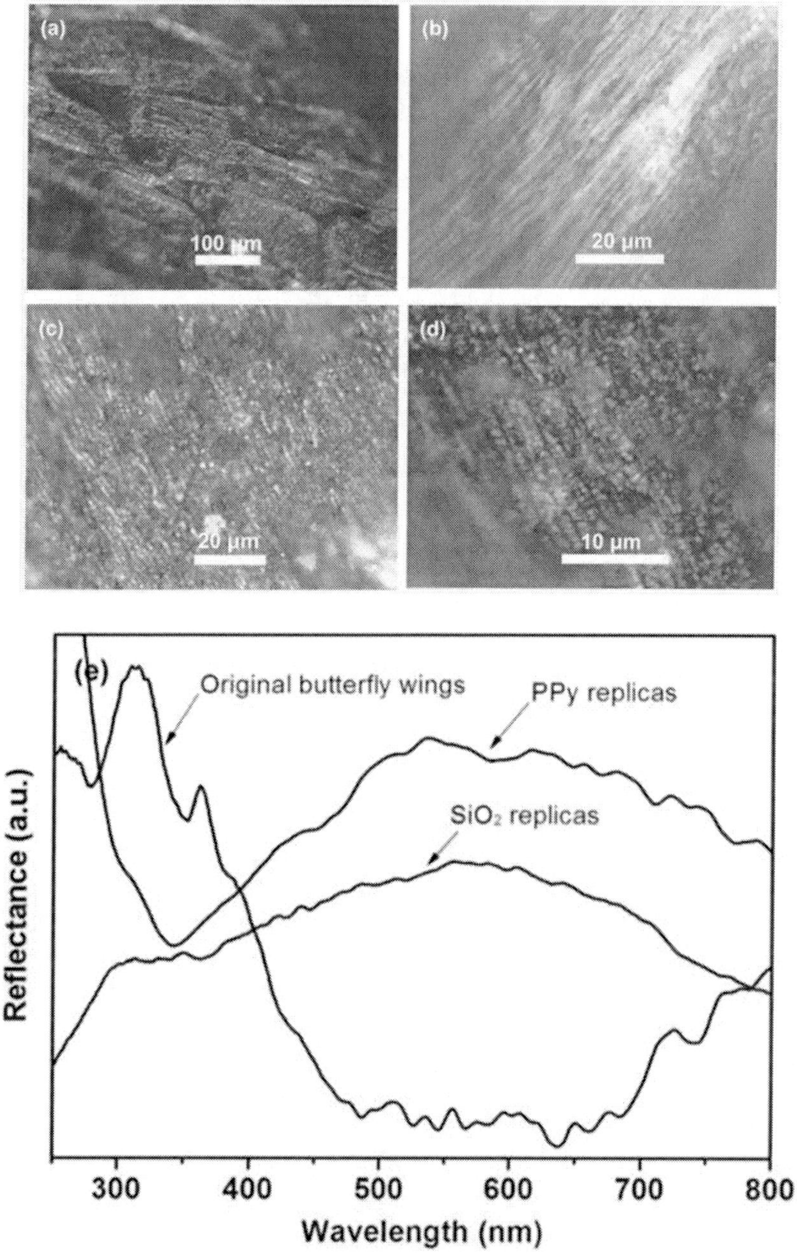

Figure 10. (a)–(d) Optical images of original butterfly wing, SiO_2 replica and PPy replica, (c) and (d) optical images of PPy replica with different magnification and (e) reflectance results of original butterfly wings, SiO_2 replica and PPy replica. Produced with permission [35].

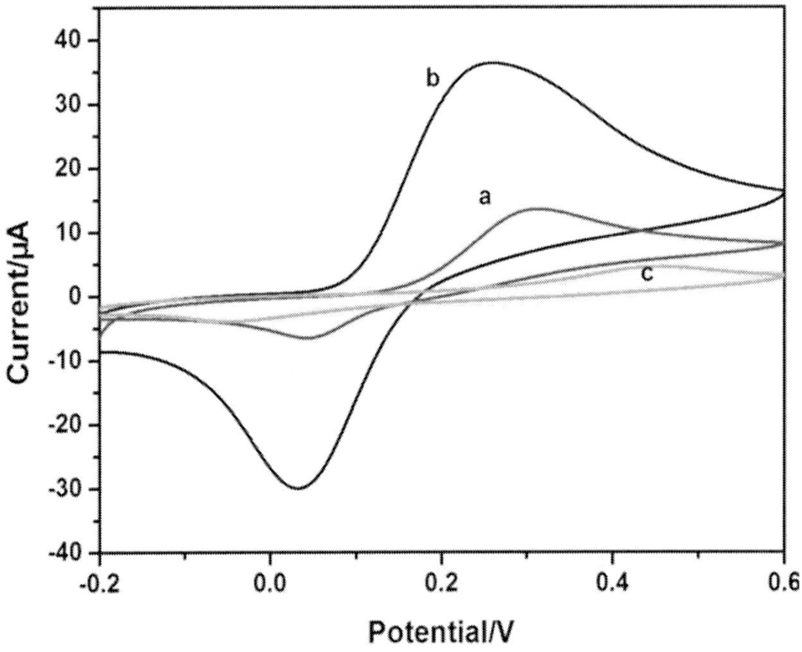

Figure 11. Cyclic voltammograms of 5.0 × 10^{-3} M DA at (a) bare GCE, (b) PPy replica–Au/GCE and (c) common PPy–Au/GCE. Produced with permission [35].

Figure 11 shows the cyclic voltammograms (CVs) of 5.0×10−3M DA in pH 7.0 phosphate buffer solution (PBS) with a scanning rate of 50 mV s^{-1}, corresponding to a bare glassy carbon electrode (GCE) (curve a), PPy replica–Au/GCE(curve b) and common PPy–Au/GCE (curve c), respectively. The bare glassy carbon electrode showed a potential gap about ΔEp = 0.275 V between the redox peaks, with redox peak current of 13 μA, illustrating that DA could exchange electrons with the electrode. After incubating the electrode with powder (PPy–Au/GCE), the redox peak currents decreased to 4 μA, with potentials shifting to −0.05 and 0.45 V. An obvious increase of the peak to peak separation was observed, suggesting the presence of PPy retarded the interfacial electron-transfer kinetics of DA at the electrode. Interestingly, the potential gap of PPy replica–Au/GCE drops to ΔEp = 0.222 V, with cathodic potential at 0.032 V and anodic potential at 0.254 V. This indicates that in PPy replica–Au/GCE, the direct electron transfer between DA and electrode is greatly enhanced as compared with both GCE and PPy–Au/GCE.

The hierarchical structures of PPy replica must provide a favorable microenvironment for DA to exchange electrons with the electrode and facilitate the rate of electron transportation. It has been demonstrated that the biosensor characteristics are related to the effective surface area of the working electrode [39]. PANI composite inverse opal has been reported to present much higher electrocatalytic efficiency for the oxidation of reduced β-nicotinamide adenine dinucleotide (NADH), compared with that of an unpatterned film [40].

Because the response of a biosensor depends on the interaction of an analyte with an immobilized capture ligand on the sensor surface, increasing the surface area of the sensor will improve the sensitivity through a higher density of the capture ligand.

The PPy replica thus prepared possessed a larger effective surface area, as a result of hierarchical structures and ordered porosity, and thus providing great accessibility for the biomolecules with higher accumulation ability and catalytic efficiency. Furthermore, the nano gold particles would disperse not only on the surface but also inside the pores of PPy replicas owing to the interconnected hierarchical structures. Thus, there is no surprise to see that the average peak current at PPy replica–Au/GCE was about 3- and 9-fold higher than that at the bare GCE (curve a) and PPy–Au/GCE (curve c), respectively.

CONCLUSION

In summary, we have demonstrated several routes to produce photonic structures from *Morpho* butterfly wings, including a sonochemical method, a sol–gel impregnation then calcination method and a two-step templating and in situ polymerization approach. A variety of photonic structural semiconductors TiO_2, SnO_2 and SiO_2 were obtained after exposing the butterfly wing templates to a suitable metal oxide precursor under ultrasonic irradiation, followed by calcination at an elevated temperature. All the samples can reflect iridescent visible light, suggesting that the simple sonochemical process does replicate not only the outer but also the inner and more complex surfaces of the hierarchical structure of the butterfly wings.

The results confirm that the interesting chromaticity of the reflected light can be adjusted simply by tuning the replica materials. The efficiency of the sol-gel impregnation approach was proved by copper-doped tungsten oxide with photonic crystal structures by using butterfly wings as a hard-template. The relationship between morphology and sensor properties of these replicas was investigated.

The photonic crystal Cu-doped WO_3 replica sensor has a much higher response sensitivity as well as a higher selectivity compared to the non-PC Cu-doped WO_3 replica sensor due to the different morphologies of these two sensors. The two-step templating and in situ polymerization approach was successfully used in the replication of PPy with photonic crystal structures from *Morpho* butterfly wings. The preliminary response of the replicas to dopamine (DA) as a biosensor was investigated and it was found that the PPy replicas showed nearly 10 times better sensitivity than that of PPy powder, attributing to the controlled porosity and hierarchical structures inherited from *Morpho* butterfly wings.

Thus, these reported biomimetic techniques provide effective approaches to produce photonic crystals from nature with three dimensional networks, which may open up an avenue for many passive and active infrared devices, especially high performance optical components and sensors.

EXPERIMENTAL SECTION

1. Synthesis of TiO_2, SnO_2, SiO_2 Replicas with Photonic Structures

The *Morpho* butterfly wings we chose as biotemplates were purchased from Shanghai Natural Wild-Insect Kingdom Co., Ltd. Analytic grade reagents HCl, NaOH, alcohol, tetraethyl. orthosilicate (TEOS), $SnCl_2{\cdot}2H_2O$ and $TiCl_4$ were purchased from Shanghai Chemical Co. The detailed processing is described as below: the wings pretreated with 6 %

HCl and 10 % NaOH were carefully dipped into ethanol/water/precursor solutions and sonicated at room temperature for a period of 2.5–3 h by a high-intensity ultrasonic probe (Ti horn, 20 kHz, 100 W cm^{-2}). The ethanol/water/precursor solutions have the molar ratio ethanol:H_2O:$TiCl_4$= 35:11:1 for the TiO_2 replica; ethanol:H_2O:TEOS:HCl = 3:12:1:0.03 for the SiO_2 replica and ethanol:H_2O:$SnCl_2 \cdot 2H_2O$ = 100:100:1 for the SnO_2 replica. Finally, the sonicated butterfly wings were taken out and washed three times with ethanol and dried overnight under vacuum.

The prepared sample is denoted as TiO_2-butterfly, SnO_2-butterfly or SiO_2-butterfly. Calcination was performed at 500 °C for 3 h, then the chitin substrates were removed by reaction with air, leaving metal oxide in the form of ceramic butterfly wings. The resultant replicas are denoted TiO_2 replica, SnO_2 replica or SiO_2 replica.

2. Cu Dopped WO_3 with Photonic Structures

Phosphotungstic acid 44-hydrate ($H_3PO_{40}W_{12}$) and $CuCl_2$ were purchased from Shanghai Chemical Co. The detailed processing is described as below: the wings pretreated with HCl and NaOH were carefully dipped into 20 wt% of $H_3PO_{40}W_{12}$ in ethanol solution with a certain amount of $CuCl_2$ (weight ratio W: Cu =1 : 0.03) and kept at 36 °C for 6 h. Then, the butterfly wings were taken out and washed with ethanol and dried overnight under vacuum. The prepared samples are denoted as Cu-W-PC-butterfly and Cu-W-butterfly, corresponding to the *Morpho* and Euploea mulciber butterfly wing templates used, respectively. Calcination was performed at 430 °C for 3 h, and the chitin substrates were removed by reaction with air, leaving metal oxide in the form of the butterfly wings. The resultant replicas are denoted as Cu-W-PC replica, and Cu-W replica, respectively. Pure WO_3 powder was prepared using the same method described above but without the presence of the butterfly wing template.

3. Synthesis of Biomorphic PPy with Photonic Structures

Pyrrole was distilled under reduced pressure and stored at low temperatures. The replication process is described as below: firstly, the wings pretreated with 6% HCl and 10% NaOH were carefully dipped into the ethanol/water/TEOS/HCl solution with the molar ratio of 3:12:1:0.03 and ultra-sonicated at room temperature for a period of 3 h using a high-intensity ultrasonic probe (Ti horn, 20 kHz, 200 W cm^{-2}). After the ultrasonic processing, the butterfly wings were taken out and washed three times with alcohol and dried in air. Calcination was performed at 450 °C for 3 h, then the chitin substrates were removed by reaction with air, leaving SiO_2 in the form of ceramic butterfly wings. Secondly, the SiO_2 replicas were placed on a microslide and dipped into 1–2 mL of pyrrole, and then in 0.1 mL of 0.21 M $FeCl_3$.

Capillary forces drew the liquid monomer into the void spaces between the silica particles. Further polymerization was carried out at 65 °C for 3 h. Finally, the microslide was carefully removed and the PPy with the SiO_2 templates inside were soaked in a 4% HF solution for 24 h to remove the SiO_2. The resultant replicas thus prepared are denoted as PPy replica. For comparison, PPy powders without biological structures were synthesized in the same way, but without SiO_2 replicas as templates.

4. Measurement of Biomolecule Sensing Properties

Cyclic voltammetry (CV) was performed on a CHI 660D computer electrochemical workstation (Chen Hua Instruments Co., Shanghai, China) with a conventional three-electrode system, which was composed of a working electrode, a platinum slice counter electrode and a saturated calomel reference electrode (SCE), respectively. A glassy carbon electrode (GCE) served as working electrode. Dopamine (DA) and Nafion (Sigma–Aldrich) were used as received without further purification. The phosphate buffer solution (PBS, 0.10 M, pH 7) was prepared by mixing NaH_2PO_4 and Na_2HPO_4. DA solution used in all experiments was freshly prepared. All solutions were prepared with doubly distilled water. Nano gold particles were introduced to the surface of the PPy replicas and the prepared PPy powder to improve the conductivity of the two materials.

The detailed process was described as below: the PPy replicas as well as PPy powders were grinded and dispersed in water evenly, 10 mL of 0.01% gold (III) chloride trihydrate ($HAuCl_4 \cdot 3H_2O$, Shanghai Chemical Company) was added into the PPy suspension and heated to boiling, 1.25 mL of 1% sodium citrate dihydrate ($C_6H_5Na_3O_7 \cdot 2H_2O$, Shanghai Chemical Company) was trickled into the boiling solution and the mixed solution was boiled for 0.5 h. The above solution was then dropped on the surface of GCE and then 5 μL of 1% Nafion/alcohol solution after it was dried. The PPy modified GCE electrodes were investigated using CV in PBS containing 5.0×10^{-3} M DA.

5. Characterization

The morphologies and microstructures of the resultant replicas were characterized using a JEOL JSM-6360LV field-emission scanning electron microscope (FE-SEM) equipped with a JED2300 energy-dispersive X-ray spectrometer (EDS) system. Transmission electron microscopy (TEM) was conducted on a JEOL 2011F. TEM specimens were prepared by grinding the synthesized samples into power with a mortar and pestle, after which the powder was dispersed in pure ethanol and picked up by holey carbon films on copper grids. Fourier transform-infrared (FT-IR) measurements were recorded with a PE Paragon 1000 spectrophotometer and the spectra were obtained from KBr pellets. The micrographs of the replicas were taken using a digital optical microscope VHX-600, Keyence.

REFERENCES

[1] Yu, AM; Zhang, HL. *Curr. Nanosci.*, 2010, 6, 206-212.

[2] Norris, DJ; Nagpal, P; Josephson, DP; Denny, NR; DeWilde, J; Stein, A. *J. Mater. Chem.*, 2011, 21, 10836-10843.

[3] (a) Cassagneau, T; Caruso, F. *Adv. Mater.*, 2002, 14, 1837-1841. (b) Fudouzi, H; Xia, YN. *Adv. Mater.*, 2003, 15, 892-896. (c) Wu, GL; Jiang, Y; Xu, D; Tang, H; Liang, X; Li, GT. *Langmuir.*, 2011, 27, 1505-1509.

[4] Aguirre, CI; Reguera, E; Stein, A. *Adv. Funct. Mater.*, 2010, 20, 2565-2578.

[5] Watanabe, K; Hoshino, T; Kanda, K; Haruyama, Y; Kaito, T; Matsui, S. *J. Vac. Sci. Technol. B,* 2005, 23,570-574.
[6] Zhang, JZ; Gu, ZZ; Chen, HH; Fujishima, A; Sato O. *J. Nanosci. Nanotechnol.,* 2006, 6, 1173-1176.
[7] Gaillot, DP; Deparis, O; Welch, V; Wagner, BK; Vigneron, JP; Summers, C. *J. Phys. Rev. E,* 78,031922-1-031922-6.
[8] Martin-Palma, RJ; Pantano, CG; Lakhtakia, A. *Appl. Phys. Lett.,* 2008, 19, 355704-5.
[9] (a) Wong, TH; Gupta, MC; Robins, B; Levendusky, TL. *Opt. Lett.* 2003, 28, 2342-2344. (b) Zhang, W; Zhang, D; Fan, TX; Ding, J; Gu, QX; Ogawa, H. *Nanotechnology,* 2006, 17, 840–844. (c) Silver, J; Withnall, R; Ireland, TG; Fern, GR; Zhang, S. *Nanotechnology,* 2008, 19, 095302-7.
[10] Huang, JY; Wang, XD; Wang, ZL. *Nano Lett.,* 2006, 6, 2325-2331.
[11] Chen, Y; Gu, JJ; Zhu, SM; Fan, TX; Zhang, D; Guo, QX. *Appl. Phys. Lett.* 2009, 94, 053901-3.
[12] Zhu, SM; Zhang, D; Li, ZQ; Furukawa, H; Chen ZX. *Langmuir.,* 2008, 24, 6292–6299.
[13] Zhu, SM; Zhang, D; Chen, ZX; Gu, JJ; Li, WF; Jiang, HB; Zhou, G. *Nanotechnology,* 2009, 20, 315303-8.
[14] Huang, JG; Kunitake, T; *J. Am. Chem. Soc.,* 2003, 125, 11834-11838.
[15] Cook, G; Timms, PL; Spickermann, CG. *Angew. Chem. Int. Edn.,* 2003, 42, 557-559.
[16] Wang, YQ; Tang, XH; Yin, LX; Huang, WP; Hacohen, YR; Gedanken, A. *Adv. Mater.,* 2000, 12, 1183–1186.
[17] Potyrailo, RA; Ghiradella, H; Vertiatchikh, A; Dovidenko, K; Cournoyer, JR; Olson, E. *Nat. Photon.,* 2007, 1, 123-128.
[18] Zhang, Y; He, XL; Li, JP; Miao, ZJ; Huang, F. *Sensors Actuators B,* 2008, 132, 67–73.
[19] Wan, Q; Wang, TH. *Chem. Commun.,* 2005, 3841–3843.
[20] Barsan, N; Weimar, U. 2001, *J. Electroceram.,* 2001, 7143–7167.
[21] Wu, NL; Wang, SY; Rusakova, IA. *Science,* 285, 1375-1377.
[22] Scott, RWJ; Yang, SM; Chabanis, G; Coombs, N; Williams, DE; Ozin, GA. *Adv. Mater.,* 2001, 13, 1468-1472.
[23] Baeck, SH; Choi, KS; Jaramillo, TF; Stucky GD; McFarland, EW. *Adv. Mater.,* 2003, 15, 1269-1273.
[24] Williams, DE; Moseley, PT. *J. Mater. Chem.,* 1991, 1, 809-814.
[25] Vlasov, YA; Bo, XZ; Sturm JC; Norris, DJ. *Nature,* 2001, 414, 289-293.
[26] Zhu, SM; Liu, XY, Chen, ZX, Liu, CJ; Feng, CL; Gu, JJ; Liu, QL; Zhang, D. *J. Mater. Chem.,* 2010, 20, 9126-9132.
[27] Wang, JY; Cao, Y; Feng, Y; Yin, F; Gao, JP. *Adv. Mater.,* 2007, 19, 3865-3871.
[28] Cassagneau, T; Caruso, F. *Adv. Mater.,* 2002, 14, 1837-1841.
[29] Yu, AM; Meiser, F; Cassagneau, T; Caruso, F. *Nano Lett.,* 2004, 4, 177-181.
[30] Saison, T; Peroz, C; Chauveau, V; Berthier, S; Sondergard, E; Arribart, H. *Bioinspir. Biomim.,* 2008, 3, 046004-5.
[31] Arslan, H; Arslan, F. *Artif. Cells Blood Substit. Immobil. Biotechnol.,* 2011, 39, 341-345.
[32] Oliveira, HP; Oliveira, AHP; Tavares, GF; Albuquerque, JJF; Andrade, CAS. *Polimeros.,* 2010, 20, 253-257.
[33] John, SW; Alici, G; Cook, CD. *IEEE ASME Trans. Mech.,* 2010, 15, 149-156.
[34] Yang, C; Liu, P. *Synth. Met.,* 2010, 160, 768-773.

[35] Tang, J; Zhu, SM; Chen, ZX; Feng, CL; Shen, YJ; Yao, F; Zhang, D; Moon, W-J; Song, D-M. *Mater. Chem. Phys.,* 2012, 131, 706-713.
[36] Liu, Y; Chu, Y; Yang, LK. *Mater. Chem. Phys.,* 2006, 98, 304-308.
[37] Johnson, SA; Ollivier, PJ; Mallouk, TE. *Science,* 1999, 283, 963-965.
[38] Yang, XL; Jin, Y; Zhu, YH; Tang, LH; Li, CZ. *J. Electrochem. Soc.,* 2008,155, J23-J25.
[39] Park SM; Lim TH; Jeon YM; Kim JG; Joo SW; Gong MS. *Sens. Actuators B: Chem.*, 2008, 132, 166-173.
[40] Tian, SJ; Wang, JJ; Jonas, U; Knoll, W. *Chem. Mater.,* 2005, 17, 5726-5730.

In: Focus on Biomimetics Research
Editor: Abele Scipione
ISBN: 978-1-62417-332-5

Chapter 4

BIOMIMETIC MOLECULAR SWITCHES

Marina Blanco-Lomas, Pedro J. Campos and Diego Sampedro
Departamento de Química,
Centro de Investigación en Síntesis Química (CISQ),
Universidad de La Rioja, Madre de Dios, Logroño, Spain

ABSTRACT

Science has always looked for inspiration in the highly effective function of biological systems. At the same time, the aspiration to mimic their functioning has been a driving force for the advancement of chemistry and material science.

Light represents a usual stimulus found in very different organisms, triggering and driving diverse biological processes. In higher organisms, the process of visual perception is extremely important. But also other processes such as photosynthesis in plants and photomovement responses imply the interaction between light and biological media to provide complex responses.

The basic concept behind such different behaviors is centered on photoresponsible proteins capable of regulating the biological activity by light. Thus, the use of light to control the activity of biomolecules, as well as to mimic their function, is an attractive strategy for both studying and controlling interactions in complex living systems. Photoactive biomolecules can be introduced into a cell and manipulated noninvasively using light.

This allows for a high degree of spatial and temporal control. Among the different types of molecules that can be used, photoswitches are a class of chemical compounds that undergo reversible photochemical behavior so that many rounds of active/inactive states may be produced. In addition to this, it is possible to find a wide range of molecules that experience a photoswitchable behavior.

In this contribution, we will describe the synthesis and photochemical properties of different families of photoswitches whose structure is based on that of natural chromophores. Also, the properties that make them attractive for the photocontrol of biomolecules will be outlined.

1. INTRODUCTION

Mankind has always looked at Mother Nature in the search of food and protection, but also as an implacable foe threatening the survival of the human race. When the most urgent necessities were covered, Nature also acted as a source of stimulation for art and recently science as well. Taking inspiration from Nature is clearly a good idea in many topics. After all, Nature has managed to create an extremely complex self-sustaining system capable of support life over billions and billions of years. Even now, there is a lot we can learn about design from many different points of view. Thus, natural systems have helped designers, engineers and thinkers over the last decades. Buildings, bridges, machines and medicine, just to mention some examples, have been developed after examining the intricacies of the natural world in which we live. Nature has aided to improve almost every aspect of human life. Our relationship with Nature's patterns and motifs provides us with both beautiful things to admire and a perfect example of the power of careful design. Both art and science could find enough material for contemplation and study.

Direct observation of Nature soon allowed to unveil certain mathematical aspects in the design and functioning of natural systems. For instance, Fibonacci (the Fibonacci sequence) tried to understand how Nature extrapolates its design. Also, artists have used since ancient times many geometrical and mathematical properties: architects from Ancient Egypt or Greece, and Renaissance artists such as Leonardo Da Vinci or Michelangelo.

In the past decades, a rich convergence has come from a wide range of scientific disciplines, including biology, chemistry, computer science, mathematics, and physics. From the overlap of these and other disciplines and the collaboration of scientists with a very different perspective a new vision of the field is emerging. Of course, this is not a simple task. Usually there is no clear line to be drawn between Nature design and human application. Besides, it took a long process of slow evolution to improve and select the adequate natural systems. Thus, our efforts to further advance in these designs or modify their properties to match human necessities constitute a task both exciting and very demanding.

Along human history, mankind has taken advantage of naturally available materials, but always has tried to adapt or improve Nature's design. For instance, silk is a natural protein fiber, some forms of which can be woven into textiles. The protein fiber of silk is composed mainly of fibroin and produced by certain insect larvae to form cocoons. The earliest evidence of silk was found in Ancient China, where a silk cocoon was found cut in half by a sharp knife, dating back to between 4000 and 3000 BC. The special features of silk made it a very valuable good along history. Specially, silk is one of the strongest natural fibers. It is made up of the aminoacid series Gly-Ser-Gly-Ala forming beta-sheets with hydrogen bonds between chains. The high proportion of glycine allows the fibers to be strong and resistant to stretching. These remarkable properties were soon recognized by Chinese and they also tried to emulate them looking for an artificial silk. [1] However, the complete understanding of the chemistry involved and the technical knowledge necessary to duplicate its properties took some more time. It wasn't until 1890's when the first artificial silk (viscose) was produced from cellulose.

Since then, a rapid growth in interest followed on how living systems synthesize materials with exceptional properties. In the last decades, scientists have drawn their bioinspiration in two main ways. On one hand, a natural system could be used once implemented in a synthetic system. Using this approach, the features of the natural system are transferred to an artificial construction. This way, its properties could be used in a new way, even completely diverse from its original one. [2] An example of this approach can be found in the recent applications of DNA for new and exotic uses. Very different functions such as using selective bonds between complementary DNA sequences to link particles to surfaces have been developed apart from its biological role. On the other hand, natural models can be emulated trying not only to use or understand how biological systems work, but also to use them in artificial systems with new or improved properties. [3] This way, we could take biological systems as a starting point and try to identify the key factors behind their structure and function. Using this information, we could build new systems with different, improved or more controllable properties.

After slow evolution, Nature has developed excellent examples of efficient machinery. A number of biomechanical machines can be used as models for the design and construction of our own synthetic prototypes. These biomolecular motors and machines are nanometric equivalents of everyday machines. From these natural machines, we could learn how to design new versions of devices capable of performing different goals or how to fine tune the properties of these systems. In the end, our ultimate objective could be to mimic Nature. Even more, we could even try to improve the properties of natural systems or at least to adapt them to our specific needs.

In the present chapter we will present some of the recent efforts trying to replicate the excellent performance of several natural devices in the context of biomimetic photochemical switches. Taking as a starting point some compounds found in Nature acting as very efficient switches, scientists have tried to mimic their behavior once included in artificial systems. In the next section we will briefly introduce some natural machines that inspired the development of artificial devices. Then, we will present some biomimetic compounds designed and synthesized to replicate the performance of natural systems. Also, some improvements on the original products will be discussed. Finally, some applications of the artificial devices once incorporated in complex systems will be shown.

2. Natural Molecular Devices

As already mentioned, different examples of nanoscale machinery are found in Nature. These natural nanomachines, as their macroscopic counterparts, require energy to perform their functions. When comparing conventional technologies with natural molecular machines, it is obvious that the latter display tremendous abilities. [4] For instance, the molecular machinery of green plants converts more energy and synthesizes a greater tonnage of organic compounds than does humanity's entire chemical industry. Besides, this conversion is performed in a truly clean way and using inexpensive raw materials. Therefore, when designing new nanosystems, one of the major objectives could be directed so that it mimics Nature's efficient machines.

The following paragraphs reveal how these molecular machines are present in a multitude of essential biological processes, such as transport of cations, synthesis of ATP, muscle contraction and vision.

During normal cell functioning, the passage of cations, across their membranes is critical. Ions such as Na^+, Ca^{2+} and K^+ control a good number of biological functions and their concentrations at both sides of the membrane should be managed. The transport of ions can be done by using carriers or ion channels. An example of an efficient natural ion channel is the light-activated proton pump bacteriorhodopsin, a small and robust membrane protein of the halophilic microorganism *Halobacterium salinarum*. [5] Bacteriorhodopsin consists of seven membrane-spanning helical structures linked by short loops on either side of the cell membrane. It also contains one molecule of a linear pigment called retinal, which is covalently attached to the nitrogen atom of the lysine residue of helix G through a protonated Schiff base (PSB). The retinal chromophore, suffers an isomerization when irradiated with light from all-*trans* to 13-*cis*, through several intermediates. This structural change is used by the Schiff base to push a single proton through the seven-helix bundle, from the inside of the cell to the extracellular medium, being subsequently reprotonated from the cytoplasm. Therefore, in this particular movement, the retinal chromophore acts as a valve inside of the cell membrane of this organism.

The ATP synthase protein [6] is a natural rotary motor that consists of two units attached to a common shaft. The complex can be conceptualized as two rotary motor units coupled together. One of these units uses free energy of ATP hydrolysis to rotate in one direction whereas the other uses the energy stored in a transmembrane electrochemical gradient to turn in the opposite direction. Thus, the ATP synthase complex is fully reversible and it may synthesize or hydrolyze ATP depending on the driving force of the movement. Besides, this complex has been reported to be extremely efficient, as the ATP synthase could couple nearly 100% of its ATP-derived energy into mechanical work.

Apart from the rotary motors, different types of biomolecular motors move in a linear fashion along a track of some kind. These natural machines are considered greatly relevant because they are involved in crucial processes, such as intracellular trafficking, cell division and muscle contraction. Eukaryotic cells require efficient systems that enable intracellular transport to supply their cellular organelles.

Among these systems, microtubule motor proteins such as kinesin can be found. [7] Conventional kinesin is a protein assembly composed of two larger protein chains, which are involved in microtubule binding and mobility, as well as two smaller protein chains, which regulate heavy chain activity and binding to cargo. Kinesin transports cargo along microtubules, which are composed of self-assembled tubular proteins in a cylindrical way. For each ATP molecule hydrolyzed during the movement, kinesin performs a walking-like motion on 8 nm steps. Moreover, the dynein family, which also moves along microtubules in a linear fashion, is usually anchored to the centrosome of the cell. [8]

Among cellular division processes, several molecular motors performing linear-like movements are also found. For instance, RNA polymerase synthesizes new RNA from a single strand RNA template, [9] and DNA helicase translates along and unwinds DNA in preparation for new DNA synthesis [10]

Among the different natural devices, one of the most remarkable examples of a molecular motor is the retinal chromophore of rhodopsin, also found in the bacteriorhodopsin previously discussed. In vertebrates, the process of vision is initiated by rods and cones, which are photoreceptor cells located in the retina of the eye. The photosensitive entities of these cells are visual pigments that consist of the appropriate apoprotein, called opsin, and a chromophore, known as 11-*cis*-retinal, that is embedded inside the opsin (Figure 1) [11]

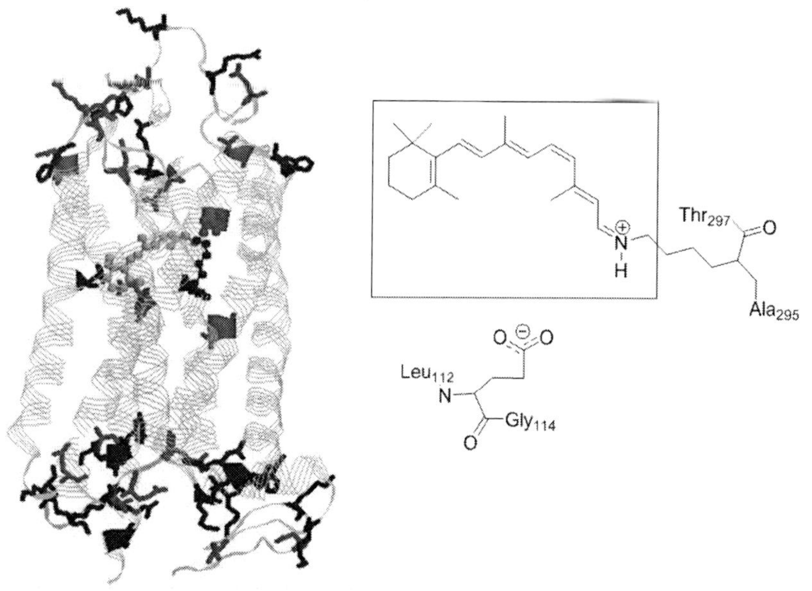

Figure 1. Rhodopsin and its chromophore (a protonated Schiff base of retinal).

While the biological aspects of the process of vision are extremely complex and are still subject to intense research, the chemical event that spreads the biological cascade is extremely simple. After receiving a photon, the retinal chromophore undergoes a *cis-trans* photoisomerization during the process of vision (Figure 2). [12]

The role of rhodopsin in the signal transduction cascade of vision is to further activate a protein, upon absorption of light. [13] In the absence of visible light, retinal presents a *cis* conformation between its carbons 11 and 12. However, when 11-*cis*-retinal absorbs visible light (λ between 400-600 nm), a selective and rapid (200 femtoseconds) photoisomerization of this double bond takes place via evolution of a single $\pi \rightarrow \pi^*$ excited state (S_1) that decays to the all-*trans* ground state (S_0) product. Picosecond time-resolved spectroscopy of 11-*cis* locked rhodopsin analogs reveals that the *cis-trans* isomerization of the retinal chromophore is the primary reaction in the process of vision, and it is followed by a conformational change in the rhodopsin protein, that activates trasducin in order to initiate the process of vision. [14]

It is known that the *cis-trans* isomerization is highly efficient in rhodopsin, presenting an *in vivo* isomerization quantum yield (Φ_{isom}) of 0.67, [15] that is essential to make twilight vision highly sensitive. In fact, a human rod cell can respond to a single photon absorption. Such an efficient photoisomerization of the retinal chromophore is characteristic in the protein environment of rhodopsin, being in contrast to the rhodopsin chromophore in solution.

Due to the high efficiency of the isomerization process of retinal *in vivo*, this process has been widely studied and used as a pattern for the design of new prototypes of molecular switches and motors activated by light.

Since the protein environment facilitates the isomerization process of retinal, the aim of these studies has been the design of synthetic compounds that reproduce in solution the features of the chromophore inside the protein. Using this approach molecular switches and motors that work efficiently can be obtained.

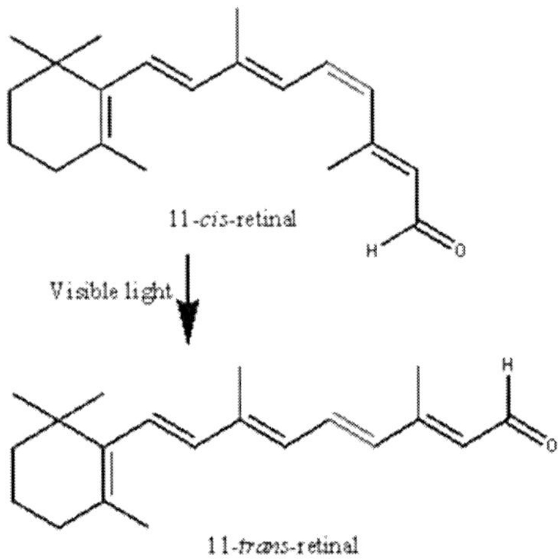

Figure 2. Photo-induced isomerization reaction of retinal.

Another striking example of biomolecules that undergo efficient photoprocesses is the chromophore of the green fluorescent protein (shortened as GFP), from *Aequorea victoria* jellyfish (Figure 3). GFP and other related proteins are intrinsically fluorescent molecules that have optical properties determined by a photoexcitable green-light emitter chromophore autocatalytically generated by the posttranslational modification of a three-amino acid sequence (Ser65-Tyr66-Gly67). [16] GFP takes advantage of the presence of a chromophore, (*Z*)-4-(4-hydroxybenzylidene)-1,2-dimethyl-1*H*-imidazol-5(4*H*)-one (*p*-HBDI), that consists of the hydroxybenzyl side chain of Tyr66 (phenolic ring) and the imidazolidinone ring formed by cyclization of the tripeptide (heterocyclic ring).

It is embedded in a protein cavity containing a number of neighboring polar and aromatic residues. These are important in establishing a hydrogen-bond network around the chromophore involving also some water molecules. Therefore, GFP fluorescence and absorption critically depend on this network and perturbation of the latter can induce marked variations in the GFP photophysics. The *p*-HBDI chromophore is the ultimate responsible for the fluorescence of this protein within a picoseconds time scale, [17] resulting in very intense anion fluorescence.

However, several studies have revealed that *p*-HBDI also undergoes competitive nonradiative processes when irradiated with light, such as a *Z*/*E* photoisomerization, [18] which contributes to reduce the luminescence quantum yield (Figure 4).

Due to its attractive emission of fluorescence, GFP has emerged in recent years as a unique fluorescent marker in molecular and cellular biology. [19] Also, in 2008, as a result of the great relevance of the luminescence of GFP, the Nobel Prize of Chemistry was assigned to three researchers that had worked in the discovery and development of the green fluorescent protein: Osamu Shimomura, Martin Chalfie and Roger Y. Tsien.

On the other hand, a wide variety of GFP mutants have been designed with different spectral characteristics that allow a very different performance. In addition, the modification of the chromophore from the natural *p*-HBDI to other derivatives permits the exploitation of other reactive channels different from the fluorescence emission. In this context, the *Z*/*E* isomerization has been also explored.

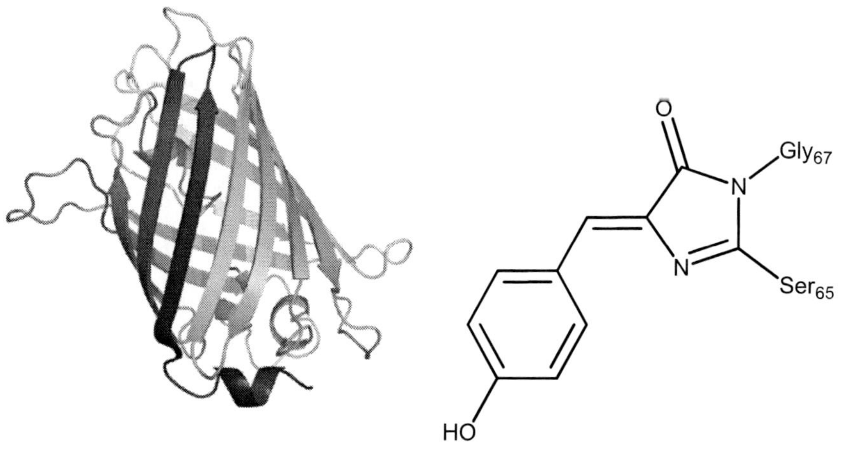

Figure 3. Green fluorescent protein (GFP) and its chromophore (*p*-HBDI).

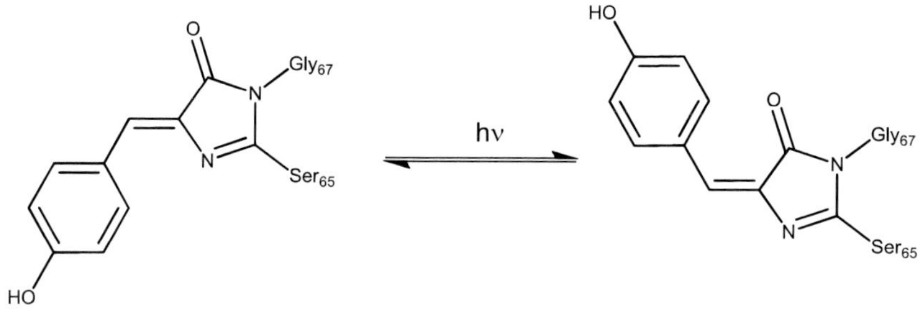

Figure 4. *Z*/*E* photoisomerization of the GFP chromophore.

3. Bioinspired Artificial Molecular Devices

3.1. Rhodopsin-Based Molecular Photoswitches

As stated previously, the high efficiency shown by the PSB retinal *in vivo* makes it an attractive target when designing new prototypes of switches that undergo *Z*/*E* photoisomerization processes. One of the pathways that lead to the design of new prototypes of molecular devices, which exhibit light-induced movement, combines the use of several *ab initio* quantum methods. Using the natural PSB-retinal chromophore as a starting point, the successful design, synthesis and characterization of a new prototype of molecular switch based on the PSB retinal chromophore was reported [20].

The design of this new prototype of molecular switch implied the modification of the chemical structure of the natural chromophore to try to mimic in solution the results of the protein *in vivo*. First, the long conjugated chain was shortened to simplify the photochemical and photophysical properties of the molecule. Then, two rings were introduced in the extreme double bonds to prevent competitive isomerizations. The resulting prototype is shown in Figure 5 (a). Furthermore, the introduction of chirality in the previous compound led to a new prototype of molecular motor (Figure 5(b)).

A molecular switch is a molecular device that can be reversibly interchanged between two different states, on and off, by means of an external stimulus. In general terms, a molecular switch is considered a bistable system. If each one of the states can be "read" non-destructively, the molecule can function, in principle, as a memory element in a digital system.

A molecular rotor is defined as a device capable of rotating in several directions with a continuous movement. We have previously described an example of a natural molecular rotor (ATP-synthase).

Finally, if the continuous rotational movement of a molecular rotor is controlled in a manner that involves unidirectional rotation, that particular molecular device is denominated a molecular motor. To design new prototypes of molecular motors three basic requirements need to be fulfilled:

- repetitive 360° rotary motion;
- consumption of energy to execute the movement; and
- control of the rotational direction, which can be done by introducing in the motor skeleton different chemical factors, such as chiral centers in the molecule, which favor the rotation in a certain direction due to asymmetry.

Different examples of natural molecular motors have been also described above.

Thus, the modifications introduced in the natural chromophore are advantageous due to the following facts:

- five member rings restrain isomerization about the –CH=NH- and terminal -CH=CH- double bonds of the structure;
- the possible existence of different conformers of the molecule is avoided as five member rings are conformationally rigid; and
- the two $-(CH_2)_2$-groups present in the molecule could be functionalized, allowing to obtain a family of compounds with a common backbone.

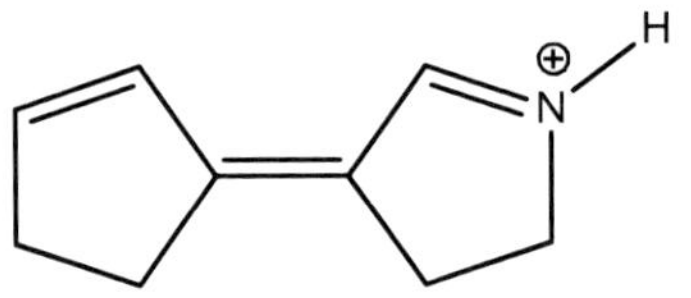

(a)

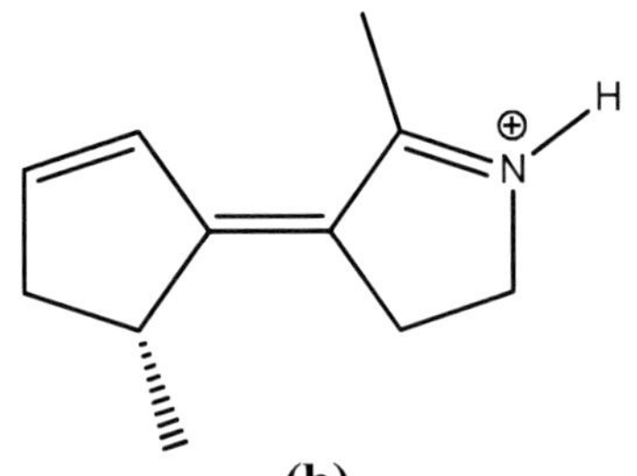

(b)

Figure 5. Prototypes of a molecular switch (a) and motor (b).

Thus, the pair of biomimetic molecular devices shown in Figure 5 were studied through CASPT2//CASSCF theoretical calculations that simulated the photochemical and thermal isomerization pathways for both systems. The results of these calculations led to the conclusion that the designed prototypes satisfied the criteria of being an efficient switch and motor. An efficient photoisomerization takes place when the photoexcited reactant evolves along a barrierless excited-state path and finally relaxes also through a barrierless path to the energy minimum of the photoproduct. Moreover, the stability of both *Z* and *E* isomers with respect to thermal isomerization is also required.

The applicability of this compound to act as an efficient switch in a photomodulable peptide was also studied by a quantum mechanics/molecular mechanics strategy. [21] The photoswitch showed in Figure 5(a) was included in an octapeptide derived from the thioredoxin reductase active site. It was demonstrated that the switch starts to isomerize immediately after photoexcitation, but it is only after the ground state is reached when the change in the peptide conformation takes place.

Although these systems showed very promising features at the theoretical level, the synthesis of compounds containing the penta-2,4-dieniminium backbone (Figure 5) has not been reported yet and their performance in real applications still has to be assessed.

Therefore, to be able to experimentally study related compounds and support the theoretical predictions, the synthesis of an analogous system shown in Figure 6 was proposed. It corresponds to the synthetically accessible *N*-alkylated structure 4-benzylidene-3, 4-dihydro-2*H*-pyrrolium (NABP) in which the terminal C=C double bond of the penta-2,4-dieniminium unit (Figure 5(a)) is replaced with a phenyl group [20].

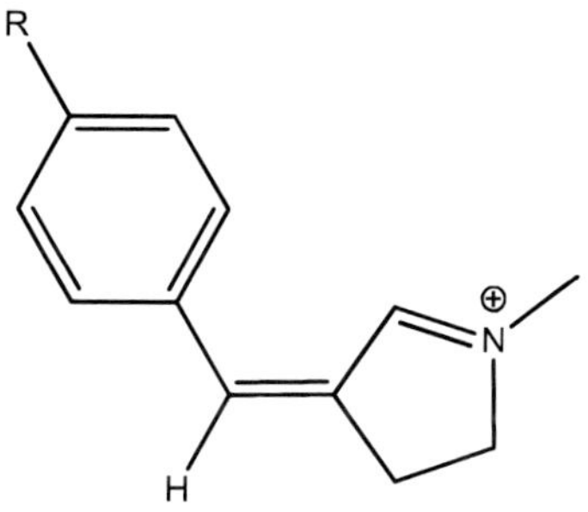

R = H, *p*-OMe, *p*-NO_2

Figure 6. *N*-alkylated benzylidene-pyrroline Schiff bases (NABPs).

After the study of these new prototypes of molecular switches using theoretical calculations and experimental methods, it was concluded that although they presented some features that were desirable for an ideal molecular switch (such as the absence of intermediates or energy barriers along the photoisomerization pathway, the presence of a simple isomerization movement, and photostability), they could not be considered efficient. The prototypes showed an extremely low isomerization quantum yield (*ca.* 10^{-3}) due to the energy that was lost in the movement of other bonds capable of rotating present in the molecule. Therefore, the behavior of these molecular switches based on the PSB retinal chromophore required improvement.

Soon after NABPs were synthesized and photochemically characterized, the synthesis of a new family of compounds based on the chromophore of the PSB retinal was accomplished (Figure 7). [22] In these switches there is a decrease of the number of torsional degrees of freedom of the backbone of NABPs and, specially, the potential competitive rotation around the benzylic single bond that might be the responsible for the low isomerization quantum yields obtained. These kind of molecular switches are called *N*-alkylated indanylidene-pyrroline Schiff bases (NAIPs). The photoinduced dynamics of these molecular switches could potentially replicate the dynamics of the PSB-retinal isomerization in rhodopsin. [23] Moreover, the *p*-methoxy NAIP derivative is a photochromic compound completing its *Z* → *E* and *E* → *Z* photocycle in picoseconds. These time scales suggest that NAIP-based motors may complete a half-rotary cycle in less than 10 ps, which means a few orders of magnitude faster than the fastest (≈6 ms for half-cycle) known biarylidene. [24] In addition to this, these switches feature a higher isomerization quantum yield (*ca.* 0.3) compared to NABPs, and increased polarity with respect to azobenzene. Besides, the performance of NAIP to act as switches in complex environments was explored through theoretical calculations. Specifically, the design and computational characterization of a functional cyclic peptidomimetic formed by a NAIP switch conjugated to a biologically active peptide has been reported. [25]

Despite the remarkable properties as molecular switches of NAIPs, some other features would be required to improve these systems. One of them is the possibility of presenting absorption in the visible region so lower energies and even solar light could be used to induce the isomerization process. So far, when R = hydrogen atom, the corresponding NAIP presents an absorption maximum of 343 nm, but if R = *p*-methoxy, the band is red-shifted to 377 nm, allowing absorbance in the near-UV region. Therefore, a different change in the substitution of the molecule would be required in order to displace the band towards the visible region.

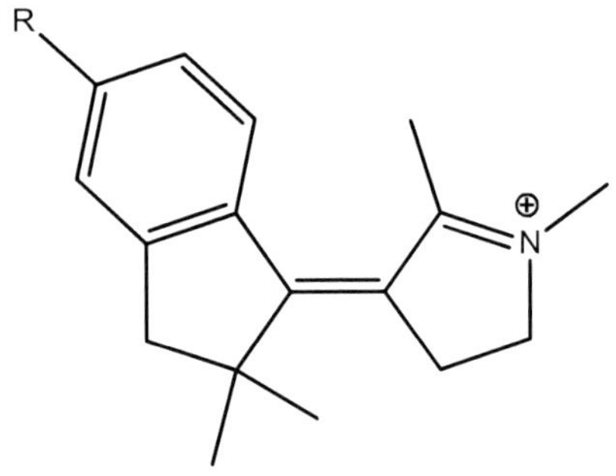

R = H, OMe

Figure 7. *N*-alkylated indanylidene-pyrroline Schiff bases (NAIPs)

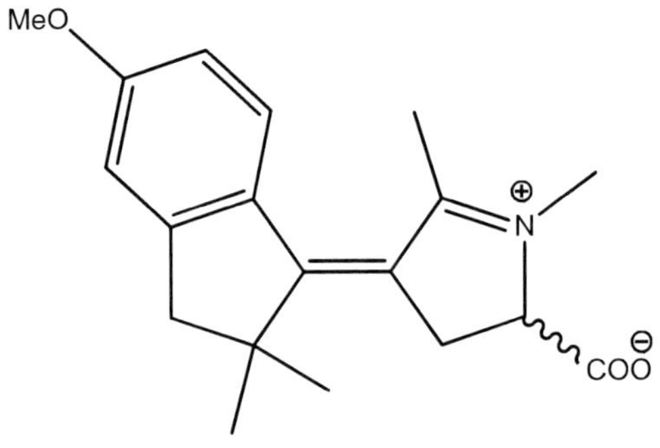

Figure 8. Structure of a NAIP switch featuring a stable zwitterionic head.

Unfortunately, an additional drawback shown by these systems is that the synthetic route that is followed is complex and hardly adaptable for switches with different substitution. Conversely, further modification of the substructure also allowed the preparation of a dipole moment switch (Figure 8). [26] Because of its large dipole moment, this switch may constitute the prototype of a generation of electrostatic switches achieving, at the single-molecule level, a light-induced dipole moment inversion of about 30 D. This dipole moment change opens up a new perspective for the light-driven conformational control of macromolecular structures determined by polar interactions.

A different family of molecular switches based on PSB-retinal was also developed by replacing the indanylidene unit by a fluorenylidene unit. [27] The resulting compounds are named fluorenylidene-pyrrolines (FPs) and *N*-alkylated fluorenylidene-pyrrolines (NAFPs) (Figure 9).

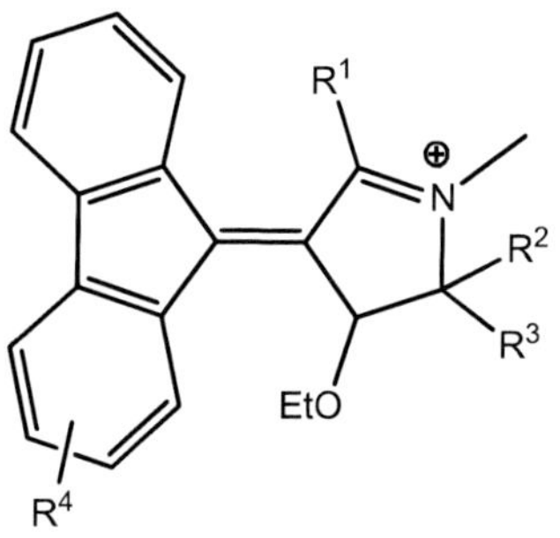

Figure 9. *N*-alkylated fluorenylidene-pyrrolines (NAFPs).

As shown in Figure 9, different substituents could be introduced with ease in almost every point of the molecule that allow for the tuning of the system properties. The synthesis of these compounds can be carried out thanks to the reaction between alkynyl Fischer carbene complexes with fluorenone imines to yield the corresponding fluorenylidene-pyrrolines (FPs).

The results of the photochemical study and characterization of FPs and NAFPs have revealed that they present really interesting features as molecular switches. [28] For instance, the methylated compounds undergo a bathochromic shift of the low energy band of approximately 70 nm in the UV spectra compared to the neutral compounds (λ_{max}= 334 nm), presenting absorption in the visible region (λ_{max}= 401 nm). This allows carrying out the photoisomerization reaction with solar light, which is a clean and renewable kind of energy. Furthermore, methylation also speeds up the photoisomerization process, so that the reaction is four times faster with NAFPs than with FPs.

In addition to this, the values of the photoisomerization quantum yields of FPs ($\Phi \approx 0.6$) and NAFPs ($\Phi \approx 0.5$) are very close to the value found for PSB-retinal *in vivo* (in rhodopsin, $\Phi \approx 0.67$), which makes these switches very efficient. Moreover, these compounds present very weak fluorescence, with a fluorescence quantum yield of 0.06. Therefore, they don't lose energy through deactivation pathways. Interestingly, the stability of these compounds after several hours of irradiation with a 125W medium-pressure Hg lamp has been checked, showing that these molecular switches are highly photostable, which is a desirable characteristic when used for technological applications.

Nevertheless these compounds present several drawbacks, such as the fact that the isomers ratio at the photostationary state is approximately 50% (*E* isomer) and 50% (*Z* isomer), because of the symmetry of the fluorenone ring. Consequently, mixtures rich in one of the two isomers cannot be obtained by means of irradiation. In addition to this, another great disadvantage that NAFPs share with NAIPs and previously reported NABPs is that the different ways of functionalizing the molecule have led to no potential applications of these molecular switches, which would be useful in order to test their behaviour in other environments.

Therefore, in an effort to avoid the drawbacks found in previous PSB-retinal switches, a new and versatile entry route to a new family of molecular photoswitches was reported. [29] These novel molecular switches based on PSB-retinal chromophore are analogous to benzylidene-pyrrolines NABP switches previously described, presenting a substituent R^1 in C-5 of the pyrroline ring (Figure 10(b)). It was anticipated that a substituent in that position could affect the photoisomerization process. Moreover, a phenyl ring in that spot could substantially increase the UV-absorption and induce a bathochromic shift due to extended conjugation.

These new switches maintain most of the good features achieved previously with other rhodopsin-based photoswitches while present several improvements, such as high isomerization quantum yield. [29] Specially, the synthetic route permits the synthesis of a family of compounds by modification of the starting material. Also, placing amino groups at both ends of the unit allowed for the linkage to a peptide. Interestingly, the cross-linked peptide showed an excellent performance as irradiating with a 400 nm light allowed to obtain a photostationary state of 7/93 *E/Z* starting from the thermal, more stable *E* isomer. Cross-linking the *E* form of the switch with the peptide led to a stabilization of the α-helical content and therefore the peptide was found to be fully helical. However, after irradiating with a 400 nm light, the switch causes a marked decrease in helicity due to the isomerization to the *Z* form. Even more, this conformational change is fully reversible to a 78/22 *E/Z* photostationary state upon irradiation with a 446 nm light. Thus, this switch can effectively photocontrol the peptide conformation in an aqueous environment using low-energy irradiation wavelengths fully compatible with biological environments. The large conformational change that occurs and the efficiency of the process make this switch an excellent candidate for peptide and protein photocontrol. This rhodopsin-like photoswitch fulfils most of the requirements needed for an efficient switch and shows how careful study of natural systems can help us in the design of new products with interesting properties. [29]

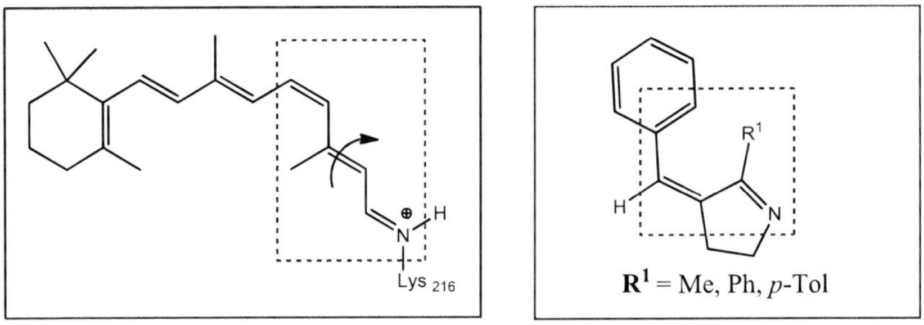

a) **b)**

Figure 10. Comparison between the structures of the protonated Schiff base of retinal chromophore (a) and the prototypes of molecular switches (b).

3.2. GFP-based Molecular Photoswitches

As discussed above, the main photophysical process in the GFP *in vivo* is fluorescence emission. However, other side-reactions are also found for the GFP chromophore, such as photoisomerization. This reactivity can be promoted by structure modifications that could lead to potential applications.

The main goal of the design of molecular switches with structure based on the green fluorescent protein (GFP) chromophore is to obtain a new family of prototypes of molecular switches that work efficiently. To this purpose, one could try to mimic the photoisomerization observed in the natural GFP chromophore. This new family of compounds has a certain structural analogy with the compounds previously shown in precedent paragraphs based on the structure of the protonated Schiff base of retinal, PSB-retinal. The structure of the GFP chromophore is shown in Figure 11, together with GFP-based photoswitches.

If we compare the structure of the GFP chromophore in Figure 11(a) with the structures in Figure 10 of the PSB-retinal chromophore, it can be inferred that there is an important difference between them related to the position of the nitrogen atom of the imine bond. In the case of the GFP chromophore, the bond sequence is –C=C-N=C-, while in PSB-retinal, the sequence is –C=C-C=N-. This modification may be relevant when the rotation of the central double bond of the designed switches takes place, due to the change in the electronegativity of the neighboring atom of that double bond.

Another structural difference is the presence of heteroatoms in the five-membered ring, apart from the nitrogen atom of the imine bond, specifically those that belong to the amide function. The oxygen atom directly bonded to the imine moiety can affect the electronic properties of the whole chromophore. Besides, the carbonyl group may also affect the rotation, as it is spatially close to the region of rotation of the C=C double bond.

When contrasting the structure of the GFP chromophore (Figure 11(a)) with the one of the new photoswitches (Figure 11(b)), it can be seen that these new compounds show subtle differences from the GFP chromophore. The main discrepancy between them is that the GFP chromophore and its derivatives present an amide function in the five-membered ring (dihydroimidazolone), while the new compounds have an ester function instead (5(4*H*)-oxazolones).

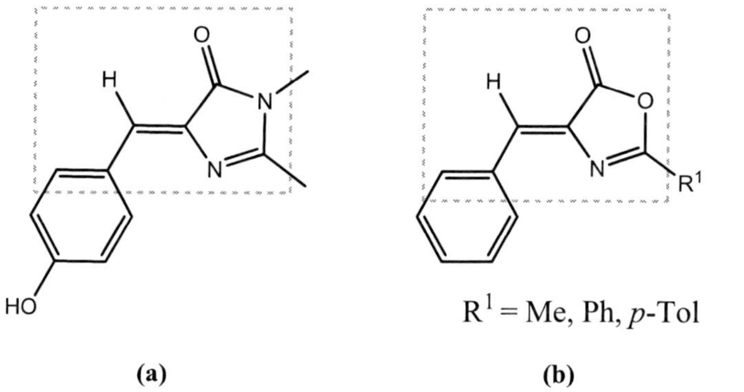

Figure 11. Comparison between the structures of the GFP chromophore (a) and the GPF-based photoswitches (b).

However, it is not complicated to reach the structure of the GFP chromophore from the synthesized structures (Figure 11(b)), as a single reaction step would lead to obtain the compounds with structure shown in Figure 11(a). [18a].

The synthesis of the new photoswitches with structure based on the GFP chromophore takes place under the classical conditions for the formation of azalactones. [30] Interestingly, a whole family of GFP-based photoswitches can be synthesized with ease and their photophysical and photochemical properties can be adjusted by the substituents present in the molecule.

These compounds have been reported [31] to feature an efficient photoswitching process. The easy and versatile synthesis allows for the preparation of derivatives with a different substitution pattern that can be used to tune the photophysical properties. This is relevant in order to be able to match the spectroscopic properties of the switches with the requirements for technological applications. Also, the main isomer present in solution can be controlled by using light or heat as the stimulus. Thus, the main isomer can be switched from *Z* to *E* and *vice versa* by changing the operating stimulus. The same effect can also be achieved through a careful selection of the wavelength of irradiation. As a result, these photoswitches show promising features that make them excellent candidates for the development of different applications.

Conclusion

Through this chapter we have presented some natural machines capable of performing efficient work. Also, we have shown that we can learn from their functioning in order to design and build our own artificial machines. Of course, it will be extremely difficult to improve Nature's performance, but we should keep in mind that trying to mimic natural systems could not only lead to new applications but also to a better understanding of how the natural world works.

Probably the most important lesson that we could learn from Nature is that the more relevant issue in any biomolecular device is the task to be performed. Efficiency is far more significant than appearance. Thus, in order to build new devices or develop new applications, we should always keep in mind that Nature could have solved the same problem before. And if that is the case, probably the solution is both simple and efficient.

The necessity to generate new photoactive materials, the design of new data storage devices or the biomedical applications of photocontrolled proteins are some of the different applications in which molecular switches could be useful. For all these cases, Nature already presents some fascinating examples. In the development of our own versions of these systems we will find that at least some paths are already opened up.

Acknowledgments

We thank the Spanish MINECO, Comunidad Autónoma de La Rioja and Universidad de La Rioja for financial support.

References

[1] D. Kaplan, W. W. Adams, *Silk Polym.; ACS Symp. Ser.* 1994, *544*, 2.

[2] I. Willner, *Science* 2002, *298*, 2407.

[3] a) S. Mann, *Biomimetic Materials Chemistry*, VCH, New York, 1996. b) M. Sarikaya, I. A. Aksay, *Biomimetics: Design and processing of Materials*, American Institute of Physics, College Park, 1993.

[4] K. E. Drexler, *Trends Biotechnol. 1999* 1999, *17*, 5.

[5] a) S. Subramaniam, R. Henderson, *Nature* 2000, *406*, 653. b) W. Kühlbrandt, *Nature* 2000, *406*, 569.

[6] A. D. Metha, M. Rief, J. A. Spudich, D. A. Smith, R. M. Simmons, *Science* 1999, *283*, 1689.

[7] M. Schliwa, G. Woehlke, *Nature* 2001, *411*, 424.

[8] H. C. Taylor, E. J. Holwill, *Nanotechnology* 1999, *10*, 237.

[9] J. Gelles, R. Landick, *Cell* 1998, *93*, 13.

[10] T. M. Lohman, K. Thorn, R. D. Vale, *Cell* 1998, *93*, 9.

[11] D. Baylor, *Proc. Natl. Acad. Sci. USA* 1996, *93*, 560.

[12] H. Kandori, Y. Shichida, T. Yoshizawa, *Biochem. (Moscow)* 2001, *66*, 1197.

[13] Y. Shichida, H. Imai, *Cell. Mol. Life Sci.* 1998, *54*, 1299.

[14] J. K. McBee, K. Palczewski, W. Baehr, D. R. Pepperberg, *Prog. Retin. Eye Res.* 2001, *20*, 469.

[15] R. A. Mathies, J. Lugtenburg, in *Handbook of Biological Physics, Vol. Handbook of Biological Physics* (Eds.: D. G. Stavenga, W. J. de Grip, E. N. Pugh), Amsterdam, 2000, p. 56.

[16] R. Heim, D. C. Prasher, R. Y. Tsien, *Proc. Natl. Acad. Sci. U.S.A.* 1994, *91*, 12501.

[17] M. Chattoraj, B. A. King, G. U. Bublitz, S. G. Boxer, *Proc. Natl. Acad. Sci. U.S.A.* 1996, *93*, 8362.

[18] a) V. Voliani, R. Bizzarri, R. Nifosì, S. Abbruzzetti, E. Grandi, C. Viappiani, F. Beltram, *J. Phys. Chem. B* 2008, *112*, 10714. b) S. Rafiq, B. K. Rajbongshi, N. N. Nair, P. Sen, G. Ramanathan, *J. Phys. Chem. A* 2011, *115*, 13733.

[19] M. Zimmer, *Chem. Rev.* 2002, *102*, 759.

[20] D. Sampedro, A. Migani, A. Pepi, E. Busi, R. Basosi, L. Latterini, F. Elisei, S. Fusi, F. Ponticelli, V. Zanirato, M. Olivucci, *J. Am. Chem. Soc.* 2004, *126,* 9349.

[21] T. Andruniów, S. Fantacci, F. De Angelis, N. Ferré, M. Olivucci, *Angew. Chem. Int. Ed.* 2005, *44*, 6077.

[22] F. Lumento, V. Zanirato, S. Fusi, E. Busi, L. Latterini, F. Elisei, A. Sinicropi, T. Andruniów, N. Ferré, R. Basosi, M. Olivucci, *Angew. Chem. Int. Ed.* 2007, *46*, 414.

[23] A. Sinicropi, E. Martin, M. Ryazantsev, J. Helbing, J. Briand, D. Sharma, J. Léonard, S. Haacke, A. Cannizzo, M. Chergui, V. Zanirato, S. Fusi, F. Santoro, R. Basosi, N. Ferré, M. Olivucci, *Proc. Natl. Ac. Sci. USA* 2008, *105*, 17642.

[24] J. Vicario, M. Walko, A. Meetsma, B. L. Feringa, *J. Am. Chem. Soc.* 2006, *128*, 5127.

[25] A. Sinicropi, C. Bernini, R. Basosi, M. Olivucci, *Photochem. Photobiol. Sci.* 2009, *8*, 1639.

[26] A. Melloni, R. R. Paccani, D. Donati, V. Zanirato, A. Sinicropi, M. L. Parisi, E. Martin, M. Ryazantsev, W. J. Ding, L. M. Frutos, R. Basosi, S. Fusi, L. Latterini, N. Ferre, M. Olivucci, *J. Am. Chem. Soc.* 2010, *132*, 9310.

[27] a) L. Rivado-Casas, D. Sampedro, P. J. Campos, S. Fusi, V. Zanirato, M. Olivucci, *J. Org. Chem.* 2009, *74*, 4666. b) L. Rivado-Casas, P. J. Campos, D. Sampedro, *Organometallics* 2010, *29*, 3117.

[28] L. Rivado-Casas, M. Blanco-Lomas, P. J. Campos, D. Sampedro, *Tetrahedron* 2011, *67*, 7570.

[29] M. Blanco-Lomas, S. Samanta, P. J. Campos, G. A. Woolley, D. Sampedro, *J. Am. Chem. Soc.* 2012, *134*, 6960.

[30] J. E. Audia, J. J. Droste, J. S. Nissen, G. L. Murdoch, D. A. Evrard, *J. Org. Chem.* 1996, *61*, 7937.

[31] M. Blanco-Lomas, P. J. Campos, D. Sampedro, *Org. Lett.* 2012, *in press, DOI: 10.1021/ol301741g*.

In: Focus on Biomimetics Research
Editor: Abele Scipione

ISBN: 978-1-62417-332-5

Chapter 5

A VOLTAMMETRIC SENSOR BASED ON A HEMIN MODIFIED MULTIWALLED CARBON NANOTUBE FOR HYDROQUINOME DETERMINATION IN PHARMACEUTICAL SAMPLES

***Débora Nobile Clausen*[1], *Eduardo Henrique Duarte*[1], *Arnaldo César Pereira*[2], *Lauro Tatsuo Kubota*[3] and *César Ricardo Teixeira Tarley*[1,3,*]**

[1]Universidade Estadual de Londrina (UEL),
Departamento de Química, Londrina, Brazil
[2]Universidade Federal de São João del-Rei (UFSJ),
Departamento de Ciências Naturais, São João del-Rei, Brazil
[3]Departamento de Química Analítica - Instituto Nacional de Ciência e Tecnologia (INCT) de Bioanalítica, Universidade Estadual de Campinas (UNICAMP),
Cidade Universitária Zeferino Vaz s/n, Campinas, Brazil

ABSTRACT

A simple, low cost and sensitive electrochemical method for the detection of hydroquinone (HQ) was developed based on a carbon nanotube paste (70%) electrode modified with hemin (20%) and mineral oil (10%). The aim of the present work was to evaluate the performance of a hemin (iron(II) protoporphyrin IX) functionalized carbon nanotube sensor for the voltammetric determination of HQ. The modified electrode demonstrated an electrocatalytic activity for the HQ oxidation in a 0.05 M Tris-HCl buffer solution (pH 7) using cyclic voltammetry. The anodic peak current of the sensor was 4 and 2.8 times higher when compared to glassy carbon and non-modified carbon nanotube paste electrodes, respectively. The best paste composition was found to be 3.5:1 (MWCNT:Hemin), considering the smaller separation between the cathodic and anodic peaks (ΔE) as well as the larger peak currents. The linear plot of anodic peak current *vs.* square root of scan rate demonstrated that the HQ oxidation was a diffusion-controlled

[*]Tel +55 43 3371 4366; fax +55 43 3371 4286; E-mail address:ctarleyquim@yahoo.com.br (C. R. T. Tarley)

process. The square wave parameters were optimized, and the best results obtained were as follows: amplitude = 70 mV, frequency = 70 Hz, and step of potential = 0.01 V. 0.15 mol L^{-1} Tris-HCl was found to be the best buffer solution condition. The analytical curve was linear in the HQ concentration range of 2.5 - 120 μmol L^{-1} (r = 0.9962, sensitivity = 0.14507 μA μmol L^{-1}, and precision = 2 %). The limits of detection and quantification for HQ were estimated to be 0.78 μmol L^{-1} and 2.60 μmol L^{-1}, respectively. The evaluation of HQ as electron mediator had insignificant effects on the sensing mechanism, showing that the proposed sensor did not act like a second-generation electrode. Therefore, peroxide was not used to measure HQ amounts. The HQ diffusion coefficient was found to be 6.34×10^{-5} cm^2 s^{-1}. The proposed method presented no significant differences at the 95 % confidence level compared to the chromatographic method. It can be employed for detecting other phenolic compounds, such as guaiacol, catechol, N-acetyl-p-aminophenol, ascorbic and uric acids; besides, it can be applied to pharmaceutical formulations.

Keywords: sensor, multiwalled carbon nanotubes (MWCNTs), hemin, hydroquinone

INTRODUCTION

Hydroquinone (1,4-dihydroxibenzene, HQ) and catechol (1,2-dihydroxibenzene, CT), two isomers of phenolic compounds, have been recognized as important environmental pollutants in accordance with the United States Environmental Protection Agency (US EPA) and European Union (EU) (Xie et al. 2006).

Both HQ and CT are used in cosmetics, pesticides, condiments, drugs and photographic dye chemicals (Ahammad et al. 2011). These compounds are extremely harmful to animals, plants and the aquatic environment even at low concentrations (Timur et al. 2003). High HQ levels in humans may cause fatigue, headache, tachycardia, kidney damage, apart from some cancer types, such as acute myeloid leukemia (Zhao et al. 2006). HQ, is one of the most common active ingredients of skin lightening creams prescribed by dermatologists.

Various methods have been employed for determining HQ and CT trace levels - *e.g.*, high performance liquid chromatography (HPLC) (Wang et al. 1999), spectrometry (Nagaraja et al. 2001), and electroluminescence (Sun et al. 2000). Electrochemical techniques, such as cyclic voltammetry (CV), differential pulse voltammetry (DPV) and square wave voltammetry (SWV), have also been used due to their simplicity, high sensitivity and viability for miniaturization analysis (Santos et al. 2007).

Since HQ is electroactive, it can be determined by these methods. The use of voltammetric sensors for the HQ determination based on ionic liquid composite film modified glassy carbon electrodes (Liu et al. 2010, Qi and Zhang 2005) and potato tissue modified paraffin /graphite electrodes (Vieira and Fatibello-Filho 2000) has been reported in literature.

The disadvantages of employing the ionic liquid composite film modified electrode are the prolonged time of film purification and preparation and a high cost of glassy carbon, whereas the drawbacks of the potato tissue modified electrode are as follows: the prolonged biosensor preparation, necessity for activation of enzymes in organic solvents and keeping them at low temperatures to avoid denaturation, and the problem of increasing the rate of electron transfer between the enzyme active sites and electrode surface.

To overcome these challenges, hemin modified multi-walled carbon nanotubes (MWCNTs) were chosen in the present work to develop a new sensor. MWCNTs exhibit superior structural, electronic and electrochemical compared to other types of carbon materials, such as glassy carbon, graphite and diamond.

Besides, they are electrically inert and do not show any voltammetric response within experimental potential windows commonly used. As a result, the MWCNTs have been widely investigated to study electrocatalysis, direct electrochemistry of proteins and electroanalytical applications, such as sensors and electrochemical biosensors (Dai 2002, Charlier 2002). MWCNT based sensors can be made through carbon paste electrodes (using Teflon-like polymers or mineral oil) (Wang and Musameh 2003, Valentini et al. 2003) or film modified electrodes (using Nafion-like materials) (Rubianes and Rivas 2003). The paste was prepared by hand mixing of a MWCNT composite and a binder (usually mineral oil) until a homogeneous phase was obtained, and then packed into the electrode cavity.

The film was made of MWCNTs dispersed in liquid composites (DMF, Nafion, etc), and the suspension was coated on the previously cleaned glassy carbon electrode. The use of surfactants also has been exploited to enhance the performance of sensor based on MWCNT (Jain and Rather 2011). The MWCNTs can also facilitate the direct transfer of enzymes and proteins, such as glucose oxidase (Gox) and hemeproteins including cytochrome c (Cyt c), hemoglobin (Hb), myoglobin (Mb) and other catalysts. Electrodes based on MWCNT paste, randomly scattered MWCNTs and MWCNTs chemically connected to a substrate have been used in protein biochemistry (Gong et al. 2003).

The observed mechanism for the facilitated electron transfer between the hemeproteins and MWCNTs can be explained considering the structural properties of the nano-sized MWCNTs, surface defects produced during treatment (Zhao et al. 2002), and direct heme-nanotube π-π communications (Davis et al. 2004).

However, the potential application of the MWCNTs in biosensors is still limited due to the following factors: (i) the nanotubes tend to aggregate in many solvents, which makes them difficult for practical use as well as for interactions with biomacromolecules, (ii) strong interactions between the proteins and nanotubes may distort the protein structure, shifting the loss in biocatalytic activity to the substrates.

Moreover, as cited above, the biosensors have been of great concern for the rate of the electron transfer between the enzyme active sites and electrode surface (Gong et al. 2005). The MWCNT based biosensors can be divided into three generations.

The first-generation biosensors are based on the electroactivity of a substrate or a product of the enzymatic reaction, whereas the second-generation biosensors use mediators to transport electrons between the enzymes and electrode, and, finally, the third-generation biosensors are based on the direct electron transfer without any mediators (Gong et al. 2005). The study of the direct electron transfer between the electrodes and biomolecules (redox proteins, oxirredutase, etc.) is a convenient and informative way to understand kinetics and thermodynamics of biological redox processes. In this context, hemin (iron(II) protoporphyrin IX) plays an important role in molecular modeling to evaluate the redox activity of hemeproteins (b-cytochrome, peroxidase) and oxygen-transporting proteins (myoglobin, hemoglobin), since it has a good stability in solutions, low molecular weight and low cost (Chen et al., 2000; Zhang et al. 2001).

Besides, it appears to be a mimetic agent for peroxidase and nitrate reductase (Mimica et al. 2001), apart from functioning as a mediator under conditions close to a native biological environment for amperometric detection of many species, such as O_2 (Zheng et al. 2002),

H_2O_2 (Zhang et al. 2001), NO/NO_2^- (Mimica et al. 2001), superoxidase (Chen et al. 2000), tryptophan and its derivatives (Nan et al. 2002).

The strong characteristic of hemin is the existence of a protoporphyrin ring that promotes its adsorption on carbonaceous materials and can be immobilized on the MWCNT surface when preparing a sensor (Turdean et al. 2006) for detecting analytes of interest in the environment (*e.g.*, toxic phenolic compounds, such as HQ).

Therefore, the rapid construction of a low cost sensor based on a MWCNT paste platinum electrode (without pre-purifying) functionalized with hemin and Mineral Oil as well as its further application for HQ determination is presented herein. Hemin has been chosen due to the fact that it has the same catalytic action as the enzymes, but, as an advantage, it does not present the above mentioned problems.

1. Methodology

1.1. Equipment

Cyclic voltammetry, square wave voltammetry and chronoamperometry experiments were performed in a conventional three-electrode electrochemical cell (capacity: 15 mL, electrodes: Ag/AgCl - reference, Pt - auxiliary, and MWCNT/Hemin/Mineral Oil - working) shown in Figure 1. Electrochemical measurements were carried out with a Tris-HCl buffer (pH 7.0) (Tris-HCl buffer) and 200 μmolL^{-1} HQ solution using an Autolab PGSTAT 101 potentiostat/galvanostat; all solutions were stirred on a Fisatom magnetic mixer.

Figure 1. Eletrochemical cell used for cyclic voltammetry, square wave voltammetry and chronoamperometry experiments.

The measurements were carried out in presence of oxygen. HPLC analyses were conducted on a Shimadzu liquid chromatograph equipped with a LC-20AT gradient pump. A UV-VIS diode array detector (at 289.0 nm) and an injector fitted with a 20 μL loop were also employed in these tests. The separation was carried out on a Shimadzu CLC-ODS(M) column (column size: 250 mm x 4.6 mm i.d., particle size: 5 μm) at 25 °C and a flow rate of 1.0 mL min^{-1} using a methanol-water mobile phase (20:80, v/v).

1.2. Reagents

All reagents - hydroquinone - HQ (Sigma, 99%), sodium dihydrogen phosphate monohydrate (Merck, 99.0 – 102.0%), Tris-HCl buffer (Invitrogen, 99.9 – 100.1%), PIPES buffer (Sigma, 99.0%), HEPES buffer (Sigma, 99.5%), hydrogen peroxide (Biotec 30%), sodium acetate trihydrate (Merck, 99.0 – 101.0%), methanol (Sigma-Aldrich, ≥ 99.9%), and MWCNTs (CNT Co. Ltd., Yeonsu-Gu, Incheon, South Korea, diameter: 10 - 40 nm, length: 5 - 20 μm, > 93 %) - were of analytical grade, except mineral oil and utilized in the experiments without purifying. Deionized water, used to prepare the solutions, was obtained from a Millipore Milli-Q system, except the HQ working and Tris-HCl buffer solutions prepared with ultra-purified water.

1.3. Preparation of the Hemin Modified MWCNT Paste Electrode

The MWCNT/Hemin/Mineral Oil working electrode was prepared in the following proportion: 70 % (MWCNT): 20 % (Hemin): 10% (Mineral Oil). The MWCNTs and hemin were weighed in a Petri dish. After that, they were well mixed for 20 minutes using a steel spatula and then macerated in an agate mortar with a pestle.

Finally, the mineral oil was added, and the three reagents were mixed until get a homogeneous paste. The paste was then packed into the cavity of the platinum electrode (area: 0.282 cm^2).

2. Results and Discussion

2.1. Performance Evaluation of the Voltammetric Sensor

To evaluate the electrochemical behavior of the MWCNT/Hemin/Mineral Oil electrode, its voltammograms were recorded and compared with those obtained for the GC and non-functionalized MWCNT/Mineral Oil electrode, as shown in Figures 2a and 2b.

As seen in these figures, there was an increase of 400 % in the anodic peak current for the MWCNT/Hemin/Mineral Oil paste electrode when compared with the GC electrode (Figure 2a and b), and of 280% when compared with the non-functionalized MWCNT/Mineral Oil electrode, showing that hemin indeed facilitated the HQ oxidation. Besides, a decrease in the peak separation between the anodic (Epa) and cathodic (Epc) peaks potential for the MWCNT/Hemin/Mineral Oil system could be observed.

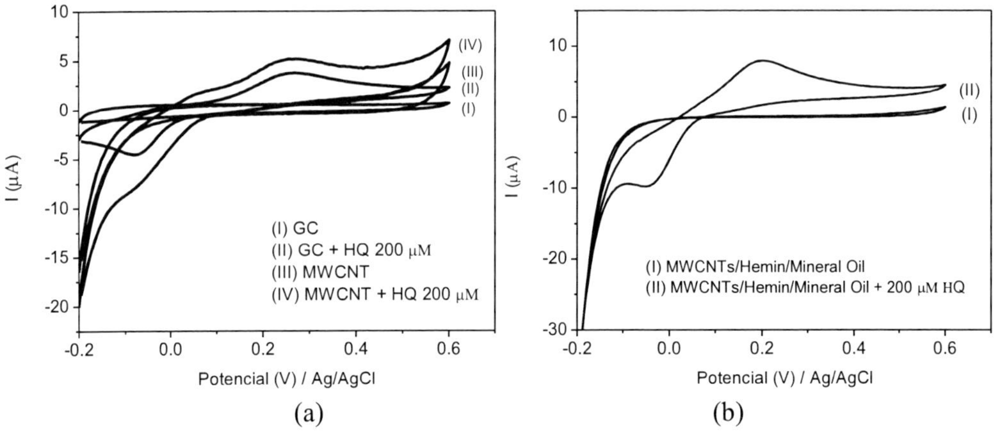

Figure 2. (a) Cyclic voltammograms obtained for the GC and MWCNT/Mineral Oil electrodes in the absence and presence of HQ, (b) Cyclic voltammograms obtained for the modified MWCNT/Hemin/Mineral Oil electrode (70:20:10) in the absence and presence of HQ. [Tris-HCl] = 0.05 mol L^{-1} (pH 7.0), [HQ] = 200 μmolL^{-1}, 50 mVs^{-1}.

2.2. Composition of the Hemin modified MWCNT Paste Electrode

After verifying the efficiency of the hemin modified MWCNT sensor for the HQ oxidation, a study was performed to elucidate the composition of the MWCNT/ Hemin/Mineral Oil paste. The following proportions were used: 35:35:30, 20:70:10, 15:75:10, 70:20:10, and 60:15:25 % (m/m/m) (Table 1).

The analysis of the paste composition presented in Table 1 shows that the increase of the hemin proportion reduced the peak separation (ΔE_p), and the lowest value of 0.2249 V was obtained for the 70:20:10 (m/m/m) proportion. However, in the proportions with higher hemin quantities, demonstrates that the accentuated increase in hemin would mask the MWCNT effect and diminish the reduction currents. Therefore, the 70:20:10 (m/m/m) proportion was chosen for further experiments.

Table 1. Difference between the anodic and cathodic peak potentials of the glassy carbon electrode, carbon nanotube and carbon nanotube modified with hemin in different proportions

Electrodes	ΔE_p (V)
GC	0.280
MWCNT	0.276
MWCNT/hemin/Nujol (20/70/10, m/m/m)	0.264
Electrodes	ΔE_p (V)
MWCNT/hemin/Nujol (60/15/25, m/m/m)	0.232
MWCNT/hemin/Nujol (35/35/30, m/m/m)	0.245
MWCNT/hemin/Nujol (70/20/10, m/m/m)	0.225
MWCNT/hemin/Nujol (15/75/10, m/m/m)	0.269

According to Ma et al (Ma et al. 2010) , the reduction of Fe(III) to Fe (II) takes place on the hemin functionalized PAMAM/MWCNT glassy carbon electrode at a potential of -0.4 V, whereas the oxidation of Fe (III) to Fe (IV) took place at 0.8 V (Kang et al. 1997). Thus, considering that the studied potential range for the MWCNT/Hemin/Mineral Oil electrode was -0.2 - 0.6 V, it can be verified that iron present in hemin structure is Fe(III).

2.3. Effect of pH

To study the effect of pH on the HQ anodic current, the pH range of 4.0 - 9.0 was chosen. For pH 4.0, 5.0 and 5.9, 0.05 mol L^{-1} acetate buffer was used, whereas for pH 7.0, 8.0 and 9.0, 0.05 mol L^{-1} Tris-HCl buffer (0.05 $molL^{-1}$) was used, as shown in Figures 3a and 3b. As observed from Figure 3a, the increase in pH promoted increments of the analytical signal up to pH 7.0. At low pH, the oxidation potential was shifted towards more positive values.

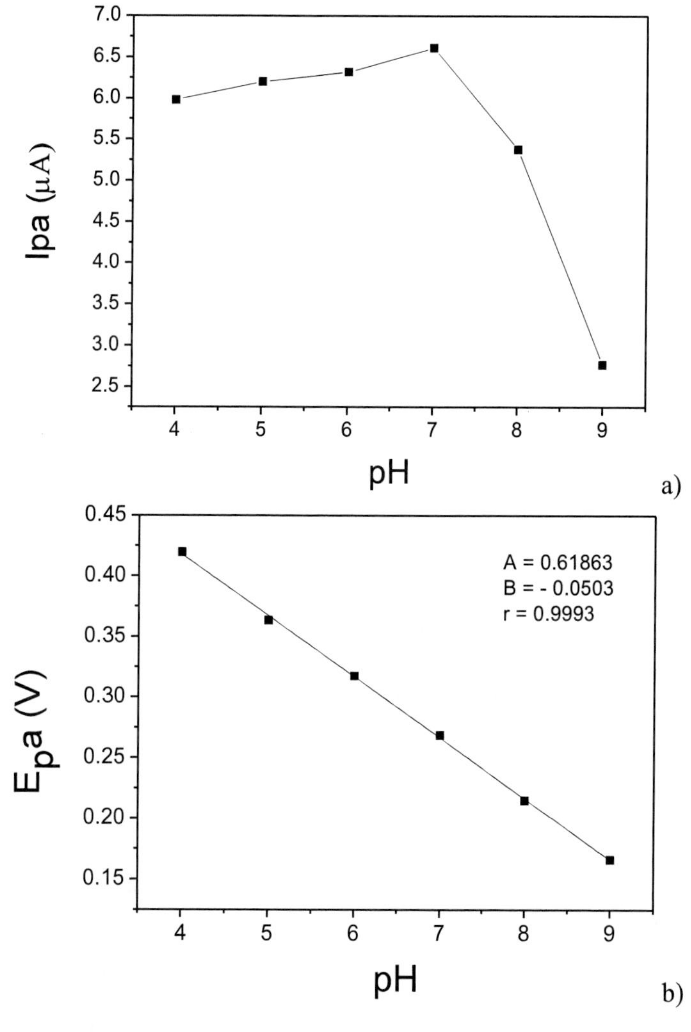

Figure 3. Ipa vs. pH (a) and Epa vs. pH (b) plots for HQ. [HQ] =200 $\mu molL^{-1}$, 50 mVs^{-1}.

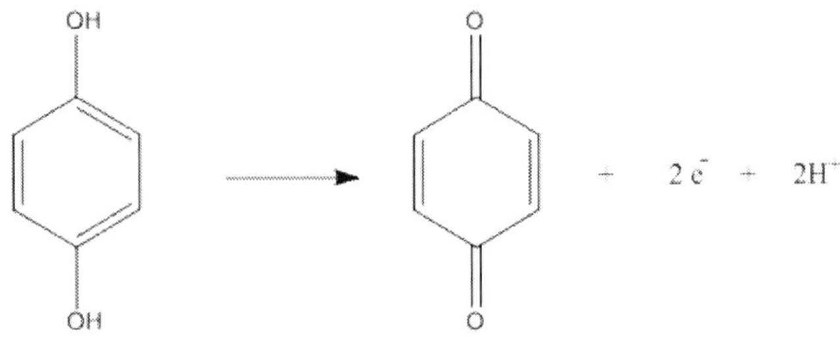

Figure 4. Mechanism of HQ oxidation in aqueous media.

As seen in Figure 4, there was a decrease in the magnitude of the HQ oxidation (Liu et al. 2010). At high pH values (8.0 and 9.0), the oxidation and reduction peaks were less evident, giving rise to two hypotheses: 1) deprotonation of carbonyl groups on the MWCNT surface, and 2) dissociation of HQ (pKa = 9.2) leading to deprotonation of the anions and thus causing electrostatic repulsion between the analyte and electrode (Qi and Zhang 2005). The HQ oxidation peak potential showed a linear variation with pH (Figure 3b), and the angular coefficient was found to be -50.3 $mVpH^{-1}$, suggesting that the total number of electrons and protons participating in the charge transfer process was the same. Since the HQ oxidation took place through the two-electron transfer, the number of involved protons was two (Liu et al. 2011).

2.4. Effect of Scan Rate

The effect of scan rate effect on the HQ voltammetric behaviour was studied under 0.05 mol L^{-1} Tris-HCl buffer solution containing 200 μmol L^{-1} HQ solution (Figure 5a). The scan rates of 20 - 500 mV s^{-1} were tested. Figure 5a demonstrates an increase in the HQ anodic peak and also a shift of the anodic potential towards more positive values with increasing of the scan rate, a typical case of a diffusion-limited mechanism. Moreover, Figure 6b indicates a linear relationship (r = 0.99572) between Ipa and $\nu^{½}$, thus confirming that the HQ oxidation process was controlled by diffusion (Teófilo et al. 2004).

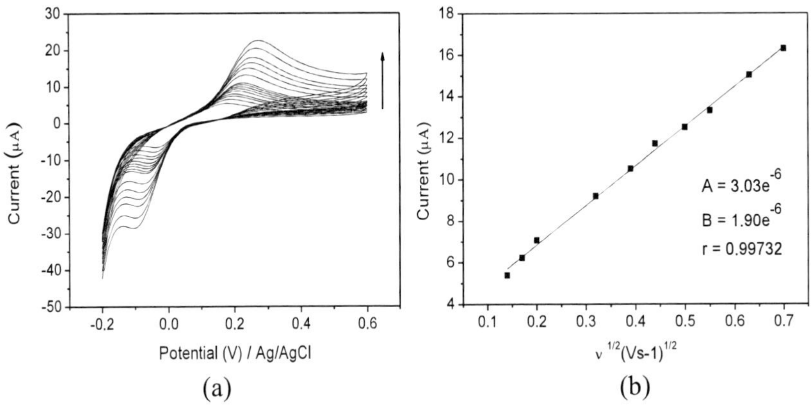

Figure 5. (a) Cyclic voltammograms of HQ at various scan rates. (b) Ipa *vs.* $\nu^{1/2}$ plot for HQ (r = 0.99732). Assays carried out at pH 7.0 with 200 μmol L^{-1} HQ.

2.5. Influence of H_2O_2 on HQ Determination

Experiments were carried out in the presence and absence of H_2O_2 by using cyclic voltammetry and chronoamperometry as well as hemin functionalized and non-functionalized electrodes. The first experiment was performed in the potential range of -0.6 - 0.6 V vs Ag/AgCl with the MWCNT/Hemin/Mineral Oil electrode and 0.05 mol L^{-1} Tris-HCl buffer employing or not 0.5 - 2.5 mmol L^{-1} H_2O_2 solutions (Figure 6a). The other experiment was performed under the same conditions, but with the MWCNT electrode in the absence of hemin (Figure 6b). As observed in Figure 6a, the reduction of Fe(III) to Fe(II) occurred on the MWCNT/Hemin/Mineral Oil electrode at -0.4 V, potential, in accordance with results of the research realized by Ma et al (2010), whereas Figure 8b, illustrates the absence of peak at -0.4 V in the case of the MWNTC/Mineral Oil electrode, thus showing that hemin exhibited catalytic properties toward the H_2O_2 reduction. Thus, as expected, it could be noted that the present sensor based on the hemin modified MWCNT paste electrode might be also used for monitoring H_2O_2 concentrations.

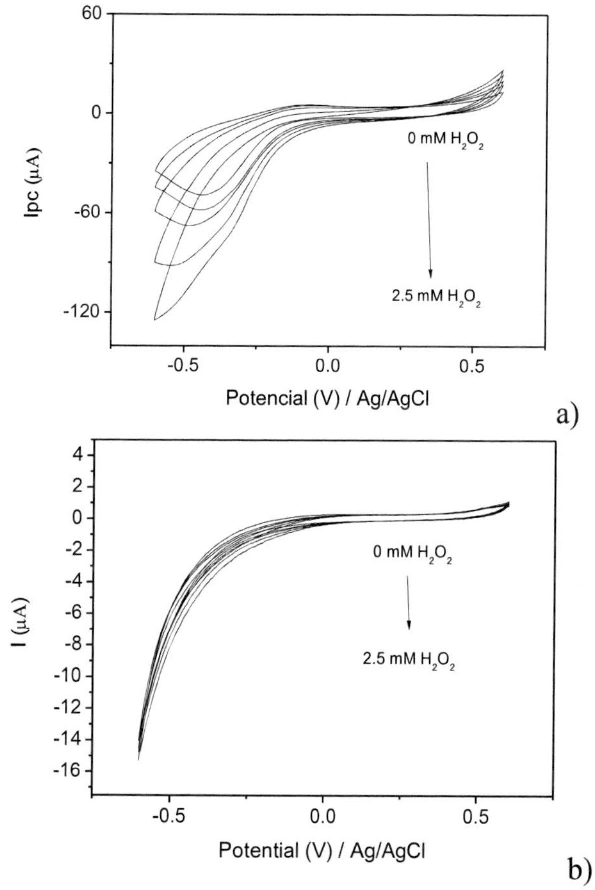

Figure 6. Cyclic voltammograms for the Tris-HCl buffer (0.05 mol L^{-1}, pH 7.0) in the absence and presence of H_2O_2 (0.5 mmolL^{-1} to 2.5 mmolL^{-1}) using the MWCNT/Hemin/Mineral Oil (a) and (b) MWCNT/Mineral Oil (b) electrodes. Scan rate of 50 mVs^{-1}.

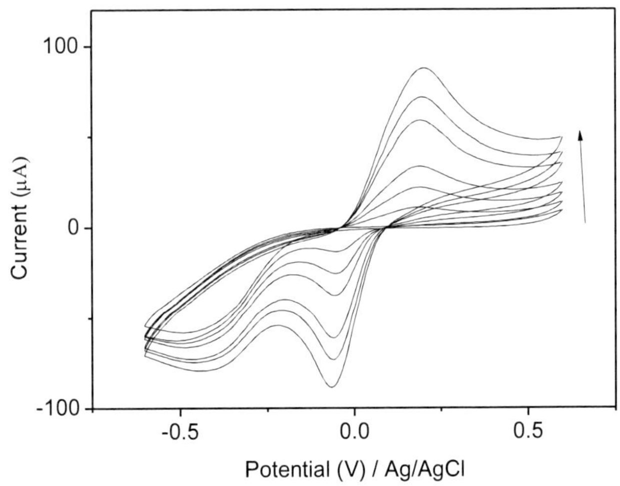

Figure 7. Cyclic voltammograms for HQ at crescent levels (0.2 mmol L^{-1} to 1.2 mmol L^{-1}) in the presence of 0.8 mmol L^{-1} H_2O_2 using the MWCNT/Hemin/Mineral Oil electrode. Scan rate of 50 mVs^{-1}. Assays carried out at pH 7.0 with 200 μmol L^{-1} HQ.

Another assay was carried out in the potential range of -0.6 - 0.6 V by using the hemin functionalized electrode and containing increasing HQ concentrations (0.2 mmol L^{-1} to 1.2 mmol L^{-1}), in 0.05 molL^{-1} Tris-HCl buffer medium and 0.8 mmol L^{-1} H_2O_2 (Figure 7).

As observed in Figure 7, the reduction of Fe (III) to Fe (II) also took place at HQ crescent concentrations and -0.4 V, thus confirming results of the study performed by Rosatto et al. (1999). However, it was observed that the increase in HQ concentration did not significantly enhance the H_2O_2 reduction currents catalyzed by hemin; in other words, HQ had no mediator effect (Gong et al. 2005). It seems that in this case, the redox process of hemin and hydrogen peroxide is independent of HQ presence.

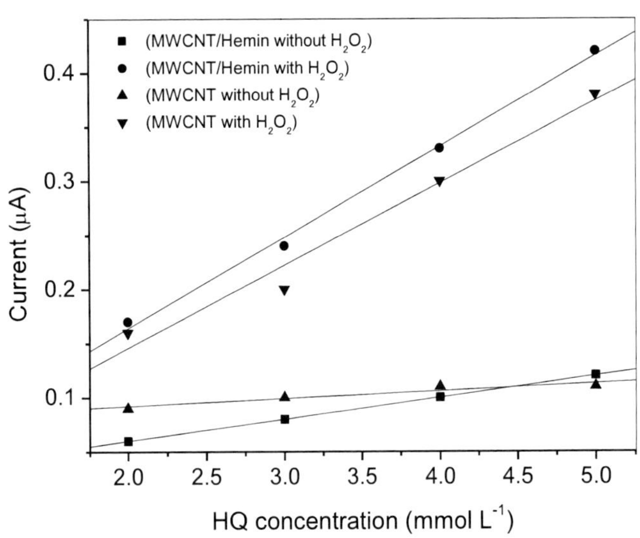

Figure 8. Analytical curves obtained from chronoamperograms for HQ in the absence and presence of H_2O_2. [HQ] = 2.0 - 5 mmol L^{-1}, [Tris-HCl] = 0.05 mol L^{-1} (pH 7.0), [H_2O_2] = 0.6 mmol L^{-1}, fixed potential at -0.05 V.

In addition, it was observed that oxidation peak of HQ increased linearly as increasing its concentration, showing again that electron transfer is not mediated by HQ.

This observation was also verified from Figure 8, which display analytical curves constructed by means of chronoamperometric experiments in the presence and absence of H_2O_2 for both MWCNT/hemin/mineral oil and MWCNT/mineral oil electrodes.

As observed, the sensitivity for both electrodes in the presence of H_2O_2 is very similar, showing that the hemin in the MWCNT paste does not catalyze the oxidation of HQ by H_2O_2.

The influence of increasing H_2O_2 concentration on oxidation/reduction peaks of HQ was also investigated. Figure 9, displays the cyclic voltammograms of HQ in the absence and presence of with increasing H_2O_2 concentrations, using the MWCNT/hemin/mineral oil and MWCNT/mineral oil electrodes. As expected, the reduction peak of hydrogen peroxide was only observed at the MWCNT/hemin/mineral oil electrode. An increase of the reduction current was observed after addition hydrogen peroxide (Figure 9a), with an insignificant increase on the cathodic current of HQ oxidized by hemin.

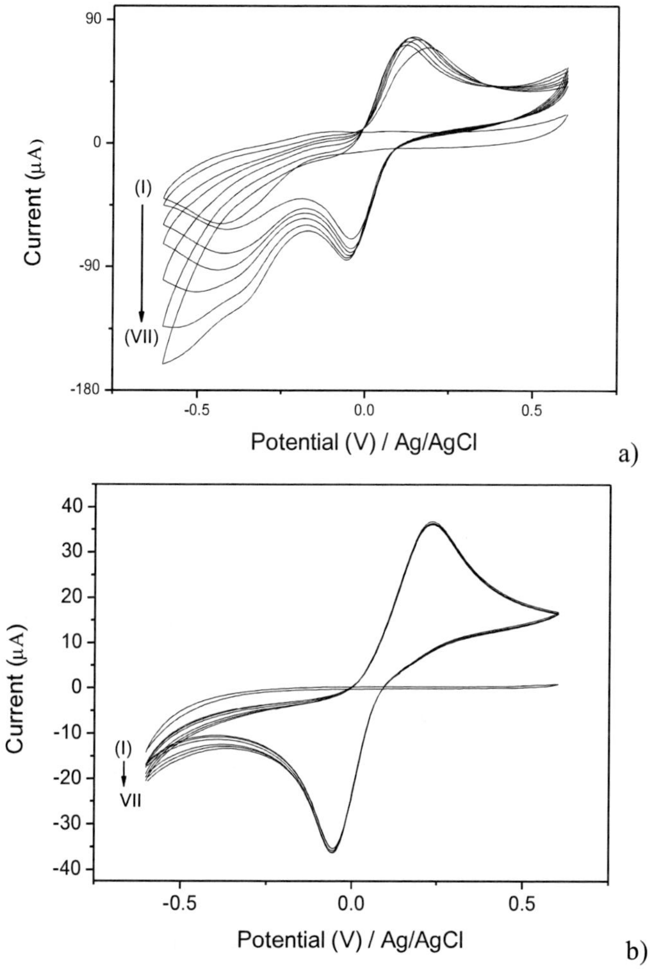

Figure 9. Cyclic voltammograms for buffer (I) and HQ at 1.0 mmol L^{-1} (II) with crescent levels (III to VII) of H_2O_2 (0.5 to 3.5 mmol L^{-1}) using MWCNT/hemin/mineral oil (a) and MWCNT/mineral oil electrodes (b). Scan rate of 50 mVs^{-1}. Assays carried out at pH 7.0.

It is worth noting that, considering the catalysis of HQ oxidation by hemin in the presence of hydrogen peroxide in the similar way as peroxidase-based sensors, it would be expected a decrease of the anodic peak of HQ with subsequent increase in the cathodic peak, whose observation was not verified.

The chronoamperograms depicted in Figure 10 present the same behavior.

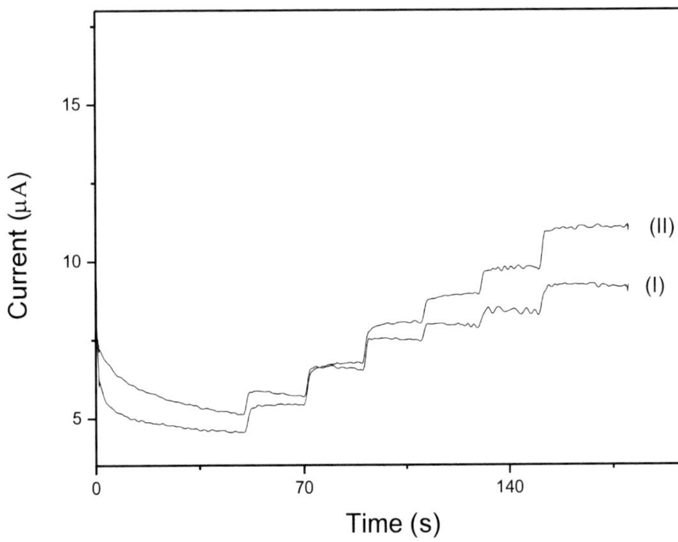

Figure 10. Current I (μA) vs time (s) amperograms for H_2O_2 in the absence (I) and presence (II) of HQ. $[H_2O_2] = 0.2 - 1.5$ mmol L^{-1}, [Tris-HCl] = 0.05 mol L^{-1} (pH 7.0), $[HQ] = 10^{-3}$ molL^{-1}, fixed potential at -0.05 V.

According to aforementioned, it could be concluded that, in fact, the HQ electron mediator effect on the sensor performance was negligible, indicating that the sensor did not act like a second-generation electrode. Hence, it was decided not to use H_2O_2 for the HQ determination.

Anyway, it must be emphasized that, in the absence of H_2O_2, the oxidation peak potential of HQ was shifted toward less positive potential and the anodic and cathodic peak currents increased in a significant way when the MWCNT was modified with hemin, as demonstrated in Figures 2 and 11, justifying herein the development of voltammetric sensor.

The enhanced performance of MWCNT in the presence of hemin probably can be associated to the low resistance of sensor and high electroactive area (0.205 cm^2) compared with the electrode without hemin (0.124 cm^2). The electroative areas of electrodes were estimated according to Randles-Sevcik equation (1) using the slope of the Ip *vs* $v^{1/2}$ plot for the $K_4Fe(CN)_6$ at 5.0 mmol L^{-1} concentration in 1.0 mol L^{-1} KCl.

$$Ip=2.69x10^5n^{3/2}AD^{1/2}v^{1/2}C \qquad (1)$$

where Ip refers to the anodic peak current, n is the electron transfer number, A the surface area of the electrode, v is the scan rate and C is the concentration of $K_4Fe(CN)_6$. Using $K_4Fe(CN)_6$ as a probe molecule, n =1 and $D = 7.6 \times 10^{-6}$ cm^2 s^{-1}.

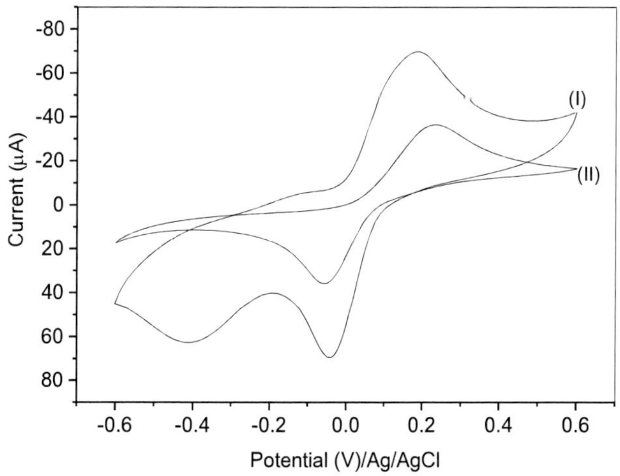

Figure 11. Cyclic voltammograms for HQ at 1.0 mmol L^{-1} in the absence of H_2O_2 using MWCNT/hemin/mineral oil (I) and MWCNT/mineral oil electrodes (II). Scan rate of 50 mVs^{-1}. Assays carried out at pH 7.0.

2.6. Determination of HQ Diffusion Coefficient

Chronoamperometric measurements of HQ at MWCNT/Hemin/Mineral Oil electrode were carried out by setting the working electrode potential at 0.25 V vs. Ag/AgCl/KCl (3.0 M) for the various concentration of HQ in 0.05 $molL^{-1}$ Tris-HCl medium (pH 7.0) (Fig. 12). For an electroactive analyte (in this case, HQ) with a diffusion coefficient D, the observed oxidation current for the electrochemical reaction at a mass transfer limited rate could be described by the Cottrell equation (Bard and Faulkner 2001).

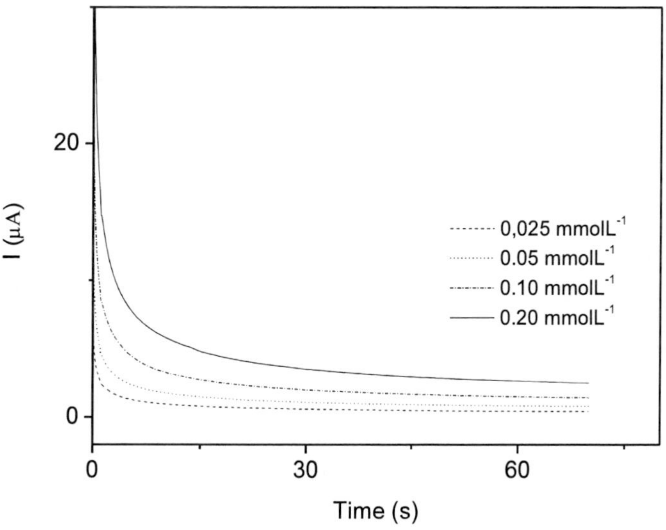

Figure 12. Current I vs. time amperograms. [HQ] = 0.025 - 0.2 mmol L^{-1}, [Tris-HCl] = 0.05 mol L^{-1} (pH = 7.0), t = 60 s, Epa = 0.25 V.

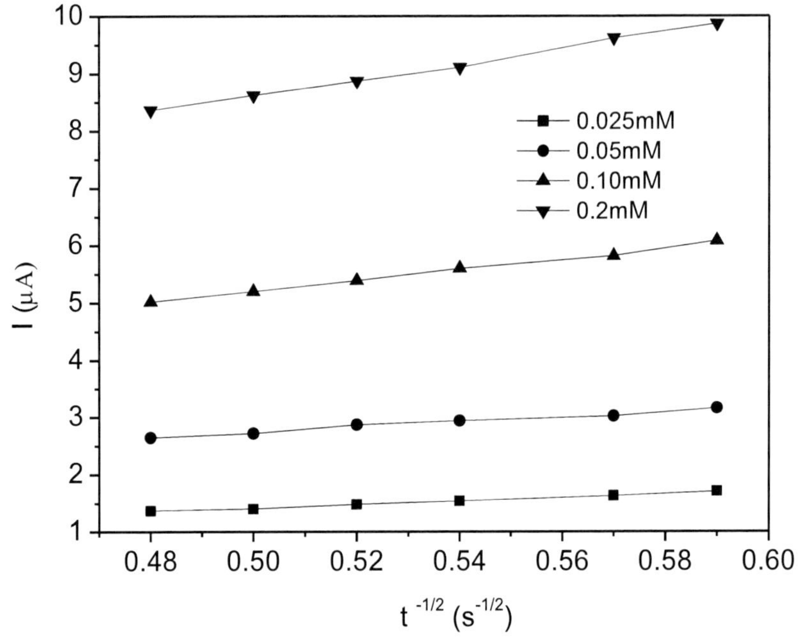

Figure 13. Current I vs $t^{-1/2}$ plot based on the chronoamperograms shown in Figure 9.

The I (µA) vs. $t^{-1/2}$ ($s^{-1/2}$) experimental plots were used for different HQ concentrations, and the mean HQ diffusion coefficient ($6.34x10^{-5}$ cm^2 s^{-1}) was obtained from the average of the straight line angular coefficients (Figure 13).

2.7. Optimization of Parameters Related to Square Wave Voltammetry (SWV)

SWV method was used to determine the concentration of HQ and the best conditions were optimized. The step potential was varied from 0.0005 to 0.04 V, whereas the amplitude (50 mV) and frequency (50 Hz) were kept constant, as illustrated in Figure 14.

As observed in Figure 14, the anodic current increased with the step potential up to 0.02 V, and decreased in the range of 0.02 - 0.03 V. Nonetheless, an enlargement and asymmetry of the peak could be seen at 0.02 V (Furlanetto et al. 2000).

Therefore, the step potential value of 0.01 V was chosen for further experiments. After having set the step potential at 0.01 V, the frequency was varied from 10 to 110 Hz, whereas the amplitude remained constant (50 mV), as shown in Figure 15.

As could be seen in this figure, the HQ oxidation current increased with the frequency up to 70 Hz and then decreased. Therefore, the frequency of 70 Hz was chosen for an amplitude variation study. The influence of pulse amplitude ranging from 10 to 130 mV is illustrated in Figure 16.

It could be observed that the HQ oxidation current increased with the amplitude up to 70 mV and then decreased. Thus, the amplitude value of 70 mV was chosen for further studies.

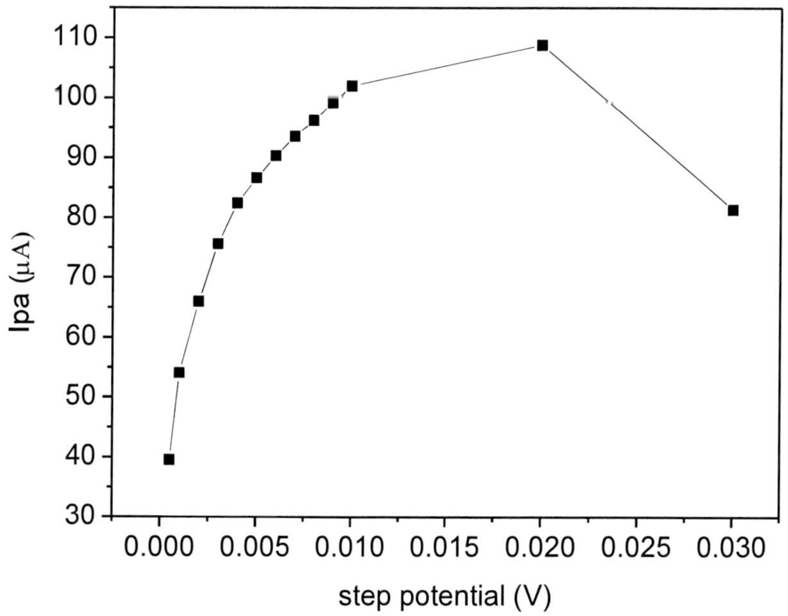

Figure 14. Anodic peak current Ipa vs. step potential variation plot. Amplitude = 50 mV, frequency = 50 Hz. [HQ] = 10^{-3} mol L^{-1}, [Tris-HCl] = 0.05 mol L^{-1} (pH = 7.0).

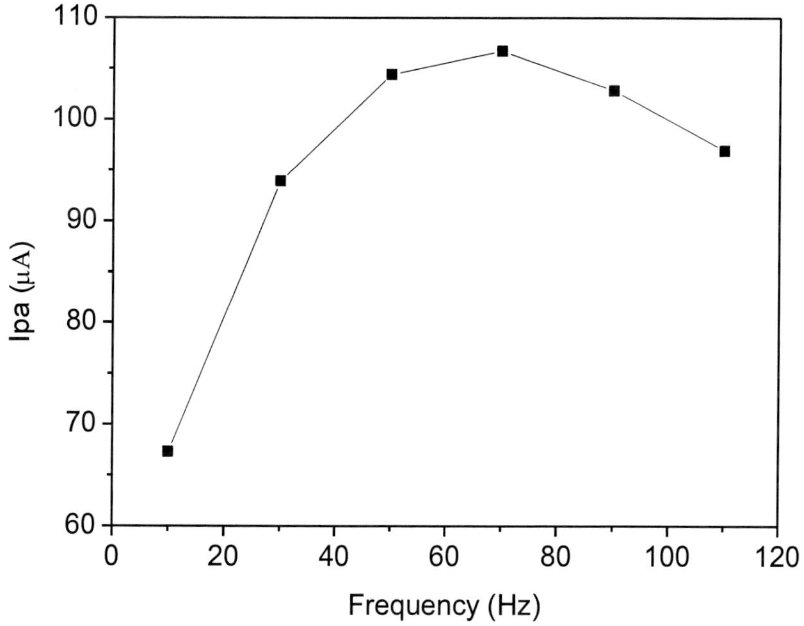

Figure 15. Anodic peak current Ipa vs. frequency variation plot. Amplitude = 50 mV, step potential = 0.01 V. [HQ] = 10^{-3} mol L^{-1}, [Tris-HCl] = 0.05 mol L^{-1} (pH = 7.0).

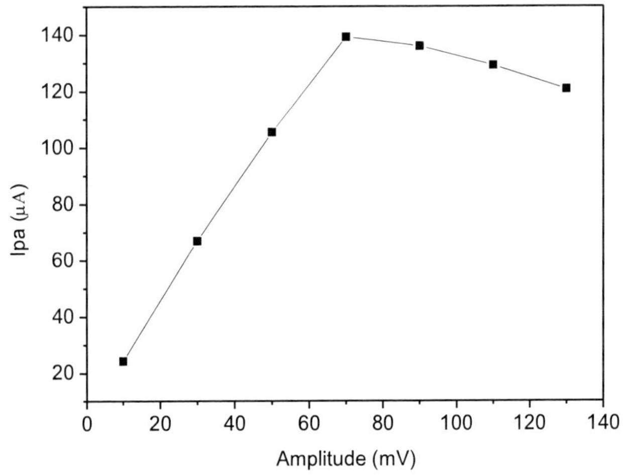

Figure 16. Anodic peak current Ipa vs. amplitude variation plot. Frequency = 70 Hz, step potential = 0.01 V. [HQ] = 10^{-3} mol L^{-1}, [Tris-HCl] = 0.05 mol L^{-1} (pH = 7.0).

2.8. Effect of Supporting Electrolyte

To evaluate the effect of electrolyte support on HQ oxidation, some solutions of Tris-HCl, HEPES, PIPES and phosphate solutions at pH 7.0 at 0.05 mol L^{-1} concentration were used under optimized conditions. Based on the analysis of the results given in Table 2, it could be observed that the anodic current response I_{pa} (μA) increased in the following order: HEPES (19.24) < phosphate (66.22) < Tris-HCl (151.10) < PIPES (163.90). Since the oxidation potential of HQ was lowest in Tris-HCl medium, this electrolyte was chosen for further experiments.

Table 2. Influence of buffer nature on anodic potential, anodic current and peak width

Buffer (0.05 mol L^{-1}, pH 7.0)	E_{pa} (V)	I_{pa} (μA)	W ½ $_{pa}$
Tris-HCl	0.31357	151.10	0.23163
HEPES	0.62576	19.24	0.54382
Phosphate	0.39413	66.22	0.19135
PIPES	0.37399	163.90	0.22156

2.9. Effect of Ionic Strength

To study the ionic strength effect on the HQ oxidation, the Tris-HCl buffer concentration was varied from 0.01 to 0.30 mol L^{-1}. The anodic peak current signal increased with the electrolyte concentration up to 0.20 mol L^{-1} and then decreased, indicating that the ions migrated at a constant velocity in the concentration range of 0.25 - 0.30 mol L^{-1} (Vieira and Fatibello-Filho, 2000) as shown in Figure 17. However, at 0.20 mol L^{-1}, the peak was obtained in lower resolution. Therefore, the concentration of 0.15 mol L^{-1} was chosen.

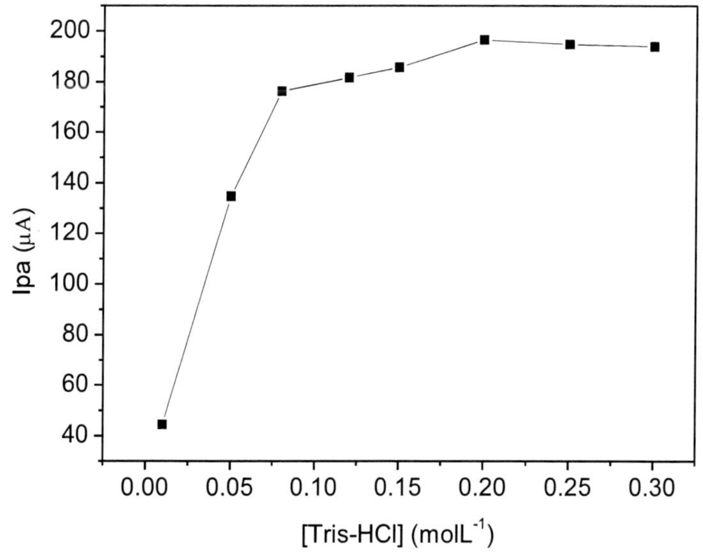

Figure 17. Influence of the concentration of electrolyte Tris-HCl in the Ipa of the HQ. Different concentrations of Tris-HCl in pH 7.0 and [HQ] = 10^{-3} molL^{-1}.

2.10. Analytical Characteristics of MWCNT/Hemin/Mineral Oil Electrode Sensor

Under optimized conditions, the proposed method presented a calibration curve for the HQ concentration in the range of 2.5 - 120 μmol L^{-1} (Figure 18). The following parameters were found: detection limit of 0.78 μmol L^{-1} (S/N = 3), quantification limit of 2.60 μmol L^{-1} (S/N = 10), and sensitivity of 0.14507 μAμmol L^{-1}. The electrode developed herein showed a satisfactory detection limit compared to electrodes developed by other researchers (Table 3). Besides, the reduced preparation time and cost for the present sensor could be also marked as advantages. The precision of sensor in terms of repeatability (n= 10) assessed as relative standard deviation - RSD) was found to be 1.33 and 1.08% for the respective concentration of HQ, 3.0 and 110.0 μmol L^{-1}. The reproducibility on the sensor preparation was also verified. For this task, sensors, prepared in two different days, were submitted to cyclic voltammetry for HQ determination.

Table 3. Comparison between the electrode developed in the present work and other electrodes

Electrode	Technique	DL (μ mol L^{-1})	Reference
GC/Poly(thionine)	DPV	0.03	Ahmmad et al
GS/MWCNT	DPV	0.75	Qi and Zhang
Paraffin/graphite (biosensor)	SWV	8.1	Vieira and Fatibello-Filho
Chitosan (biosensor)	SWV	2.6	Oliveira and Vieira
Chitosan (biosensor)	SWV	2.0	Oliveira and Vieira
MWCNT/Hemin/Mineral Oil	SWV	0.78	Present work

GC - glassy carbon, GS - graphite sleeve, MWCNT - multi-walled carbon nanotube, DL - detection limit.

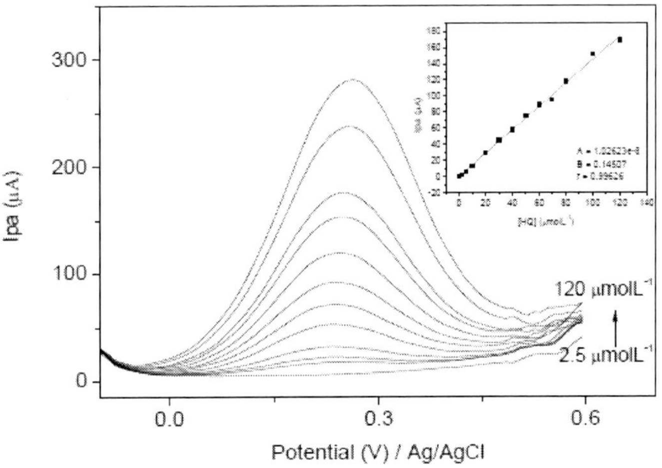

Figure 18. SWV plots for different HQ concentrations (2.5 - 120 µmolL^{-1}). [Tris-HCl] = 0.15 molL^{-1} (pH 7.0), amplitude = 70 mV, frequency = 70 Hz, step potential = 0.01 V. The analytical curve is shown inset this figure.

Figure 19 demonstrates the very similar profile of voltammetric behavior of HQ using the MWCNTHemin/Mineral Oil sensor prepared in different days.

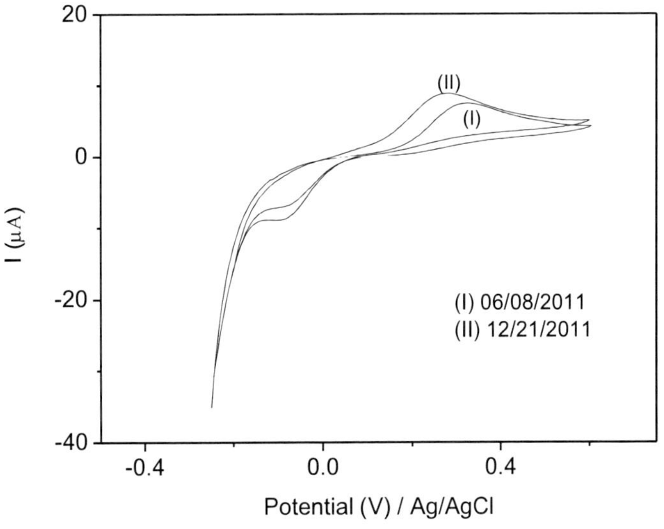

Figure 19. Cyclic voltammograms recorded for the MWCNT/Hemin/Mineral Oil (70:20:10) paste prepared on 06/08/2011 and 12/21/2011. [Tris-HCl] = 0.05 mol L^{-1} (pH = 7.0) and scan rate of 50 mVs^{-1}.

2.11. Determination of Other Phenolic Compounds by MWCNT/Hemin/Mineral Oil Sensor

The individual determination of other phenolic compounds by SWV using the MWCNT/Hemin/Mineral Oil sensor was verified under the best conditions obtained. The peak current of each phenolic compounds at 1.0 x 10^{-3} mol L^{-1} concentration was compared with anodic current signal for hydroquinone, at the same concentration.

Table 4. Interfering compounds tested in the potential range of -0.2 to 0.7V and their respective signals

Interfering compound	Epa (V)	Ipa signal (%)
Hydroquinone (HQ)	0.25	100.00
2-nitrophenol	-	-
4-nitrophenol	-	-
Guaiacol	0.66	7.23
N-acetyl-p-aminophenol	0.50	116.90
Catechol	0.36	102.17
Uric acid	0.43	87.66
Ascorbic acid	0.40	34.50
Carbamide (urea)	-	-

Table 5. HQ determination in pharmaceutical formulations using the chromatographic and proposed voltammetric methods

Sample	Voltammetry (present sensor) (g/100g HQ)	HPLC (g/100g HQ)	Relative error 1 (Er_1) (%)	Relative error 2 (Er_2) (%)
I (2g/100g)	1.92±0.07	1.98±0.11	-3.03	-4.00
II (4g/100g)	4.04±0.03	4.03±0.11	0.25	0.75
III (4g/100g)	4.03±0.02	4.18±0.29	-1.22	4.50
IV (4g/100g)	3.81±0.18	3.82±0.04	-0.25	-4.50
V (4g/100g)	4.07±0.07	4.05±0.04	0.49	1.25

Mean follows of standard deviation, 95% level confidence level (paired t test) N = 3, .
Er_1 = voltammetry (present sensor) *vs.* chromatography; Er_2 = voltammetry (present sensor) *vs.* labeled value.

As shown in Table 4, the potential of catechol was the closest to that observed for HQ. Studies with both hydroquinone and each interfering showed that only is possible to separate the peaks of uric acid and N-acetyl-p-aminophenol. On the other hand, it was not possible separate the peaks of guaiacol, cathecol and ascorbic acid. However, the electrode showed advantageous responses to the other phenolic compounds, apart from HQ, in the same potential range (-0.2 to 0.7 V). Among the compound tested, urea, 2-nitrophenol and 4-nitrophenol did not present oxidation potentials in this range of studied potential, thus not interfering with the HQ measurements.

2.12. Application of Proposed Method and Validation

The proposed MWCNT/Hemin/Mineral Oil sensor was employed for determining HQ in five cosmetics creams using the voltammetric method. Samples I - III presented commercial brands, whereas Samples IV and V were acquired at manipulation pharmacies in the city of Londrina (Brazil). 500 mg of cream samples (II to V) containing 4g/100g of HQ was weighed in a beaker of 100 mL capacity, followed by addition of 90 mL of ultrapurified water and stirring for 30 min at room temperature. Next, the mixture was transferred to a 100 mL volumetric flask and the volume was made up with ultrapurified water. From this solution, 200 μL was transferred without filtration into the electrochemical cell containing 9.80 mL of 0.15 mol L^{-1} Tris-HCl. For sample I containing 2g/100g of HQ, the preparation was similar to that mentioned above except by using 1000 mg of cream samples. The obtained results by proposed method were compared with HPLC method, as a reference technique (Garcia et al 2005). The cream samples were solubilized with a mixture of MeOH/H_2O (20:80, v/v), the same mobile phase of HPLC, and filtered three times before analyzing. The analytical curve of chromatographic determination was ranged from 6.0 up to 30 μg mL^{-1} (r=0.9935). As one can see from Table 5, the amounts of HQ determined by sensor in real samples were very similar to that labeled and there is no difference at 95% confidence level (paired t test) when compared with HPLC method. The main advantage of the proposed method is that it allows to direct determination of HQ in the aqueous phase, without pre-filtration, after solubilizing the cream samples in water.

CONCLUSION

The sensor based on the MWCNT/Hemin/Mineral Oil electrode has been proven to be more effective for the HQ determination than those based on the GC and non-functionalized MWCNT electrodes. The best anodic current response was obtained for the 70 % MWCNT : 20 % Hemin : 10 % Mineral Oil proportion, using Tris-HCl as buffer (pH 7.0). The scan rate study indicated that the HQ oxidation process was controlled by diffusion. The SWV experiments showed that the amplitude of 70mV, frequency of 70 Hz and step potential of 0.01 V were found to be the best conditions. The analytic curve for HQ was linear in the range of 2.5 - 120 $\mu molL^{-1}$ (r = 0.9962), and relative standard deviation were 1.33% and 1.08% for the solutions containing 3.0 $\mu molL^{-1}$ and 110.0 $\mu molL^{-1}$, respectively (n = 10). The sensor developed herein presented some advantages, such as low cost, reduced paste preparation time, paste composition reproducibility, low limits of quantification and detection and catalytic behavior towards HQ oxidation. The proposed method can be used to detect others phenolic compounds and also be applied to pharmaceuticals formulations, since it presents no significant differences at the 95% confidence level compared to the chromatographic method.

ACKNOWLEDGMENTS

The authors would like to thank the CNPq, Fundação Araucária do Paraná, CAPES, FAPEMIG and INCTBio for their financial support and fellowships.

REFERENCES

Ahammad, A. J. S., Rahman, M. M., Xu, G. R., Kim, S. and Lee, J. J.; *Electrochim. Acta*, 2011, 56, 5266.

Bard, A. J., Faulkner, L. R.; *Electrochemical Methods Fundamentals and Application*, 2nd e., Wiley, New York, 2001.

Charlier, J. C.; *Acc. Chem. Res.*, 2002, 35, 1063.

Chen, J., Wollenberg, U., Lisdat, F., Ge, B. and Scheller, F. W.; *Sens. Actuators B*, 2000, 115.

Dai, H.; *Acc. Chem. Res.*, 2002, 35, 1035.

Davis, J. J., Coleman, K. S., Azamian, B. R., Bagshaw, C. B. and Green, M. L. H.; *Chem. Eur. J.*, 2004, 9, 3732.

Furlanetto, S., Orlandini, S., Aldini, G., Gotti, R., Dreassi, E. and Pinzauti, S.; *Anal. Chem. Acta*, 2000, 413, 229.

Gong, K., Yan, Y., Zhang, M., Su, L., Xiong, S. and Mao, L.; *Anal. Sci.,* 2005, 21, 1383.

Jain, R. and Rather, J.A.; *Colloids Surf. Biointerfaces,* 2011, 83, 340.

Kang, T. F., Shen, G. L. and Yu, R. Q.; *Anal. Chim. Acta,* 1997, 354, 343.

Liu, X., Ding, Z., He, Y., Xue, Z., Zhao, X. and Lu, X.; *Colloids Surf. B: Biointerfaces*, 2010, 79, 27.

Liu, Z., Wang, Z., Cao, Y., Jing, Y. and Liu, Y.; *Sens. and Actuators B*, 2011, 157, 540.

Ma, Q., Ai, S., Yin, H., Chen, Q., Tang, T.; *Electrochim. Acta*, 2010, 55, 6687.

Mimica, D., Zagal, J. H. and Bedioui, F.; *J. Electroanal. Chem.*, 2001, 497, 106.

Mimica, D., Zagal, J. H. and Bedioui, F.; *Electrochem. Commun.*, 2001, 3, 435.

Nagaraja, P., Vasantha, R. A. and Sunitha, K. R.; *J. Pharm. Biomed. Anal.*, 2001, 25, 417.

Nan, C. G., Fena, Z. Z., Li, W. X., Ping, D. J. and Qin, C. H.; *Anal. Chim. Acta*, 2002, 452, 245.

Oliveira, I. R. W. Z., Vieira, I.C.; *Quim. Nova,* 2006, 29, 932.

Oliveira, I. R. W. Z., Vieira, I.C.; *Enzy. Microb. Technol.,* 2006, 38, 449.

Qi, H. and Zhang, C.; *Electroanalysis*, 2005, 17, 10.

Rosatto, S. S., Kubota, L. T. and Neto, G. O.; *Anal. Chim. Acta,* 1999, 390, 65.

Rubianes, M. D. and Rivas, G. A.; *Electrochem. Commun.*, 2003, 5, 689.

Santos, A. S., Pereira, A. C., Sotomayor, M. D. P. T., Tarley, C. R. T., Duran, N. and Kubota, L. T.; *Electroanalysis*, 2007, 19, 549.

Sun, Y. G., Cui, H. H., Li, Y. H. and Lin, X. Q.; *Talanta*, 2000, 53, 661.

Teófilo, R. F., Reis, E. L., Reis, C., Da Silva, G. A. and Kubota, L. T.; *J. Braz. Chem. Soc.*, 2004, 15, 865.

Timur, S., Pazarlioglu, N., Pilloton, R. and Telefoneu, A.; *Talanta*, 2003, 61, 87.

Turdean, G. L., Popescu, I. C., Curulli, A. and Palleschi, G.; *Electrochim. Acta*, 2006, 51, 6435.

Valentini, F., Amine, A., Orlanducci, S., Terranova, M. L. and Palleschi, G.; *Anal. Chem.*, 2003, 75, 5413.

Vieira, I. C. and Fatibello-Filho, O.; *Talanta*, 2000, 52, 681.

Xie, T., Liu, Q., Shi, Y. and Liu, Q.; *J. Chromatography A*, 2006, 1109, 317.

Wang, J. and Musameh, M.; *Anal. Chem.*, 2003, 75, 2075.

Wang, L. H. and Kuo, Y. P.; *Chromatography*, 1999, 49, 208.

Zhang, Y. L., Zhang, C. X. and Shen, H. X.; *Electroanalysis*, 2001, 13, 1431.

Zhao, Y., Zhang, W., Chem, H., Luo, Q. and Li, S.S. F. Y.; *Sens. Actuators B*, 2002, 87, 168.

Zhao, G., Li, M., Hu, Z., Li, H. and Cao, T.; *J. Mol. Catal. A*, 2006, 255, 86.

Zheng, N., Zeng, Y., Osborne, P. G., Li, Y., Chang, W. and Wang, Z.; *J. Appl. Electrochem.*, 2002, 32, 129.

In: Focus on Biomimetics Research
Editor: Abele Scipione
ISBN: 978-1-62417-332-5

Chapter 6

TRAJECTORY FORMATION OF BIOMIMETIC ROBOT ARMS WITH REDUNDANCY

Tadashi Kashima
Tomakomai National College of Technology, Japan

ABSTRACT

It is an important issue for the design of biomimetic robots to consider not only that their appearance should resemble a human arm, but also that the arm should move along human-like trajectories. The human arm trajectories in reaching movements have been widely investigated, and it is known that the hand paths lie along either a roughly straight or a simple curved line, and the tangential velocity profile of the hand is always bell-shaped. One of the most effective methods for controlling biomimetic robots along such trajectories is to apply human arm trajectory formations. However, most of trajectory formations proposed to date are limited to either planar motion or the arm model within three degrees of freedom. Accordingly, it seems to be difficult to apply existing trajectory formations to the human arm model with redundancy since the human arm consists of a shoulder with three degrees of freedom and an elbow with one degree of freedom even when the motion of a wrist is frozen.

This study investigates the trajectory formation of biomimetic robot arms with redundancy. Here, a redundant arm model with three joints at the shoulder and one joint at the elbow movable in a three-dimensional space is considered. The basic trajectories of the shoulder and elbow joints are first determined by the minimum jerk criterion in joint angle space. Then, an idea for a time adjustment of the joint motion of the elbow relative to the shoulder has been brought. This time adjustment supposed at the beginning or the end of movement can be numerically determined so as to demonstrate the property of human arm trajectories. Also, a systematical scheme for formulating the required trajectory without complex analytical calculation has been developed. Consequently, the hand trajectory derived from the joint angle trajectories can be uniquely produced once the initial and target states of the arm and the movement duration are given. The trajectories produced under the proposed trajectory formation are evaluated by the trajectories observed by experiments. The results show that the proposed trajectory formation is an advantageous scheme for biomimetic robots to control along human-like trajectories.

INTRODUCTION

One of the important issues in the design of biomimetic robots is to provide a control capability to enable them to move like a human arm. The human arm is controlled smoothly in reaching movements, and the trajectories during the movement have specific properties. To be more specific, human hand paths describe either a straight or a simply curved line according to movement conditions, and the tangential velocity profile of the hand is always bell-shaped [1]-[4]. In addition, the hand path and velocity profile have altered little even if the movement duration or the hand held load is changed [3],[5]. An effective method for controlling biomimetic robots along such trajectories is to apply the human arm trajectory formation. In order to understand how such human arm trajectories are planned in the central nervous system, many studies have been undertaken and feasible trajectory formations have been proposed. These studies have been conducted mainly in dynamic and kinematic coordinates. Trajectory formations in dynamic coordinates have been investigated by generally defining performance criteria characterizing human movements such as torque change [6], effort [7], commanded torque change [8] and acceleration with constraints [9]. More complex trajectory formations by adding the characteristics of muscles have also been investigated [10]-[12]. The optimal trajectories minimizing these criteria are in good agreement with human arm trajectories if dynamic parameters such as the movement duration and the mass of each part of the arm are properly set up. In fact, the optimal trajectories are sensitively influenced by the change in the dynamic parameters, even if the two hand positions in a two-point reaching movement are kept constant. Thus, it seems to be difficult for the trajectory formations in dynamic coordinates to demonstrate the invariant properties to change of dynamic parameters. Moreover, the numerical calculation for searching the optimal trajectory becomes extremely difficult even under specific movement conditions in the two dimensional space. Consequently, such criteria result in difficult problems for the trajectory formation of biomimetic robots.

In contrast, it is possible to formulate human-like trajectories which satisfy the invariant properties if we assume the trajectory formation in kinematic coordinates. The trajectory formations in kinematic coordinates can be mainly considered either in Cartesian space or in joint angle space. As regards the trajectory formation in Cartesian coordinates, the typical performance criteria are defined by a derivative of the hand position, including a jerk [2, 13]. The trajectory produced consistently shows a straight hand path. However, curved hand paths are occasionally observed in human arm movements. Besides, it is difficult to apply these trajectory formations to a redundant arm model since the hand position must be represented only in three dimensions. Such curved paths can be produced by supposing a linear relationship between the shoulder and elbow angles in joint angle space. One of the trajectory formations which promise a linear relationship between joint angles is to define the performance criterion by the minimum angular jerk. The hand trajectories produced under the pure minimum jerk of joint angles do not necessarily demonstrate the human arm trajectories [8], [14, 15]. However, most human hand paths in a sagittal plane can be duplicated, when the linear relationship between the joint angles are firstly set up and a specific time difference of the joint motion of the elbow relative the shoulder has been adequately provided [16]. A hand trajectory cannot be produced unless a human trajectory, which can be duplicated, is given. In spite of this problem, the trajectory formation defined by joint angles possesses a rare feature in its ability to apply to the human arm model with any degree of freedom according to the number of the joint angles. Thus, the trajectory formation specified in joint angle space has potential ability to succeed the trajectory formation of biomimetic robot arms with redundancy.

There are few trajectory formations applicable to three dimensional arm movements. In this study, the trajectory formation in kinematic coordinates has been investigated for biomimetic robots with four degrees of freedom movable in a three dimensional space. In the beginning, the minimum jerk trajectories for all joint angles are analytically produced. Then, the angular trajectory of the elbow joint must be modified by a time difference of the joint motion of the elbow relative to the shoulder in order to reach the maximum hand velocity at the midpoint of the movement. Here, the condition for deciding the time difference cannot be given by an equation which is described in our previous literature [17] since it becomes extremely complex for the biomimetic robot arm with redundancy. Accordingly, a systematical scheme based on the Denavit-Hartenberg method has been developed for numerically determining the time difference. As a result, the trajectory for demonstrating human arm movements can be uniquely produced once a pair of the initial and target states of an arm with the movement duration are given.

Experiments have been conducted for movements in a sagittal plane and a three dimensional space. When the movement is limited to a plane motion, the dynamic behavior of the angles of the shoulder and elbow joints can be measured with high accuracy and high resolution by gyro sensors. In experiments in a three dimensional space, only the path and the velocity of the hand are analyzed by use of visual information from video cameras because the angular velocities around each axis for the redundant arm model cannot be independently measured by gyro sensors especially for a sphere joint at this moment. Finally, the trajectories produced under the proposed trajectory formation are evaluated by the trajectories observed in experiments not only in a sagittal plane but also in a three dimensional space. According to the results, the advantages of this trajectory formation for biomimetic robot arms with redundancy have been discussed.

TRAJECTORY FORMATION

Most of trajectory formations proposed until now cannot be applied to a redundant arm model which is allowed to move in a three dimensional space. On the other hand, the trajectory formation based on joint angles in kinematic coordinates has an important advantage in its ability to apply any degree of freedom to an arm model, according to the number of the joint angles. The model of biomimetic robot arms with redundancy is shown in Figure 1, and this model consists of a shoulder with three degrees of freedom and an elbow with one degree of freedom. Here, the local coordinate systems are defined according to the Denavit-Hartenberg method, and the hand position can be systematically determined as the origin with reference to the end coordinate frame $O_4 - x_4 y_4 z_4$. Besides, the link parameters for the arm model are listed in Table 1. The invariant property to change the hand-held load or movement duration can be performed by assuming the trajectory formation designed in kinematic coordinates. In this study, the minimum angular jerk criterion is assumed in order to reasonably characterize the human arm trajectories. Then, the performance criterion becomes

$$J = \frac{1}{2}\int_0^{t_f} \sum_{i=1}^{4} \dddot{\theta}_i(t)^2 \, dt \ , \tag{1}$$

where $\theta_i(t), i = 1, \cdots, 4$ are the shoulder and elbow angles as shown in Figure 1. Besides, t_f is the movement duration. Here, a set of necessary conditions for J to have a satisfactory value becomes

$$\theta_i^{(6)} = 0, \quad i = 1, \cdots, 4 \ . \tag{2}$$

In reaching movements, the initial conditions of a movement can be given as

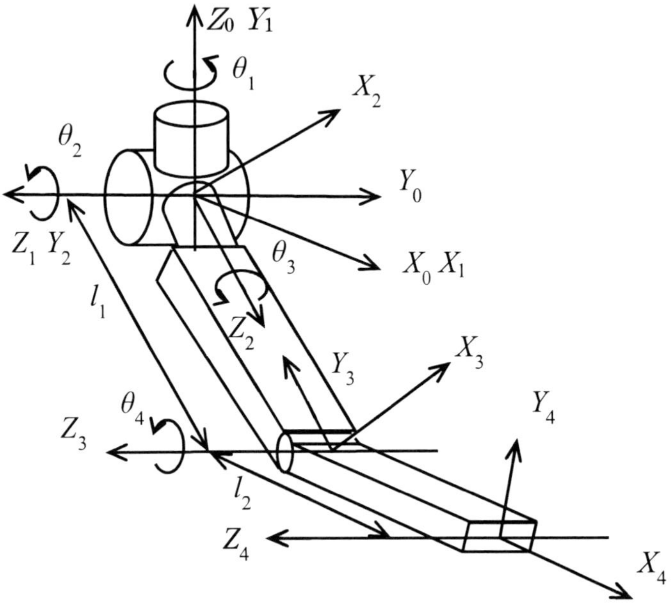

Figure 1. Arm model with four revolute joints.

Table 1. Link parameters for the arm model

Link number	α_i	a_i	d_i	θ_i
1	90°	0	0	θ_1
2	90°	0	0	θ_2
3	-90°	0	l_1	θ_3
4	0°	l_2	0	θ_4

$$\begin{cases} \theta_i(0) = \theta_{is}, \dot{\theta}_i(0) = 0, \ddot{\theta}_i(0) = 0 \\ \theta_i(t_f) = \theta_{if}, \dot{\theta}_i(t_f) = 0, \ddot{\theta}_i(t_f) = 0 \end{cases} ,i = 1,\cdots,4 , \tag{3}$$

where θ_{is} and θ_{if} represent the joint angles at the beginning and the end of the movement. The angular trajectory to satisfy with Equations (2) and (3) can be independently calculated as

$$\theta_i(s) = \theta_{is} + (\theta_{is} - \theta_{if})(15s^4 - 6s^5 - 10s^3), \quad i = 1,\cdots,4 , \tag{4}$$

where $s = t/t_f$ is the normalized time. In the study for human arm movements in the planar motion, the hand trajectories under the minimum angular jerk for the shoulder and elbow joints without any constraint does not demonstrate human arm trajectories [8].

Especially, the hand velocity of the produced trajectory does not necessary show a bell-shaped profile with the maximum value in the vicinity of the halfway point at the midpoint of movement, which is observed in human arm trajectories. However, most of hand paths can be duplicated by providing an adequate time difference between the elbow and shoulder joint motions which are defined under a linear relation of angular velocities [16]. Here, the hand trajectory cannot be produced without an observed trajectory to be duplicated, but this idea brings in possibilities that any human arm trajectory can be described under an appropriate time difference between the elbow and shoulder joint motions. In this study, a time adjustment scheme of the motion of the elbow joint is newly developed so as to demonstrate human arm trajectories. The first case is when the onset of the motion of the elbow joint relative to the shoulder is delayed, and we also set the second case where the motion of the elbow joint stops before the motion of the shoulder joint. In the first case, we suppose that the motion of the elbow joint was delayed by time *d* with respect to normalized time *s*. Now, the angle of the elbow joint θ_4 in Equation (4) can be replaced by

$$\theta_4(s) = \theta_{4s} + (\theta_{4s} - \theta_{4f})(15p^4 - 6p^5 - 10p^3) \tag{5}$$

where

$$p = \begin{cases} 0, & 0 \le s < d \\ \dfrac{s-d}{1-d}, & d \le s \le 1 \end{cases} . \tag{6}$$

In the second case, when the elbow joint stops at a time *d* before the shoulder joint stops, *p* in Equation (5) becomes

$$P = \begin{cases} \dfrac{s}{1-d}, & 0 \le s < 1-d \\ 1, & 1-d \le s \le 1 \end{cases} . \tag{7}$$

As a result, the joint angles $\theta_i\ ,i=1,2,3$ are the function of the normalized time *s*, and the elbow joint angle θ_4 is the function of *s* and *d*. Here, the hand position can be systematically represented by the joint angles according to kinematics. In the local coordinate systems in Figure 1, a matrix representing the location frame *i* relative to *i*-1 can be determined by

$$A_i^{i-1} = \begin{bmatrix} \cos\theta_i & -\sin\theta_i \cos\alpha_i & \sin\theta_i \sin\alpha_i & a_i \cos\theta_i \\ \sin\theta_i & \cos\theta_i \cos\alpha_i & -\cos\theta_i \sin\alpha_i & a_i \sin\theta_i \\ 0 & \sin\alpha_i & \cos\alpha_i & d_i \\ 0 & 0 & 0 & 1 \end{bmatrix}, \tag{8}$$

where α_i, a_i and d_i are the link parameters defined in Table 1, and θ_i is the joint angle. By use of these matrices, the relationship between the hand position $\boldsymbol{p}_0$ represented with reference to the base coordinate frame $O_0 - x_0 y_0 z_0$ and the origin $\boldsymbol{p}_4$ represented with reference to the end coordinate frame $O_4 - x_4 y_4 z_4$ can be expressed by

$$\boldsymbol{p}_0 = A_1^0(\theta_1)A_2^1(\theta_2)A_3^2(\theta_3)A_4^3(\theta_4)\boldsymbol{p}_4\ , \tag{9}$$

where

$$\boldsymbol{p}_0 = \begin{bmatrix} x & y & z & 1 \end{bmatrix}^T \text{ and } \boldsymbol{p}_4 = \begin{bmatrix} 0 & 0 & 0 & 1 \end{bmatrix}^T.$$

Here, the derivative of the hand position can be described by

$$\frac{d\boldsymbol{p}_0}{ds} = \left(\frac{dA_1^0(\theta_1)}{ds} A_2^1(\theta_2)A_3^2(\theta_3)A_4^3(\theta_4) + A_1^0(\theta_1)\frac{dA_2^1(\theta_2)}{ds} A_3^2(\theta_3)A_4^3(\theta_4) + A_1^0(\theta_1)A_2^1(\theta_2)\frac{dA_3^2(\theta_3)}{ds} A_4^3(\theta_4) + A_1^0(\theta_1)A_2^1(\theta_2)A_3^2(\theta_3)\frac{dA_4^3(\theta_4)}{ds}\right)\boldsymbol{p}_4 \tag{10}$$

Furthermore, the second derivative can also be analytically given by differentiating Equation (10) with respect to *s*. Thus, the components of the hand velocity $\dot{x}$, $\dot{y}$ and $\dot{z}$ are derived from Equation (10), and those of the hand acceleration $\ddot{x}$, $\ddot{y}$ and $\ddot{z}$ are derived from the derivative of Equation (10). As the acceleration of the hand in a three dimensional space is

$$\frac{dv}{ds} = \frac{d(\sqrt{\dot{x}^2 + \dot{y}^2 + \dot{z}^2})}{ds} = \frac{\dot{x}\ddot{x} + \dot{y}\ddot{y} + \dot{z}\ddot{z}}{\sqrt{\dot{x}^2 + \dot{y}^2 + \dot{z}^2}}, \tag{11}$$

the condition that the maximum velocity exists at the midpoint of the movement becomes

$$[\dot{x}\ddot{x}+\dot{y}\ddot{y}+\dot{z}\ddot{z}]_{s=0.5}=0\ , \tag{12}$$

where

$$\dot{x}\ddot{x}+\dot{y}\ddot{y}+\dot{z}\ddot{z}=\frac{d\boldsymbol{p}_0}{ds}^T\cdot\frac{d}{ds}(\frac{d\boldsymbol{p}_0}{ds})\ .$$

Equation (12) is the function of the generalized time s which is set as s=0.5 and the time difference d. In this study, the time d satisfying Equation (12) is numerically searched for by the Newton-Raphson method. Consequently, the arm trajectory of biomimetic robots can be uniquely produced once the initial and target states of the arm and the movement duration are given.

RESULTS AND DISCUSSION

Experiments which consist of two categories have been conducted to quantitatively evaluate the proposed trajectory formation. In the first category, a subject moves his arm in a sagittal plane, and the trajectories of the upper arm and the forearm are directly measured by use of two gyro sensors fixed on the upper arm and the forearm of the subject. Then, the trajectories are produced and evaluated under the movement conditions observed by the experiments.

In the second part, the human arm trajectories in a three dimensional space has been analyzed by use of still images from two video cameras since the joint angles of the sphere joint of the shoulder cannot be independently measured at this moment. The produced hand trajectories in Cartesian space are mainly evaluated by means of the observed trajectories.

A. Trajectories in a Sagittal Plane

The experiments under four conditions to cover the basic movements have been conducted. Here, the trajectories are analyzed on the basis of signals from gyro sensors since angular velocities are directly measured and the resolution is high. The details of the experimental procedure, including the signal processing, are described in the literature [17]. Subsequently, not only the angles of the shoulder and the elbow joints at start and stop but also the movement duration are identified from the experimental data. These data are utilized as movement conditions in order to produce the trajectory according to the proposed trajectory formation. But, the angles of θ_1 and θ_3 in Figure 1 are fixed as 0 rad when the trajectory formation described in the previous chapter is applied. The trajectories under four movement conditions in forward movements are shown in Figure 2.

In this experiment, the trajectories are analyzed on the basis of signals from gyro sensors in order to precisely examine dynamic behavior of joint angles. Thus, the behavior in joint angle space is illustrated in the first three rows of each movement.

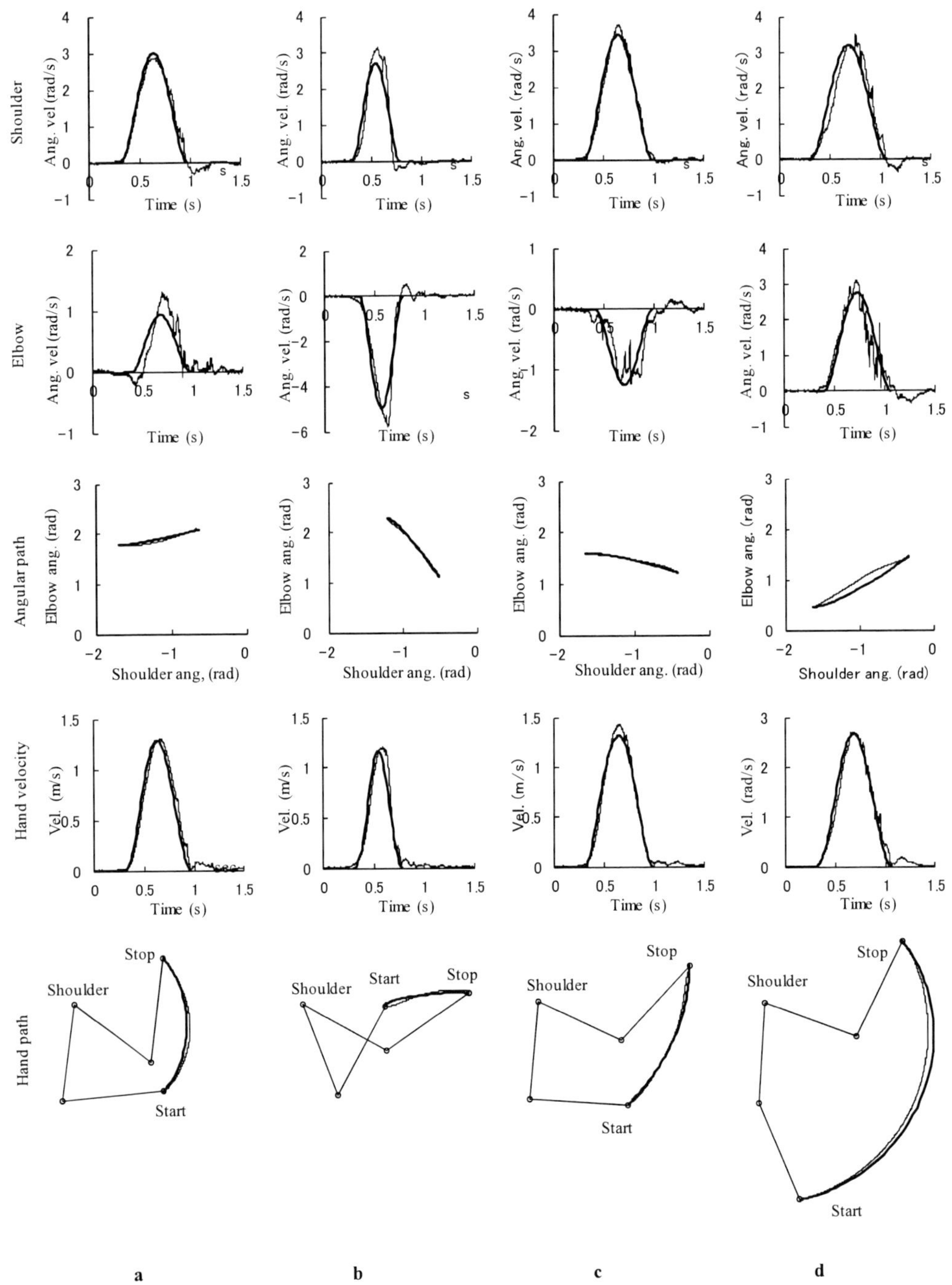

Figure 2. Trajectories in forward movements in a sagittal plane. Trajectories in forward movements given in experiments(thin line) and trajectory formations (heavy line). The first row and the *second row* show angular velocities of shoulder and elbow. The *third row* exhibits angular paths. The *forth row* and *the fifth row* represent hand velocities and hand paths. **a** to **d** represent trajectories under condition 1 to 4, respectively.

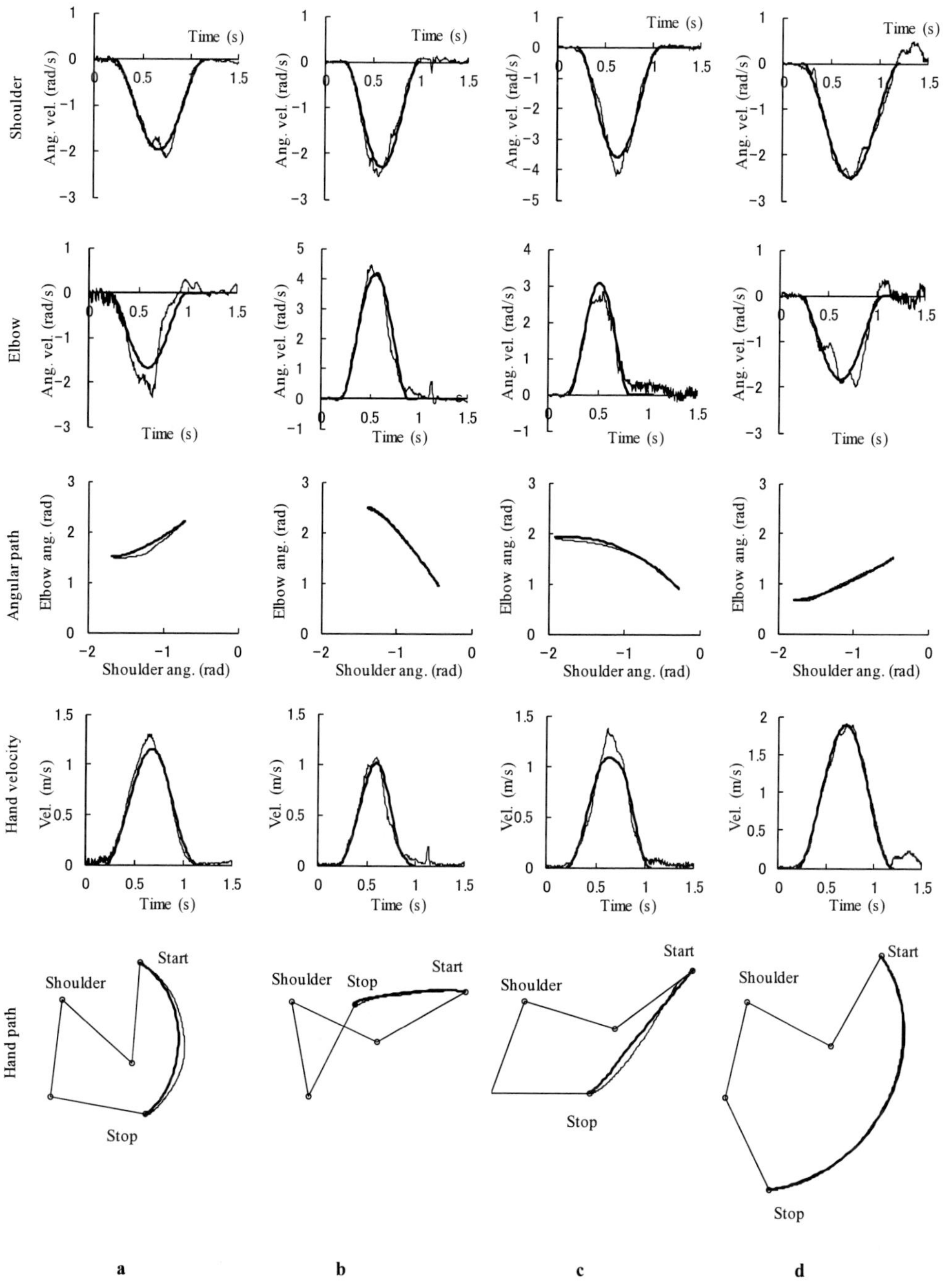

Figure 3. Trajectories in backward movements in a sagittal plane. Trajectories in backward movements given in experiments(thin line) and trajectory formations (heavy line). See Figure 2 legend for description.

Then, the hand velocity and the hand position at every sampling time are calculated by use of the angular displacement and the angular velocity, and they are shown in the last two rows. Then, the proposed trajectory formation is applied to each movement and the produced trajectories are also displayed. In this trajectory formation, it is necessary to set the time of start and stop of the movement. In addition to the results of the forward movements, those of the backward movements in which the initial and target hand positions are exchanged are shown in Figure 3. In this experiment, the subject is instructed to move a hand without concerning himself fine accuracy of the target point for performing natural movements. As a result, a pair of hand positions of the forward movement is not, in a precise sense, the same as a pair of positions of backward movement.

As regards the performance in joint angle space, the angular velocity of the observed trajectory is basically bell-shaped in the forward and backward movements. Besides, Figures 2d and 3d show that the angular velocities of the shoulder and elbow joints slightly overshoot near the termination time around the target value of 0 rad/s especially in the movements of condition 4. This phenomenon is called "zero velocity crossing" [18] and it notably appears in the movements when the high velocity is required. The angular velocity profiles of the produced trajectory well agree with those of the observed ones in all movements except this phenomenon of zero velocity crossing.

As can be seen from Figure 2, the elbow joint motion in most of the movements starts after the shoulder joint motion does, though the time difference cannot be clearly recognized in the movement of condition 2. On the contrary, Figure 3 exhibiting the trajectories of the backward movements show such an interesting feature that the shoulder and elbow joint motions simultaneously start, and the elbow joint motion stops before the shoulder joint motion does.

In the experiments, the initial hand position in the forward movement is adequately determined out of a pair of two hand positions. Accordingly, notation of either forward or backward does not make sense in itself. However, it is an important fact that behavior of the angular velocities of the elbow joint is mostly reversed between the forward movement and the backward movement for a pair of the hand positions. Thus, the time difference between the elbow and shoulder joint motions shows distinctive features of the human arm trajectories.

As can be seen from Figures 2 and 3, it is possible to recognize the time difference of the joint motion of the elbow relative to the shoulder though the angular velocities fluctuate around 0 rad/s with not only high frequency but also low frequency. So, the approximate value of time difference is measured and indicated in Table 2.

In addition to the analysis of the observed data, the time difference can be numerically obtained in the process of this trajectory formation. Hence, the time difference calculated in the trajectory formation is also listed in Table 2. In this Table, its index is quantitatively expressed by the ratio of the time difference to the movement duration and it changes from 0 % to 20 % in the forward movements. As the time difference indicates positive values, it turns out that a time delay of the joint motion onset of the elbow relative to the shoulder comes into existence. On the other hand, the time difference in all backward movements indicates negative value, and its ratio to the movement duration is more than 10 % in every movement condition. Consequently, the time difference between the motions of the elbow and shoulder joints basically shows an opposite phenomenon in the forward and backward movements. Also, the angular paths in the third raw of Figures 2 and 3 are slightly curved because of the time difference of the joint motion of the elbow relative to the shoulder. These results support the proposed trajectory formation which well reflects the property of the human arm trajectories. The hand trajectory in Cartesian space is one of the most important issues for the trajectory control of biomimetic robot arms. The hand trajectory can be characterized by a hand velocity profile and a hand path.

As can be seen from Figures 2 and 3, the hand velocity of the observed trajectories always show a bell-shaped profile with the maximum value in the vicinity of the midpoint of movement.

The produced hand velocity profiles nearly overlap the observed ones either in the forward movement or the backward movement. As for hand paths of the observed trajectories, they are curved in the movements of condition 1 and condition 4, and are almost straight in the movements of condition 2 and condition 3. The produced trajectories well characterize both kinds of properties Meanwhile, a slight difference between the produced and observed trajectories is recognized in the hand paths of condition 4 in the forward movement (Figure 2d) and those of conditions 1 and 3 in the backward movements (Figures 3a and 3c). However, judging from the experimental results observed for plural subjects [3],[10], this variation can be considered within the range of the individual difference. Consequently, it is reasonable to evaluate that the trajectories numerically produced under the proposed trajectory formation well demonstrates the human arm trajectories.

Table 2. Movement duration and time difference. The *third* column and the *fourth column* indicate movement duration(MD) and observed time difference(OTD) measured in experiments. Here, minus in OTD means the case that the joint motion of the elbow stops before that of the shoulder does. The fifth column shows the rate of absolute value of observed time difference to movement duration. The last column denotes the time difference of produced trajectory for reference

Kind of motion	Condition	Movement duration [s]	Observed time difference [s]	Ratio= |OTD|/MD	Simulated time difference [s]
Forward	1	0.66	0.16	0.24	0.12
	2	0.48	0.0	0.0	0.05
	3	0.68	0.12	0.18	0.11
	4	0.76	0.1	0.13	0.09
Backward	1	0.93	-0.21	0.23	-0.15
	2	0.78	-0.1	0.13	-0.09
	3	0.85	-0.2	0.24	-0.24
	4	1.0	-0.18	0.18	-0.14

B. Trajectories in a Three Dimensional Space

In addition to the investigation in a sagittal plane, the human arm trajectories are analyzed in a three dimensional space. As described in detail, the dynamic behavior of the joint angles during movement takes an important role in the proposed trajectory formation. However, gyro sensors, especially for a sphere joint, cannot independently measure the angular velocity around each axis for the redundant arm model.

This is the reason why the analysis is limited to fundamental movements, and only the hand path and the velocity profile are analyzed. In experiments, two video cameras are utilized, and the positions of the hand and the elbow in a three dimensional space are analyzed. Then, the hand velocity is approximately obtained by differentiating the polynomial of a fitted curve along a hand path with respect to time. As the arm model illustrated in Figure 1 has four degrees of freedom, it has redundancy for locating a hand position.

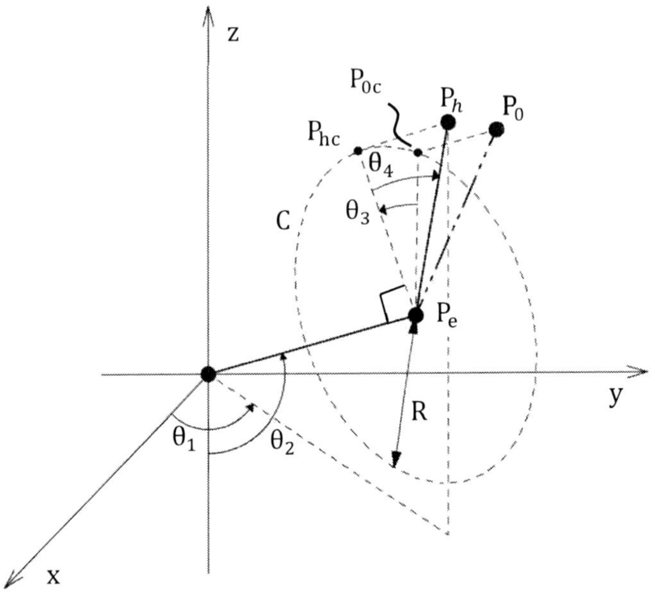

Figure 4. Angles of the arm model.

While the positions of the hand and the elbow corresponding to joint angles can be calculated by Equation (9), the inverse kinematics for the redundant arm model has to be developed in order to calculate every angle from an arm state. The angles in any state of the arm are shown in Figure 4. Here, $\boldsymbol{p}_h = (x_h, y_h, z_h)^T$ and $\boldsymbol{p}_e = (x_e, y_e, z_e)^T$ indicate the hand position and the elbow position, respectively. The shoulder angle around z axis can be simply given as

$$\theta_1 = \tan^{-1}(y_e / x_e) \,. \tag{13}$$

Next, the angle from the z axis becomes

$$\theta_2 = \cos^{-1}(-z_e / l_1) \,, \tag{14}$$

where l_1 is the length of the upper arm. Here, the elbow angle θ_4 is the angle when the shoulder, the elbow and the hand are on the one plane, and it can be described as

$$\theta_4 = \sin^{-1}(l_1^2 + l_2^2 - (x_h^2 + y_h^2 + z_h^2)/(2l_1 l_2)) \,, \tag{15}$$

where l_2 is the length of the forearm. As regard the angle θ_3, a circle C with radius R in Figure 4 is drawn on the plane which is normal to the upper arm and passes the elbow joint. Here, $\boldsymbol{p}_0$ is the position when θ_3 is 0 rad, and $\boldsymbol{p}_{0c}$ is the position which is perpendicularly projected to the circle C. Similarly, $\boldsymbol{p}_{hc}$ is the projected position of the hand position $\boldsymbol{p}_h$. Then, the distance between the positions $\boldsymbol{p}_{0o}$ and $\boldsymbol{p}_{hc}$ yields

$$|\boldsymbol{p}_{hc} - \boldsymbol{p}_{0c}|^2 = 2(l_2 \cos\theta_4)^2 - 2(l_2 \cos\theta_4)^2 \cos\theta_3 \ , \tag{16}$$

Consequently, the angle θ_3 becomes

$$\theta_3 = \cos^{-1}(1 - |\boldsymbol{p}_{hc} - \boldsymbol{p}_{h0}|^2 / (2l_2{}^2 \cos^2\theta_4)) \ , \tag{17}$$

where

$$\begin{aligned} &|\boldsymbol{p}_{hc} - \boldsymbol{p}_{0c}| = |\boldsymbol{p}_h - \boldsymbol{p}_0| \\ &= \sqrt{2l_2{}^2 - 2l_2(\cos(\theta_2 + \theta_4)((x_h - x_e)\cos\theta_1 + (y_h - y_e)\sin\theta_1) + (z_h - z_e)\sin(\theta_2 + \theta_4))} \end{aligned}$$

Thus, every angle can be calculated if the positions of the shoulder and the elbow are given. In this study, the positions at the beginning and the end of the movements are analyzed by the still images of the video camera. At the same time, the movement duration is given by counting the still images sampled at 60 Hz.

In this experiment, two movement conditions are prepared. Under the first condition, the hand travels a short distance and a straight hand path is expected. Conversely, a curved hand path is expected under the second condition with a long travel distance. The observed and the produced hand trajectories under two movement conditions are shown in Figures 5 and 6. As regards the hand paths of Figure 5, all paths of the observed trajectories are roughly straight as shown in Figures 5a, 5b and 5c.

The produced hand paths well demonstrate these human hand paths although the path on a y-z plane in Figure 5c is slightly different from the observed hand path. The hand velocity of the observed trajectory in Figure 5d shows a bell-shaped profile in the same way as the velocity profile in the planar motion.

The time to reach the peak of hand velocity in the produced trajectory is a little sifted to the left from that in the observed trajectory, but the maximum velocity of the observed and the produced hand velocities indicate almost the same value. With respect to the curved trajectory in Figure 6, the human hand path with a convex curve is well represented by the produced trajectory in a three dimensional space as shown in Figure 6a. This curved path is clearly exhibited in the path on a x-z plane in Figure 6b. Conversely, it is interesting that the hand path of the observed trajectory on a y-z plane in Figure 6c shows almost a straight line. However, a slight difference between the observed and the produced hand paths are recognized immediately after starting movement. The time to reach the maximum velocity is a little different between the observed and the produced trajectories, but the velocity profiles of both trajectories are bell-shaped as shown in Figure 6d. Thus, slight difference between the observed and the produced trajectories is recognized in the path on a y-z plane of front view for both typical human arm movements. However, it is possible to evaluate that the trajectories produced under the proposed trajectory formation generally demonstrate properties of human arm trajectories in a three dimensional space.

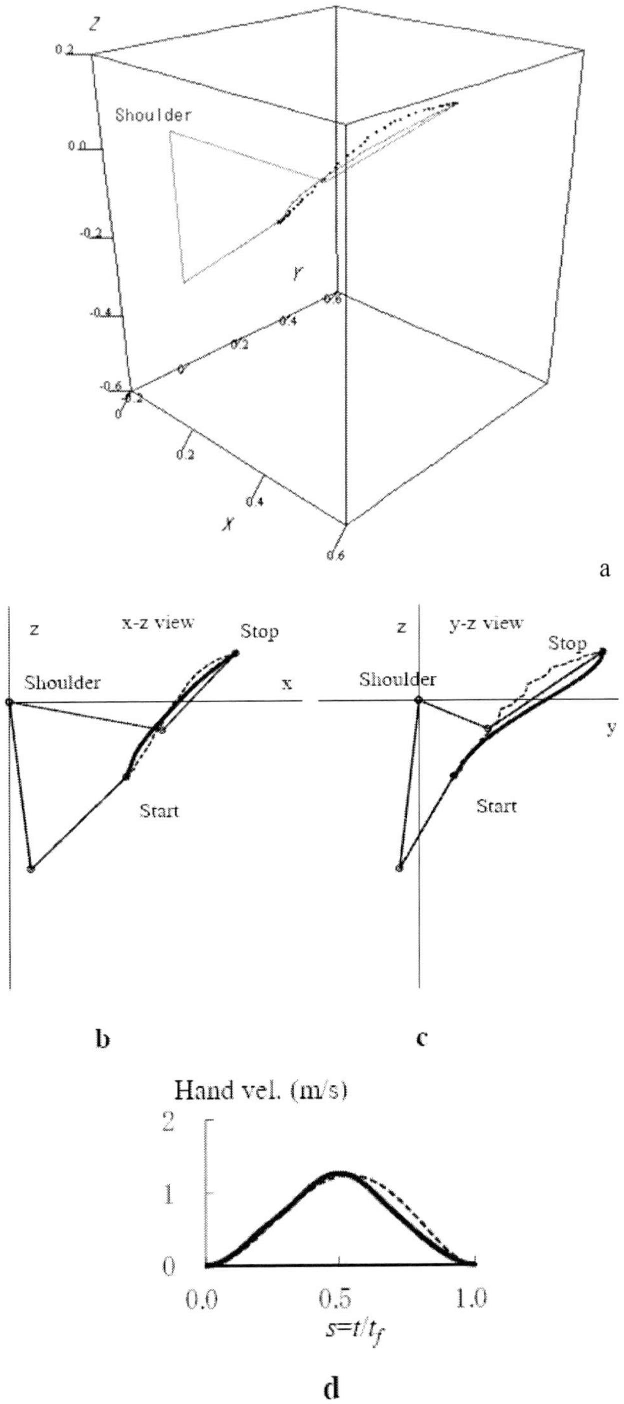

Figure 5. Trajectory under condition 1. Trajectories in experiments(broken line) and trajectory formations (heavy line).

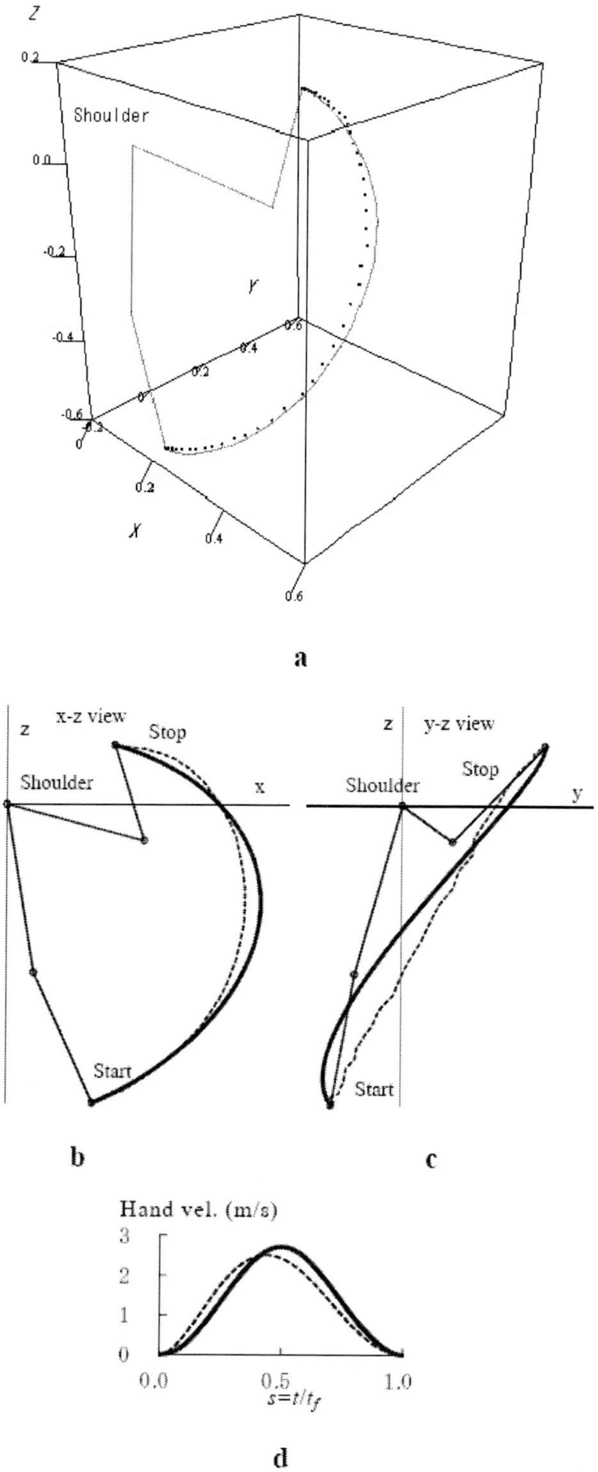

Figure 6. Trajectory under condition 2. Trajectories in experiments(broken line) and trajectory formations (heavy line).

Conclusion

In this study, the trajectory formation for controlling biomimetic robot arms along human-like trajectories has been investigated in kinematic coordinates. The trajectories produced have been compared with experimental results which are observed for the movements in both a sagittal plane and in a three dimensional space. The results in planar movements showed not only that the time difference between the motions of the elbow and shoulder joints are empirically identified but also that the produced hand trajectories are in good agreement with human trajectories.

With respect to the investigation in a three-dimensional space, a redundant arm model with a structure similar to a human arm has been considered, and the scheme of trajectory formation has been newly developed. The produced hand trajectories mostly agreed with the human arm trajectories, but a slight difference has been recognized for the hand paths in front view. However, this difference does not seem to be a crucial disadvantage for this proposed trajectory formation, because the produced trajectory well illustrated the property of human arm trajectories according to the movement conditions. Consequently, these results indicate that the proposed trajectory formation is effective for controlling biomimetic robots along human-like trajectories.

References

[1] Abend W, Bizzi E, Morasso P (1982). Human Arm Trajectory Formation. *Brain* 105, 331-348.
[2] Flash T, Hogan N (1985). The Coordination of Arm Movements: An Experimentally Confirmed Mathematical Model. *J. Neurosci* 5, 1688-1703.
[3] Atkeson C, Hollerbach J (1985). Kenematic Features of Unrestrained Vertical Arm Movements. *J. Neurosci* 5, 2318-2330.
[4] Hollerbach J (1990). Planning of Arm Movements. In *Visual Cognition and Action*, Osherson D, Kosslyn S, Hollerback J(eds.), MIT Press, 183-211.
[5] Goodman S, Gottlieb G (1995). Analysis of Kinematic Invariances of Multijoint Reaching Movement. *Biol. Cybern* 73, 311-322.
[6] Uno Y, Kawato M, Suzuki R (1989). Formation and Control of Optimal Trajectory in Human Multi-Joint Arm Movement. *Biol. Cybern.* 61, 89-101.
[7] Lan N (1997). Analysis of Optimal Control Model of Multi-Joint Arm Movements. *Biol. Cybern.* 76, 107-117.
[8] Nakano E, Imamizu H, Osu R, Uno T, Gomi H, Yoshioka T, Kawato M (1999). Quantitative Examinations of Internal Representations for Arm Trajectory Planning: Minimum Commanded Torque Change Model. *J. Neurophysiol.* 81, 2140-2155.
[9] Ben-Itzhak S, Karniel A (2008). Minimum Acceleration Criterion with Constraints Implies Bang-Bang Control as an Underlying Principle for Optimal Trajectories of Arm Reaching Movements. *Neural Computation* 20, 779–812.
[10] Alexander R (1997). A Minimum Energy Cost Hypothesis for Human Arm Trajectories. *Biol. Cybern.* 76, 97-105.

[11] Kashima T, Isurugi Y, Shima M (2000). Analysis of a muscular control system in human movements. *Bioi. Cybern* 82, 123-131.

[12] Kashima T, Isurugi Y, Shima M (2002). An optimal control model of a neuromuscular system in human arm movements and its control characteristics. *J. Artificial Life and Robotics* 6, 205-209.

[13] Richardson M, Flash T (2002). Comparing Smooth Arm Movements with the Two-Thirds Power Law and the Related Segmented-Control Hypothesis. *J. Neurosci.* 22, 8201-8211.

[14] Wada Y, Kaneko Y, Nakano E, Osu R, Kawato M (2001). Quantitative examination for multi joint arm trajectory planning - using a robust calculation algorithm of the minimum commanded torque change trajectory. *Neural Networks* 14, 381-393.

[15] Hermens F, Gielen S (2004). Posture-based or trajectory-based movement planning: a comparison of direct and indirect pointing movements. *Exp. Brain Res.* 159, 340-348.

[16] Hollerbach J, Atkeson C (1987). Deducing Planning Variables from Experimental Arm Trajectories: Pitfalls and Possibilities. *Biol. Cybern* 56, 279-292.

[17] Kashima T, Iwaseya M (2011). Trajectory control of biomimetic robots for demonstrating human arm movements. *Artif Life Robotics* 16, 338-342.

[18] Plamondon R, Feng C, Woch A (2003). A kinematic theory of rapid human movement, Part IV a formal mathematical proof and new insights. *Biol. Cybern* 89, 126-138.

[19] Kashima T, Isurugi Y (1998). Trajectory formation based on physiological characteristics of skeletal muscle. *Biol. Cybern* 78, 413-422.

INDEX

C

D

E

F

N

O

P

T

U

V

W

X

Y

Z

ratio
Veritas

CASSELL CARAVEL BOOKS

1. KNIGHTS OF THE CRUSADES
2. HEROES OF POLAR EXPLORATION
3. NELSON AND THE AGE OF FIGHTING SAIL
4. ALEXANDER THE GREAT
5. RUSSIA UNDER THE CZARS
6. CAPTAIN COOK AND THE SOUTH PACIFIC
7. JOAN OF ARC
8. EXPLORATION OF AFRICA
9. SHAKESPEARE'S ENGLAND
10. MARCO POLO'S ADVENTURES IN CHINA
11. THE VIKINGS
12. CAESAR
13. THE FRENCH REVOLUTION
14. CORTES AND THE AZTEC CONQUEST
15. THE UNIVERSE OF GALILEO AND NEWTON
16. LEONARDO DA VINCI
17. THE BATTLE OF WATERLOO
18. PHARAOHS OF EGYPT
19. MOUNTAIN CONQUEST
20. BUILDING THE SUEZ CANAL
21. THE SPANISH ARMADA
22. THE HOLY LAND IN THE TIME OF JESUS
23. RUSSIA IN REVOLUTION
24. DISCOVERERS OF THE NEW WORLD
25. CHARLEMAGNE
26. THE SEARCH FOR EARLY MAN
27. FERDINAND AND ISABELLA
28. CHARLES DARWIN AND THE ORIGIN OF SPECIES

A CASSELL CARAVEL BOOK

FERDINAND AND ISABELLA

By the Editors of
HORIZON MAGAZINE

Author
MELVEENA McKENDRICK

Consultant
J. H. ELLIOTT
Professor of History
King's College, University of London

Cassell · London

First published in Great Britain 1969
Printed and bound in Italy by Arnoldo Mondadori Editore
S.B.N. 304 93411 9

PRADO, MADRID: PHOTO, DOMINGUEZ RAMOS

In a painting done about 1490, St. Dominic (right, holding a lily) and St. Thomas Aquinas join Ferdinand and Isabella, their son, John, and daughter Isabella in prayer at the throne of the Virgin and Infant Jesus.

FOREWORD

Ferdinand and Isabella were an unlikely pair to bring greatness to Spain. Only a few years before their wedding in 1469, they even had been somewhat unlikely to rule their separate kingdoms. Between Isabella and the throne of Castile stood an older half-brother and his daughter; likewise, Ferdinand's older half-brother was next in line for the throne of neighboring Aragón. Few people would have guessed that the teen-age couple marrying in secret would become the renowned monarchs of a united Spain. To achieve this unprecedented unity and strength, Ferdinand and Isabella waged a successful civil war, initiated a devastating religious persecution, led a bitter campaign against the Moors, launched foreign wars, and forged intricate marriage alliances for their children with half of Europe.

Today the famous rulers are most often remembered for the epochal voyage of a Genoese adventurer in their service, Christopher Columbus. To them, however, the historic landfall of October, 1492, was only a secondary event of the year. The preceding January they had accepted the surrender of Moslem Granada, thus ending centuries of Islamic rule in their peninsula. And later that year they had ordered the expulsion or forced baptism of Spain's Jewish minority, a cruel and fanatical crusade undertaken in an excess of zeal for their Catholic faith.

Europe, in the century of Ferdinand and Isabella, was also awakening to the glories of a new age, the Renaissance; and the Spain of the Catholic Kings —as Ferdinand and Isabella came to be known—was not untouched by this brilliant revival of learning. Manuscript illuminations, paintings, sculpture, wood carvings, and architecture reproduced on the following pages reveal the inspired artistry of the era.

Some of Ferdinand and Isabella's successors would bring greater power to Spain, as the wealth from Columbus' "Land to the West" poured into the nation's coffers and as its brave sons built out of faith and brutality one of history's mightiest empires. But the memorable achievement of creating a nation out of rival kingdoms and of raising an entire people to new heights of grandeur remains that of Ferdinand and Isabella alone.

THE EDITORS

RIGHT: *The top of a scepter of state belonging to Isabella offers a splendid example of the Spanish silversmith's art in her century.*

GRANADA CATHEDRAL: PHOTO, ORONOZ

COVER: *At their supreme moment of triumph, Ferdinand and Isabella enter captured Granada, as shown on a 1522 polychrome bas-relief in the Cathedral's Royal Chapel.*

GRANADA CATHEDRAL: PHOTO, ORONOZ

FRONT ENDSHEET: *A fifteenth-century Flemish tapestry brought to Spain presents Christ as King of a brilliant Renaissance court.*

TARRAGONA CATHEDRAL: MAS

TITLE PAGE: *A splendidly caparisoned knight decorates a history of the Spanish Order of Santiago, a group headed by Ferdinand.*

The Order of St. Iago, ARCHIVO MUNICIPAL, BURGOS: MAS

BACK ENDSHEET: *Immortalized in wooden polychrome statues, the Catholic Kings kneel at their burial place in Granada Cathedral.*

GRANADA CATHEDRAL: PHOTO, ORONOZ

CONTENTS

I

EVELYN HOFER

DIVIDED SPAIN

One night at the beginning of October, 1469, a small party of merchants set out across the mountains above Saragossa in the Spanish kingdom of Aragón. Their destination was the town of Valladolid in the neighboring kingdom of Castile and León. Although their mules were loaded with boxes and bales, these men were no ordinary merchants. The shabbily dressed boy of seventeen who tended the mules was in fact their young master, Ferdinand, King of Sicily and heir to the crown of Aragón. In Valladolid he was to meet and marry the eighteen-year-old Infanta Isabella, heiress of Castile.

Ferdinand traveled in disguise and at night for two reasons: first, because Isabella's half-brother, King Henry IV of Castile, and most of his nobles were eager to prevent the wedding; and second, because Ferdinand's father, old King John of Aragón, could afford neither the men nor the money his son would have needed to enter hostile Castile in a manner becoming his rank.

The journey was a long and dangerous one. The merchant caravan, winding its way through the wild sierra country that divided Aragón and Old Castile, had to keep a constant lookout for the robber bands that infested the mountains. Once safely through, the travelers followed the curving river Duero across the empty, rock-strewn plains of the Castilian tableland. Whenever they stopped at some lonely inn, Ferdinand, to avoid suspicious glances from the innkeeper and fellow guests, waited on the other members of the caravan and fed and watered their mules.

At Burgo de Osma, a walled town perched high above the Duero, he was suddenly forced to reveal his identity. Some of Isabella's partisans challenged him from the battlements. Whatever reply he invented on the spur of the

This simple chapel, which crowns a rocky outcrop high in the tableland of central Castile, was already old when Ferdinand of Aragón crossed such barren hills as these on his way to marry Isabella of Castile in 1469.

BRAUN, *Civitates orbis terrarum*, 1576: NEW YORK PUBLIC LIBRARY

Sure-footed mules were the chief means of transportation over the winding, hilly roads of medieval Spain. In this detail from a sixteenth-century engraving of Jerez de la Frontera, in Andalusia, two muleteers drive a laden pack train toward the town. The man at right carries a huge wineskin that may be filled with the wine to which Jerez has given its name—sherry.

moment was obviously not satisfactory. One of the sentries hurled a large stone, which narrowly missed killing the future King of Spain. When Ferdinand revealed that he was their princess' bridegroom, however, the townsfolk quickly made up for their hostility, and the mayor himself escorted the Prince on the next stage of his journey, to Dueñas in León. Here Ferdinand was greeted by a band of nobles who were loyal to Isabella, and he could abandon his disguise at last.

Six days later, on October 15, he rode into Valladolid to meet his bride for the first time. Isabella was waiting for him in the private house where she had been staying. With her were the Archbishop of Toledo, a strong supporter of the proposed match, and her lifelong friend, a girl her own age named Beatrice de Bobadilla. Isabella had dressed very simply for the occasion and her only ornament was a magnificent gold and ruby necklace, which Ferdinand had sent her as a present. The two young people talked together for two hours, and then Ferdinand returned to Dueñas, eighteen miles away.

By arranging to marry Ferdinand, Isabella was deliberately defying her brother the King, who had other plans for her marriage, plans that dismayed her. She had risked a great deal by coming to Valladolid and asking Ferdinand to meet her there. But she was cautious by nature and had already insisted on writing to her brother, asking for his approval. The marriage, she now said, could not go ahead until she received his answer. However, the Archbishop, who was alarmed that an unfavorable reply might upset everything, managed to persuade her that the wedding should not be postponed. Isabella then raised another objection. Since she and Ferdinand were second cousins, they could not marry without a dispensation from the Pope. A letter requesting this would have to be sent to Rome.

By now the suspense was unbearable. Fortunately, King John, Ferdinand's wily father, had foreseen such an obstacle, and he and the Archbishop between them had forged a papal bull of dispensation that authorized Ferdinand to marry any female cousin, if he so wished. This was now shown to the unsuspecting Isabella, and the Archbishop proclaimed joyfully that the marriage could proceed.

On October 19 the couple were quietly married by the Archbishop of Toledo in the same house where Isabella had been staying. There was none of the pomp, pageantry, and splendor usually associated with royal weddings. Only after they were safely married could the cloak of secrecy be thrown off, and even then they had to borrow the money to pay for the week-long celebrations that followed.

This furtively contracted marriage was to prove one of the most important events in Spanish history. It meant that when Ferdinand and Isabella in turn acceded to their separate thrones, the kingdoms of Aragón and Castile—which formed the greater part of what we know as Spain—would be united. Although each kingdom was to remain separate, the same heir would inherit them both; this would inevitably, if not immediately, mean the integration of the two. Upon unity depend power and strength, and it was as a united country that Spain was to become one of the foremost powers of Europe in the sixteenth and seventeenth centuries.

The concept of Spain as a nation emerged during the reign of Ferdinand and Isabella. But the task of molding a collection of sovereign states into a more or less united whole was by no means an easy one. It was to take all Isabella's resolute and selfless devotion and all Ferdinand's military ability and diplomatic guile.

The land to which they were born had a long and turbulent past, so turbulent that it is impossible to appreciate the extent of their achievements without first realizing exactly how much they had to contend with.

In a woodcut illustration from a popular fifteenth-century romance of chivalry, Oliver, a mythical Castilian prince, is married in a quiet ceremony reminiscent of Ferdinand and Isabella's wedding.

Oliveros de Castilla, 1491: HISPANIC SOCIETY OF AMERICA, NEW YORK

From earliest times, when the land was occupied by the primitive Iberians who gave their name to the peninsula, Spain suffered a series of invasions, reconquests, and civil wars. Phoenicians, Celts, Greeks, Carthaginians, and Romans occupied the country in turn. In the fifth century A.D., as the Roman Empire crumbled under barbarian attacks, successive waves of Suevians, Alans, Vandals, and Visigoths poured into Spain, adding to the existing mixture of races.

In 711 came the most fateful invasion of all, when the Arabs crossed the Strait of Gibraltar and established the

BIBLIOTECA NACIONAL, MADRID, RES. 204 FOL 171 R: PHOTO, ORONOZ

Spain's distant past is glimpsed in two medieval manuscript illuminations. Olives and grapes, introduced by Greek traders in the sixth century B.C., *soon became mainstays of Spanish agriculture. At right, Christians gather in the fruits of the earth. At left is the climax of the centuries-long domination of Spain by Carthage—Hannibal's army crossing the Alps to attack Rome. The artist's ignorance of elephants led him to symbolize the huge animals that terrified the Romans by one rather timid-looking beast, unfortunately daubed with blood.*

religion and culture of Islam throughout Spain. The Visigothic King Rodrigo disappeared in battle, and his few remaining followers fled to the mountains in the far north. Within eight years the Arabs dominated the entire peninsula—a situation that was to continue for another five hundred years. When Isabella married Ferdinand in 1469, the kingdom of Granada was still under Arab control, and the Spain they inherited had been shaped by the centuries of combined co-existence and warfare with the Moors, as the Arabs in Spain are usually called.

BIBLIOTECA NACIONAL, MADRID, VIT. 14-2 FOL 209 R: PHOTO, ORONOZ

BIBLIOTECA NACIONAL, MADRID, VIT. 19-2 FOL 19 V; PHOTO, ORONOZ

Rodrigo "the Unfortunate," the last Visigothic king, is shown in this miniature from a sixteenth-century genealogy of the kings of Spain. Comparison with the flourishing family tree on page 24—taken from the same manuscript—reveals some artful details with which the painter has reinforced the unhappy Rodrigo's depression.

It was from the remote and rugged land of Asturias in the north of Spain that the Christians launched the movement to win back their country: the Reconquest. A Visigothic chieftain named Pelayo was elected as Rodrigo's heir, and in 718 Pelayo led the Christians to their first victory over the Moslems in the deep mountain gorge at Covadonga. As a result, they managed to secure the tiny kingdom of Asturias for themselves and gradually added to it by more conquests. (See map, pages 52–53.)

By the beginning of the tenth century, Asturias had grown big enough to be divided by one of its kings into smaller kingdoms, each ruled by one of his sons: Asturias proper; Galicia and the area that was to become northern Portugal; and the territory of León. The eastern region of León, which gradually assumed a dominant role in the war against the Moors, was itself to become a separate province known as Castile, because of its numerous frontier castles. Still farther to the east, other small kingdoms had sprung up as more territory was regained from the Moslems: first, Catalonia, retaken by Charlemagne; then, Navarre, established by Basque warriors just south of the Pyrenees; finally, Aragón, the area between the two.

It is essential to remember that what is rather grandly called the Reconquest was in its early stages little more than intermittent guerrilla warfare among a kaleidoscopic array of states. This splintering of Christian Spain occurred partly because the northern mountain ranges made natural territorial divisions, separations that were further reinforced by the Roman system of administration. By the time the Moors invaded Spain, these divisions had hardened into a petty state system in which each region jealously preserved its independence and treated all outsiders with hostile suspicion. An alliance against the Moors was almost as great a threat to their independence as was conquest by the Moors themselves. Both sides, in fact, were divided by internal struggles, and when it suited them, Christians and Moors actually fought on the same side. Until the eleventh century the Reconquest was in no way a religious or even a national crusade. The Christians fought to regain their lost territory and nothing more. In times of peace there was frequent trading and intermarriage with the Moors. It even became fashionable for Christian lords to wear the long, flowing Moorish dress and to adopt such Moslem customs as sitting on floor cushions instead of on chairs.

For centuries internal quarrels prevented the Christian kingdoms from making much headway against the Moors.

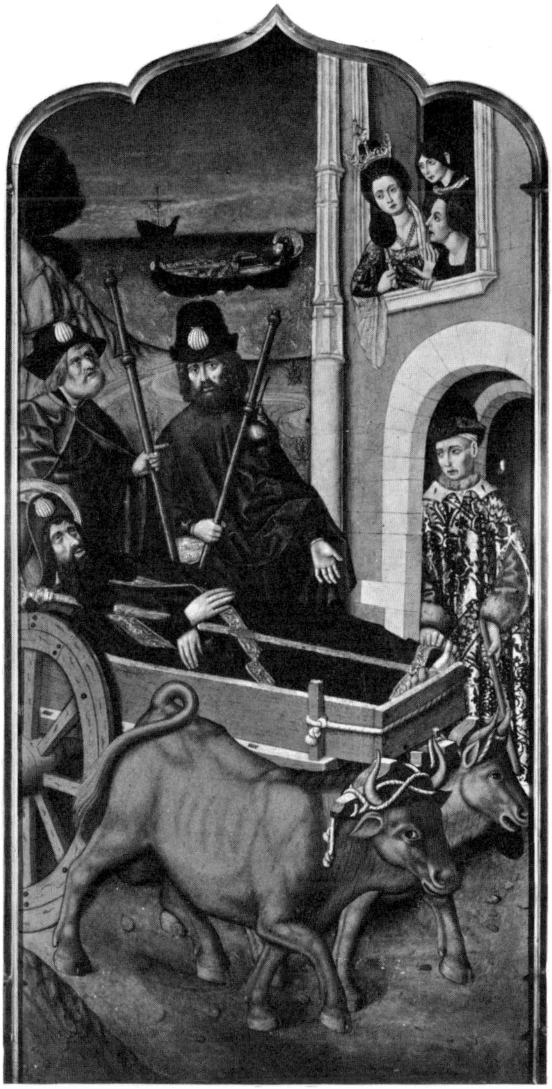

PRADO, MADRID

Santiago—St. James—has been venerated as Spain's patron saint since the ninth century, when a body believed to be his was discovered in Galicia. This retablo, *or altarpiece, retells the legendary arrival of the saint's body by boat from the Holy Land (rear) and its subsequent journey by oxcart to Compostela for burial. During the Middle Ages, St. James' shrine there attracted countless pilgrims, each of whom, like the disciples at left, wore Santiago's scallop-shell emblem in his hat.*

After the beginning of the ninth century, however, Christian Spain fought under the protection of a patron saint. The remains of a body believed to be that of James, son of Zebedee, were discovered in Galicia, and from that time, so soldiers claimed, the peaceful fisherman from the Sea of Galilee had led the Christians to victory in their battles. With St. James' scallop-shell banner proudly displayed, the Christians in 1085 recaptured Toledo—an important step forward, since the city had been the ancient Visigothic capital. But within a year, a new Moorish invasion brought the Christian armies to a virtual halt; even St. James' shining armor and flashing sword could not prevent their terrible defeat by the Moors at Alarcos, in 1195.

وتنحل الفص والجبالة والقشر والذبالة انها لضغث على ابالة فانصاعت تنقض مذرجها

وتنشد مذرجها فلما دانتني قرنت بالرقعة درهما وقطعة وقلت لها ان رغبت في المشوف المعلم

واشري الى الدرهم فوحي بالسر المدهم وان ابيت ان تشري فخذي القطعة وابيري

نالت الى استخلاص البدر التم والابلج الهم وقالت دع جدالك وسل عما بدالك فاستطلعت

طلع الشيخ وبلدته والشعر وناسج بردته فقالت ان الشيخ من اهل سروج وهو الذي وشى

BIBLIOTHEQUE NATIONALE, PARIS: B. N. FONDES ARABES 5847 FOL 94 V

The Christian army was almost totally destroyed, and after the battle the vast plain with its fringe of low, barren hills was covered with fallen soldiers and horses. The Moors burned the town of Alarcos to the ground and laid waste the land as far as the walls of Ávila, one hundred sixty miles to the north. The disaster, however, finally shocked the Spanish Christians out of their preoccupation with internal affairs. It also focused the attention of the outside world upon the struggle with the heathen that was taking place in Spain.

As a direct outcome of the defeat at Alarcos, a large-scale rendezvous was arranged at Toledo in 1212. Knights and men-at-arms flocked there from all over northern Spain, to be joined by crusaders from France, England, and other parts of Europe. Once assembled, they headed south. Their aim was to strike at the heart of Andalusia, the territory the Moors had ruled for five centuries.

To get there, the Christians had to cross the Sierra Morena, the towering bastion of rock that cuts off the lush south of Spain from the arid wastes of the central plateau. Although it was summertime, it was bitterly cold up in the mountains. The sheep and goat tracks that had to serve as paths were steep and treacherous and often impassable to knights and horses in full armor. To the disgust of the Spaniards, many of the foreign companies turned back. It is doubtful, however, whether the rest of the army would have survived the ordeal without the help given them by shepherds who knew the passes.

Eighth-century Spain, divided by different inherited cultures and creeds, was an easy prey for Moorish invaders from North Africa. The warlike spirit of these ardent Moslems is evident in the defiant pose and blaring trumpets of the warriors at left, from a thirteenth-century Persian manuscript. In the detail at right, from a Spanish chronicle of the same date, a Moorish chieftain holds a war council during the Reconquest of Spain by the Christians, a task completed by Ferdinand and Isabella.

ESCORIAL: MS. T-1-1 FOL CLXXXI

CONVENT OF LAS HUELGAS, BURGOS: SKIRA

It is extraordinary that the Moors, who surely must have learned of the Christians' approach through their spies, did not ambush their enemies while they were still in the mountains. If they had done so, the Moors certainly would have been victorious, and the history of Spain might have taken a very different course. As it was, perhaps overconfident of their superior strength, they allowed the Christians to pour down from the mountains onto the great plain of Tolosa. There, on July 16, 1212, the two sides met in pitched battle and the Moors were defeated. The Christians had won the decisive victory of the Reconquest: Las Navas de Tolosa.

For the Moors it was the beginning of the end. In 1236, St. Ferdinand, King of Castile and León, captured Córdoba, and in 1248 he took Seville, with its prosperous harbor. The only Moorish territory left on Spanish soil was the little kingdom of Granada, which was to resist Spanish pressure for another two hundred years.

Although Moorish domination of the peninsula was at an end, the states that made up Christian Spain were far from united. Years of fighting an enemy common to all but rarely resisted by all had reinforced each state's basic sense of independence, self-reliance, and hostility to its neighbors. Each province had developed its own customs, laws, and institutions and even its own version of the Latin speech that had been common throughout the peninsula in the days of the Roman occupation.

By the middle of the fifteenth century, however, many of the less powerful states had been absorbed, and the peninsula was divided into five kingdoms: Portugal; Navarre; the Moorish kingdom of Granada; Castile; and Aragón. The last two shared between them the territorial bulk of the peninsula. Castile was the greater, comprising Galicia, Asturias, the Basque provinces, León, Old and New Castile, Estremadura, Andalusia, and Murcia. In other words, the kingdom stretched from the Bay of Biscay to the Strait of Gibraltar, from Corunna in the northwest to Cartagena in the southeast.

The kingdom of Aragón, made up of Aragón itself, Catalonia, and Valencia, formed a solid block in the east. But although Aragón was smaller than Castile, it possessed

BIBLIOTECA NACIONAL, MADRID, VIT. 19-2 FOL 34 V: PHOTO, ORONOZ

In a miniature from the same genealogy as those on pages 16 and 24, St. Ferdinand, clad in gold-bedecked sixteenth-century armor, is handed the huge key of a town that is probably Seville, taken by him from the Moors in 1248.

This ten-foot-high banner was captured from the fleeing Moors at Las Navas de Tolosa in 1212. A superb example of Moorish art, it combines Arabic prayers with intricate geometric motifs woven in silk and gold threads.

BIBLIOTECA NACIONAL, MADRID: VIT. 17-3

Charles of Viana, dispossessed by his father in favor of Ferdinand, is portrayed here by a fifteenth-century artist so sympathetic to the slighted prince's cause that he has even given his head a halo.

what Castile did not—foreign territory. For centuries, while Castilian kings and nobles led the armies of the Reconquest, Aragón had taken advantage of its long coastline to trade with other Mediterranean countries and interfere in their affairs. In the course of time, Aragón succeeded in conquering the Balearic Islands, Naples, Sardinia, and Sicily.

Castile and Aragón were in many ways very different: in their social structure, their economic affairs, their legal systems, and their political aims. As Castile expanded, much of the territory recovered from the Moors had been handed over to the nobles who had helped to win it back, and many of them had grown extremely powerful—too powerful for any but the strongest monarch to control. With their wealth, their castles, and their companies of fighting men, the Castilian nobles ran their huge estates like private kingdoms. In Aragón, on the other hand, there had always been a strong parliamentary tradition, and since the country's fortunes depended so directly on its overseas commerce, the merchant class had the most influence.

There existed between the two kingdoms an inborn suspicion, an ill-concealed enmity that was to survive even the reign of Ferdinand and Isabella. The secrecy in which the pair had to get married is proof of their countries' hostility. The jealousy with which each guarded its sovereignty is shown by the terms of their marriage treaty, which stipulated that each monarch, in the event that both succeeded to their thrones, was to retain control of his or her own kingdom. Only a male heir could inherit both their crowns. Yet the marriage of Ferdinand and Isabella was to mean that the very names of Castile and Aragón would one day be replaced by the all-embracing "Spain."

The circumstances that led to the permanent union of Castile and Aragón were curious ones. Ferdinand, born on March 10, 1452, was the second son of the enormously vital and intelligent King of Aragón, John II. Between Ferdinand and the throne of Aragón stood his much older half-brother Charles, who was, through his mother, John's first wife, also the heir to Navarre. When Charles' mother died, John married a Castilian noblewoman of remarkable spirit and energy. Ferdinand, the son of an aging father and an ambitious mother, became the natural favorite. By 1460 court politics had alienated Charles so completely from his father that the King imprisoned his elder son and proclaimed Ferdinand his heir.

His action threw the whole realm into such turmoil that John was forced to release his elder son. In 1461, however,

Charles died, at the age of forty. The cause of his death was officially said to be a fever, but suggestions of poison have often been made and may quite possibly be true. Charles' death brought renewed revolts to Catalonia, where the insurrection continued for another ten years. The rebels did not finally capitulate until three years after Ferdinand's marriage, but the ending of the civil war left him at last undisputed heir to his father's lands.

Isabella's chances of succeeding to the throne had seemed even smaller than Ferdinand's. She was born on April 22, 1451, the only daughter of another John II, of Castile. John was not an evil man, but he lacked both the inclination and the ability to govern well. He loved to dance and joust, to compose poetry and play the lute, leaving the affairs of his kingdom in the much more capable hands of his favorite, Álvaro de Luna. But this was not an age in which monarchs could afford to delegate responsibility. The Castilian nobles were constantly jostling for power, wealth, and land. A weak king and a powerful fa-

The simplified family tree below shows that Ferdinand and Isabella had the same great-great-grandparents and that each was the child of a father's second marriage (indicated by figure 2). Premature deaths of Charles of Viana, Ferdinand's half-brother, and Alfonso, Isabella's brother, made each next in line for a throne, although Isabella had to fight Juana la Beltraneja for Castile's crown.

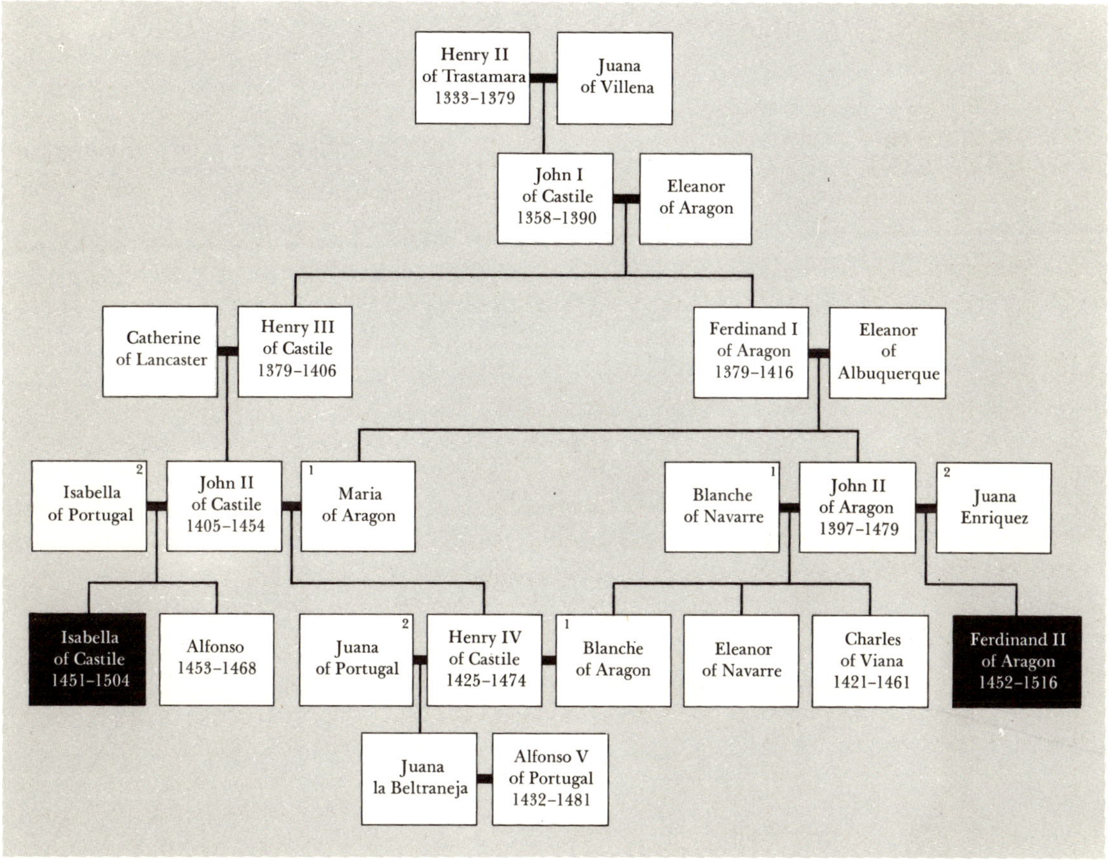

vorite were a combination guaranteed to encourage anarchy. Castilian prestige fell at home and abroad.

Eventually the reign of John II came to an ignominious end. When the King's first Queen died, Álvaro de Luna chose Isabella of Portugal as John's new wife, hoping through the grateful Queen to open up another avenue of influence over the King. His error of judgment proved disastrous. Resentful of his ascendancy and his arrogant treatment of her husband, the new Queen successfully plotted his downfall. Álvaro de Luna was arrested and executed, after a summary trial, in the public square of Valladolid on June 2, 1453. A year later John himself died, fully conscious of the appalling state in which he had left his country and regretting that he "had not been born the son of a mechanic, instead of King of Castile."

Isabella was only three years old when her father died, and her half-brother, Henry, succeeded him. Unfortunately for Castile, Henry IV proved to be an even more disastrous king than his father. The promise he had shown as a prince disappeared when he reached power. The generosity that earlier won him the nickname The Liberal soon turned out to be an extravagance that sapped the country's economy. As a result of his apathy and mismanagement, crime and rebellion thrived, and the corruption at court set the pattern for all levels of society. Short of money to meet his vast expenses, Henry tried to increase the taxes that already imposed an unfair burden upon the poor. Worst of all, he allowed the coinage to be debased, and the cost of living increased astronomically.

Henry's personal life brought the reputation of the Cas-

The family tree of John II of Castile shows him mounted (top), with his queens, Maria of Aragón and Isabella of Portugal, guarded by St. Vincent, at right. Below are Maria's children, Henry IV and his sisters; and at bottom, Isabella's, Alfonso and Isabella (Elisabet).

MUNTADAS COLLECTION, MUSEO DE ARTE CATALUNYA, BARCELONA: MAS

The lovely, uncluttered countryside of fifteenth-century Spain forms the background of this detail from a Catalonian painting of the Epiphany. A group of gaily clad knights spur down a road between a church (left) and a moated castle, where the artist's fancy has run riot among the turrets.

tilian monarchy lower still. As a youth he had led a licentious life, but when he married Blanche of Aragón, he proved unable to produce an heir. After twelve years the marriage was annulled. In 1455, soon after his accession, he married Juana of Portugal. By the time another six years of childlessness had gone by, Henry had been dubbed The Impotent. When Queen Juana at long last gave birth to a daughter, in 1462, Beltrán de la Cueva, Henry's favorite, was commonly believed to be the father, and the King became the laughing stock of Spain. For her entire life the unfortunate girl was known as La Beltraneja, after her supposed father.

Doubts about the paternity of Juana la Beltraneja finally gave rise to outright civil war. A large confederacy of nobles

DUKE OF VILLAHERMOSA COLLECTION, MADRID: MAS

This portrait of John II of Aragón, painted many decades after the King's death, nevertheless conveys something of the energy that characterized Ferdinand's determined and ambitious father.

already in rebellion against King Henry refused to acknowledge the baby as his heiress. At a mock ceremony outside the walls of Ávila, they insolently dethroned an effigy of Henry, stripping it one by one of its crown, its scepter, and its royal vestments. They then proclaimed as their king Henry's half-brother Alfonso, who with his older sister Isabella had been brought up in seclusion by their widowed mother, Isabella of Portugal.

By this time, Isabella was fourteen. Several suitors had already been proposed for her, including Ferdinand of Aragón, his brother Charles, and the King of Portugal, Queen Juana's brother, who was twice Isabella's age. This last distasteful match she evaded with the diplomacy that was to be one of her most valuable attributes as Queen, saying that "the infantas of Castile could not be disposed of in marriage without the consent of the nobles of the realm."

On July 5, 1468, the whole course of her life was changed when her brother, Alfonso, was found dead in his bed. Since he was a healthy boy of fourteen, it seems probable that he was murdered. In any case, his death meant that Isabella and Juana la Beltraneja were sole rivals for the throne.

With characteristic good sense, Isabella, instead of challenging her half-brother, Henry, arranged a reconciliation with him. Its outcome was that she was legally proclaimed heir to the throne of Castile and León, instead of her "niece" Juana. Henry promised never to compel her to marry against her wishes, while Isabella, for her part, promised not to marry without his consent. As heir, her marriage, of course, acquired new importance. A brother of Edward IV of England asked for her hand. So, too, did the French Duke of Guienne, heir apparent to Louis XI. But Isabella held out for Ferdinand of Aragón.

She had never seen Ferdinand, but he was young and reputed to be charming, good looking, and a fine horseman. In addition, he was her second cousin and a member of the same dynastic house, the Trastámaras. Union with him, therefore, had a certain appropriateness that the other proposed matches did not.

Most of the Castilian nobles, however, objected. A faction once again rose in support of Juana la Beltraneja's claim to the throne; another faction persuaded Henry that an alliance with Portugal would be more profitable. So successful were they that the King even threatened Isabella with imprisonment if she did not accept the Portuguese match. But no opposition could change Isabella's mind once it was made up. Since her brother had violated their agreement by trying to compel her marriage, she felt herself free to marry against his wishes. Accordingly, she sent word to Aragón that she would accept Ferdinand. King John of Aragón, as aware as Isabella's advisors of the importance of uniting the two kingdoms, joyfully ratified the agreement.

To pretend that the marriage was a love match would be pointless. It was a political match that worked out brilliantly. Isabella's determination to marry Ferdinand can best be explained by crediting her with the ability to see in the marriage the possibility of a united Spain. She was fully aware that, for a princess, duty must come before personal happiness, but fortunately for her, personal inclination and political expediency coincided. Even more fortunately, her marriage turned out a happy one.

United both by political expediency and by personal inclination, Ferdinand and Isabella took pleasure in decorations that stressed their joint claim to rule. A page of a devotionary made about 1500 for one of their children is ornamented with their crowned initials: F for Fernando and Y for Ysabel.

MUSEE CONDE, CHANTILLY, MS. 604/1339 FOL IV PEC 731: PHOTO, GIRAUDON

The couple who stood before the Archbishop of Toledo in Valladolid that memorable day made a handsome pair. Isabella was of medium height, with the ample flesh that in middle age runs to fat. Her skin was pale, her hair fair with copper tints, and her eyes were the striking blue-green of the house of Trastámara. Her face, although its features were regular, was not strictly beautiful, but her level gaze and sweetness of expression made it an attractive one. She was courteous, dignified, intelligent, and extremely devout.

Ferdinand, for his part, was generally considered a comely young man. Although his hair was already becoming sparse, he had a fine forehead and bright, lively eyes. His strong body was toughened by war, hunting, and jousting; his skill in the saddle was renowned. His most noticeable personal quality at this time, before responsibility and power had properly shaped his character, was his equanimity—an equanimity that he was never to lose for long, even in periods of the greatest stress. His education compared unfavorably with Isabella's, but he had a great deal of native intelligence. Where she was forthright, however, he was to prove devious. As a result, they complemented each other perfectly.

There is no doubt that he and Isabella became devoted to each other. The court life of the time made minor infidelities almost inevitable, and Ferdinand was no exception. The Queen kept a jealous eye on his affairs: ladies-in-waiting whom he took as mistresses were quietly dismissed; at least three illegitimate daughters were relegated to convenient convents. But these were casual, momentary lusts. That his basic emotional commitment was to Isabella, and hers to him, cannot be questioned. When Isabella died, she ordered that, upon her husband's death, she be buried with him, so that "our bodies may symbolize and enjoy beneath the ground the close relationship that was ours when we were alive." Ferdinand's will likewise stipulated that he should, whatever the circumstances, be buried with the Queen.

If Isabella had any doubts about the wisdom of her choice when, against all opposition, she committed herself to Ferdinand that day in Valladolid, they were needless ones. Their marriage was to prove a successful personal union and a triumphant political one.

Artfully enclosed in the initial letter of a 1484 legal patent, Ferdinand and Isabella, youthful but regally dignified, hold each other by the arms.

La cronica de España.

II

SPAIN UNITED

AMERICAN NUMISMATIC SOCIETY

Ferdinand and Isabella replaced Henry IV's debased currency with coins, like this excelente, *showing the joint rulers face to face.*

King Henry of Castile was not pleased by his half-sister's choice of a husband. When Ferdinand and Isabella notified him of the marriage and asked for his blessing, his only answer was that "his ministers must be consulted." In the meantime, on the other side of the Pyrenees, the crafty Louis XI of France realized that the eye he had been keeping on Spain had not been quite watchful enough. In the hope of gaining control of Catalonia, Louis, a master of intrigue, had made several treaties with Castile and Aragón in turn. He had even proposed that Isabella marry his brother. Now, suddenly, he was faced with the possibility of a united Castile and Aragón that would form a rival power bloc near at hand. Louis immediately urged King Henry to reject Isabella and acknowledge the rights of his daughter, Juana la Beltraneja. He made his proposal more attractive by offering as a husband for Juana the same brother whom Isabella had previously refused.

Henry obligingly proclaimed that Isabella had forfeited her rights to the throne by marrying against his will. The claims of the rival princesses were more than enough to set the unruly barons at one another's throats. Old feuds were revived, armies were raised, towns were burned down, and crops destroyed. Isabella spent most of her time riding from town to town in the hope that a personal appearance would create enthusiasm and support for her cause. For months on end she could not even count on the help of her husband. Aragón and France were fighting desperately for possession of Roussillon and Cerdagne, in the northeastern Pyrenees. Louis had taken advantage of the Catalan civil war to annex these two counties from Aragonese rule, and Ferdi-

Ferdinand and Isabella's coat of arms (left), from a 1493 Spanish history, stresses their equal majesty by quartering the castles and lions of her Castile and León with the bars and eagles of his Aragón and Sicily.

VALERA, *La Cronica de España*, 1493: BIBLIOTECA NACIONAL, MADRID

nand was forced to help his father in the bitter struggle to win them back.

The situation was brought to a head on December 11, 1474, for that night Henry died in his palace at Madrid. Mismanaging his affairs to the very end, he had failed to indicate his successor. Isabella, therefore, immediately proclaimed herself Queen of Castile. Two days later, in Segovia, she was led by the city's dignitaries to the main square. There, on a raised platform, she received their homage. Then, with an unsheathed sword of state carried upright before her, its hilt pointing to the sky in the traditional Castilian way, the twenty-three-year-old Queen went in procession to the nearby cathedral to give thanks to God for His goodness.

Isabella's claim to the throne was technically a legal one. The Castilian Cortes, or parliament, had refused to revoke its recognition of her when ordered to do so by Henry after her marriage. Once she was proclaimed Queen, the Cortes met in Segovia to swear allegiance to her. Ferdinand also hastened there from Aragón, and his exact position in Castile was established. This was a rather delicate problem, since, as a Trastámara, he, too, had a claim to the Castilian throne. But it was decided that Isabella alone was heir to Castile and that therefore she alone was its monarch. Tactfully she assured Ferdinand that "since you and I by God's grace are so much in harmony, there can be no distinction between us. Whatever has been decided, you are still, as my husband, King of Castile and in Castile your wishes will be law." As far as her kingdom of Castile was concerned, however, Ferdinand was always, in theory at least, to remain a consort king.

To the supporters of young Juana, recognition of Isabella by the Cortes was irrelevant. Since her illegitimacy had never been proved, to all intents and purposes La Beltraneja was the late King's daughter. Toward the end of May, 1475, Juana in her turn claimed the throne.

The bitter war of succession that followed was a critical point in Spanish, and even in world, history. Juana's most powerful ally was her uncle Alfonso V, King of Portugal, who hoped to win Castile for the thirteen-year-old princess and then marry her. Had he succeeded, Castile would have been united with Portugal in the west instead of with Aragón in the east. Isabella would have been merely the consort Queen of Aragón. Since Aragón's interests had always been directed toward the Mediterranean, it is hardly likely that its rulers would one day have listened to the proposals of a visionary sailor who thought he could sail westward across the Atlantic Ocean to China.

When Portuguese troops poured into Spain in May, 1475, Ferdinand and Isabella, with relatively few soldiers, were quite unprepared for war. Fortunately for them, the Portuguese forces stopped in Plasencia, close to the border, for the formal betrothal of Juana and the uncle who was thirty years her senior. The ceremony was celebrated with such pomp that it was almost a week before the army was able to continue its march. At Arévalo, barely thirty miles from Segovia, the Portuguese encamped to wait for Castilian reinforcements.

Ferdinand and Isabella put the extra time to good use. Once again, Isabella rode across the country, enlisting support. She was expecting their second child that fall—the first had been a girl but now there was another chance of a prince to become their joint heir. Neither her condition, the scorching heat of a Castilian summer, nor the throat-clogging dust of the rough roads that linked town to town could smother her energy and her spirit. Such endurance seems almost unbelievable, but Isabella never lacked courage and determination in times of need. While she exhorted nobles and peasants to follow her to Valladolid, Ferdinand scoured the northern regions for troops. By early July their joint exertions had put Ferdinand in command of some fifty thousand men. Later that month, Isabella was to pay for her incredible endeavors with a miscarriage.

In the campaign that followed, Ferdinand played a part that cannot be overestimated. His was the leadership, the military skill, and the strategic resourcefulness. He sent to Aragón for military advisors. He plotted, planned, and bargained. Not least, he was a constant example of unceasing energy and cheerful confidence. The money so urgently needed for the campaign was obtained, probably at Ferdinand's suggestion, in an extraordinary way. The Castilian clergy agreed to lend the royal treasury half the silver and gold plate owned by the churches throughout the kingdom. At first, the pious Isabella protested, but her political instincts soon got the better of her religious qualms, and she authorized the loan.

The campaign dragged slowly on. Squadrons of Isabella's light cavalry from Estremadura and Andalusia made destructive raids into Portugal, and Portuguese enthusiasm for Juana's cause began to wane. In addition, the Portuguese armies gradually lost most of their Castilian supporters. Castilians and Portuguese were always uneasy

PASTRANA: MAS

In this fifteenth-century tapestry detail, a regally helmeted Alfonso V of Portugal marshals his men during a 1471 Moroccan campaign —a battle that was much more fruitful than his invasion of Castile four years later in support of his fiancée and niece, Juana.

GRANADA CATHEDRAL: PHOTO, ORONOZ

Symbol of the regality for which Isabella so desperately fought is this elegantly simple silver crown, which the Spanish Queen once wore.

allies, and even when the war first broke out, Portuguese intervention had served chiefly to induce the majority of Castilians to rally to Isabella's flag.

The most important battle of the campaign was fought in March, 1476. The Portuguese King had led an army to relieve Zamora, on the river Duero, which was besieged by Ferdinand's men. Through sheer military incompetence Alfonso camped on the bank of the river opposite the town, and the Castilians at once seized the chance to cut off his supplies and forced him to retreat. Scarcely able to believe his luck, Ferdinand followed in pursuit. The Castilians caught up with the Portuguese in the late afternoon, on a wide plain a few miles outside the city of Toro. In spite of the failing light, they attacked, and within a few hours the Portuguese were overwhelmingly defeated.

Isabella was at Tordesillas, farther up the river, waiting for news. When word of the victory arrived, she celebrated her triumph by walking barefoot in procession to the church of St. Paul's to give thanks.

DUKE OF ALBA COLLECTION, MADRID: PHOTO, ORONOZ

The soldiers clashing above appear in a manuscript Bible prepared between 1422 and 1433 for the Grand Master of the Knights of Calatrava. Though meant to depict King Saul's Old Testament victory, it more accurately depicts fifteenth-century Spanish combat, with a plumed royal helmet like the one on page 27.

The psychological effect of the victory was enormous. More and more towns threw in their lot with Isabella. In her determination to subdue all the remaining opposition, she moved her headquarters close to the Portuguese border, so that she could supervise operations in person. When her advisors pointed out the danger, she is said to have replied: "It is not for me to think of the dangers and exertions I experience in my own cause, nor by untimely fears to discourage my friends, with whom I am resolved to remain until I have brought the war to an end."

The end came, finally, in 1479 when King Alfonso of Portugal signed a treaty on behalf of himself and Juana, renouncing all rights to the Castilian throne. Juana retired to the convent of Santa Clara at Coimbra, and Isabella, at twenty-eight, was at last the undisputed Queen of Castile.

In January of the same year, John II of Aragón had died in Barcelona at the age of eighty-two, and Ferdinand succeeded him. Nearly ten troubled years had passed since Ferdinand and Isabella secretly plighted their troth to each other in Valladolid. Now, as joint sovereigns of Castile and Aragón, they stood on the threshold of greatness. The future of Spain was theirs to shape.

Once she had won her throne, Isabella's most pressing task was to make sure that she kept it. This meant bringing to heel the Castilian nobles, including those who had supported her. They had, after all, rebelled against the previous monarch, Henry IV, for her sake. If it suited them, they were quite likely to rebel against her for the sake of somebody else. Since most people in Castile were weary of civil war, Ferdinand and Isabella knew they would not lack support in their campaign to curb the power of the aristocracy. One grateful chronicler wrote that the King and Queen had come "to restore these kingdoms and rescue

Historia de Vespasiano, 1499, BIBLIOTECA NACIONAL, LISBON: MAS

EVELYN HOFER

From castles like the fifteenth-century one in Estremadura below, the nobility wielded great power. The 1499 Spanish woodcut at left —supposedly Roman Emperor Vespasian leveling Jerusalem—may be a sly comment on Ferdinand and Isabella's razing of such castles.

them from the tyrannical government to which they have for so long been subjected." Ferdinand and Isabella both saw it as their mission to restore peace and order under the authority of the crown.

Backed by the Cortes and by the towns, the King and Queen undertook a series of measures that would restrict the power and privileges of the landed aristocracy. Many of the baronial castles from which Castile derived its name were deliberately destroyed, and a statute was passed that forbade the building of new ones. In Galicia alone fifty fortresses were razed. This proved to be an extremely effective move; without their castles to fight from and flee to, the nobles were poorly equipped for war. Dueling, which had been a licensed form of murder among aristocrats, was made a treasonable offense, and as a supplementary move, Ferdinand and Isabella made the nobles swear to keep the peace among themselves. (In the past, the barons' rank had always exempted them from the normal processes of the law.) In addition, they were stripped of the semiroyal ceremonials and privileges that they had gradually usurped over the years. As a final measure, the sovereigns began a policy of granting positions at court and in the administration on the basis of merit rather than wealth and rank.

Naturally, the nobles objected, and their leaders wrote a collective letter of complaint to Ferdinand and Isabella. In return they received such a stern, uncompromising rebuke that they let the matter rest. The King and Queen, by firmness, prudence, and skillful wooing of the Cortes and the municipalities, had already managed to invest the crown of Castile with an authority it had not possessed for hundreds of years.

One of the severest blows suffered by the Castilian aristocracy was financial. In 1480, the year after the war of succession had ended, the Cortes realized that it was essential to restore the kingdom's economy. To avoid increasing general taxation—which would put the heaviest burden on those who could least afford it—half the revenues usurped by the nobles in the previous sixteen years were to be restored to the Crown. The Queen's confessor, Hernando de Talavera, was given the task of carrying out this scheme, and such was his efficiency that the Crown gained an additional annual income of 30,000,000 maravedís, equivalent to $1,500,000 in modern currency. By far the greater part of this sum came out of the nobles' pockets.

A second great blow affected not only the nobles' pockets but their pride as well. By the end of the twelfth cen-

: comer. beuer. calçar. vestir. visitar. consolar. enterrar.

tury, the Reconquest of Spain had acquired some of the characteristics of a religious crusade. In imitation of the monastic orders of knighthood that had been created to fight for the protection of the Holy Land, three knightly orders had been established in Castile: those of Santiago, Calatrava, and Alcántara. Although they were intended to embody the religious and secular ideals of chivalry, over the years these orders had acquired immense wealth and power. Their members were subject only to the authority of their grand masters and officers, and as religious knights, they enjoyed the normal privileges of the clergy as well as those of the aristocracy. Membership, and particularly ranking membership, was therefore the ambition of all nobles, and the elections held to appoint knights and officers were a constant source of strife. In Isabella's view, the power of these orders was an obvious threat to her authority.

In 1476 the Grand Master of Santiago, the principal order, died, and the Queen seized this opportunity to intervene. Unwilling to entrust her mission to anyone else, she raced on horseback from Valladolid to the convent at Uclés, two hundred miles away, where the election of the new master was being held. After an exhausting three-day journey over bad roads, she arrived at Uclés after dark and in pouring rain. Bursting in upon the astounded electors with her clothes and hair still dripping with water, she announced that she wished her husband to be elected Grand Master. Such an important order of knights, the Queen cunningly added, deserved no one less eminent than the King himself at its head.

Isabella had her way, but Ferdinand, cautious as ever, allowed the electors' candidate, Alonso de Cárdenas, to occupy the mastership as his delegate. A valuable precedent had been established, and when, in 1487, the Grand Master of Calatrava died, Ferdinand was able to assume the office immediately without a murmur of protest. Seven years later he took over the order of Alcántara, and when Cárdenas died in 1499, Ferdinand succeeded him. No sooner did the King and Queen have control of all three orders than they reformed them, diverting much of their vast revenues into the royal treasury. By the time the grand mas-

BIBLIOTECA NACIONAL, MADRID, RES 204 FOL 115 V: PHOTO, ORONOZ

As part of their traditional privileges, Spanish nobles insisted upon a total obedience from their vassals that they themselves were unprepared to show their king. In a fifteenth-century illustration for a Latin text, vassals with a request get a stern, negative answer.

DUKE OF ALBA COLLECTION, MADRID: PHOTO, ORONOZ

The frontispiece of the Calatrava Bible shows the Grand Master, wearing the order's red cross insignia, enthroned above a line of knights performing the seven deeds of charity. Beneath them, more members receive a manuscript from a kneeling man who may be the rabbi who translated the Bible.

BIBLIOTHEQUE NATIONALE, SERVICE PHOTOGRAPHIQUE, PARIS: MS. ESPAGNOL 36 FOL 48 V

Rough justice was nothing new in Spain. An illustration for a fourteenth-century romance depicts an execution in open country, as was usual. Yet the condemned man, though stripped and strung to the gallows, is about to be saved.

terships were permanently annexed to the Crown of Castile in the next reign, one of the greatest potential sources of anarchy had been neutralized. Membership in the orders was an honor to be bestowed by the Crown.

The barons were not the only cause of lawlessness and unrest in Castile. Decades of constant warfare, economic inflation, and irresponsible administration had produced a restless society with no respect for law and order. Crime went unpunished; cities as well as the countryside were infested with robbers. As a contemporary writer put it, "men seem to have lapsed into a state of savagery." Ferdinand and Isabella coped with this chaos by introducing an efficient police force.

Many Castilian towns in the Middle Ages had financed vigilante groups known as *hermandades*, or brotherhoods, which looked after the interests of the towns and maintained law and order within them. However, since the *hermandades'* jurisdiction extended only to the areas closest to each town, large areas of countryside had degenerated into lawless badlands. At the very beginning of her reign Isabella decided to create a national band of vigilantes called the Santa Hermandad, or Holy Brotherhood. It would be controlled by the Crown and would exercise justice impartially without being swayed by local prejudices. Its main task was to patrol the highways and open countryside, to suppress violence and theft, and to arrest criminals who had escaped from the cities. Its troops were recruited from the ordinary householders of Castile, and each village and town contributed its quota of men. When the Santa Hermandad was called out, a group would pursue the criminals to the edge of its boundaries, where the chase would be taken up, if necessary, by the neighboring group.

Since the Santa Hermandad could try as well as arrest prisoners, organization of the brotherhood drastically restricted the absolute authority the nobles hitherto had held over their own lands. Predictably, the barons resented the idea of being legally responsible to the Crown through a group of their own dependent vassals. Ferdinand and Isabella countered opposition by appointing a standing army of two thousand men—under the command of Ferdinand's half-brother Alonso of Aragón—who were always ready, properly mounted and armed, to suppress insurrection. This professional force naturally commanded more respect from the nobles than did the local, more amateurish groups.

The Santa Hermandad came to exercise a reign of terror over the Castilian countryside. It was dealing with a savage society and its measures were correspondingly savage. When a criminal was caught, the least he could expect was torture and probably mutilation. A man found guilty of stealing from 500 to 5,000 maravedís (worth approximately twenty-five to two hundred fifty dollars), for example, lost a foot. Very often bandits were captured, tried, and immediately executed, all on the same spot. This barbarity proved effective. Just over twenty years after the Santa Hermandad was founded, the King and Queen were able to abolish its Supreme Council and its expensive salaried officers. Punishments became less harsh, and appeals against the Santa Hermandad's sentences were permitted. Whereas in the past its name alone had inspired fear and awe, it now lost much of its former power and prestige and dwindled into an ordinary rural police force. But it had fulfilled its purpose.

BIBLIOTECA NACIONAL, MADRID, VIT. 25–7 FOL 18 R: MAS

The Santa Hermandad's methods of reducing crime, described by the sovereigns' own physician as "terrifying and horrible vivisections," were effective indeed. Above, a bandit is about to lose his head.

Robber gangs no longer terrorized the countryside, and their hideouts in the mountains were destroyed.

In her burning desire to make her kingdom a safe and peaceful place to live in, Isabella often intervened personally in the process of reform. The year after her coronation in Segovia, the people of that city rose up against their governor. When Isabella heard the news, she ordered her horse to be saddled and rode immediately to Segovia. Intrepid as ever, she addressed the angry crowds: "Tell me what your grievances are and I will do all in my power to satisfy them; for I am sure that what is in your interest must also be in mine and in that of the city as a whole." Her calmness and obvious sincerity had their effect, and she was able to persuade the people to return to their ordinary occupations while she conducted a full inquiry.

Isabella's readiness to travel anywhere and everywhere in the pursuit of her duties was one of her remarkable features. Even pregnancy did not deter her, and each one of her five children was born in a different city. Since the

BIBLIOTHEQUE ROYALE DE BELGIQUE, BRUSSELS: MS. 7 FOL 78 R

Something of fifteenth-century court life can be glimpsed in the scene above of a royal party entering a Belgian city. Although most of the ladies ride demurely in a wagon, one prefers the saddle—as did Isabella. The Spanish artist who depicted the birth of the Virgin at right came closer to visualizing the ceremony of a royal confinement than he did Mary's humble birth.

DUKE OF ALBA COLLECTION, MADRID: PHOTO, ORONOZ

The Old Testament's Queen Esther—a model for Isabella in her concern for justice—dictates laws to royal scribes in this Calatrava Bible illustration.

BRITISH MUSEUM: MS. ROY 10E4 FOL 187 R

When Isabella launched her reform of Spain's corrupt and decadent religious orders, some four hundred Andalusian friars emigrated to North Africa rather than give up their concubines. Above, in a miniature illustrating an earlier papal decree, an errant monk and his lady friend share the stocks.

Queen traveled around, the court did too. It was never permanently fixed in any one city. During 1477 and 1478, for instance, Isabella made her headquarters at Seville in Andalusia, and there, on June 30, 1478, the bells rang out to celebrate the birth of a son, Prince John—their second child in eight years and an heir at last for both Castile and Aragón. Throughout her stay in Seville, the Queen, as usual, held audiences every Friday. Seated on a throne on a high dais draped with cloth of gold, she presided over lawsuits that required special attention, listening to anyone who wished to put his case before her. On these occasions she showed she was merciful and generous as well as intelligent and just. Once the war of succession was over, it was not unusual for Isabella, either alone or with Ferdinand, to make such working visits to at least four towns every year.

Under this constant care, Castile gradually emerged from its medieval chaos. As the nobles lost power, the Crown gained it. Justice became a meaningful word, and the laws of Castile, which over the centuries had become a tangle of often contradictory provincial codes, statutes, and ordinances, were revised. A code was compiled that would have national application. With the whole administrative system running more effectively, the royal revenues increased without the need for additional taxes. The currency was reformed, and the private mints permitted by Henry IV were suppressed. Roads and bridges were built to improve communications, and merchants were actively encouraged to find markets for the wool that Castile's great migrant flocks produced in such quantities. The Consejo Real, or Royal Council, which was the central governing body of Castile, was refashioned so that it would be controlled not by powerful lords but by lawyers, townsmen, and members of the lesser nobility.

All this was done under the close supervision of the King and Queen. Isabella's attention to detail, in particular, became a byword. It is said that a piece of paper one day fell out of her sleeve; on it, written in her own hand, was "The post of town crier must be given to so-and-so because he is the man with the loudest voice."

Even the Church did not escape reform. Although Castilian sovereigns had once had almost complete authority over ecclesiastical appointments, papal control had gradually increased, and it had become common for foreign clerics who had never before set foot in Spain to be allotted high-ranking posts by the Pope. Ferdinand and Isabella regained the right to make their own appointments by threatening Pope Sixtus IV with a large-scale investigation of the many abuses of the Church. They used their new power to promote to high positions those ecclesiastics who by their scholarship and integrity were deemed worthy of them.

The Pope's concession did not deter Isabella from the project that was perhaps nearest to her heart—the reform

BRITISH MUSEUM: MS. ROY 15E3 FOL 269 R

From Spain's vast migrant flocks of sheep came the wool that was a mainstay of the growing nation's economy. In the 1468 miniature at right the shepherds receiving the news of Jesus' birth are unmistakably Spanish, with their tooled leather waist pouches, sandals, bagpipes, and full wineskin by the fire. The sparse trees and barren hills over which the angels hover are also reminiscent of Spain. At left in another miniature of the late fifteenth century, men dye cloth in a huge vat heated beneath so as to keep the new colors fast.

PIERPONT MORGAN LIBRARY, NEW YORK: MS. M450 FOL 4 V

TOMICH, *Historias . . . dels Reys de Aragó*, 1534: NEW YORK PUBLIC LIBRARY

ANDRES DE LI, *Reportorio de los Tiempos*, 1493: HISPANIC SOCIETY OF AMERICA, NEW YORK

The sun beams happily down upon a Spanish lord and lady secure behind the moat of their castle at left. The woodcut illustrated a 1534 history of Aragón's rulers. The peasant cutting wood above appeared in a 1493 almanac of the seasons and represented December's preparation for winter's approach.

of religion in Spain. The fifteenth century was an age of religious decline all over Europe, a situation that was eventually to lead to the Protestant Reformation. Under Isabella's leadership, Spain became the first country in which measures were taken to root out corruption in the Church. The man she entrusted with the reform was Francisco Jiménez de Cisneros, an austere Franciscan friar. Named by Isabella Archbishop of Toledo, and thus Primate of the Spanish Church, Cisneros set about reforming the monastic orders, enforcing rules, and if necessary, expelling serious offenders.

Eventually opposition to the reform gave way, and at a crucial moment in its history, the Church in Spain was purged of its disreputable elements. By their remarkable foresight, Ferdinand and Isabella had anticipated the inevitable attacks on the Spanish Church and therefore avoided serious upheaval. Had they not acted when they did, Spain would have felt the impact of the Protestant Reformation much more severely.

In their efforts to curb the grandees, as the high Spanish noblemen were called, Ferdinand and Isabella had needed and received the support of the towns. No statute compelled the King and Queen to summon the nobles and the clergy to meetings of the Cortes, and they saw to it that after about 1480 the parliament was attended mainly by the town representatives. Once the nobles were under control, however, and the good will of the municipalities was no longer so essential, Ferdinand and Isabella decided to clamp down on the towns in their turn.

During the Reconquest the towns that sprang up one by one as the Christians advanced had been granted large areas of conquered territory and often their own charters by the kings of Castile. The towns therefore enjoyed a large degree of independence. Most of them, again as a legacy of the Reconquest, were heavily fortified and could conceivably offer very effective opposition to the monarchy. Ferdinand and Isabella adopted measures to prevent such opposition. The most important of these was the appointment to every main town of a *corregidor*—a royal official with administrative duties and judicial powers who was to watch over the affairs of his community and to be held responsible for them to the Crown itself. In this way the sovereigns gained such tight control over the towns that they need expect little further opposition from the town-dominated Cortes of Castile.

And thus it proved. The Crown at last was in complete command. The three estates of Castile—the nobility, the clergy, and the municipalities—were firmly subjected to royal authority. After the long years of anarchy, Castile was a united kingdom.

Candid comments on Spanish life at the time of Ferdinand and Isabella appeared in woodcuts illustrating a 1491 book entitled Mirror of Human Life. *Good and bad lawyers are contrasted in the scenes below. At left, an honest lawyer arbitrates among knights from an open book of the law; at right, a corrupt practitioner accepts bribes of money and goods.*

BOTH: AREVALO, *Espejo de la Vida Humana*, 1491, BIBLIOTECA NACIONAL, MADRID

In the measures they took to counteract internal disorder, Ferdinand and Isabella were never motivated by selfish gain. They had visions of greatness for Spain and they realized quite clearly that a country torn by civil strife has no time for greatness. They realized that not only the greatness but also the very well-being of Spain depended on a settled Castile, which was by far the largest and most populous kingdom of the Iberian Peninsula. So firmly did Ferdinand believe this that in the early years of his reign he devoted his attention entirely to his wife's kingdom of Castile. His own kingdom of Aragón, which was internally much more peaceful during this period, Ferdinand left in the hands of a viceroy.

It says much for the personal qualities of the monarchs that they were able to fulfill their aims without undue violence and deception. Isabella, in particular, impressed everybody with her dignified sincerity, her forthright behavior, her extraordinary capacity for action, and the great personal interest she took in the affairs of her kingdom. However, where Isabella inspired both respect and love in Castile, Ferdinand inspired only respect. He was, after all, Aragonese and a more cautious and uncommitted person than Isabella. But he had the great virtue of being successful, and this brought its own reward in loyalty. In addition, of course, he was a man in a man's world and could command an obedience that Isabella alone might have found impossible to enforce. Together, and only together, they imposed on their country a unity that their predecessors would scarcely have believed possible and at which contemporaries never ceased to express their amazement. The sight of Spaniards from all over Spain fighting in unison at the siege of Granada inspired this comment from an Italian, Peter Martyr:

> Who would have believed that the Galician, the fierce Asturian, and the rude inhabitants of the Pyrenees, men accustomed to deeds of atrocious violence and to brawl and battle on the lightest occasions at home, should mingle amicably, not only with one another, but with the Toledans, the La-Manchans, and the wily and jealous Andalusians; all living together in harmonious subordination to authority, like members of one family.

Aragonese and Castilian kings each shared legislative power with a Cortes, which was summoned chiefly to grant the sovereigns' financial requests. In the 1495 miniature at right, a king of Aragón presides over the nobles, clergy, and town deputies who comprise the "three estates" of his Cortes.

Constituciones de Catalunya, 1495, ARCHIVO DE LA CORONA DE ARAGON, BARCELONA: MAS

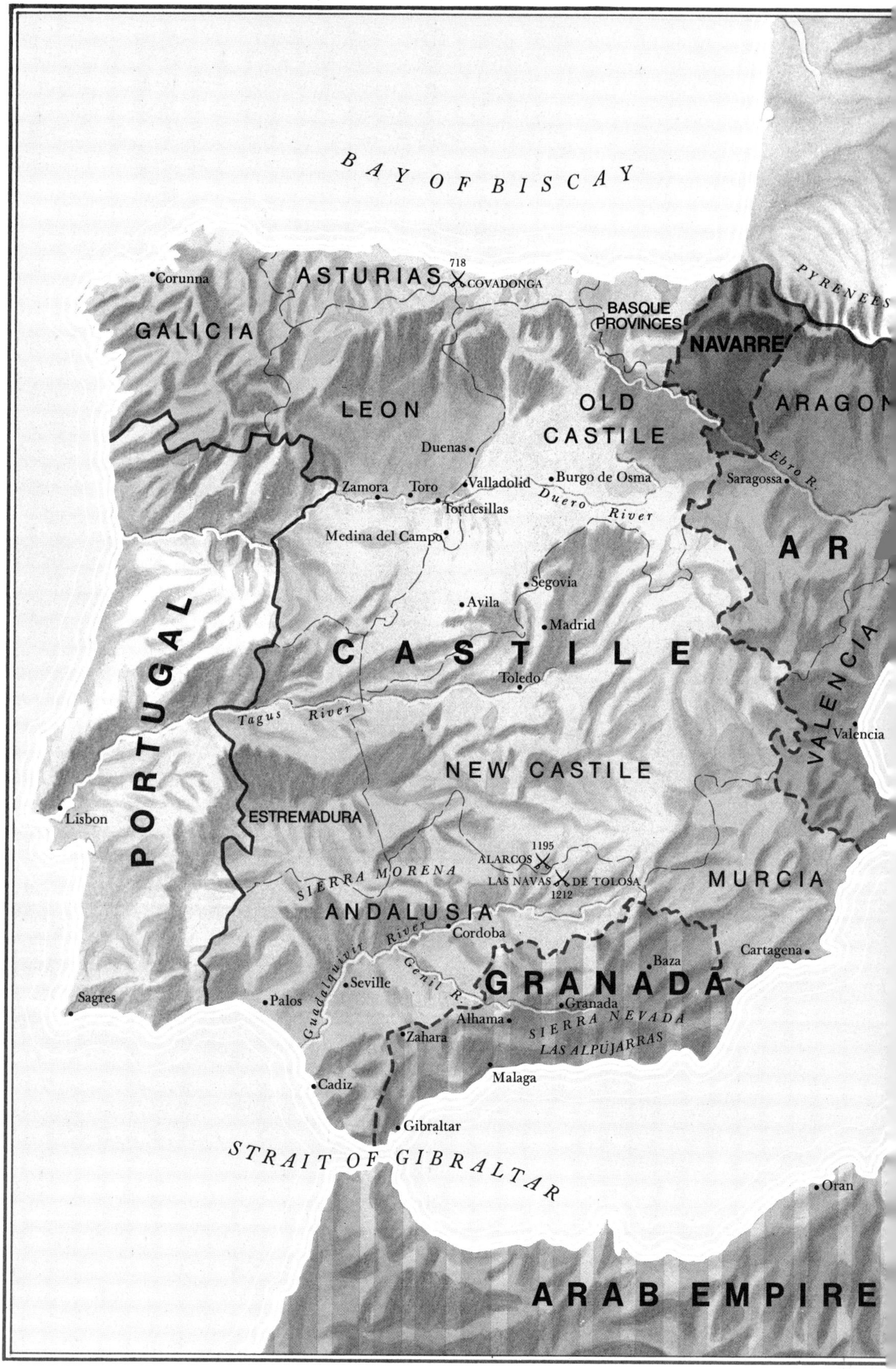

BAY OF BISCAY
Corunna
ASTURIAS
718
COVADONGA
GALICIA
BASQUE PROVINCES
NAVARRE
PYRENEES
LEON
OLD CASTILE
ARAGON
Duenas
Ebro R.
Saragossa
Zamora
Toro
Valladolid
Burgo de Osma
Duero River
Tordesillas
Medina del Campo
AR
Segovia
Avila
Madrid
PORTUGAL
CASTILE
VALENCIA
Toledo
Tagus River
Valencia
NEW CASTILE
Lisbon
ESTREMADURA
1195
ALARCOS
LAS NAVAS DE TOLOSA
1212
SIERRA MORENA
MURCIA
ANDALUSIA
Cordoba
Guadalquivir River
Genil R.
Cartagena
Baza
GRANADA
Seville
Sagres
Palos
Granada
Alhama
SIERRA NEVADA
Zahara
LAS ALPUJARRAS
Malaga
Cadiz
Gibraltar
STRAIT OF GIBRALTAR
Oran
ARAB EMPIRE

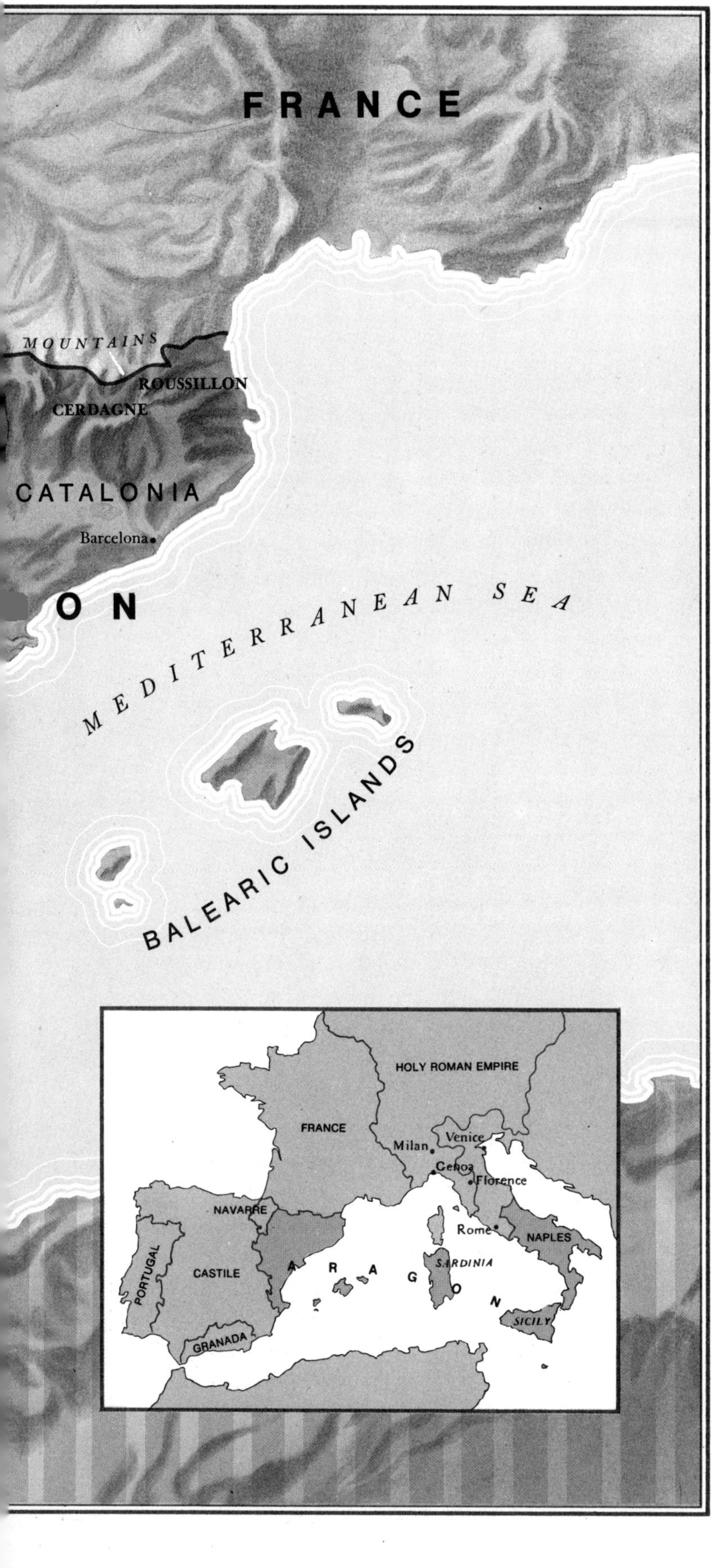

THE SPAIN OF FERDINAND AND ISABELLA

Isabella finally won undisputed rule over Castile in 1479, the same year Ferdinand succeeded to the throne of neighboring Aragón. The Iberian Peninsula they then sought to unify was divided into five separate kingdoms, as shown on the map at left: Portugal; Castile; a small, independent Navarre straddling the Pyrenees; Aragón; and—in the south—Granada, the sole remaining possession of the North African Moslems who had once dominated the entire peninsula. (The location of three major battles in the centuries-long struggle with the Moors is indicated with crossed swords.) Aragón's Mediterranean interests are reflected in the inset map of Southern Europe. Ferdinand's kingdom stretched from mainland Spain, through the Balearic Islands and Sardinia, to Sicily. The Italian kingdom of Naples—united to Aragón earlier in the fifteenth century—was ruled by Ferdinand's first cousin King Ferrante I. Not until 1504 would Ferdinand add Naples to his own extensive Mediterranean realm.

III THE

TOLEDO CATHEDRAL: PHOTO, ORONOZ

CONQUEST OF GRANADA

In January, 1492, church bells rang out joyfully all over Europe. In London's St. Paul's Cathedral and in the cathedrals of many other capitals priests sang *Te Deums*. In Rome the Pope and his cardinals walked in procession to St. Peter's to attend a special high mass. All Europe was celebrating the news that on the second day of that month Granada, the last Moorish stronghold on Spanish soil, had surrendered to the armies of Ferdinand and Isabella. In Spain itself the people were ecstatic. Their King and Queen had given them back their country.

In the minds of Spaniards down to the present day no event has been more closely associated with Ferdinand and Isabella than the conquest of Granada, which restored the city to Christianity after nearly eight centuries of Moorish rule. From the beginning of her reign Isabella had been attracted to such an undertaking for two reasons. As an extremely devout Christian, full of crusading zeal, she longed to destroy the last remnants of Moslem power in Spanish territory. As a monarch whose chief political aim was unity, she resented the presence on Spanish soil of a potentially hostile kingdom of different race and religion. With her usual political intuition, she also realized that a common crusade against the Moors was the most effective means of creating a sense of nationalism among the peoples of Spain and of winning their enthusiastic support.

Crusading was to some extent in the air. In 1453, to the horror and dismay of Christendom, Constantinople had fallen to the Moslem Turks. Christian armies made frantic efforts to stem the Moslem advance into Europe. In Spain, Isabella's half-brother, Henry IV, organized six military expeditions into Granada between 1455 and 1457, but he

The decisive sieges in the long Granada campaign are depicted on a series of choir stalls carved in 1495 for Toledo Cathedral. At left, Spaniards assault a town with lombards as three men (right of center) scale the walls.

achieved nothing. Nevertheless, these campaigns fired the imagination of his countrymen, and in 1464 large numbers of Castilians had to be restrained from leaving Spain in order to join a European crusade against the Turks.

Isabella, remembering this, had no sooner won the struggle for the Castilian throne than she turned her attention southward to the Moorish kingdom of Granada. Its King, Mulay Hassan, was a proud, impetuous man who all his life had harbored secret dreams of one day reconquering Spain. In 1476 Ferdinand and Isabella sent a demand for payment of the annual tribute that Mulay Hassan's predecessors had agreed to give to Castile. Contemptuously the Moorish King replied that "the mints of Granada coined no longer gold, but steel." At the time, the King and Queen were very angry but were too involved in the war of succession to exact any punishment, and Mulay Hassan was encouraged by what he took to be a sign of weakness. On the night of December 26, 1481, his armies attacked the unsuspecting Christian fortress of Zahara on the Andalusian frontier. Those who resisted were killed; every other man, woman, and child was carried off into slavery in Granada.

ARCHIVO HISTORICO NACIONAL, MADRID

"From childhood [Ferdinand] had been brought up in wars, in which he underwent much hardship and personal danger," wrote an early chronicler. On this seal from a document issued in 1481, the King is shown in martial pose astride a royally caparisoned warhorse.

Christian Spain was appalled at this flagrant breaking of the truce between Granada and Castile. But in a way the disaster was a relief, because it brought hostilities out into the open and allowed the King and Queen an opportunity to retaliate. They now had their excuse for undertaking a campaign to drive the Moors from their land forever.

In 1482 they opened their attack. A Spanish spy reported that the town of Alhama, situated on a hill of rock in the heart of enemy territory, could easily be taken by surprise and overcome. The Marquis of Cádiz, Don Rodrigo Ponce de León, set off for Alhama with a large force. Just before dawn on February 28, a small group of men managed to scale the battlements of the castle that dominated the town. They killed the guards and opened the castle gates to let in the army. But when they attempted to take the town itself, the citizens, armed with crossbows and harquebuses, fought from street to street. Not until dusk did the Spaniards break through and begin to ransack the town of its riches—wrought gold and silver, delicate Moorish jewelry, inlaid furniture, and embroidered silks.

The battle of Alhama, though, was not yet won. Twenty-five miles away, in Granada, Mulay Hassan hastily gathered together an army. Five days later he appeared before the gates of Alhama with more than fifty thousand men and laid siege to the town. Meanwhile, Ferdinand, who had

DUKE OF ALBA COLLECTION, MADRID: PHOTO, ORONOZ

The decade of fighting to oust the Moors from Granada climaxed a century of strife within Castile itself. As always, the chief victims were the common people, who lost homes, livestock, and often their own lives to the rival armies. Above, a kneeling woman intercedes for her family and flocks.

foreseen that this might happen, was riding day and night from the north to bring help. Isabella stayed behind to collect money, reinforcements, and supplies. Then, in spite of the fact that she was heavily pregnant with her fourth child, she too marched south to join Ferdinand at Córdoba. There many of her advisors warned her that it was foolish to try holding a town so near the Moorish capital. "Glory," she replied, "is not to be won without danger." As it happened, there was little danger. Wisely, Mulay Hassan retreated as the Spanish army approached, and the exhausted garrison joyfully opened the town gates to their fellow Christians.

Immediately the two monarchs, both acting in the way most typical of their separate characters, set about consolidating their position in Alhama. Isabella ordered the three main mosques to be purified and consecrated as churches, for which she herself provided the bells, crosses, and furnishings. Ferdinand, turning his mind to less spiritual matters, arranged for the provisioning and strengthening of the garrison, then conducted forays into the fertile plain between Alhama and Granada itself, destroying crops and orchards.

The loss of Alhama was a terrible blow to the morale of the Moors. Many saw it as an omen of greater disasters. A ballad written at the time, lamenting Alhama's fall, warned Mulay Hassan that one day he would lose all Granada. Before that day arrived, however, many fortresses were to be won and lost.

After Alhama the war against Granada became a methodically conducted campaign of sieges. The mountainous

BIBLIOTECA NACIONAL, MADRID: VIT. 26-2 FOL 205 R

Warfare between Moslems and Christians was a fact of European life from the seventh to the sixteenth century. In an illustration from a fourteenth-century Byzantine manuscript, Christians besieged in Edessa in 1032 resist a two-pronged attack by Arab bows and spears. Arrows were still the chief Moslem weapon in the Granada campaign but had added potency because of poisoned tips.

terrain did not allow for full-scale formal battles in the medieval fashion, with vast armies drawn up facing each other across an open plain, their foot soldiers ready to advance, their cavalry ready to charge. Most of the Moorish towns were built on commanding positions that allowed them to defy the enemy besiegers for months at a time, and good artillery was essential if the war was not to exhaust the finances of the Christian monarchs. Good artillery was something the Spaniards did not have, and Isabella sent to France, Germany, and Italy for men, equipment, and advice. Slowly, her army acquired heavy lombards, primitive cannon large enough to fire balls of marble and iron weighing up to 165 pounds. Transporting these guns from siege to siege was a heartbreaking task. Crossing the sierras was often difficult for a man on horseback; for the oxen or mules harnessed to a train of heavy guns it was well-nigh impossible. Nevertheless, it was done. A large body of soldiers preceded the artillery, hacking through undergrowth, clearing boulders, and bridging streams, to make rough-and-ready tracks for the clumsy guns and their sweating teams.

As the Christian army marched from one town to the next, it laid waste the surrounding country. An organized group of thirty thousand men devoted all their time to destroying farms and burning, flattening, or uprooting the olive trees, the orange groves, and the vineyards that covered the countryside. The fertile *vega*, or plain, of Granada—one of the most spectacular sights in Spain when covered by luscious crops and backed by the snow-capped range of the Sierra Nevada—soon became a desolate, blackened waste. But however cruel, this destruction was necessary. By cutting off the Moors' supplies of food, the Spanish army could decisively shorten the war. Once the Moslems were defeated, the land would have time enough to recover.

Meanwhile, Ferdinand's Mediterranean fleet patrolled the stretch of sea between Spain and North Africa to prevent the Moors from sending for or receiving aid and supplies. Granada was now almost completely isolated by the Christians.

Slowly but inexorably, the Spaniards made inroads upon the Moorish kingdom. As each town fell, Moslem territory shrank. Spanish successes attracted mercenaries and volunteers from all over Europe, soldiers of fortune who swelled the ranks of the militia levied by Ferdinand and Isabella from Castile and Andalusia. Throughout the campaign, the King rode at the head of his army, while the Queen carried out the duties of supreme financier and quartermaster. In addition, she constantly visited the camps to encourage the soldiers and established field hospitals for the sick and wounded. Six large tents, known to the soldiers as the Queen's Hospitals, were set up near the front lines and supplied with nurses and surgeons and such dressings and medicines as were available, all at Isabella's own expense. Four hundred covered wagons, armed against enemy attack, were set aside as ambulances.

This handsomely decorated morion or cabasset—an open helmet with an upturned brim permitting a soldier to sight along his barrel—became the favorite headgear of sixteenth-century Spanish infantry.

THE JOHN WOODMAN HIGGINS ARMORY, WORCESTER, MASS.

The war was not all stark and grim. It was carried on with the pomp and grandeur thought appropriate to a religious crusade. The nobles set up camp in silken tents, and richly liveried servants offered them food on gold and silver plate. Although the King and Queen normally lived very simply, they too showed a sense of occasion when necessary. Whenever news of another victory reached the Queen in their headquarters at Córdoba, she would go in procession to the cathedral for a service of thanksgiving, accompanied by all the ecclesiastical and secular dignitaries of the city. When Ferdinand himself arrived triumphant from a battle, he would be escorted—still dusty with the grime of war—to the cathedral, and there he would prostrate himself before the altar.

Not all the battles ended in victory for the Christians. In March, 1483, Don Alonso de Cárdenas, Grand Master, by Ferdinand's gracious consent, of the Order of Santiago, set

ESCORIAL: PHOTO, BRADLEY SMITH

In confused flight after a major Spanish victory, Moorish soldiers crowd into the forecourt of a Granadan town while more disciplined troops march past in the distance. This detail from a fresco in the Escorial was painted in 1587 and depicts a Reconquest battle that took place in 1431. The Christians attacking the high-walled town in the foreground; the women grouped on the hill above, awaiting their fate; the shepherd and his flock caught up in the confusion; and the fight to the death going on behind the palisade (lower right) are all reminiscent of the battles of the later conquest of Granada.

off on a raid into the region of Málaga with three thousand cavalry and a company of infantrymen. In the mountains north of Málaga they were attacked by Moorish cavalry. Cárdenas' men tried to retreat but were so burdened with plunder acquired en route that they lost their way and were trapped in a deep ravine. Their only chance of escape was to go back the way they had come, but in the dark they missed the track. When the sun rose, they found themselves in a valley commanded on every side by Moorish archers and soldiers. The Christian army panicked. Some threw away their arms; others, it is said, fainted from terror and fatigue. Frightened and confused, large groups were herded together like sheep by mere handfuls of Moors. Some strayed as far as Málaga and were captured by Moorish women working in the fields. Many escaped only after wandering for days in the mountains, keeping themselves alive on berries and wild herbs. In all, more than eight hundred were killed and almost two thousand were taken prisoner. For Ferdinand and Isabella this was one of the most ignominious disasters of the whole war.

In spite of such reversals, however, the Christians advanced. Their cause was enormously helped by the fact that the kingdom of Granada was torn apart by internal feuds. Mulay Hassan's two sons, Boabdil and Yusuf, rebelled

ESCORIAL, MS. T-1-1 FOL CLXV: PHOTO, BRADLEY SMITH

A fate similar to that of Cárdenas' unlucky party in 1483 has befallen the Christian prisoners in this thirteenth-century miniature. Closely guarded by Moors on horseback and on foot, they are herded into captivity along with plundered cattle, goats, and sheep.

As the fifteenth century wore on, siege equipment became more and more elaborate, as can be seen in an illustration from the Bible of the Knights of Calatrava. At left, a modernized version of the age-old catapult lobs cannon balls toward a distant castle while, at right, an ingenious form of hoist enables a crossbowman to shoot down onto unprotected battlements.

against their father and fled to Guadix, some thirty miles from Granada, where Boabdil was proclaimed King. After a fierce struggle, Granada too went over to Boabdil, and Mulay Hassan and his brother El Zagal ("The Valiant") were forced to retreat to Málaga. War broke out between the two halves of the kingdom, each vying with the other for success in the continuing war with the Christians. It was El Zagal who led the Moors to the crushing victory over the army of Don Alonso de Cárdenas. Boabdil realized that, to remain the Moorish champion, he would have to surpass his uncle's feat. With an army of ten thousand men, he laid siege to the town of Lucena in April, 1483, but succeeded only in getting himself captured by a Spanish foot soldier.

Ferdinand seized this heaven-sent opportunity to exploit the division amongst the Moors. In return for his freedom, Boabdil had to swear allegiance to Ferdinand and undertake to help the Spaniards. Boabdil was to prove an unreliable ally, but because he needed the Christians' help to preserve his throne, he could not afford to break the truce and reopen hostilities with them.

While the two Moorish factions were at each other's throats, Ferdinand and Isabella decided to exert themselves to the utmost and attempt to take Málaga. This important south coast city was the key to Granada. Ships from Málaga had been the only vessels able to pierce the tight Spanish naval blockade of the coast and maintain some degree of communication between Granada and North Africa. If Málaga fell, the capital must eventually collapse. In May, 1487, therefore, the Christians launched a mighty attack by land and sea upon the city's fortifications.

Málaga was not easy to approach. It occupied a very strong natural position, with a mountain range to the north and the sea to the south providing formidable barriers to an attacking enemy. In addition, El Zagal, who had succeeded Mulay Hassan as leader of the opposing Moorish faction, had seen that it was well fortified and ready for a siege. The city had large supplies of artillery and ammunition, and with its regular garrison reinforced by local volunteers and by a corps of African mercenaries, Málaga was well able to resist the Spaniards.

For three months Ferdinand's heaviest guns pounded cannon balls at the walls on the land side. For three months the Moors fired back. Francisco Ramírez, a famous military engineer from Madrid, devised and constructed great wooden towers on rollers. Loaded with soldiers, the towers could be pushed close enough to the ramparts for the men they carried to drop down onto the battlements. But the Moors captured the soldiers and destroyed their machines. In the meantime, the Moorish fleet put out to sea and engaged the Spanish ships, trying to draw them within range of the city's guns.

Eventually hunger and exhaustion got the better of courage. The citizens of Málaga offered to surrender on the same honorable conditions that had been granted to Moorish cities in the past. The sovereigns' terms, indeed, had been unusually liberal: the defeated population was given the choice of leaving, with their personal possessions, or of becoming Spanish subjects, with the right to continue practicing their old laws, customs, and religion. To the Málagans' consternation, however, Ferdinand refused to accept their surrender on those same terms. Again they offered and again he refused. This time, he swore that if they now harmed a single hair of a Christian, he would put every man, woman, and child in the city to the sword. The Moors had no choice but to surrender unconditionally.

Clerics were essential members of the army led by Ferdinand and Isabella against the Moors. Above, a bishop consecrates armor and pennons to God's service. At right, amid the colorful tents of a field camp, trumpeters summon the Castilian troops to hear priests (center) say Mass before a battle.

RIGHT: ESCORIAL: PHOTO, BRADLEY SMITH. LEFT: BIBLIOTECA NACIONAL, MADRID, VIT. 18-9 FOL 228 R: PHOTO, ORONOZ

Historia de Vespasiano, 1499, BIBLIOTECA NACIONAL, LISBON: MAS

The interest in classical learning that spread through Spain in Isabella's reign brought publication of such books as the history of Vespasian, from which this scene is taken. The "Roman" soldiers, however, are drawing water for a siege camp typical of the 1490's.

On August 18, 1487, Ferdinand and Isabella made a state entry into Málaga. Shortly afterward, the city's entire population was driven into the great courtyard of the fortress and the sentence of slavery was passed on every one of them. One third of the slaves were to be exchanged for Christian prisoners in North Africa; one third were to be the property of the state, to help defray the expenses of the war; and the remainder were to be disposed of as presents at the sovereigns' whim.

In an apparent softening of this severe sentence, Ferdinand held out a crumb of hope to the wretched Málagans. They could purchase their freedom if they could raise a huge ransom within nine months, and he suggested they do this by surrendering all their personal effects and by asking for money from the Moors in Granada and overseas. As the wily Ferdinand had foreseen, however, the ransom demanded was too much for the Málagans to accumulate in the time available; consequently, they stayed slaves, and the Crown gained the whole of the city's treasure.

Ferdinand's actions seem both harsh and deceitful to modern eyes, but his decision to make an example of Málaga was strategically sound: an act of terrifying severity might well frighten the Moors into surrender and thus shorten the war.

The tactic succeeded. Boabdil, deciding to cut his losses, offered to surrender Granada to the Christians. In return, he asked for the overlordship of four towns that were still under the control of El Zagal. Since these would have to be taken by the Spaniards before they could pass to Boabdil, this meant that the Spanish armies would in fact be furthering his own war against his uncle. Nevertheless, Ferdinand and Isabella agreed, and in the spring of 1489 their troops laid siege to El Zagal's capital, Baza.

The Moors resisted fiercely, and the siege dragged on for five months. As autumn approached, Ferdinand ordered more than a thousand clay and timber huts to be built as winter quarters for his armies. When this town and all the surrounding roads were washed away in a cloudburst, the morale of the Christians fell so low that the King sent for Isabella. She had already been forced to pawn her jewels to meet the expenses of a war that seemed as if it would never end. Nevertheless, she arrived at the camp to cheer and comfort her troops, and her presence had an overwhelming effect. Three days later Baza offered to capitulate. Subtle as ever, Ferdinand treated the townspeople so generously that their mayor consented to go to El Zagal

BOTH, GRANADA CATHEDRAL: PHOTO, ORONOZ

Still to be seen in the Granada chapel where Isabella lies buried is the silver chest ornamented in gold (left) in which she kept her jewels. The illuminated page from her missal (below) is discolored by centuries of exhibition on the anniversary of Granada's fall.

and persuade him to surrender—not only Baza but also the rest of his territory—to the Spaniards. This El Zagal did, declaring, "What Allah wills, he brings to pass in his own way. Had he not decreed the fall of Granada, this good sword might have saved it, but his will be done!"

Now that El Zagal had given up the towns requested by Boabdil, Ferdinand and Isabella demanded the surrender of Granada, as agreed. Instead, the unreliable Boabdil declared that he was going to fight for what little remained of the Moorish kingdom of Granada and backed up his words with raids into Christian territory. Ferdinand and Isabella decided to finish once and for all with the Moors. In the spring of 1491, their army, fifty thousand strong, camped on the *vega* below the city of Granada itself, to remain until Spain was wholly theirs.

At this final stage the Christians made elaborate preparations so that nothing should go wrong. The surrounding countryside was ravaged and its farms and villages were wiped out. Isabella arrived to supervise arrangements for provisioning and equipping the Spanish army and to inspect the troops. Something, however, did go wrong. One night, Isabella asked a maid to remove a candle whose glow was preventing her from sleeping. The maid put the candle too near some draperies, and the silk-walled royal tent caught fire. The flames spread so rapidly that most of the camp was destroyed, but fortunately no one was hurt. Instead of replacing the charred tents and pavilions, the Spaniards built a whole new city with large buildings, roads, and squares on the same site. Such was the enthusiasm and fighting fitness of the troops that the city took a mere three months to erect. When they begged the Queen to let them call the city after her, Isabella asked that they give it instead the significant name of Santa Fe—the Holy Faith.

EVELYN HOFER

The sight of Santa Fe, proving as it did the seriousness of the Spaniards' intentions, profoundly discouraged the citizens of Granada. They put up a token resistance, challenging the Christians to meet them in equal combat, in groups or one by one, in true chivalric fashion. But in the face of such massive opposition, the Moors had little confidence in their own fortifications and resources, and in October, 1491, Boabdil opened secret negotiations with the Christians.

Hernando de Zafra, the royal secretary, and Gonzalo de Córdoba, a military leader who spoke Arabic, were chosen to represent the King and Queen. Long conferences were held, usually at night, to discuss possible terms of surrender.

Looming darkly against the glittering peaks of the Sierra Nevada, the Alhambra—with its crenelated towers and archways under tiled roofs—is as superb a sight today as it was in 1492 when Ferdinand and Isabella took Granada. Later additions include the church at center and the palace to its right.

In sharp contrast with the vengeance Ferdinand had taken on the unfortunate people of Málaga, the terms eventually agreed on were liberal in the extreme. In return for the surrender of the city and all its artillery and fortifications, the Moors would be allowed to practice their religion freely, to be judged by their own laws, and to retain their traditional dress, language, and customs. Their property would remain their own, and they would continue to pay taxes at the same rate as they had to their Moorish sovereigns. For three

A magnificently carved altarpiece in the Royal Chapel at Granada commemorates crowning moments in the sovereigns' holy war to liberate Spain. In this detail, a grim-faced Boabdil, with two soldiers leading his horse, leaves Granada forever to hand over the keys of the city to the Spaniards. Behind him, bound Moorish hostages file through the Alhambra's gates.

years, indeed, they would not be taxed at all. Lastly, despite his treachery, Boabdil was granted a tiny kingdom under the authority of the Castilian Crown, in the wild mountain region called Las Alpujarras.

The terms, in fact, could not possibly have been more generous, and it is probable that at the time Ferdinand and Isabella had every intention of keeping them. Their hold over southern Spain was still precarious, and very wisely they hoped to instill in the war-weary Moorish population of Granada a sense of gratitude toward the kingdom's new Christian rulers.

On the morning of January 2, 1492, Boabdil and fifty of

GRANADA CATHEDRAL: PHOTO, ORONOZ

his cavaliers rode out through the gates of his beautiful palace, the Alhambra, and down through the lovely gardens that lined the hill to the banks of the river Genil, where Ferdinand and Isabella were waiting. Boabdil tried to dismount and kiss Ferdinand's hand in token of homage but the King gently restrained him, whereupon Boabdil kissed Ferdinand's sleeve instead. Sadly, he gave the King the keys of the Alhambra; then, taking his leave of the conquerors, he set off across the *vega* to his substitute kingdom in Las Alpujarras.

Solemnly, a carefully picked detachment of the Christian army marched into the city and up to the Alhambra. Soon a great silver cross was raised upon its highest tower. At the sight of the cross, the priests in Isabella's entourage, waiting by the river, burst into a *Te Deum*, and tears of joy streamed down every face. When the royal standard fluttered out beside the cross, the heralds shouted, "Granada, Granada for Ferdinand and Isabella." "Castile, Castile," the crowd roared back, and the grandees crowded to kiss Isabella's hand as Queen of Granada.

Four days later, at Epiphany, Ferdinand and Isabella made their state entry into the city. In a glittering parade of armored troops, beplumed knights, and richly robed priests, they went in procession to the largest mosque, where they offered up prayers of thanksgiving. They then took possession of the Alhambra and in its splendid throne room received the homage of the inhabitants of the conquered city.

After 781 years, Christianity once more reigned supreme in Spain. It was the end of an era. It was also in many ways a beginning. For the first time, men from every region, forgetting their petty quarrels, had joined together to fight as a united country under a single leader, in far greater numbers and in longer campaigns than ever before. Ferdinand's armies had learned military tactics, discipline, and obedience. Soon the Spanish infantry, trained in the wars against the Moslems, was to become famous all over Europe. In addition, now that the crusading Spaniards had finally defeated the Moorish empire on their own soil, they would decide to carry their victories into the heart of Moslem territory, into North Africa itself. Spain, in fact, was poised at the beginning of a new career.

But of even greater significance for the future of Spain was the fact that standing among the cheering crowds who witnessed the triumphant entry of Ferdinand and Isabella into Granada was a man called Christopher Columbus.

MUSEO DE EJERCITO, MADRID: PHOTO, ORONOZ

The hilt of Boabdil's sword displays the intricately wrought patterns that made Moslem metalwork famous and treasured works of art.

ALCAZAR OF SEVILLE: PHOTO, BRADLEY

IV

SPAIN AND COLUMBUS

To Europeans of the late Middle Ages the Orient, like outer space today, was an area of mystery that inspired the wildest speculations. Tales were told of rivers filled with gold and cities paved with precious jewels, of strange beasts and weird humans, men with dogs' heads and men with no heads at all, men with hairy bodies and men with one foot.

So firm a hold did the Orient have on the European imagination that even after traders and travelers had penetrated into Asia as far as Peking, accounts of their experiences did nothing to dispel the aura of fable and fantasy. On the contrary, they often added to it, for the pioneer explorers soon realized that the more wonderful their tales, the more respect and admiration they inspired. The Orient became the great goal of European exploration. Its spices, silks, gold, and jewels lured merchants eager for profit; adventurers hoped to win land and become fabulously wealthy; the pious longed to convert the heathen.

The great age of Renaissance discovery and exploration was opened by the Portuguese in the fifteenth century. Their brilliant young prince, Henry the Navigator, had dreamed of reaching India by sea, since the overland route was blocked by the Moslem Turks. On the rocky headland of Sagres, jutting out into the Atlantic from the southern tip of Portugal, Prince Henry founded the world's first school of navigation. There he gathered together the best mariners, navigators, cartographers, and astrologers (as astronomers were then called) that Europe could offer. From this base the Portuguese, in the little ocean-going vessels called caravels, began to inch their way down the west coast of Africa.

COLUMBUS, *Book of Privileges*, 1502: NEW YORK PUBLIC LIBRARY

Columbus' coat of arms, granted by the grateful sovereigns, bears the castle and lion of Castile as well as islands (left, below) and anchors symbolizing his discoveries.

Protecting a group of American Indians with her mantle, the Virgin blesses Spanish navigators and the craft they used at the time of the New World's discovery. Columbus (in gold) and the two Pinzón brothers kneel at right.

The men of the fifteenth century were sailing into an unknown that only the imagination had penetrated. Until 1433 no captain had dared take his ship beyond Cape Bojador on the northwest coast of Africa, for fear of entering the Ocean of Darkness, where the sea boiled and white men turned black. Using the compass and charts developed by Prince Henry's navigators, Portuguese sailors passed Cape Bojador, reached Cape Verde and the mouth of the Gambia —and returned to tell their story. But progress was slow. By the time Henry died in 1460, his caravels had still to round the southern tip of the African continent. Eighteen years later, Bartholomeu Dias was blown past the Cape of Good Hope in a storm; but even he was unable to carry on to India because his crew threatened to mutiny unless he returned home.

Prince Henry's great-nephew John II of Portugal was as eager as The Navigator to encourage his countrymen's exploration, and in 1484 he appointed a committee to deal with matters of discovery. In that year there appeared before this committee a tall, well-built man of striking appearance. Contemporaries never failed to mention the fairness of his skin, the brilliant red of his hair, and most notably, the piercing blue of his eyes. He had a startling proposition: he would reach the Orient for Portugal, not by sailing around Africa and across the Indian Ocean, but by sailing *west*. Since the world was round, as most informed people then realized, there was no reason at all, he argued, why a ship sailing due west from Europe should not reach the eastern coast of Asia. The man, of course, was Christopher Columbus.

BIBLIOTECA NACIONAL, MADRID

The Middle Ages' most popular travel book was Sir John Mandeville's Book of the Marvels of the World, *which embroidered a basic fabric of facts with such extravagant fictions as these mythical creatures, supposedly to be found in the mysterious Orient.*

It is more or less certain that Columbus, born in 1451, was the son of a Genoese woolen weaver. He too may well have been a weaver before he began making trading voyages in the Mediterranean in his early twenties. In 1476 he embarked on a trading mission to Lisbon and England. His fleet was attacked by Franco-Portuguese warships off the coast of Portugal, but Columbus managed to escape by swimming ashore. He stayed in Lisbon, married a Portuguese girl, and began earning his living as a mapmaker and dealer in books. Whenever possible, he went to sea.

As early as 1474, while he was still in Genoa, Columbus had begun corresponding with an eminent Florentine geographer, Paolo Toscanelli, about the possibility of sailing westward to Cathay and Cipango, the medieval names for China and Japan. Toscanelli had even sent the enthusiastic young man a chart showing the course a ship should take

BIBLIOTHEQUE NATIONALE, SERVICE PHOTOGRAPHIQUE, PARIS: MS. FR. 2810 FOL 222

Far to the East, medieval Europeans believed, lay the Garden of Eden, once the home of Adam and Eve. From a fountain within it poured the Rivers of Paradise: the Indus, Tigris, Euphrates, and Nile. In this illustration from a 1395 edition of Marco Polo's travels, enterprising seamen attempt to reach the Earthly Paradise by ship, a task bound to be frustrated by contrary winds and weather.

across the Ocean Sea, as the Atlantic was called. Soon, Columbus' "great and noble ambition to pass to where the spices grow," as Toscanelli put it, became an obsession. Years later he was to describe it to Ferdinand and Isabella as a fire burning within him. Avidly he studied the writings of scholars and travelers whose views about the extent and condition of the world's surface fitted in with his dream. He himself was far from being an expert cosmographer, and his fixation on the idea of a western passage led him to deny the claims of logic and ignore the opinions of contemporary experts. As a result, he greatly underestimated the size of the earth and overestimated the width of the Asian continent. In other words, he believed that the eastern coast of Asia was much nearer the western coast of Europe than it really is.

In believing this, Columbus was following accepted theory. The great Greek geographer Ptolemy had estimated the circumference of the earth at 18,000 miles, 7,000 miles short of the true circumference. This meant that the distance between the Canary Islands and Hangchow, in China, ought to be only 5,000 miles. Such authorities as Marco Polo and Toscanelli himself more or less agreed with Ptolemy. Marco Polo placed Japan 1,500 miles off the

Asian mainland and consequently only 3,500 miles away from Europe. Columbus, however, had misinterpreted him and optimistically estimated the distance between the Canaries and *China* as 3,550 miles. In fact, of course, it is just under 12,000 miles.

The idea that there was a vast continent blocking the way to Asia never occurred to Columbus. Although North America had already been visited several times by the Norsemen, it is extremely doubtful whether news of their discoveries had filtered down to southern Europe. Columbus expected, obviously, to come across various islands on his travels, but there is no doubt that the "Land to the West" that he set out to find was, to his mind, an integral part of Asia.

The Portuguese committee rejected Columbus' proposal, although they did not dismiss the idea itself. It was an understandable decision. Here was a group of nautical

This 1457 mapmaker's view of the world, based on Marco Polo's description of his journeys and looking much like a modern football, remained fairly standard in Europe until Columbus' discovery of America. Europe is recognizable at left, but east of the Red Sea nearly all is highly fanciful.

Epitomizing the wondrous world Columbus hoped to reach when he set sail in 1492 is this fifteenth-century view of Peking. Also based on Marco Polo's descriptions, the imperial city appeared as a detail on Fra Mauro's 1459 map of the Far East. At center are the towers of Kublai Khan's fabled palace.

experts being asked to entrust a naval expedition to a comparatively inexperienced sailor, who had never been in command of anything in his life. If Portugal's rulers should wish to embark on such a project, they had any number of well-seasoned master mariners to call upon; why resort to a foreigner?

Disgusted by his treatment in Portugal, Columbus decided to look elsewhere for aid and patronage. In 1485 he set off with his five-year-old son Diego to visit the court of Castile.

Instead of making the journey to Castile by land, Columbus chose to travel from Lisbon in a ship that was bound for Palos, on the Río Tinto in southwestern Castile. As the vessel sailed up the estuary of the Tinto, he noticed the Franciscan monastery of La Rábida, standing high above the river on a promontory. He realized at once that it would be an ideal place to leave his young son; he could not then guess that La Rábida would be his greatest source of encouragement in the difficult years to come.

By the time he and Diego had walked the dusty miles from Palos to the monastery, the little boy was hungry and thirsty and Columbus' first words, when the monastery gate swung open, were to ask for a piece of bread and a drink of water. As it turned out, not only were the friendly friars willing to give him alms and lodging, but Fray Juan Pérez, their prior, also lent a sympathetic ear to his grandiose schemes. Either in the monastery itself or as the result

BIBLIOTHEQUE NATIONALE, PARIS: MS. LAT. 1176 FOL 132

Franciscan friars, typical of those who befriended Christopher Columbus and his son Diego in Spain, are shown parading ceremoniously through a French town in this roundel from an illustration in a fifteenth-century manuscript.

of an introduction given him by Pérez, Columbus met Fray Antonio de Marchena, a Franciscan with a lively interest in navigation and a high reputation as an astronomer. "He alone never treated my ideas with ridicule," Columbus later wrote. His gratitude was justified, for no sooner had Fray Antonio heard what the would-be explorer had to say than he promised to help him.

The friar realized that it would be impossible to take the enterprise straight to the King and Queen. They were far too involved in the struggle to win back their own land from the Moors to attend to schemes for finding distant lands by an unknown and highly improbable route. Instead, Fray Antonio arranged introductions for his friend to the Duke of Medina-Sidonia, Castile's wealthiest grandee, and the Duke of Medina-Celi, one of the country's most influential noblemen.

Each duke in turn considered financing and equipping a fleet for Cristóbal Colón, as Columbus became known in Spain. The Duke of Medina-Celi, after entertaining the explorer as his guest for more than a year, finally decided that a venture of such great proportions demanded nothing less than royal support. Judging that Columbus stood a better chance of impressing the enthusiastic, intuitive Isabella than her cautious, calculating husband, Medina-Celi applied to the Queen. The order came back that Columbus should hasten to the court.

Early in 1486 Columbus arrived in Córdoba, the court's headquarters that year. In a room of the Alcázar the thirty-four-year-old visionary unfolded his dreams and hopes for the first time before Isabella, who was only a few months older than he.

The Queen listened to his persuasive words with an interest that was more than her usual gracious courtesy. Tremendous benefits could result from the discovery of a western route to Asia. Not only would her country hold the key to a storehouse of wealth and riches—including the almost legendary Moluccas, the Spice Islands that produced the world's supply of cloves and nutmeg—but, in addition, an alliance with the Great Khan would vastly increase Castile's power and importance in Europe. The enterprise would also strike a blow at the prestige of Portugal, which hitherto had monopolized the field of maritime exploration. These considerations appealed to Isabella the stateswoman.

Just as important for a woman of her extreme piety were the advantages that Columbus could win for Christendom. Since the fall of Constantinople in 1453, the Ottoman Turks had been steadily advancing into a divided, ineffectual, and increasingly frightened Europe. A western route to Asia would make it possible for the Christians to take the Turks from the rear in a pincer movement and save Europe by overthrowing their empire once and for all. Perhaps an even greater appeal to Isabella's crusading zeal, however, was the thought of the countless millions of benighted Asian natives who might then be won over to Christianity.

Yet even the political and religious ambitions of a queen have to succumb to economics. The Crown of Castile could not afford to finance an expedition for Columbus. Every maravedí Isabella could scrape together was being poured into the war against Granada. Not until the whole of Spain was Christian once again would the country be able to direct its attention and its resources elsewhere.

GRANADA CATHEDRAL: PHOTO, ORONOZ

Some sense of the splendor and rigid formality of Renaissance court life, into which Columbus suddenly found himself thrust, is seen in this bas-relief of a mace-bearer, in full livery, from a retablo at Granada. The mace was a ruler's ensign of authority, in this case Ferdinand and Isabella's, since the figure wears their coat of arms on his doublet.

At the same time, the Queen was reluctant to allow Columbus to go off and interest some other monarch in his project. She gave instructions that he be well looked after, and she appointed her confessor, Hernando de Talavera, as head of a commission to examine Columbus' proposals.

Contrary to legends that have risen since, the commission did not challenge Columbus' view of the world as a globe; in fact, most of them shared his belief. What they did challenge was his estimate of the distance across the Ocean Sea between Europe and Asia. They were sure the sea could not be that narrow (they were, of course, absolutely right), and they remained unconvinced. The commission adjourned without reaching any conclusions.

The years between 1486, when the commission was first set up, and 1492, when the Enterprise of the Indies was finally accepted, were difficult ones for Columbus. Later, he spoke of them with great bitterness. The self-elected discoverer of "Land to the West" became a shabby hanger-on at court, continually pestering the preoccupied monarchs with requests for further interviews. For a proud and sensitive man it must have been an agonizing ordeal.

Twice his hopes were raised only to be dashed to the ground. The commission of 1486 eventually advised the King and Queen to reject Columbus' project, since "his promises and offers were impossible and vain and worthy of rejection." Their reasons for this advice included the argument that the voyage between Europe and Asia would take three years and the contention that—even if it were possible to sail to the other side of the globe—it would be impossible to sail back because no ship could sail uphill. On another occasion Columbus was turned down at the last stage of negotiations. After countless delays and disillusions, he had decided to make Castile pay handsomely for the benefits he brought back. He demanded the ranks of ad-

Among the possessions that Columbus took with him when he sailed from Palos in August, 1492, was a 1485 Latin edition of Marco Polo's travels. Notes in the margin are believed to have been made by the discoverer himself, who apparently used the sketch of a hand to mark such interesting passages as "Here are found pearls in great quantity" and "shoally seas."

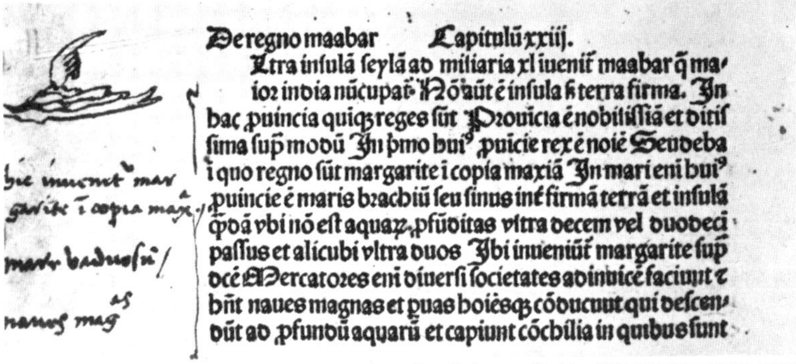
De regno maabar Capitulũ xxiij.
Ltra insulã seylã ad miliaria xl iuenit maabar q maior india nũcupat. Nõ aũt ẽ insula s terra firma. In hac puincia quinqꝫ reges sũt. Provicia ẽ nobilissiã et ditissima sup modũ. In pmo hui puicie rex ẽ noiẽ Seudeba i quo regno sũt margarite i copia maxiã. In mari eni hui puincie ẽ maris brachiũ seu sinus int firmã terrã et insulã qdã vbi nõ est aquaꝝ pfũditas vltra decem vel duodeci passus et alicubi vltra duos. Ibi inveniũt margarite sup dcẽ. Mercatores eni diversi societates adinvicẽ faciunt et hñt naues magnas et puas hoiẽsqꝫ cõducunt qui descendũt ad pfundũ aquarũ et capiunt cõchilia in quibus sunt

POLO, *De constituciones . . . Orientalum*, 1485: COLUMBIAN LIBRARY, SEVILLE

TOLEDO CATHEDRAL: PHOTO, ORONOZ

At the time Columbus was seeking their aid, the monarchs were preoccupied with the conquest of Granada. An incident of the campaign, the attempted assassination of Ferdinand and Isabella in their camp outside a besieged city, appears on the 1495 Toledo choir stall above. As the rulers quietly read in their tent, a Moor claiming to be a prophet (left) steals toward them. Guards (right) rush to seize him.

miral and of viceroy of any lands he discovered and a tenth part of all the gold, silver, and jewels they might yield. Discussions were broken off at once; the rewards he asked were ridiculously, alarmingly high.

At one stage, tired of waiting for Castile to make a decision, Columbus reopened negotiations with Portugal. Yet again his hopes were shattered when Bartholomeu Dias rounded the Cape of Good Hope, and King John II understandably decided that one sea route to the Orient was enough for Portugal. In an attempt to interest Henry VII of England or Charles VIII of France in the great enterprise, Columbus' brother Bartholomew took ship for the north while Columbus himself returned to Spain.

How Columbus supported himself during these years of frustration is a mystery. He was probably based in Córdoba, but he traveled around in the wake of the court, and he was certainly present at the sieges of Málaga and Baza. From time to time, he must have visited his son and his friends at La Rábida. His spare hours were no doubt devoted to learning as much about ocean navigation as was possible for a man living on land. Seeking confirmation of his theories, he read and reread his geography books, noting and underlining every mention of Asia, its fertile soil, its fabulous riches, its spices and silks, and its great cities. Not for one moment, obviously, did he lose faith in his dream.

Libre . . . dels fets maritimes, 1502: BIBLIOTECA NACIONAL, MADRID

Looking very sturdy in a facsimile from a maritime tract dated 1502, this tiny caravel—reminiscent of the ones used by Columbus on his famous voyage—must have seemed to her crew slim protection against the vast, watery unknown.

Columbus did not entirely lack friends and supporters. Juan Pérez, the prior of La Rábida, had once been the Queen's confessor, and late in 1491 he made a special journey by mule to the military camp of Santa Fe to beg the Queen to grant Columbus an audience. Isabella was preoccupied with arranging the surrender of Granada, and although she did agree to see Columbus once again, the interviews ended badly. Angry and depressed, Columbus set out for France to seek another patron. Fortunately, he had another good friend at court: Luis de Santángel, Ferdinand's Keeper of the Privy Purse.

When Santángel heard that the monarchs had firmly bid Columbus farewell forever, he hurried to the Queen and told her "he was astonished to see that her Highness, who had always shown a resolute spirit in matters of great pith and consequence, should lack it now for an enterprise of so little risk, yet which could prove of so great service to God and the exaltation of his Church, not to speak of very great increase and glory for her realms and crown." By way of more concrete encouragement, the courtier offered to finance the expedition himself.

The Queen was so much affected by Santángel's words that she at once sent a servant on horseback after Columbus. Her messenger overtook the explorer at a village some four miles from Santa Fe. Columbus, full of excitement at the dramatic unexpectedness of the summons but inevitably wary after his previous disappointments, turned his mule's head round and plodded back. After an hour's audience with the King and Queen, the enterprise was accepted on Columbus' terms. Three months later, on April 17, 1492, their agreement was formally sealed, and the Genoese adventurer walked out of the audience chamber a gentleman, an admiral, a viceroy, and a governor-general of lands not yet discovered.

What made both Isabella and Ferdinand decide to support Columbus? They could have had no inkling that when he left Granada, scene of their greatest triumph to date, he was setting out to lay a new world at their feet. But it is almost certain that both monarchs, and in particular Isabella, saw the enterprise as part of their glorious vision of an expanding, crusading Spain. Alarmed by Columbus' excessive demands, Isabella had momentarily forgotten her queenly mission to make her reconquered country preeminent in Europe. Santángel's direct and forceful speech, reminding her of this, stung the Queen into action.

How had the wily and suspicious Ferdinand been convinced that Columbus was indispensable to the expedition? After all, Spain, as well as Portugal, possessed expert navi-

Palos' busy port provides the setting for this broadly romanticized version of Columbus' departure, by a sixteenth-century engraver, Théodore de Bry. At right, the navigator bids farewell with a flourish to a stiffly formal Ferdinand and Isabella, as his crew (left, background) take longboats to the ships.

India occidentalis, 1590: NEW YORK PUBLIC LIBRARY

The earliest-known representation of Columbus' discovery of America (right) provides a highly imaginative view of the occasion, complete with stylized trees and huts and a passive race of natives, clad only in fig leaves, warmly welcoming the explorer. At home, Ferdinand (left, foreground) directs all across a miraculously narrowed Atlantic. Below is a sixteenth-century plan for the defenses of Santo Domingo, the oldest city in the New World. Ships appear in the harbor at right.

DATI, *La lettera dell'Isole . . .*, 1493: NEW YORK PUBLIC LIBRARY

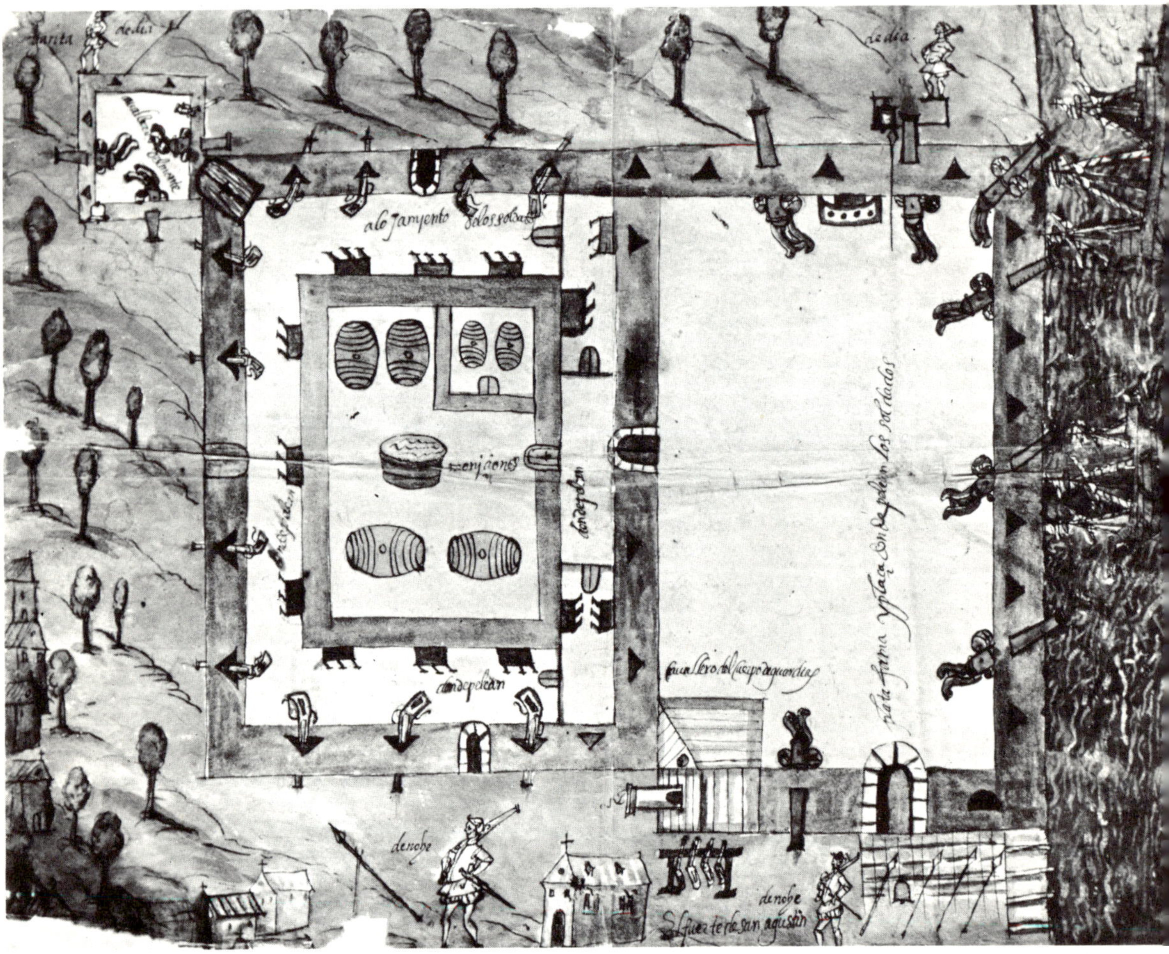

BRADLEY SMITH

gators, and Columbus had far less claim than they to be admiral of a naval expedition. He had, however, one quality that they had not: he was Genoese, a native of one of Europe's wealthiest trading cities. Ferdinand very probably saw Columbus as a means of tapping the vast resources of Genoese capital and concluded that there was nothing to lose and everything to gain by supporting this particular adventurer.

In the spring of 1492 Columbus returned to Palos, the scene of his landing in Spain seven years earlier. The little Atlantic fishing port was ordered to supply him with two fully equipped caravels for a year at its own expense, and Columbus chartered a third, larger vessel to serve as the expedition's flagship.

It took three months to man and fit the fifty-ton *Pinta*; the *Niña*, ten tons lighter; and the clumsy, hundred-ton *Santa María*. Columbus easily persuaded Martín Alonso Pinzón, head of Palos' chief shipping family, to join the expedition as co-leader, but finding seamen proved much more difficult. The Ocean Sea was regarded by ordinary mariners as unnavigable, and the men of Palos did not wish to be the first to show that it was not. They might never reach land, or if they did, they might never be able to come back. They might meet whirlpools, terrible reefs, or storms of gigantic ferocity. They might even meet enormous monsters of the deep that would drag them and their flimsy caravels down to a watery grave.

Eventually, however, thanks partly to an order suspending all civil and criminal processes against men who signed on with Columbus, and partly to the lure of Oriental treasure, ninety men were found to sail the three ships. As dawn broke on August 3, 1492, Columbus, armed with a letter of recommendation from the monarchs of Castile to the Great Khan and the other lords of India and the East, set out on one of the most momentous voyages the world has ever known.

Two months followed, months of fair wind and fine weather, of false alarms and sea mirages, with Columbus faking his entries in the log book so that his men should not become alarmed at the great distance they were from home. On October 12, when all on board the ships were almost completely disheartened, the lookout on the *Pinta*'s forecastle shouted the longed-for words: "*¡Tierra! ¡Tierra!*" The expedition had reached San Salvador, or Watlings Island, as it is also called, one of the smaller islands of the Bahamas. A new continent had been found.

Although it was sixty miles up the Guadalquivir River from the Atlantic, Seville became the principal Spanish port for trade with the New World. Wealth from the Indies soon turned it into the busy, prosperous city depicted above

MAS

by Alonso Sánchez Coello, a late-sixteenth-century court painter. The Giralda, a magnificent Moorish minaret completed in 1195, looms next to Seville Cathedral on the horizon, as caravels and galleys vie for space along the waterfront.

WILLIAM L. CLEMENTS LIBRARY, UNIVERSITY OF MICHIGAN

The Spaniards' relentless search for New World riches led them to enslave and slaughter thousands of Indians in the Caribbean, Mexico, and Central and South America. A 1582 edition of a Dominican friar's exposé of this cruelty included this scene of the conquerors beating and torturing chained natives being treated as beasts of burden.

After visiting other small islands, and discovering Cuba and Hispaniola, where the *Santa María* was wrecked on a coral reef, Columbus returned to Spain on board the *Niña*. In April, 1493, he presented himself at court in Barcelona. Splendidly dressed, he rode into the city at the head of a cavalcade of officers, servants, and six Indian natives he had brought back to be baptized. The Indians were dazzlingly decked out in their finest clothes, with gold bracelets on their wrists and ankles; they carried pearls, gold masks, strange fruits, and exotically feathered parrots to present to the King and Queen. When Columbus entered the throne room, which was packed with eager courtiers in their richest clothes, Ferdinand and Isabella rose to greet him, then made him sit beside them to relate the marvels of his journey. No greater honor could have been bestowed upon the explorer.

The story of Spain and Columbus was not yet over, and it was to prove a sad one. The trouble was that the rulers of

Spain, like Columbus and the men who sailed with him, expected that the discovery of the land they believed to be Asia would yield immediate results. But the coasts of "the Indies" were not lined with spices, jewels, and gold for the taking; the true riches of America, of its soil and climate, would have to be worked for.

In the years between 1493 and 1500, Columbus made two more voyages to the Caribbean, discovering many more islands, among them Puerto Rico, Jamaica, and Trinidad, and establishing permanent Spanish colonies on Hispaniola. During his third voyage, the explorer's vessels sighted the coast of what is now Venezuela. As his tiny ships bobbed on the dangerous breakers formed by the waters pouring from the delta of the Orinoco, Columbus realized that so vast a river must rise deep within a continent. Excited by having found mainland "Asia" at last, he made for Hispaniola.

The Spaniards' settlements there had not prospered. Finding little gold, the discontented colonizers had resorted to open warfare with the natives they had originally come to convert and to quarreling among themselves. The disease-stricken Spaniards who made their way home proved a bad advertisement for the "fabulous Indies," and few new recruits could be found. Columbus was a fine seaman but he had no gifts as an organizer, and he resented taking time from exploration to administer the affairs of his troublesome viceroyalty. Feeling against him both in the Indies and at court ran so high that the King and Queen sent a new chief justice, Francisco de Bobadilla, to investigate the situation. Bobadilla, obviously prejudiced against Columbus, used his supreme authority to arrest the former viceroy and send him to Spain for trial. In November, 1500, Columbus arrived at Cádiz, in chains.

BENZONI, *La Historia*, 1572: NEW YORK PUBLIC LIBRARY

Europeans were fascinated by Indian customs, such as sleeping in hammocks—a practice of Arawaks of Venezuela and the Caribbean.

When Isabella heard of his plight, she ordered him to be set free and his rights restored. But obviously, both she and her husband had lost some of their confidence in Columbus. Out of pity and gratitude they sent him on a fourth and last voyage in 1502. They were not, however, prepared to reinstate him as governor of the Indies. The new voyage was to be one of exploration and discovery, nothing more.

On this voyage Columbus explored the coastal region of Central America (which he again took for Asia) and touched the fringes of its great Aztec and Mayan civilizations. He was later marooned for twelve months in Jamaica and almost abandoned by his men. In November, 1504, he returned to Spain a broken man, crippled with arthritis,

TOLEDO CATHEDRAL: PHOTO. ORONOZ

bowed and white-haired at the age of fifty-three. On November 26, Isabella died, and Columbus' last hope of being reinstated as viceroy died with her. The colonies in the Indies were beginning to flourish without him, and Ferdinand was not a man to put sentiment before gain.

Although Columbus' continual entreaties could not persuade Ferdinand to accord him all the financial privileges to which he thought he was entitled, the explorer ended his life as a wealthy man, since he duly received a share of the revenues from the Indies. Nevertheless, he was profoundly discontented. He wanted his honors and titles, he wanted his rights and exemptions, he wanted to secure his family's future. Above all, he resented other, younger explorers taking advantage of his discoveries. Struggling painfully around Spain on his mule in the wake of the court, he alternately demanded and begged justice from Ferdinand. It was all in vain; he did not even live to see his son Diego appointed governor of the Spanish colony on the island of Hispaniola in 1509.

Christopher Columbus died in Valladolid on May 20, 1506, believing to the end that he had landed on the continent of Asia. Of the true importance of his discovery he had no idea. In 1493, the year of his triumphant return from the New World, Pope Alexander VI issued a bull that proclaimed the partition of all the newly discovered lands between Portugal and Spain. The line of demarcation, as finally fixed, ran from pole to pole three hundred seventy leagues west of the Azores. All discoveries east of this line were to belong to Portugal; all discoveries to the west, to Spain.

Columbus' dream and Isabella's breadth of vision had begun the most extraordinary, the most splendid, and the most terrifying chapter in Spanish history. Between them, they gave Spain a fabulously rich and powerful empire, although the discovery that led to Spain's greatness also carried within it the seeds of the country's decay.

To the world Columbus gave a new continent, and the world rewarded him by naming it after another man—the adventurer who first suggested that the land was not, as Columbus believed, a part of Asia—Amerigo Vespucci.

RUGE, *Des Zeitalters der Entdeckungen*, 1881: NEW YORK PUBLIC LIBRARY

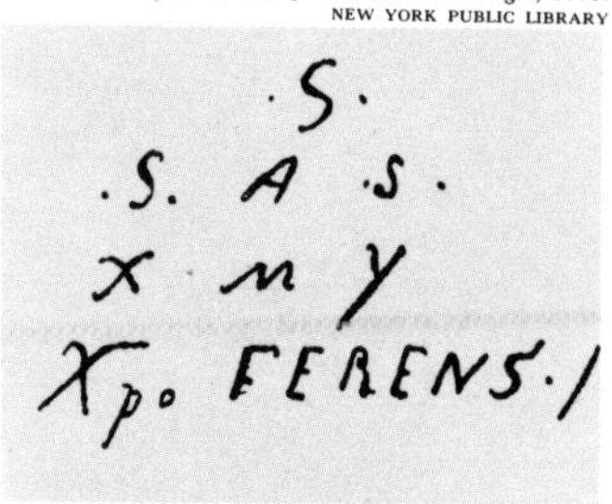

Columbus' monogram—"I am Christopher, servant of the Most High Saviour, Christ, Son of Mary"—combines both Latin and Greek initials. Here it appears on a letter to the Catholic Kings, discussing navigation, and sent from Granada on February 6, 1502.

The six-foot-tall ciborium at left, ordered by Cardinal Cisneros for the Toledo Cathedral, contains gold found by Columbus on his voyages. The precious receptacle for the Host is typical of the elaborate filigree work of Spanish silversmiths that led to plateresque architecture (see page 104).

V

A RENAISSANCE COURT

GRANADA CATHEDRAL: PHOTO, ORONOZ

Four tiny faces peer from a circle of exquisite metalwork adorning a reliquary, once Queen Isabella's mirror. After her death, the looking glass was replaced by the glass case at center for a religious relic.

As the banner of the Cross was raised at Granada in January, 1492, Ferdinand and Isabella and all their court wept for joy. The following year they gazed with excitement and curiosity upon Columbus and his strange trophies from the Indies. These two events were turning points for Spain and the rest of the world, yet they are only a part of the story of Ferdinand and Isabella. It is impossible to understand their reign and appreciate them as people without also knowing how they lived from day to day and how life at their court was profoundly affected by the cultural changes that were then sweeping Europe.

Partly because of the restless energy of the sovereigns and partly because of such practical considerations as food supplies and sanitary arrangements, the court of Ferdinand and Isabella was constantly on the move. Dotted all over Spain were royal palaces and castles where the courtiers would stay when the monarchs had business in the surrounding area. If there was no royal *alcázar* (a Spanish version of the Arabic word for "castle") conveniently at hand, the sovereigns would be entertained by one of their nobles or accommodations in a monastery would be arranged for them. But if nightfall caught the royal travelers unaware, they were quite prepared to stay at a roadside inn, and when there was not even an inn within sight, up went a forest of tents in the middle of the countryside.

On its travels the court stayed in some magnificent buildings. Not all the castles of Castile were stern, forbidding guardians of its frontier. Some of the city *alcázares* were both pleasant and comfortable, with delicate, intricately wrought architecture and cool, green gardens and fountains in the Moorish fashion. After 1492 the King and Queen had

Looking younger than they actually were when a Flemish artist completed this romantic tapestry, Ferdinand and Isabella (left) reign in Renaissance style, surrounded by richly costumed courtiers in a lush pastoral setting.

LERIDA CATHEDRAL: MAS

COURTESY H. F. BAERWALD

Perched on a cliff, 262 feet above its moat, and boasting a commanding view of the countryside, this castle at Segovia was the Queen's favorite. A fairy-talelike fortress, with light, unornamented walls and fanciful turrets, the structure, begun in the eleventh century, was altered somewhat by John II of Castile in the fifteenth century.

at their disposal the loveliest royal residence in Europe: Granada's exquisite Alhambra and its no less beautiful summer palace, the Generalife, both triumphs of Moorish art and inspiration. Isabella's favorite home was neither of these buildings, but rather the turreted castle of Segovia. Perched on the edge of a cliff, it offered a magnificent view of a countryside that was particularly dear to her.

From the outside, these buildings were certainly worthy of their royal occupants, but inside they were remarkably plain. Not only would it have been impractical to fill with luxurious furniture and hangings a residence that the King and Queen might visit only once in five or six years, but also the standards of interior decoration and domestic comfort were austere ones. The Queen lived with a minimum of furniture, no carpets, few hangings, and very little gold and silver plate, and she encouraged her nobles to do the same.

One royal residence, however, was furnished in a style rather closer to the popular idea of the luxury befitting a lord of the land. This was the palace of Almazán in Soria, in the heart of Old Castile, which Isabella chose as the home of her only son, Prince John. Its large public rooms were furnished in the style of the period, with sparse, massive furniture, long trestle tables, cushioned benches, and huge open fireplaces. But the young Prince's own suite revealed the lavish touches of a devoted mother. There were silver mirrors and basins; a manicure set; a silver ewer and bowl for the Prince to use when washing his hair; and large, carved chests to hold his velvet and brocaded clothes. On his dressing table were jars of incense, rosewater, and such perfumes as musk, benjamin, and rosemary. Isabella saw to it that he enjoyed the luxuries she herself had been denied in the remote castle where she had spent her childhood.

BIBLIOTECA NACIONAL, MADRID: PHOTO, ORONOZ

Handsome little Prince John is believed to be the subject of this miniature, used as marginal ornamentation in a lavish missal made for Cardinal Cisneros after 1503.

The monarchs were continually surrounded by retinues of servants, ladies- and gentlemen-in-waiting, chaplains and confessors, doctors and teachers. By 1497 there were eleven hundred people in attendance at court—yet Ferdinand and Isabella managed to lead comparatively austere lives. They ate very sparingly and drank little wine, but they required the servants who waited on them to be well mannered and attractively dressed. The monarchs themselves wore simple clothes. On occasions, however, Isabella indulged in the visual splendor of majesty. When in 1474 the Burgundian ambassador arrived to present Ferdinand with the coveted order of the Golden Fleece, Isabella received him wearing a skirt of ruffled green velvet, with an overdress of crimson brocade held at the neck with her famous ruby and pearl necklace. Later, presiding over a tourney held as part of the celebrations, she changed to a magnificent robe of crimson and cloth of gold and wore her gold crown. Even on such occasions, however, she refrained from exorbitant expenditure. Nothing outraged her prudent soul more than the constant competition among her courtiers that resulted in their spending far more than they could afford on personal finery.

The unusually active part they took in their kingdoms' affairs meant that Ferdinand and Isabella led extremely busy and exhausting lives. But they did not spend all their time on the battlefield, on the road, or on their thrones; they had their recreations. Ferdinand loved to hunt and ride, and all his life he excelled at these sports. Isabella's

BIBLITECA DE PALACIO REAL, MADRID: PHOTO, ORONOZ

extraordinarily strenuous way of life brought action and exercise enough. In addition, she spent long hours on her knees in prayer or confession.

In spite of her decisiveness and her remarkable capacity for action, Isabella was a woman of great serenity. Some of her leisure was spent sitting quietly and sewing, embroidering, or spinning with her ladies-in-waiting. Sometimes they would intone prayers as they worked; sometimes they would sing or listen to the court musicians, of which Isabella kept a permanent establishment of forty. On other occasions a *juglar* (minstrel) entertained them with the newest frontier ballad inspired by the struggles with the Moors, or a courtly poet recited a ditty of praise dedicated to the Queen or his latest lament of frustrated love. If Isabella was in a particularly lenient mood, she and her ladies might even dance together.

Very few other amusements were available, for by later standards the court was distinctly unsophisticated. The

TOLEDO CATHEDRAL: MAS

Ferdinand, like the noble shown above with his servant, falcons, and dog, loved to hunt. Isabella's long hours at silent prayer are reflected in the miniature at left, illustrating her own Book of Hours. As angel musicians hover, the Queen kneels in solitary devotion before the Virgin and Child.

HISPANIC SOCIETY OF AMERICA, NEW YORK

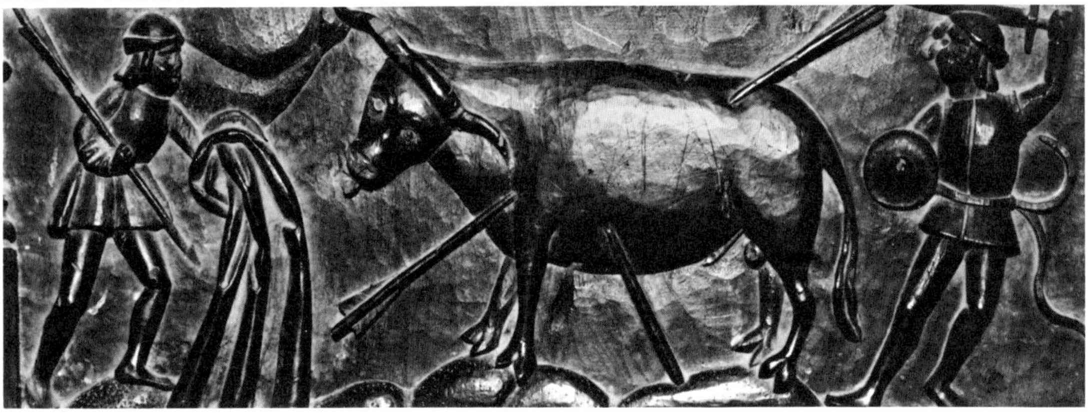

An early version of a bullfight, later Spain's national sport but one to which Isabella sorely objected, is depicted in this bas-relief from a choir stall. Unlike modern bullfighters, these heroes of the ring seem well-armed against their unhappy opponent, one with a spear and the traditional cape, the other with sword and shield.

theatre as we know it did not exist. The only forms of drama were short religious pieces and brief snatches of humorous dialogue, usually between two peasants or shepherds, that often preceded these miracle and nativity plays. The Church, in fact, with its services, its plays, and its splendid processions, was the chief source of color and spectacle. On special occasions, however, there were elaborately staged tournaments. And a form of bullfighting was growing ever more popular, although Isabella disliked the sport intensely and tried to prohibit it.

Reading was the Queen's favorite way of spending her few spare hours of leisure. In the Alcázar at Segovia she kept a personal library of over two hundred fifty volumes, carefully catalogued and well thumbed. Her interests ranged from expositions of the Scriptures and lives of the saints, through histories of Spain and the classics of Greek and Roman history and philosophy, to the legal codes and ordinances by which Spain was governed. Several books on medicine and astrology reflected Isabella's curiosity about subjects that sovereigns as a rule gladly left to their scholars. The presence in her library of treatises on the arts of hunting and hawking no doubt resulted from her desire to share in the sports beloved by Ferdinand.

The Queen's choice of reading matter was by no means limited to the informative and instructive. She possessed a collection of songbooks, both Spanish and French. Her taste for the creative and romantic led her to read the works of Spain's medieval poets and to enjoy the adventures of King Arthur's knights and their search for the Holy Grail and such classical epics as Virgil's *Aeneid*. She even had on her shelves a collection of Oriental fables and the scandalous tales of the Italian writer Boccaccio.

These books, of course, were the accumulation of a lifetime. Many of them Isabella would never have read but for the lively interest she took in the new learning of the Renaissance that was spreading across Europe from Italy. Such was her eagerness, in fact, to keep Spain and herself abreast of all that was going on in the world that, as a middle-aged woman, she set herself—and accomplished—the difficult task of learning Latin well enough to converse with foreign ambassadors to her court. Aware of the defects of her own education, she resolved to give her children every cultural benefit. Her heir and the apple of her eye, the frail and precocious Prince John, received the best instruction that could be bought for a prince of the Renaissance; his sisters, Isabella, Juana, Maria, and Catherine, also were brought up to be highly educated, enlightened young women. To tutor them, the Queen brought from Italy two

COLEGIATA, TORO: MAS

Robed in a sumptuous gown of handsome Spanish brocade, her shining hair hanging freely about her shoulders, the lady of this detail from a fifteenth-century Flemish painting, done in Spain, is said to be Isabella. Typical of so many of her portraits, it shows her at prayer, her fingers gracefully marking her place in an open missal.

BIBLIOTECA NACIONAL, MADRID, VIT. 17-1: PHOTO, ORONOZ

Elio Antonio de Nebrija, the famous Spanish scholar of Ferdinand and Isabella's time, appears in the frontispiece (right, at center) to a late-fifteenth-century edition of his work Latin Institutions. *In the left foreground, the Master of Alcántara reads along as Nebrija expounds on his grammar.*

of the most eminent humanists of the age: Antonio and Alessandro Geraldini. To the arts and skills familiar to their mother—reading, writing, sewing, weaving, spinning, a little history and geography, genealogy, heraldry, and scriptural knowledge—the four future queens added Latin. Their study of appropriate works of history, drama, poetry, and philosophy gave them a remarkable fluency in that tongue. The youngest, Catherine, who became the wife of Henry VIII of England and a champion of female education in her adopted country, was regarded by Erasmus and Sir Thomas More as "a miracle of feminine learning."

Under Isabella's influence the Renaissance flourished in Spain. In the field of the plastic arts, however, its full flowering was to be delayed half a century. With the estab-

In the Spanish retablo painting at right, the Virgin Mary as a girl (shown with a halo) displays her tapestry sample to an approving teacher. The palatial setting more accurately suggests the instruction in domestic arts given Renaissance ladies of noble birth, such as Isabella and her daughters.

CHURCH OF SAN FRANCISCO, VILLAFRANCA DEL PENEDES: MAS

lishment of a regular flow of bullion from the Spanish empire in America came the capital that financed a spate of churches and palaces incorporating the latest styles in architecture. They housed superb canvases, intricate ornaments, and decorative furnishings that were the marvel of Europe. Even though painting, sculpture, and architecture were still relatively undeveloped, the reign of Ferdinand and Isabella saw an astonishing revival of interest in what are now known as the humanities.

The printing press—the greatest invention of the age—had made its first appearance in 1450, the year before Isabella was born. Early in her reign, at a time when the Paris parliament was refusing licenses to would-be printers in France, the farsighted Queen gave Spanish printers

HISPANIC SOCIETY OF AMERICA, NEW YORK

This 1502 woodcut showing an author presenting Ferdinand and Isabella with his work reflects the great interest they took in the printed word, allowing printers special exemptions and permitting books to enter Spain duty-free.

special tax exemptions to encourage them in their trade. Later, to facilitate an even greater spread of reading matter, Isabella removed all duties on books entering the country from abroad. Along with the books, a continuous stream of foreign scholars and humanists flowed into Spain, and chairs of Hebrew and Greek were established at the medieval university of Salamanca, which rapidly became one of the chief centers of European learning.

A leading influence on this vital new movement was the man who had been the Queen's confessor and remained her friend and advisor after his elevation to the Archbishopric of Toledo: Francisco Jiménez de Cisneros. The Archbishop shared Isabella's high ideals and burning faith; like her, he saw religious reform as being inseparably linked with educational and cultural innovations. Very much in the spirit of the new humanism, he believed that his priests and monks would lead richer spiritual lives and purer physical lives only if they were better equipped to understand the theological sources from which Church doctrine had evolved. In 1497, Cisneros engaged Pedro de Gumiel, the renowned Spanish architect, to submit a plan for a new university at Alcalá de Henares—a university that would embody the spirit of Isabella's Spain.

The difficulty of attracting scholars and professors to the new foundation was overcome by granting special tax exemptions to all who joined it. Among the first professors was Elio Antonio de Nebrija, a man of encyclopedic learning. Nebrija had been educated in Italy, but in 1473 he returned to Spain, where he rapidly became one of the leading lights of Salamanca University. By the time Cisneros persuaded him to join the faculty at Alcalá, he was a scholar of international fame, with a Spanish grammar (the first vernacular grammar to be published in Europe) and a Spanish-Latin dictionary to his credit. His daughter, Doña Francisca de Nebrija, also lectured to the students at Alcalá—just as another woman, Doña Lucía de Medrano, taught the students at Salamanca. In no other country did women participate so fully in the new learning—but after all, Spain had Europe's only reigning queen.

After Isabella died in 1504, Cisneros carried on alone the task of fostering culture and learning in Spain. The

The dramatic Pietà at right, by Flemish artist Rogier van der Weyden, was but one of many foreign works in the Spanish monarchs' collection. Such acquisitions testify to unrestricted cultural exchanges during their reign, and unhappily, to a dearth of major, distinctively Spanish painters.

GRANADA CATHEDRAL: PHOTO, ORONOZ

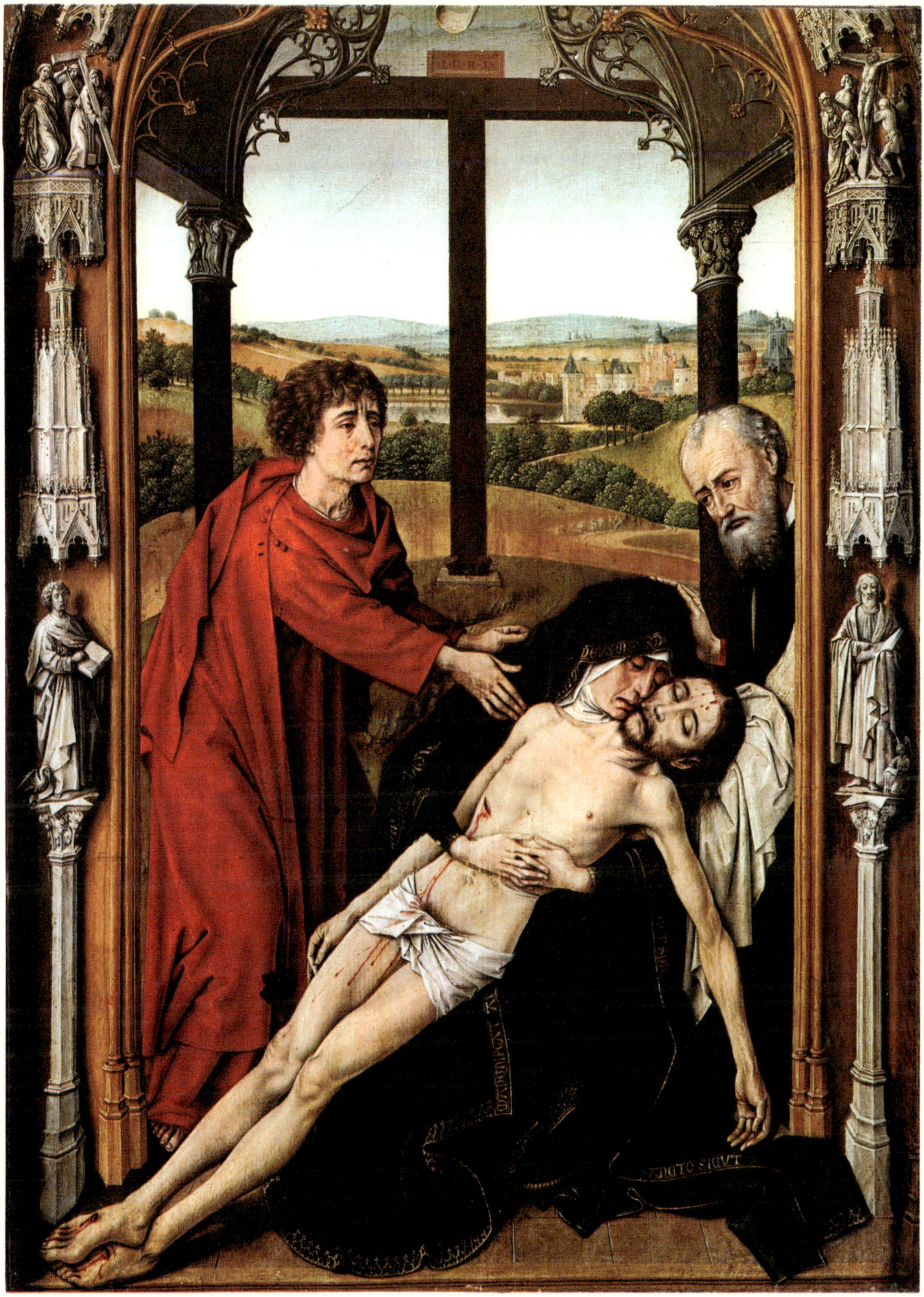

Queen had been one of his greatest inspirations; he had admired and respected all she stood for, and he was determined to keep alive the new way of life she had symbolized. The concern they had shared about the low level of Biblical studies in Spain led Cisneros to undertake a monumental project: producing the world's first Polyglot Bible. Over the centuries the original scriptural texts had become highly inaccurate. Cisneros' idea was to print side by side the Hebrew, Aramaic, Latin, and Greek versions of the Old Testament and the Latin and Greek texts of the New.

COURTESY H. F. BAERWALD

The impressive portal at the University of Salamanca provides a classic example of the profusely but delicately ornamented plateresque style of architecture, prominent in Spain in the early 1500's. In the detail above, a medallion of the Catholic Kings is below the escutcheon of their heir, Charles V.

At Alcalá University the Archbishop collected a group of nine scholars that included Nebrija; Demetrius Ducas of Crete, whom Cisneros had invited to Alcalá as professor of Greek; and three learned Jewish converts. The best printers were brought from France and Germany, and no expense was spared in collecting Biblical manuscripts from far and near. The enterprise of publishing the Polyglot Bible took fifteen years, and its total cost eventually reached 50,000 gold ducats, some $935,000 in modern currency. The last volume was completed only four months before Cisneros died in 1517, but even in those four months the Archbishop had already started planning an edition of the complete works of Aristotle. With his death, however, the project was abandoned, and along with it died the impulse toward organized scholarship in Spain.

Literature, as well as scholarship, thrived during the reign of Ferdinand and Isabella. For the first time the sons of burghers and the minor gentry began to enter a literary world that the aristocracy and the clergy had hitherto monopolized. The classical and Italian Renaissance writers were translated into Spanish. Most poets tried their hand at the new techniques these translations made available and they all found a seemingly inexhaustible subject of inspiration for their allegories and sonnets: love.

At court, relations between the sexes were governed, on the surface at least, by the rules of courtly love, a conception of human love first expressed in the twelfth century by Provençal troubadours. This love was based on suffering that ennobled and refined the spirit. Since suffering was an essential ingredient of true love, the beloved one had to be unattainable—that is, married to someone else. Courtly love, in fact, was impossible between husband and wife, and undoubtedly the prevailing system of arranged marriages made this a realistic state of affairs. On the other hand, it was accepted that the noble heart could not long be untouched by passion. As a result, devotion, or a pretense of

EVELYN HOFER

Magnificent lacelike stonework, characteristic of the plateresque style, entwines a score of tiny cherubs and heraldic figures along a balcony of the Colegio de San Gregorio, founded by Ferdinand and Isabella.

A precious souvenir of Renaissance courtly love is this 1470 songbook, with a lady and her suitor strolling across its page. Closed, the book forms one heart; open, two hearts, joined. The song above: "My sweet lady do not abandon me."

BIBLIOTHEQUE NATIONALE, PARIS

devotion, to some lady was expected of every young courtier. Custom equally required that each young gallant should express this devotion in poems of impassioned longing and frustration. The songbooks and sentimental novels of the age are full of the antics and lamentations of lovers in the grip of their real or pretended passion.

Toward the end of the fifteenth century it became the fashion to use religious terms and symbols to express the ravages of love. The sufferings of the lover were likened to those of Christ on the Cross, and the poet called his lady *mi dios* (my god). (The present-day use of the word "adore" derives from the practice of courtly love.) Even Isabella did not escape. Soon after she became Queen, a poem written in her honor proclaimed that her beauty was so great that had she been born before the Virgin Mary, *she* would have been chosen as the Mother of God. Such blasphemy did not go unnoticed; churchmen in their pulpits and in their writings called down the wrath of Heaven upon all those who thus exalted woman and love.

Despite the vogue of courtly love and the themes and poetic forms of the Italian humanists, Spanish writers did not neglect the traditional poetry of their own country. On the contrary, as courtly poets began writing original versions of *romances* (ballads) and *villancicos* (lyrical songs), which hitherto had been the stock in trade of popular minstrels, a valuable link was formed between the traditional and the artistic styles in Spanish literature.

The influence traditional poetry exercised in Spain pinpoints a vital difference between the Spanish Renaissance and the movement in Europe as a whole. Spain took what it needed from the Renaissance and rejected the rest. While Italian literature had definite nonreligious, even pagan, tendencies, in Spain there was a marked increase in religious poetry and a revival of religious scholarship. In northern Europe the impetus toward religious reform took such an anticlerical tone that it threatened Church doctrine itself; Archbishop Cisneros' reform of the Spanish Church was never anything but orthodox.

Isabella, an intelligent and in many ways enlightened woman, was desperately eager for Spain to participate in all the advantages of the new learning. At the same time, she considered Spanish unity and well-being to be inseparable from religious unification and orthodoxy. With these nothing, not even the Renaissance, must be allowed to interfere. Under her influence the Spanish Renaissance took a course that was to make a Spanish Reformation impossible.

PIERPONT MORGAN LIBRARY, NEW YORK: MS. 450 FOL 126 V

A slyly irreverent artist of 1468 provided this charming tribute to his country's favorite sport—bullfighting—as ornamentation for a missal. The bull, emerging from a pomegranate, a popular fruit among the Spaniards and symbol of Granada, is playfully pierced with arrows by two impish children.

VI CHRISTIANS,

BIBLIOTHEQUE NATIONALE, SERVICE PHOTOGRAPHIQUE, PARIS: MSS. FR. 20.068 FOL 125

JEWS, AND MOORS

The year 1492 is a landmark three times over in Spanish history. It was the year of the final conquest of Granada, which ended Moorish rule in Spain. It was also the year Columbus discovered the "Land to the West," which was to become a vast Spanish overseas empire. By themselves, these two events would be enough to mark the end of one era and the beginning of another.

Perhaps no less momentous to the shaping of the country Spain was to become was the edict of March 31, 1492, that forced every Jew in Spain to choose either converting to Christianity or leaving the country forever. The cruelly hard decision between baptism and exile had to be made within four months. A decade later, on February 12, 1502, all the Moors who still lived in Castile were ordered, in their turn, to make the same choice. Moors and Jews alike, they knew no other home; for centuries their families had formed an essential part of the fabric of Spanish society. Now they had to make the agonizing choice between their religion and their country.

In issuing these edicts, Ferdinand and Isabella hoped to establish Spanish unity, at a time that seemed particularly propitious. To the monarchs the decrees represented the only practical solution to the problem of co-existence in a society composed of three separate races and three of the world's great faiths: Christian, Moslem, and Jewish. Unfortunately, the result of these drastic measures was to be not a genuinely unified country, but a closed society in which racial purity was taken as the sole guarantee of a man's true worth.

There had been a time when the Jews, Moors, and

A popular symbol of the Church as a citadel besieged by the Seven Capital Sins (all ladies) and heretics blindfolded against the True Faith appears in the fifteenth-century French manuscript illumination at left. In Spain the defense of Christianity led to the cruel persecution of the Inquisition.

STADTBIBLIOTHEK, BERNE: MSS. HIST. HELV. I, 1-3, P. 44

A late-fifteenth-century Swiss manuscript depicted the ritualistic murder of a Christian child, the crime of which Jews throughout Europe were so often accused. People actually believed that the victim's blood (which the man at left has come to collect in a pot) was used in Hebrew religious rites.

Christians of the Iberian peninsula had lived together in comparative harmony. The first large-scale persecution of the Jews took place under the Visigoths in the seventh century. Under the liberal conditions of the first four hundred years of Moorish rule, the Jewish population prospered. However, in the twelfth century, Moslem fanatics began persecuting the members of other faiths, and Jews and Christians were forced to flee to Christian territory in the north.

Accepted as useful civilian members of a society whose chief concern was the Reconquest of Spain, the Jews continued to flourish. As town after town fell to the advancing Christians, the *mudéjares* (Moors living under Christian rule) found that they too were allowed to live their own lives and follow their own customs. Christians were expressly forbidden to oppress members of either faith or interfere with their religious observances. In towns, mosques, synagogues, and churches catered to the different spiritual needs of inhabitants. In Moorish Spain, once the flame of Moslem fanaticism had died down, the same conditions prevailed.

In this tolerant atmosphere, Spain became one of the main centers of culture and scholarship in early medieval Europe. Jewish, Moslem, and Christian scholars worked alongside each other as teachers and collaborated in the production of distinguished historical, legal, and scientific manuscripts. "In the commercial sphere," as one historian has expressed it, "no visible barriers separated Jewish, Christian, and Saracen merchants. . . . Christian contractors built Jewish houses and Jewish craftsmen worked for Christian employers. Jewish advocates represented gentile clients in the secular courts. Jewish brokers acted as intermediaries between Christian and Moorish principals. As a by-product, such continuous daily contacts inevitably fostered tolerance and friendly relationships."

Despite this remarkable degree of liberalism, the tolerance was never complete, and the Christians made every effort to preserve the invisible barriers that kept the races socially apart. Intermarriage was discouraged by the threat of severe penalties. Paradoxically, the Jews fared much worse than the Moors, because they occupied a more important place in the social structure. Under Christian rule, most Moors earned their living as agricultural laborers, craftsmen, and traders, all minor roles in the country's society. The Jews, on the other hand, were not only traders and craftsmen but also physicians, merchants, tax-collectors, treasury officials, and money-lenders—all pivotal positions in Christian Spain. They managed the revenues and estates of the aristocrats, lent money to the Crown for military enterprises, and even became royal counselors.

As a result, this prosperous and powerful minority gradually aroused envy and jealousy among the Christians. As tax-collectors merely carrying out their jobs, they were hated; as money-lenders they were despised by the very people who came to borrow; as doctors they were accused of murdering their patients. As a race they were resented by the Christians for appearing to shirk manual labor and choosing instead professions that would bring them quick, easy profits.

The anti-Semitic hysteria went to such incredible lengths that stories were told of Jews stealing Christian children and crucifying them as part of their religious rites. This terrible slander was current throughout Europe in the Middle Ages. Not until the eighteenth century did the Church, after an exhaustive inquiry, state categorically that not one of these child-murder scares had any basis whatever in reality.

Anti-Semitism received official backing as early as the thirteenth century, and in 1371 the Cortes approved the decision of an earlier council that all Jews should identify themselves by wearing a round yellow patch over their hearts. The ruling was not rigorously enforced because of opposition from powerful Spanish Jews, but to Christian Spaniards it was the signal for giving their anti-Semitism free rein. The year 1391 saw a series of massacres in which the Jewish quarters of the country's largest cities—Toledo, Valencia, Barcelona, and Seville among them—were looted and then completely destroyed. That June over four thousand Jews were murdered in Seville alone.

Large numbers of Jews who escaped death accepted Christianity, becoming what were known as *conversos*. Their forced conversion did nothing to solve their problems; on the contrary, it made them worse. Many became Christians in name only and lived in constant dread of discovery. Even those who were fully converted were regarded with suspicion, and conversion did not necessarily mean integration into Christian society—even for the second, third, or fourth generation of a *converso* family. However often a *converso* went to Mass, however complete his renunciation of Jewish customs and habits, however ignorant he might be of the Hebrew language and the forms and procedures of the Jewish religion, he was considered a potential enemy, who was all the more dangerous for being ostensibly a friend.

BIBLIOTHEQUE NATIONALE, PARIS: N. A. FR. 24.541 FOL 35

Anti-Semitism gave rise to many weird legends. In the one depicted above, a Jewish father—upon learning that his son attended Mass with Christian companions—puts the boy in an oven. The Virgin Mary, however, appears at left to rescue the child from his father's cruel, unnatural punishment.

בעל הבית ובני ביתו שאומרים ההגדה

These persecutions and conversions, like those that were to follow, were carried out in the name of religion. But while religious fanaticism was a strong element, it was not the sole cause. Underlying it was a racial hatred based on envy, jealousy, and fear. As one contemporary observer remarked, "All this was caused by a thirst for robbery rather than by devotion."

In the fifteenth century the numbers of *conversos* grew as the persecution continued. In 1412 a law was passed in Spain depriving Jews and Moors of the right to hold office, to hold titles, to bear arms, to hire Christians to work for them, or to follow the trades of carpenter, tailor, butcher, or grocer. They were not to change their residences, nor might they eat, drink, bathe, or even talk with Christians. They were to wear rough clothing and distinguishing badges and they were to wear their hair and beards long. This piece of legislation alone was responsible for thousands of conversions.

As more and more rich, eminent, and powerful Jews accepted Christianity, more and more of them married into established Christian families. Inevitably, those who were gifted rose to important positions at court, in the administration, and even in the Church. Four of Isabella's bishops and three of her royal secretaries were from *converso* families. So too were her former confessor, Hernando de Talavera, Archbishop of Granada, and the grand masters of the military orders of Santiago and Calatrava. In Aragón few noble families were without some Jewish blood, and Ferdinand himself is said to have had Jewish forebears.

The old aristocrats of Spain were alarmed at this challenge to their monopoly of power by members, even converted members, of a despised race. After centuries of fighting the infidel Moors, nobility had become identified with pure Spanish blood and with the true faith. Many nobles owed their titles, lands, and wealth solely to the services they had rendered their race and their religion. Belief in their own racial and religious purity had become a way of life, and any challenge to it they regarded as a threat to their pre-eminence in society.

This political jealousy further reinforced the religious

HISPANIC SOCIETY OF AMERICA, NEW YORK: B 241 FOL 99 V

The lion of León and the castles of Castile ornament a 1491 Toledo Hebrew Bible, a pathetic touch of patriotism only the year before Jews were expelled from Spain.

Quite contrary to the savage practices rumored to take place in Jewish synagogues, the typical Hebrew service—as shown in this fourteenth-century picture of Spanish Jews—consisted of old men quietly praying under an eternal light. Women, when allowed, sat upstairs, apart from the men.

ERMITA DE SAN BARTOLOME, VILLAHERMOSA: MAS

The most dastardly act of which Jews were accused by many of their Christian brethren was the profanation of the Host—believed by Catholics to be the living body of Christ. Illustrations, like this fourteenth-century retablo detail showing Jews boiling the wafer to destroy its power, did much to propagate anti-Semitic sentiments.

reasons for Christian hatred of the *conversos*. After several bloody race riots in Toledo and Valladolid, all *conversos* were forcibly expelled from Córdoba in 1473. In the outbreak of murders that followed, the *converso* Constable of Castile, a man of the highest eminence, was killed in front of a church altar at Jaén, in southern Spain.

While the populace resorted to violence, influential Old Christians, as those of pure Spanish descent were called to distinguish them from the New Christians or *conversos*, adopted subtler but no less effective methods. After 1449, cities began passing laws that forbade *conversos*, as well as Jews, to hold public office or to give testimony against Old Christians in court. This was the situation when Isabella came to the throne.

Christian Spain, it must be said, was not unanimous in this denunciation of the *conversos*. Alfonso Carrillo, the Archbishop of Toledo who had married the King and Queen, was one of those who favored racial integration. "Divisions bring great scandals and schism," he wrote in 1481, "and divide the seamless garment of Christ, who, as the Good Shepherd, commanded us to love one another in unity and obedience to Holy Mother Church. . . . From which it is obvious how culpable are those who . . . create different lineages, some calling themselves Old Christians and others calling themselves New Christians or Conversos." But a few enlightened and well-intentioned men had no chance of prevailing against the prejudice and self-interest of the majority of Spanish society.

Ferdinand and Isabella were subjected to constant propaganda from the anti-*conversos*. One of the most fervent and persistent of these was Alfonso de Hojeda, prior of the Dominican monastery of St. Paul in Seville. During the fifteen months the Queen spent in the city between 1477 and 1478, Hojeda preached to her constantly on this subject. The devout Isabella was accustomed to taking spiritual advice from the fathers of the Spanish Church. One of her early confessors, the fanatical Tomás de Torquemada, had urged her, should she ever come to the throne, to undertake the sacred mission of stamping out heresy. It is important to remember, also, that the child-atrocities then attributed to the Jews were sincerely credited by the greater part of the population. An added consideration was that from time to time groups of *conversos* were found to be practicing their forbidden religion in secret. It is not difficult to understand why the King and Queen decided that action of some sort was needed.

UNIVERSITY LIBRARY, BOLOGNA, MS. 2197: PHOTO, E. BRIGHT

Jews and Arabs tended to dominate the medical field in fifteenth-century Spain. A Hebrew translation of a tenth-century Arab physician's medical canon includes this open-air dispensary. From its jars and baskets of curatives, a chemist (right) mixes a potion, while the physician rolls up his sleeves to administer the drug.

In 1478 Ferdinand and Isabella formally asked the Pope's permission to introduce into Castile an Inquisition that they hoped would stamp out heresy and enforce religious orthodoxy.

The machinery of the Inquisition was not entirely new to the peninsula. Aragón had had an Inquisition since 1238, although by the time of Ferdinand and Isabella, it had all but ceased to function. The old Aragonese body, however, had been controlled directly by Rome, whereas the new Castilian Inquisition was to develop into an instrument of the Spanish government. It was answerable (insofar as it was answerable at all) not to the Pope but to the Crown, which appointed its officers.

The Pope granted permission for a Castilian Inquisition in 1478, but Isabella was reluctant to subject her people to such severe repressive measures, and it was not until 1480 that two inquisitors and an advisor were appointed and began their investigations.

The Spanish Inquisition was the Old Christians' trump card in their attack on the *conversos*. Its official aim was to uncover all so-called Christians who had relapsed into their former religious practices and by means of penances to bring them back into the Church. This is certainly what Isabella intended it to be: a means of solving social and racial unrest by enforcing uniformity. In reality, of course, the Inquisition, at least in its initial stages, was chiefly an instrument of racial persecution.

JEWISH THEOLOGICAL SEMINARY OF AMERICA, NEW YORK

Church merciful and Church militant both find expression in this standard of the Inquisition. The olive branch of peace balances the crusader's sword, pointed upward to denote victory, and both are under the protective arms of the Cross. The motto: "To carry out God's will toward the Jews."

The introduction of an Inquisition was the first step in the monarchs' solution to Spain's religious problem. Only the baptized came under its jurisdiction, since by definition only the baptized could be guilty of heresy. It could not interfere with practicing Jews and Moslems. The Old Christians found this a fairly satisfactory state of affairs for the time being. Those Jews and Moslems who had not accepted Christianity were by now an inferior social minority group, easily recognizable, and officially discriminated against. They represented no direct threat to Old Christian superiority and caused a very insignificant amount of social unrest. Not until the sovereigns decided that their country's strength depended on absolute unity of race and creed would the Crown's full attention be turned upon them.

The Inquisition started its operations immediately in Seville by publishing an edict that requested all citizens to report to the authorities anybody and everybody they suspected of heresy. Since accusations could be made anonymously, it was not long before they began to pour in. So numerous did the victims become that the tribunal of the Inquisition was forced to move to a fortress outside the city of Seville, the only building in the area that was large enough to accommodate its prisoners.

The procedure of anonymous accusation became one of the most insidious and unjust aspects of the Inquisition's machinery. Legend, misinformation, and prejudice have built up around the Inquisition an aura of horror, cruelty, and ruthlessness, of refinements of torture and secret-police methods. This reputation is not entirely justified. But nothing could be more regrettable than this process whereby a man could languish for months in prison, awaiting trial, without even knowing what he was accused of or who had accused him. The Inquisition's defense of anonymous accusation was that it enabled a poor man, for example, to denounce a wealthy and powerful one without fear of reprisal. But the temptation to get rid of enemies in this way was too great, and the process was considerably abused.

The Inquisition was aware of this danger and did its best to prevent such abuses. Nevertheless, the atmosphere of fear and suspicion that surrounded its operations was directly due to the knowledge that a person could be denounced, and hence ruined, by a relative, friend, or acquaintance, for reasons as flimsy as a fit of pique.

Those who wanted to report a *converso* had no difficulty finding pretexts for doing so. Any behavior connected with the Jewish way of life was considered evidence of heresy.

GRANADA CATHEDRAL: PHOTO, ORONOZ

As a token of his friendship and in reward for the great works they had accomplished for God and the Church, Pope Sixtus IV sent the Spanish monarchs this jewel-studded silver reliquary, said to contain a splinter of the True Cross.

While many *conversos* found it comparatively easy to renounce all outward adherence to their faith, they naturally found it extremely difficult to forget a whole way of life and adopt a new one. Even Old Christians who accidentally did something traditionally associated with the Jewish life were in danger of being denounced by some suspicious witness. It was proof of Judaism if a man sat down to eat with Jews, ate the meat of specially slaughtered animals, or drank a certain Jewish drink; if he put on clean clothes on the Sabbath or neglected to light his fire the previous evening; if he washed a corpse in warm water or turned his face to the wall to die. It was certain proof if he gave his children Jewish names. Since Jews had long been forbidden to give their children Christian names, recent *conversos* were in an impossible situation.

The immediate result of the Inquisition's operations in Seville was that more than four thousand Andalusian families decided they could not afford to await denunciation and took flight. Few of them got very far, for anyone who harbored refugees was given two weeks to turn them over to the Inquisition, on pain of excommunication, and the threat was extremely successful.

In Seville itself a group of influential *conversos* decided to make a stand. Among them were several magistrates and members of the clergy, and their leader, Diego de Susán, was one of the town's wealthiest citizens. They realized that they were the victims of race hatred and that nothing—wealth, influence, property, even priestly robes—could save them from persecution. If they were to act at all, it would have to be at once.

Powerful *conversos* from the surrounding countryside joined them; they raised money, provisions, and arms. If they had taken control of Seville, they might have eliminated the Inquisition in its earliest stages and thereby changed the course of Spanish religious policy. But they were betrayed—by Diego de Susán's own daughter, Susanna, who had an Old Christian lover she feared might be harmed in the fighting.

All were arrested; those directly implicated as well as most of the rich and influential (hence dangerous) *conversos*

At right, Spanish artist Alonso Berruguete sums up the Inquisition's barbarity. Its judges go about their fanatical work (one so callous, he even sleeps), heedless of the agonies of the condemned (right), nailed as well as tied to the stake, waiting to burn. Saint Dominic symbolically presides.

PRADO, MADRID

Condemned heretics wear san-benitos *decorated with the fiery jaws of hell as they proceed to their doom in this sixteenth-century German engraving of an auto-da-fé. The victims, flanked by officers of the Inquisition beseeching them to repent, receive final sentencing in the structure at right and are at last released to secular authorities and taken to the stake (right, above). Bones of dead heretics and effigies of escapees are also paraded through the long and gruesome spectacle.*

of Seville. On February 6, 1481, six prisoners were paraded through the streets in a solemn procession for the first auto-da-fé—"act of faith"—in the history of the Spanish Inquisition. First Alfonso de Hojeda delivered yet another eloquent sermon denouncing heresy. Then, because the Inquisition did not officially execute its victims, the prisoners were handed over to the secular authorities to be burned at the stake. Most of the others received one of the Inquisition's lesser punishments: imprisonment, confiscation of goods, exile, scourging, or service as a galley slave. But a few days later, another large group was burned, including Diego de Susán himself. According to legend, Susanna, filled with horror and remorse, ended her life in poverty and shame.

The discovery of this group of militant "heretics" was taken as justification for setting up tribunals throughout the country. Seven more inquisitors were appointed, all Do-

BIBLIOTHEQUE NATIONALE, PARIS: GIRAUDON

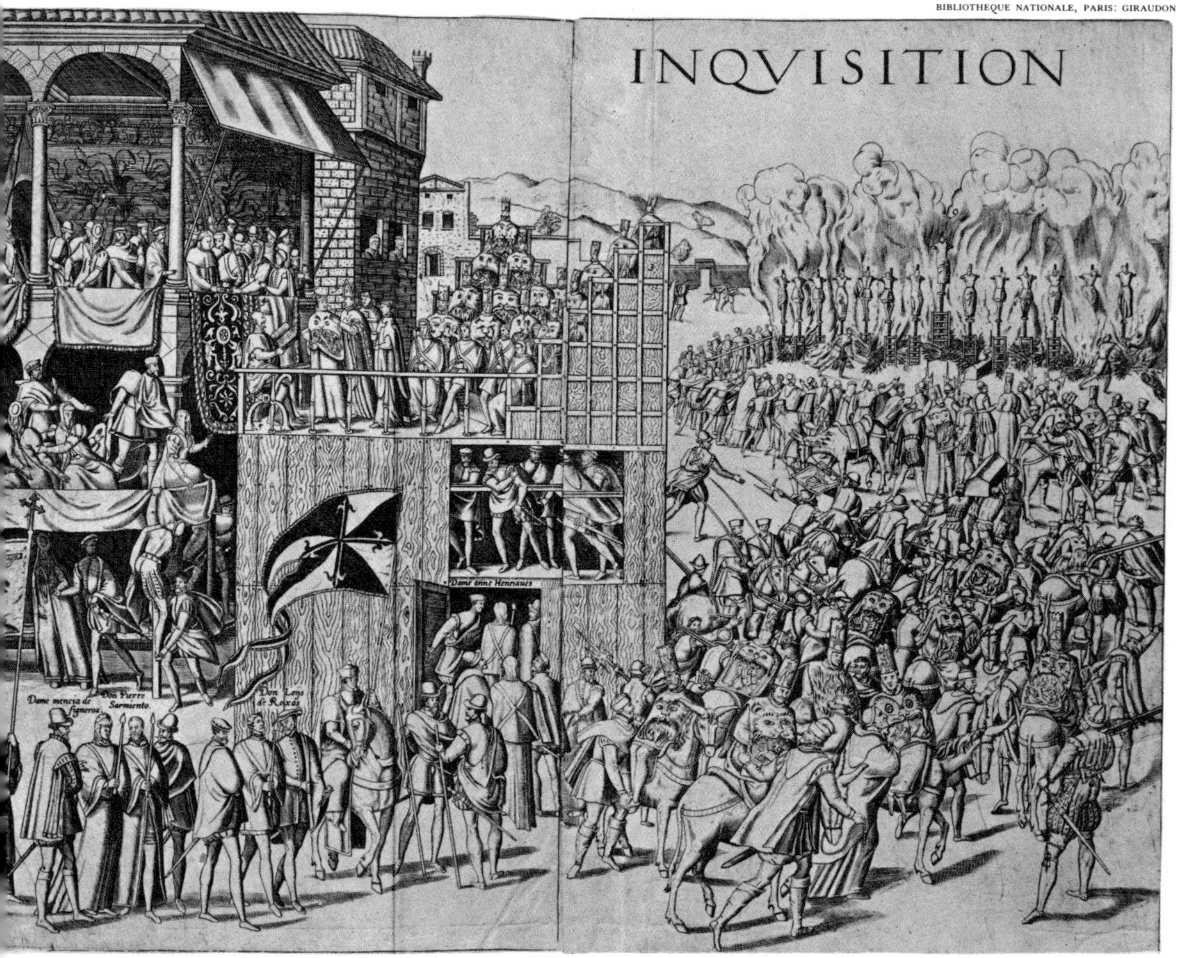

minicans. One of them, appropriately enough, was Tomás de Torquemada, Isabella's early confessor. Torquemada's appointment in 1483 as the first Inquisitor General marked the beginning of one of the most savage periods in the Inquisition's history. In the first eight years of the Seville tribunal seven hundred people are believed to have been burned and more than five thousand punished. Since anyone sentenced to punishment automatically forfeited all his property—even if he was "reconciled" to the Church after performing his penance—the victims' families were reduced to beggary. When Isabella questioned the need to impose such suffering, Pope Sixtus IV silenced her scruples and told her to proceed with the "great work of purification."

With the problem of the *conversos* under control, attention was focused on the orthodox Jews. The old stipulation that they wear distinctive badges was enforced once more; then laws were passed restricting Jews to their ghettos. But the only really effective way of limiting *converso* infiltration was to remove its source: Spanish Jewry itself. Gradually, more and more laws were passed expelling the Jews from various parts of Spain; finally, in 1492, the monarchs issued their edict of general expulsion or baptism. Henceforth no practicing Jew could live on Spanish soil.

Jewish money had financed the conquest of Moorish Granada; Jewish money was soon after to finance Columbus' epic voyage; but even Spain's great financiers were forced to abandon their religion or go. One, Abraham Senior, chose to stay, even though he was one of a handful of Jews who had official permission to take possessions in the form of gold or silver with him. His friend the theologian Isaac Abrabanel, who had been one of the monarchs' foremost financial advisors, chose to go.

In all, an estimated 150,000 Spanish Jews accepted

MUSEO DE BELLAS ARTES, VALENCIA: PHOTO, BRADLEY SMITH

In a painting depicting the Spanish desire for religious unity, Saint Vincent Ferrer, beloved Dominican preacher of the late 1300's, is seen directing the burning of Jewish and Moorish books—religious as well as secular—all judged heretical. Jews (bearded) and Moors look on helplessly while the word of God, here the New Testament, ascends from the flames: "One God, One Faith, One Baptism."

exile. Forbidden to take gold and silver out of the country, they sold their property for anything transportable—a vineyard for some linen, a house for an ass. They sailed from the coast of Spain in overcrowded and unseaworthy ships; many of them were driven back by storms and forced to accept baptism. Those who reached the coast of Africa risked murder and robbery. A rabbi later reported their fate: "The Turks killed to take out the gold which [the refugees] had swallowed to hide it; some of them hunger and the plague consumed and some of them were cast naked by the captains on the isles of the sea; and some of them were sold for men-servants and maid-servants in Genoa and its villages and some of them were cast into the sea."

Expulsion solved the Jewish problem, but it enormously aggravated the *converso* problem by creating at one fell swoop thousands and thousands of forcibly baptized Jews. In their desire for unity, strength, and stability, Ferdinand and Isabella had made a potential "fifth column" of the whole of remaining Spanish Jewry. The racists, the religious bigots, the Old Christians, and the Inquisition were quick to realize this; and it took them a long time to forget.

Heresy, or the disposition toward heresy supposedly inherited with *converso* blood, was an ineradicable stigma. During an auto-da-fé, heretics were forced to wear a smock-like penitential garment known as a *sanbenito*. Those who had been "reconciled" to the true faith wore a yellow one with diagonal crosses on it, and part of their punishment was to continue wearing this mark of shame for an allotted time. Those who were to be executed or "relaxed to the secular arm," as it was called, wore a black *sanbenito* covered with paintings of demons and hell-fire. After the heretic's death or once his penance was over, his *sanbenito* with his name attached was hung up in his parish church and remained there until it crumbled into dust. Even then, it was often replaced by a new one so that his infamy might be perpetuated and his family never be free of the blot on their name.

The sovereigns' primary aim in expelling the Jews was social stability rather than religious unity in itself—it was another ten years before they touched Spain's other religious minority, the Moors. After the fall of Granada, Ferdinand and Isabella solemnly guaranteed that those Moors who chose to stay in Spain would be allowed free exercise of their religion and customs. Yet within a decade, the relentless progress of the Inquisition and the ruthless zeal of Francisco Jiménez de Cisneros brought about a complete

TOLEDO CATHEDRAL: PHOTO, ORONOZ

Like the Jews, Moors, too, were victims of Spanish nationalism and bigotry and finally had to choose between baptism and expulsion. Above, sad Moorish prisoners, despairing and uncertain of the future, huddle aboard a Spanish ship carrying them to exile in Africa.

reversal of these liberal conditions, and the Moors in their turn were ordered to choose between conversion and expulsion.

The people of Granada were fortunate in having Hernando de Talavera as their first Archbishop. A naturally tolerant and gentle man who was himself of *converso* descent, Talavera realized that forcible conversion meant nothing. Instead, he undertook an evangelizing campaign that he hoped would lead the Moslems to accept Christianity of their own free will. He even learned Arabic so that he could preach to them in their own language, and he arranged for Arabic translations of parts of the Bible. His enlightened efforts met with some success, but progress was inevitably very slow.

TOLEDO CATHEDRAL: MAS

In this portrait, Archbishop Cisneros wears a richly embroidered chasuble on which St. Peter holding the keys of Heaven is shown. His tonsured, or shaved, head is a symbol of humility and poverty.

In 1499 Archbishop Cisneros came to Granada. Like most of the Spanish clergy, Cisneros believed that much greater efforts ought to be made; he took over the missionary movement from Talavera and started proselytizing in an extremely active and forceful way. His methods were effective: on December 18, 1499, he baptized three thousand Moors in a mass ceremony.

Cisneros' zeal did not stop at baptism. Carried away by his enthusiasm, the man who had arranged such contributions to scholarship as the founding of Alcalá University and the production of the Polyglot Bible now determined to root out the sources of Moslem teaching. The libraries of Granada had long been renowned because they contained so many outstanding scholarly works, exquisitely decorated and bound by Moorish craftsmen, who were the finest in Europe. Cisneros ceremoniously committed to a huge public bonfire all the Arabic books and manuscripts he could lay his hands on—not only copies of the Koran, the bible of the Moslem faith, but thousands of other works that were quite unrelated to theology. A few rare manuscripts on medicine he saved for the library at Alcalá, but Arabic schclarship in Spain was virtually destroyed.

Infuriated by the Archbishop's high-handed disregard of their rights, the Moors rose in revolt. It was several days before Hernando de Talavera and the city governor—an equally honorable and popular man—were able to persuade them to return to their homes and occupations.

News of the rebellion reached the court, which was then in Seville. "So we are like to pay dear for your Archbishop, whose rashness has lost us in a few hours, what we have been years in acquiring," Ferdinand remarked scathingly to the Queen. Isabella, horrified by what had happened, immediately wrote to "her Archbishop" demanding an explanation. Cisneros hurried to Seville to assure his sovereigns that everything had been done in the interests of religion. He even convinced them that, through their rebellion, the Moors of Granada had forfeited their former rights. Like the Jews, they should be given the alternatives of conversion or exile.

Most of the Granadans opted for conversion—no doubt they had heard tales of the sufferings undergone by the exiled Spanish Jews. Some braved the dangers of fleeing to North Africa. Others joined the Moors who lived in the wild mountains of Las Alpujarras, to the south of Granada, and tried to establish an effective resistance. But they had little chance of resisting the combined strength of Castile and Aragón; methodically and brutally they were subdued by the royal armies.

As a result of a series of largely political measures, Isabella's ideal of a religiously united kingdom was now within her grasp. To complete this service to God and further the spiritual welfare of her people, she issued the order of February 12, 1502, that all remaining Moors in Castile should accept baptism or exile. The vast majority, as usual, chose baptism. For the first time in its history Castile became an entirely Christian realm, and the Inquisition was charged with the spiritual welfare of thousands of *moriscos* (Christian Moors).

Isabella's religious fervor was responsible for this final step in Castile. In Ferdinand's own kingdom of Aragón the Moors were tolerated until 1525. Either because he was more moderate than Isabella, or more probably, because he did not care to remove a productive source of cheap labor for his nobles, Ferdinand warned the Aragonese inquisitors to leave the Moors in peace. The sovereigns' behavior on the Moorish issue, in fact, sums up the essential difference in their characters. Ferdinand, unlike Isabella, was prepared to put financial considerations before moral principles; as a contemporary saying expressed it: "*Mientras más moros, más ganancia*"—"the more Moors, the more profit."

For Spain Isabella is a symbol of past greatness; subsequent generations in other countries, however, have found it hard to understand or to forgive her religious policy. Nor can its motivation be explained simply as religious bigotry. The Queen was certainly a bigot; dogmatic and even narrow-minded prejudices were inevitable in a person of her great piety and zeal. As a conscientious leader, she was convinced that her subjects would be better off if they were

GRANADA CATHEDRAL: PHOTO, ORONOZ

Many Moors, like the ladies in this mass baptism scene from a retablo in Granada, chose conversion over exile—but hardly in the spirit of true converts. Here they come to the sacrament with acquiescence, garbed in traditional costume, their veils before their faces in typical Arab modesty.

devout Christians like herself. As a faithful servant of God, she was eager to extend His kingdom.

Most of the measures she took were in fact prompted by aims that were not primarily religious. She wanted Spain to be strong; and strong it could not be while divided by racial unrest. The Old Christians could not be persuaded to love and accept their rivals; their rivals must therefore be dealt with. It was as simple as that. The Jews were expelled; the *conversos* were kept under control and made to conform. Once these repressive measures were well under way, her religious ideals came to the surface with the expulsion of the Moors; yet probably even this was prompted largely by patriotic motives. As the threat of Turkish Islam grew, many thought it dangerous for Spain to harbor practicing Moslems, possible allies in any Turkish attempt to re-establish Moslem rule in Spain.

METROPOLITAN MUSEUM OF ART, NEW YORK: GIFT OF J. PIERPONT MORGAN, 1917

This Moorish belt decoration of copper and gilt enamel provides a splendid example of the fine art born of the merger of Spanish and Moorish cultures prior to the Inquisition. Although elaborate in pattern, its fineness of line and attention to detail give it an exquisitely delicate appearance.

The sovereigns' treatment of Spain's Jewish and Moorish minorities helped create qualities of intolerance, self-satisfaction, and arrogance that were all too often to characterize Spanish policy in the centuries to come. In criticizing their behavior, however, it is important to bear in mind not only their motives but also certain facts about their times. First, Spain was not the only country in Europe to resort to expelling the Jews; England and France had already done so. Second, Spanish society had been formed by a set of circumstances that were complex as well as unique. Third, the whole of Europe was about to enter a period of terrifying religious intolerance and persecution. And last, the dreaded Inquisition founded by Ferdinand and Isabella was not, in fact, quite as dreadful as its opponents have asserted.

Undoubtedly, the Inquisition spawned its sadists and fanatics; it was often barbarically unjust. But in theory, and most of the time in practice, its aim was religious orthodoxy, not gratuitous suffering. It imposed penances, not punishments. Its judgments and its treatment of prisoners were on the whole less harsh than the sentences imposed by secular courts. As for its victims: 25,000 people are thought to have been sent to the stake by the Spanish Inquisition during its three and a half centuries of existence. To keep this terrible and tragic figure in perspective, however, we should not forget the witchcraft scare that raged through Europe in the sixteenth and seventeenth centuries (and was treated with remarkable sanity by the Inquisition). In Germany alone, in the space of one hundred years, *100,000* so-called witches were burned to death.

BIBLIOTECA ESTENSE, MODENA, MS. A. X. 2. 14: PHOTO, E. BRIGHT

By the late 1400's, Europe had become a battleground for the politically ambitious, and Mars (in roundel), god of war, was the symbol of the age. In this miniature of the period, an army assaults a city in Italy, a hotbed of warring city-states and the prize of foreign nations, including Spain.

VII

INTRIGUES AND ALLIANCES

"A prince still reigning whom it would not be fitting for me to name never talks of anything but peace and good faith, yet had he ever observed either he would several times have lost his credit and his estates." So says the Italian observer Machiavelli in his realistic summing-up of the ethics of international politics at the turn of the sixteenth century. The prince he referred to was Ferdinand of Aragón.

It was in the field of European politics and diplomacy that Ferdinand's true interests lay; this was the arena for which his gifts best suited him. Isabella had a more provincial attitude; her main concern was for the political, spiritual, and cultural welfare of Castile. When she and her countrymen let their gaze wander beyond the boundaries of Spain, they looked to the west, not to the north or the east. Aragón, on the other hand, had a long history of participation in European affairs, and Ferdinand was not slow to follow the tradition established by his predecessors. Under his guidance, however, this participation was to be on a much larger scale than ever before.

The European scene at the end of the fifteenth century was both complex and fascinating. Centralized authority was beginning to replace the feudal anarchy of the Middle Ages. New techniques in industry and mining, in shipbuilding and transportation had contributed to an economic revival, and the recent invention of the printing press made it possible to communicate effectively over extensive areas of territory for the first time.

With internal stability and strength, of course, came international power politics. New-found vigor and energy inevitably brought a desire for power, importance, and territorial expansion. Europe became a battleground on which the leading nations fought and schemed against one another for pre-eminence. The aims of each were entirely selfish, and treaties were signed and broken without the slightest compunction. In the constantly changing pattern

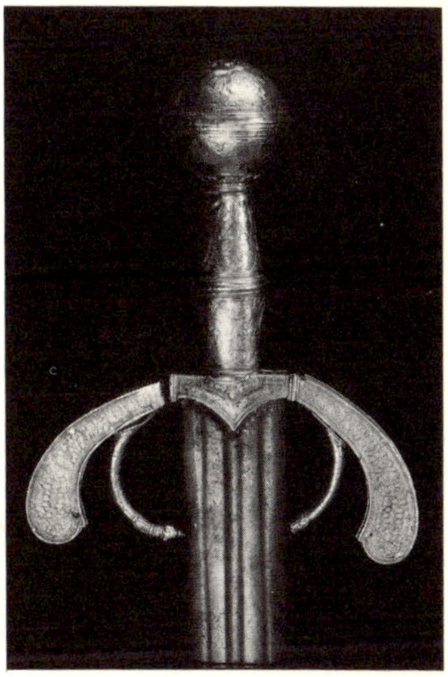

GRANADA CATHEDRAL: PHOTO, ORONOZ

A golden hilt adorns Ferdinand's sword of state. It was used to bestow honors and to enhance his appearance at regal functions and not as a weapon in his campaigns.

What could not be accomplished by war was sometimes accomplished by marriage, as in the unhappy union of the Infanta Juana and Archduke Philip of Burgundy, heir to the Holy Roman Empire. This miniature, done about 1500, is from a devotionary given Juana as a wedding present. In it Queen Isabella, Juana's mother, accepts the prayer book from the artist, as three of her daughters look on.

of political alliances, today's friend was most probably yesterday's enemy.

There were four major European nations: France, largest and most powerful of all; England, by now France's traditional enemy; Burgundy-Austria, ruled by the Hapsburg Holy Roman Emperor; and as the outcome of the union of Castile and Aragón, Spain. As a result of this union and its sovereigns' capable government, Spain had become, for the first time, a European power to be reckoned with. Ferdinand, however, knew that his country's position was not fully secure and that the kaleidoscopic diplomacy of the age could, without warning, leave Spain standing completely alone. He proposed to prevent this by establishing close ties with as many countries as possible. The most obvious and convenient way to do this would be to place the royal children, by means of marriage, in various strategic spots. What else, after all, were royal children for?

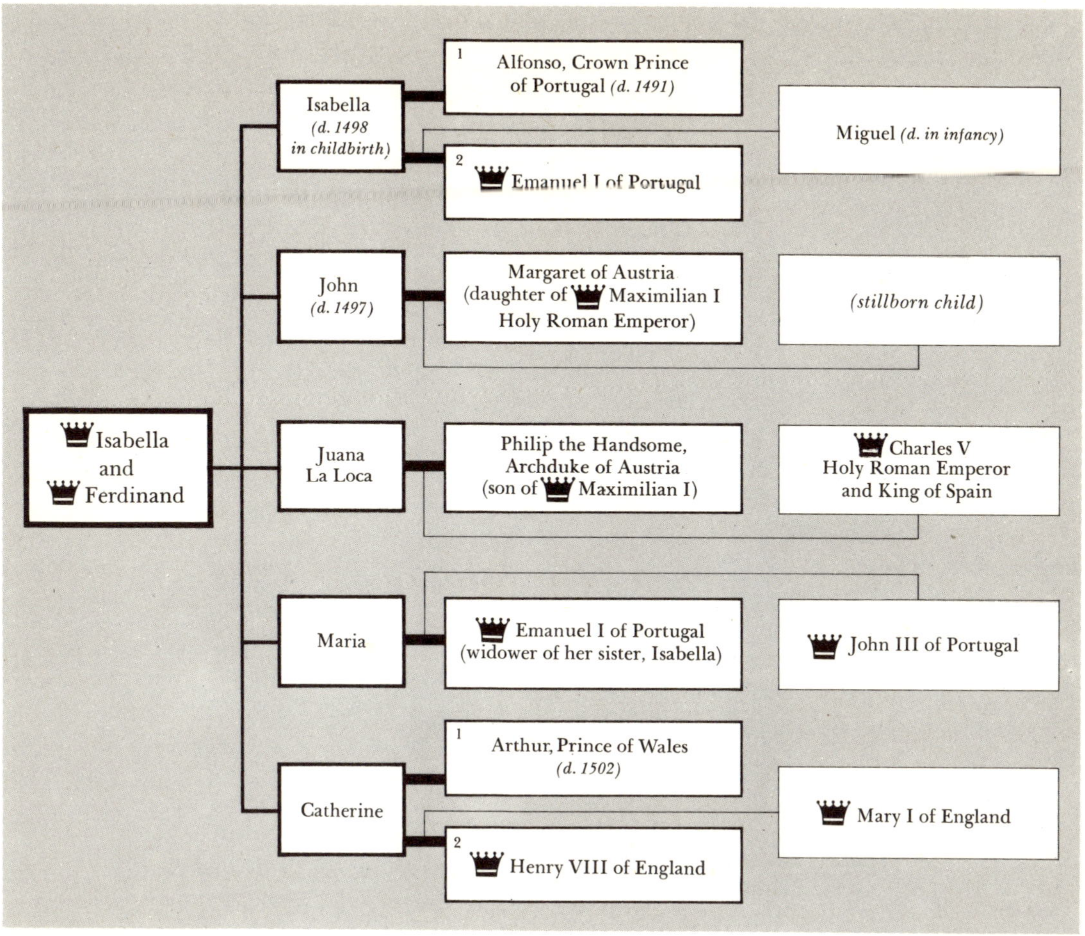

The family tree above shows the series of marriages that brought alliances with Portugal, England, and the Holy Roman Empire—but not always happiness—to the five offspring of the Catholic Kings. Grandchildren, by their daughters Juana, Maria, and Catherine, ruled Spain, the Holy Roman Empire, Portugal, and England.

For their five children, Ferdinand and Isabella arranged a series of marriages that were, in varying degrees, advantageous to their country. They could not know that these marriages would bring personal grief and suffering to their children. Even if they had known, it is doubtful whether the sovereigns would have altered their policy in the least. Ferdinand certainly would not have done so. Even Isabella, for all her warm maternal feelings, with her high concept of a queen's duties, probably would have hesitated only slightly before sadly consenting to sacrifice her children to the welfare of Spain. Her daughters had been brought up to be queens, and queens they would be, whatever the cost either to them or to her.

The first Infanta to go was Isabella, the eldest. In 1490 the twenty-year-old Princess was married to Alfonso, son and heir to King John II of Portugal. There were two reasons for the alliance: first, it might create a totally united peninsula (a dream that Ferdinand and Isabella never entirely relinquished), and second, it lessened the chance of Portugal's reopening hostilities with its neighbor. Ferdinand had no wish to enter full-scale European politics with a potential enemy at his rear.

The alliance, however, was pitifully short. A few months after the wedding, Alfonso was killed in a riding accident, and Isabella returned to Castile a sad and forlorn young widow. Four years later, Alfonso's uncle Emanuel, who had succeeded John II as King of Portugal, asked for her hand. Isabella, who shared much of her mother's devotion to religion, was reluctant to marry again, but parental pressure proved too great. For a second time, she agreed to make a Portuguese alliance in order to promote the interests of Spain. However, she imposed two conditions: the marriage must be celebrated very quietly, and all Jews were to be expelled from Portugal. Such was King Emanuel's eagerness to marry her that he accepted them both.

Shortly before Isabella's second marriage, a double alliance was arranged between the house of Hapsburg and the house of Trastámara. The Spanish heir, Prince John, then almost seventeen, was to marry Princess Margaret, whose father, the Emperor Maximilian, ruled Austria and

This royal feast from the Calatrava Bible depicts an episode in the life of Belshazzar, from the Book of Daniel, but its pomp and splendor seem more typical of fifteenth-century Spain than of early Babylon. It was at such a banquet that Prince John, no doubt, dined to celebrate his marriage to Margaret of Austria. Like King Belshazzar, who died shortly after this famous feast, John did not long outlive the celebration of his marriage.

DUKE OF ALBA COLLECTION, MADRID: PHOTO, ORONOZ

ARCHIVO DE SIMANCAS, VALLADOLID: PHOTO, ORONOZ

At left, a double-marriage agreement between the Hapsburg and Trastámara families, betrothing Princess Margaret to Prince John and Princess Juana to Philip of Burgundy, gives no small hint of the intent of such unions. The most important names, those of the monarch-parents—Ferdinand and Elizabeth (Latin for "Isabella") and Maximilian—are unusually prominent, while the names of the betrothed remain inconspicuous.

Burgundy and was the acknowledged head of the Holy Roman Empire. Archduke Philip of Burgundy, Maximilian's son and heir to the empire, was to marry John's younger sister, the Infanta Juana.

The promise of becoming empress of Austria and Burgundy made Juana's future seem even brighter than her sisters'; to have as her husband a man who was considered to be the handsomest prince in Europe seemed an extraordinary bonus. In 1496, when she was almost seventeen, she sailed north, accompanied by a splendid retinue of high-born nobles and their ladies, to meet and marry Philip in Flanders, one of the provinces controlled by the dukes of Burgundy. Some months later, the fleet that had borne Juana north returned to Spain with Princess Margaret on board. In April, 1497, she was married to Prince John at Burgos in a brilliant ceremony attended by Ferdinand and Isabella, all the grandees of Castile and Aragón, and representatives of the house of Hapsburg.

MUSEE CONDE, CHANTILLY, MS. 604/1339 FOL 3 V: GIRAUDON

Hopeful of inheriting a Spanish kingdom from his wife, Philip of Burgundy adopted the above coat of arms as a Prince of Castile.

As she sat and watched the Archbishop of Toledo marry her beloved son to the Austrian princess, Isabella must have experienced one of the happiest moments of her life. Up to now, fate had smiled on her, her family, and her kingdom. Granada had been conquered, a western empire had been discovered, and Spain had become a stable and prosperous country. By her side was a husband for whom she cared deeply. Their daughter Isabella was about to become Queen of Portugal; Juana was already Archduchess of Burgundy; and John had just married the most eligible princess in Europe. Their youngest child, Catherine, was betrothed to Arthur, Prince of Wales, son and heir to Henry VII of England. Isabella's sole cause for sadness was the inevitable physical remoteness of her two eldest daughters.

A queen who had worked so selflessly and unstintingly for the good of her country certainly deserved personal happiness. But life, with devastating irony, had stored up tragedy for the years when Isabella no longer had the energy to withstand it. The last seven years of her life dealt her a series of crushing blows. The first, and perhaps the worst, came only six months after her son's marriage. While she and Ferdinand were attending the wedding of Isabella and Emanuel of Portugal, Prince John fell dangerously ill with fever. Ferdinand rushed to his son's bedside but arrived there only to see him die, on October 3, 1497. A few months later, John's widow gave birth to a stillborn child.

Ferdinand and Isabella had lost not only their beloved son but also their last hope that the crowns of Spain would be inherited by a Spaniard. "Their Highnesses do their best to hide their sorrow," wrote the observant Peter Martyr, "but we can all see that they have suffered a mortal blow. When they sit quietly together in the patio we often catch them looking at each other, their eyes full of a grief too deep for words."

But life and politics had to continue. With their common heir gone, Ferdinand and Isabella were forced to reconsider their marriage policies. Hearing of his brother-in-law's death, Philip of Burgundy impudently claimed the title of "Princes of Castile" for Juana and himself. The monarchs at once hurried to have their eldest child, Queen Isabella of Portugal, proclaimed heiress to Castile. The young Queen was about to have a child. As far as Castile was concerned, the baby's sex was unimportant, but only the male succession was recognized in Aragón. If the kingdoms of Ferdinand and Isabella were to remain united, therefore, a grandson was essential.

On August 23, 1498, a son, Miguel, was in fact born to Isabella of Portugal. Within an hour his mother died, and Ferdinand and Isabella had lost yet another child. Even the consolation they derived from the birth of a grandson was soon denied them, for before the frail little Miguel reached his second birthday, he too was dead.

It seemed as if the policy of marriage alliances was beginning to founder, but the sovereigns were determined to retain the connection with Portugal. In 1500, King Emanuel asked for the hand of his sister-in-law Maria, the only Infanta not yet spoken for, and the Pope granted a dispensation. As it turned out, Maria was the only one of the family to lead a full and happy life; she gave her husband eight healthy children, heirs to the throne of Portugal. Unlike Miguel, however, they were never to be heirs to Castile. Ahead of Maria in the succession was Juana, the third child of Ferdinand and Isabella, who was already the mother of one son, Charles.

Immediately after Miguel's death in July, 1500, Juana was sent for by her parents. As the new heiress to Castile, she must receive the nobles' oath of loyalty. Isabella had not seen her daughter for four years, but a certain amount of information had filtered back from Flanders to Spain. It was a worried and uneasy mother who awaited Juana in Toledo, for it was said that the Archduchess showed definite signs of being mentally disturbed.

On her arrival with Philip, Juana threw royal restraint aside and embraced her parents most affectionately. But Isabella's joy at seeing her again was soon marred by her daughter's revelations about her marriage. It seemed that Juana was passionately in love with her handsome young husband but that he returned her passion only with indifference. Tormented by jealousy and incapable of coping with his infidelities, she could not bear to let Philip out of her sight.

After a few months in Spain, Philip decided suddenly to return home. He did not like Spain; he did not agree with his father-in-law's anti-French policies; and he was tired of Juana's constant reproaches. He deliberately left her behind on the excuse that she was soon to have her fourth child and the winter journey would be dangerous.

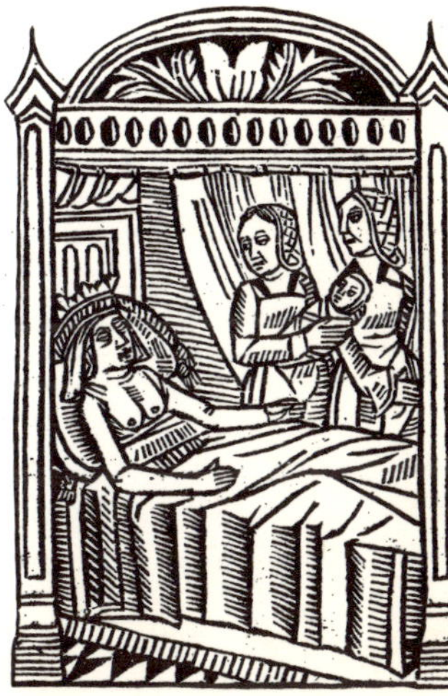

Oliveros de Castilla, 1499: HISPANIC SOCIETY OF AMERICA, NEW YORK

Reminiscent of the tragic fate of Isabella of Portugal, this woodcut from a Spanish translation of a French romance shows an unnamed queen of Castile dying in childbirth. Bare-breasted and fatigued, she beckons to a lady in waiting to hand her the swaddled infant.

Once her second son, Ferdinand, was born, Juana tried to follow Philip north. But Spain was at war with France, and it was impossible to travel. Any delay was intolerable to the insanely jealous Juana; she completely lost control of herself. Pathetic scenes of hysteria and violence followed, including one in which she fled from the royal residence near Medina del Campo in her nightdress and clung to the railings outside for a day and a night, heedless of protests from her attendants. Her mother finally managed to persuade the desperate girl to go indoors and rest. Despite their fear and foreboding, the troubled parents had no choice but to allow her to rejoin Philip in Flanders, as soon as it was safe to do so. But the damage was done. Henceforward, Juana became known as Juana la Loca—Juana the Mad.

It is an irony of history that this marriage, for all its tragedy and pathos, would turn out to be extraordinarily successful from a political and dynastic point of view. The son already born to Philip and Juana, who had remained in Flanders during his parents' visit to Spain, was to unite the kingdoms of Spain and Austria and inherit the title of Holy Roman Emperor. As Charles V, he would rule more of Europe than any monarch since Charlemagne, and his Hapsburg descendants would govern Spain for another two hundred years.

Aragonese warships tow their defeated French enemy into the Bay of Naples in a painting of an early Spanish-French clash over Italy. Castel Nuovo, home of the Aragonese rulers of Naples, stands left of center, an impressive bulwark against future invasions.

One nation had been conspicuously absent from Spain's network of marriage alliances: France. In deciding to ally Spain with Austria, Ferdinand in effect established an alignment that would set the pattern of power in Europe for two centuries to come. The immediate result, however, was that, as Spain entered European warfare on a large scale, France was its natural first enemy.

The chief bone of contention between the two countries was Italy, a far more worthwhile prize than the frontier territory over which they had long squabbled and fought. Italy at this time was not a unified kingdom but a cluster of independent city-states. The most powerful of these were the republics of Venice and Florence, the duchy of Milan, the kingdom of Naples, and the Papal See.

Milan at this time was governed by Duke Lodovico Sforza, who had made himself regent during the minority of his nephew, Giangaleazzo, and had later refused to surrender his power. Giangaleazzo's father-in-law was the powerful King of Naples, and Lodovico, anxious to divert Neapolitan attention from the situation in Milan, offered to support a long-standing French claim to the Neapolitan throne. King Charles VIII of France accepted his help, and in 1494 French armies marched into Italy.

France's activities dismayed Ferdinand. For more than

BIBLIOTHEQUE NATIONALE, PARIS: VELINS MS. 724

This allegorical depiction of Lodovico Sforza's regency in Milan for his nephew, Giangaleazzo, shows Sforza as a mulberry tree (his emblem) supporting a tender sapling (the nephew). A banner entwining them reminds the boy to rejoice at his uncle's support.

MUSEO DI SAN MARTINO, NAPLES: SCALA

half a century Naples had been ruled by a branch of the house of Aragón, and he had no desire to see his cousins dispossessed. Nor did he relish the idea of having such redoubtable neighbors as the French next door to his own kingdom of Sicily. In 1493 Ferdinand and Charles VIII had signed the Treaty of Barcelona, which restored Roussillon and Cerdagne to Aragón and stipulated in return that Spain should not contract an alliance against France with any power other than the Papal See. This last proviso, a commonplace of diplomatic treaties, offered Ferdinand a perfect loophole. He blandly told the indignant Charles that, since the kingdom of Naples was a fief of the Church, his duties toward Naples would have to override his obligations toward France.

The Italian states had neither the strategy nor the equipment to resist the French advance. By December, 1494, Charles had occupied Rome; in February, 1495, the

VERARD, *Mer des Histoires*, c. 1503: BIBLIOTHEQUE NATIONALE, PARIS

The fury of the Italian wars is captured in a woodcut of the 1495 Battle of Fornovo. Outnumbered three to one, Charles VIII (in a crowned helmet at far right) breaks through the enemy line to lead his French troops to safety.

French were in Naples. Pope Alexander VI very cannily reminded Ferdinand of his openly acknowledged obligations toward Naples. By way of more concrete encouragement, he granted Ferdinand and Isabella a certain percentage of Castilian Church revenues and conferred upon them, in advance recognition of their projected services to the Papal See, the title of *los reyes católicos*, the Catholic Kings. By this title they have been known ever since. With or without such bribes, Ferdinand would have intervened in Italy, but of course, he eagerly accepted them. The Treaty of Barcelona was torn up in front of Charles' very nose by one of Ferdinand's more intrepid ambassadors, and Spanish soldiers crossed from Sicily into southern Italy to begin the liberation of Naples from the French.

Thus Aragón entered into nine years of warfare, dragging Castile along with it. On paper the Spanish resources were no match for the French, but the Moorish wars had developed a well-trained and resilient infantry. In the course of the Italian wars this infantry, under the command of Gonzalo de Córdoba, a veteran of the Granada campaign, became one of the most formidable land forces in Europe. Gonzalo's first Italian battle, early in 1495, was the last he ever lost, and this one failure was caused only by a lack of co-ordination between his forces and their allies. After this one defeat, Gonzalo proved insuperable. Relying on speed and tactics to make up for his lack of strength, Gonzalo won back the greater part of the kingdom of Naples in less than a year. For himself he won a brilliant reputation and an impressive title: The Great Captain. In 1496, Naples capitulated and Gonzalo rode into Rome in triumph to receive the Pope's blessing.

A peace treaty was signed with France in 1498. But before long, trouble erupted again in Italy. Supported by Venice, Florence, and the Papal See and led by a new king, Louis XII, the French took over the duchy of Milan. Open hostilities between Spain and France broke out once more in Naples, and in 1503 Gonzalo de Córdoba captured its capital from the French for a second time.

Still the fighting dragged on. In December, 1503, French and Spanish forces met in a fierce pitched battle on the banks of the river Garigliano in central Italy. On a stormy, pitch-black night when the river was in flood, Gonzalo gave his men the order to attack. Soldiers on both sides, weakened by fatigue, disease, and lack of food, stood up to their knees in mud. The Spanish troops had not been paid for months, but they rallied bravely to their "Great Cap-

When the Infanta Catherine sailed north to England in 1501 to marry Arthur, Prince of Wales, she took with her, as a wedding gift, this portrait of her father, Ferdinand.

Before her marriage to England's Crown Prince, Catherine of Aragón was painted as Mary Magdalene, holding the jar of ointment with which to anoint the body of Jesus.

tain." The sheer weight of the French cavalry made a Spanish victory seem improbable, but the spirit of the famous infantrymen was unbeatable. The sun rose on a waterlogged battlefield strewn with the corpses of several thousand French soldiers and their horses. The wars in Italy were over for the time being.

In terms of direct achievement the Italian wars seem singularly fruitless. In other ways, however, they were immensely significant. For the first time Spain had fought as a nation outside its own frontier—and had won. Only twenty years earlier it had been a divided country plunged in anarchy; by annihilating the might of France, Spain had earned a place among the leading nations of Europe.

The path to supremacy was laid not only in military terms but in the field of diplomacy as well. The constantly shifting position in Italy offered valuable opportunities for establishing friendly relations with countries that shared Spain's interest in curtailing the power of France. Ferdinand did not fail to make use of them. Ambassadors as well as brides were sent abroad to provide valuable links and information. He was the first ruler outside Italy, where diplomacy was born, to use a resident ambassador for the purpose of collecting military intelligence, that is, for espionage. It was Ferdinand who laid the foundations of the intricate and complex diplomatic network through which his successors kept their fingers so effectively on the pulse of Europe.

At home, Spanish successes were overshadowed by approaching tragedy. The Queen was in failing health and alone, but for Ferdinand. Young Isabella and John were dead. Juana was growing more disturbed every year. Even Maria, in Portugal, was too far away for the sick Queen to visit. Catherine, the daughter most like Isabella herself in appearance and character, had sailed north to England in 1501. England, like Spain, was growing in importance, and marriage to its heir was an even better proposition than it had been when the marriage contract was originally made. Nevertheless, Isabella was reluctant to relinquish this last and dearest child. When she eventually did, her duties prevented her from accompanying the fifteen-year-old bride to the farthest limits of Spain, as she had escorted each of the others. The Queen never saw Catherine again; fortunately, she never knew the indignities her daughter was to suffer in England. After Prince Arthur's early death, Catherine of Aragón became the unfortunate first wife of his brother, Henry VIII, who divorced her in 1533.

MUSEO DE ARTE MODERNO, MADRID: PHOTO, ORONOZ

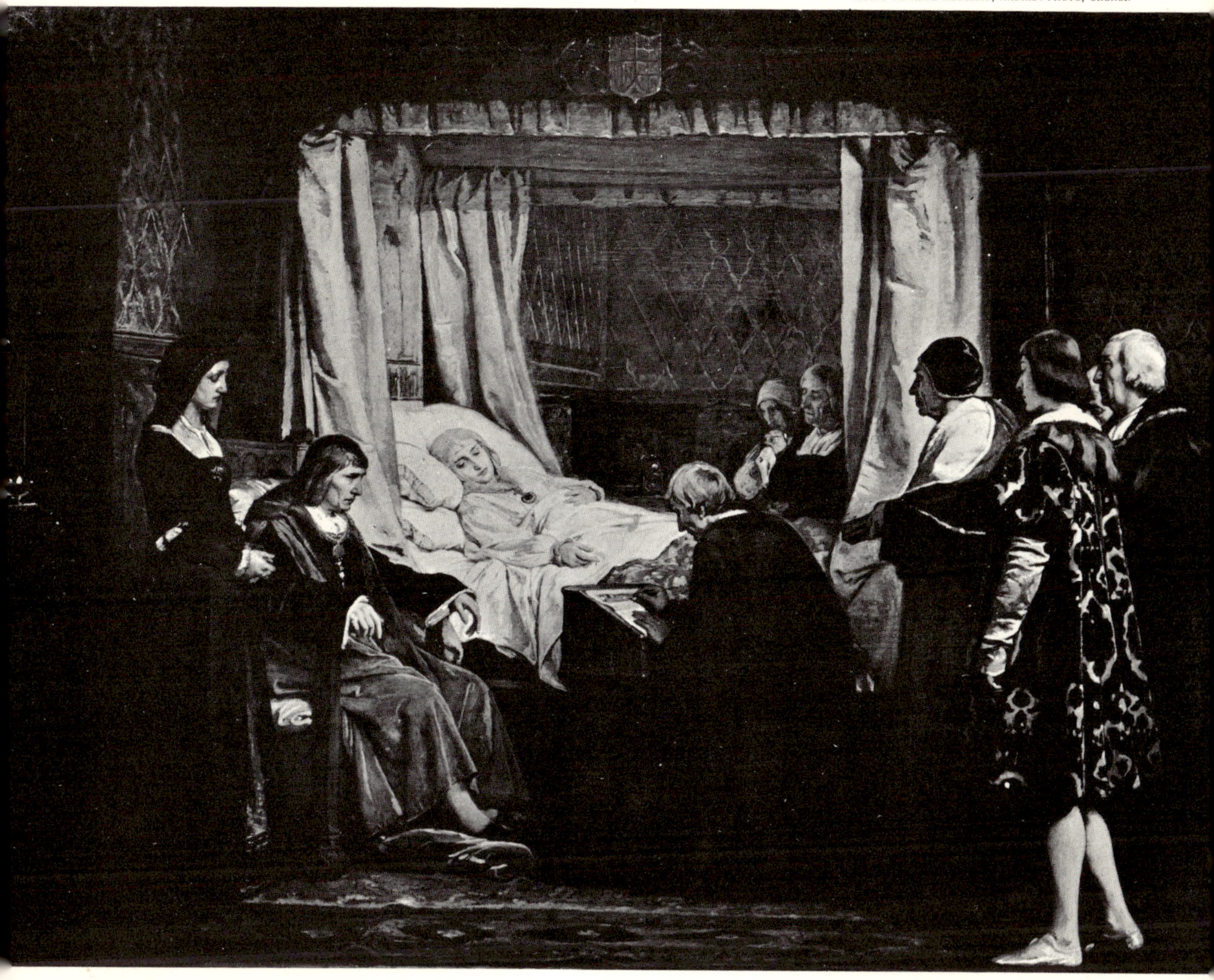

A late-nineteenth-century Spanish artist, Eduardo Rosales, painted this sentimental version of Isabella's death in 1504. A melancholy Ferdinand sits at the head of her bed, as the expiring Queen dictates her last will to a scribe.

As Isabella sank deeper into grief, she appeared less and less in public and took to wearing black robes over a religious habit. Ferdinand, more resilient, continued much as usual but spent increasingly less time in the gloom-ridden presence of his wife. In July of 1504, both monarchs came down with fever, and although Ferdinand recovered quickly, in Isabella the disease revealed more alarming symptoms. Throughout the summer she lay weak and listless on her bed or on her couch. She listened to reports of what was happening to her beloved Spain. She continued to handle as much official business as she could. Above all, she prayed.

On October 12 she made her will, with its touching proclamation of her love for Ferdinand. Her daughter Juana was named as legal heir, but Isabella provided for Juana's possible "incapacity" by appointing Ferdinand as

BOTH: MUSEES ROYAUX DES BEAUX-ARTS, BRUSSELS

Castile's ill-fated monarchs Philip and Juana appear on side panels of a fifteenth-century Flemish triptych.

regent of Castile until her four-year-old grandson Charles would be old enough to rule. Just before noon on November 26, 1504, Isabella died, at the age of fifty-three. She had reigned for thirty years; her children were gone; her job was finished. Spain had entered a political arena where there was no place for monarchs of her sincerity and integrity. The medical cause of her death is not certain, but if anybody ever died of a broken heart, it was Isabella of Castile.

"My hand falls powerless by my side, for very sorrow," wrote Peter Martyr on the day she died. "The world has lost its noblest ornament." At Isabella's orders, her body was taken to Granada, which she had always considered as the finest jewel in her crown.

When the Queen died, the Spain of the Catholic monarchs died with her. Ferdinand survived his wife by twelve years, but they were sad, gray years, in which the newfound unity of Spain was threatened. Ferdinand lacked Isabella's claim upon Castilian loyalty, and with the Queen gone, the nobles of Castile openly showed their distrust and dislike of him. They set aside Isabella's wish that he act as regent of Castile and sent instead to Philip of Burgundy, asking him to rule in Juana's place.

Ferdinand was determined not to lose the power that he had enjoyed for thirty years, and he resented the ignominy of being discarded by Castile, the kingdom that owed him so much. He had one card left to play: if he himself were to have a son, by law the throne of Aragón would go to his heir, not to the sons of Juana and Philip. In this desperate attempt to secure his position, Ferdinand reversed his policies of the last twenty years. He married again, and his bride was French: the eighteen-year-old niece of his old enemy Louis XII. The French King, eager to break up the enormous inheritance that would otherwise devolve on Juana's son Charles, gladly consented.

Fate, however, had decreed that Castile and Aragón should be one, and neither Ferdinand's diplomatic maneuvering nor the antagonism of the Castilian nobles was to prove capable of separating them. In 1506 Philip and Juana arrived in Spain, and Ferdinand was forced to surrender the sovereignty of Castile. But Philip the Handsome merely sampled the joys of kingship. Two months after his accession, he caught a chill after taking part in a strenuous ball game at Burgos. The chill became a fever, the fever grew worse, and six days later the twenty-eight-year-old King was dead.

His death drove the last vestiges of sanity from Juana's mind. She refused to allow his body to be buried, and after it had been embalmed, she refused to be separated from it. For several years the tragic Queen of Castile wandered around Spain with her husband's coffin. Finally, Ferdinand managed to lock her away—still with her beloved corpse—in a fortress at Tordesillas. She entered its gates at the age of twenty-nine, and there she remained until she died, forty-six years later. When death mercifully claimed her, the daughter of the Catholic monarchs and mother of two emperors was a ragged, wizened old crone. Toward the end, she crouched in a filthy, pitch-dark room, refusing all comforts and whining of the handsome young prince whose heart she had never touched.

Juana's obvious incapacity brought her father back to the throne of Castile for another ten years. As soon as Ferdinand was re-established as regent, Cisneros urged him to undertake a crusade against Oran. From this city and other bases on the North African coast, the Moors constantly raided southern Spain, killing, looting, burning, and carrying off hundreds of Christian slaves. Shortly after Isabella's death, the Christians had succeeded in capturing Mazarquivir, a busy port that served as the hideout for several bands of Moorish pirates, and this town became the base for the operations against Oran.

Cisneros himself accompanied the troops. Before the battle, he addressed them from his mule, promising them rich spiritual and material rewards for fighting bravely. The day ended with four thousand Moors dead and many more taken prisoner—a tribute to Cisneros' eloquence. With the taking of Oran, the Spanish crusade in Africa was under way. Ferdinand lacked Isabella's religious enthusiasm, however, and to him the campaign was only one of many projects. Spanish armies were never to expel the Moors from North Africa.

As the undisputed ruler of Spain, Ferdinand was free to turn his attention once again toward Italy. A country so divided within itself, promising such rich rewards, could hardly expect to be left in peace, and possession of Naples naturally involved Ferdinand in Italian politics. In 1508 Spain and France joined the Papal See and Emperor Maximilian in forming a coalition whose purpose was to acquire and divide the possessions of the wealthy republic of Venice. In the process, Ferdinand acquired five cities.

Then the maneuvering began again. The Pope, Ferdinand, and the Venetian republic formed a Holy League

To commemorate his 1509 victory over the Moors at Oran, Cardinal Cisneros commissioned the monumental fresco above for a Toledo chapel. As infantry scale the walls (left foreground) and cavalry skirmish on a hill (right background), the Cardinal himself (extreme right) prepares for a triumphal entry into the city once his troops have captured it.

TOLEDO CATHEDRAL: PHOTO, ORONOZ

against France, and in 1512 Ferdinand persuaded first England and then Maximilian to join. Once the French armies were routed, however, the League fell to quarreling over the division of France's Italian holdings, and in 1515, despite Ferdinand's warnings, France persuaded England and Venice to sign a mutual defense treaty. Back and forth went the struggle, with Ferdinand constantly plotting fresh ways of gaining control of Italy.

At home, too, he was rounding off his kingdom. Since the union of Castile and Aragón, the tiny kingdom of Navarre, which straddled the Pyrenees between France and Spain, had been fighting a losing battle to remain inde-

GRANADA CATHEDRAL: PHOTO, ORONOZ

pendent. In such an age, however, no country could afford to scorn alliances, and in 1512 Navarre negotiated a treaty with France.

This happened to be the year in which Ferdinand had persuaded Henry VIII of England to join him in an attack on France, and for this to succeed, Spanish troops had to be able to cross through Navarre without hindrance to link up with their English allies. One of Ferdinand's spies informed him of the terms of the Navarrese treaty with France before it was signed, and the King decided not to negotiate, but to act. His armies marched into Navarre, took its capital, Pampeluna, without a fight, and conquered most of the country within two weeks. The following year, all the states of Navarre swore allegiance to the triumphant Ferdinand, and Spain's borders finally ran the length of the Pyrenees.

The Catholic King reigned over Aragón and Castile until he died. As he grew older, he conceived a profound dislike for his grandson Charles, who was being educated in Flanders under solely Hapsburg influence yet who would inherit Spain's vast possessions. At times, he plotted to make Juana's second son, Ferdinand, who had been brought up in Spain, heir to Aragón and his Italian acquisitions, if he himself should fail to produce an heir by his young French wife. But it was not to be. Germaine bore Ferdinand only one child, who died almost at once. On his deathbed, the King's advisors persuaded him not to weaken his empire by dividing it between Charles and young Ferdinand. Spain was safe.

Ferdinand died on January 23, 1516, aged sixty-three. As he had requested in his will, his body was buried at Granada, side by side with that of Isabella, where it rightly belonged.

The King's last will settled his entire territorial possessions on Juana and her heirs and also named two regents to govern them: Cardinal Cisneros in Castile, and the Archbishop of Saragossa, Ferdinand's own illegitimate son, in Aragón. On September 17, 1517, the seventeen-year-old Charles V landed in Spain to claim his inheritance, and the Catholic monarchs' vision became a reality at last. From their once-divided kingdoms of Castile and Aragón had come the Spain of modern Europe.

Side by side in the Royal Chapel of Granada Cathedral rest marble effigies of Ferdinand and Isabella (left) and Philip and Juana (right). Their actual remains are in a crypt beneath, placed there in 1521 by Charles V.

TOLEDO CATHEDRAL: PHOTO, ORONOZ

The surrender of Marbella to the Catholic Kings during the Granada campaign is shown on a Toledo Cathedral choir stall. As trumpets sound, a Moor submits to Ferdinand, who rides in triumph into the walled city.

ACKNOWLEDGMENTS

The Editors would like to thank the following individuals and organizations for their valuable assistance:

The Duke of Alba, Madrid
Archivo Historico Nacional, Madrid
Archivo MAS, Barcelona—Mrs. Monserrat Blanch de Alcolea
Biblioteca Nacional, Manuscript Division, Madrid—D. Ramon Paz
Hispanic Society of America, New York
Mrs. Evelyn Hofer, New York
Jewish Theological Seminary of America, New York—Miss Anna Kleban
The Pierpont Morgan Library, New York
New York Public Library, Rare Books Division—Mrs. Maud Cole
Oronoz, Madrid
Skira, New York—P. Gregory
Miss Bianca Spantigati, Rome
Maps and family trees by Francis & Shaw, Inc.

FURTHER REFERENCE

Readers interested in the history of Spain, her arts, and her colonies in the New World will find exhibits at the Bowes Museum, Barnard Castle; the British Museum, the National Gallery, and the Victoria and Albert Museum (also Islamic art), all in London; Stoke-on-Trent City Museum and Art Gallery (also Islamic). The New World is represented in Birmingham City Museum and Art Gallery; Camborne Public Library and Museum; and the British Museum, London. In Spain, there are important collections at the Escorial; the Biblioteca Nacional, the Museo Naval and the Museo del Prado, all in Madrid; the Museo de Historia de la Ciudad, Barcelona; the Museo Arqueológico Provincial, Seville; and Granada and Toledo cathedrals.

For those who wish to read further on Spain under Ferdinand and Isabella and their successors, the following books are recommended:

Colón, F. *The Life of the Admiral Christopher Columbus by His Son Ferdinand.* Translated by B. Keen. New Brunswick: Rutgers UP, 1959.

Elliott, J. H. *Imperial Spain: 1469–1716.* Arnold, 1963.

Kamen, H. A. *The Spanish Inquisition.* Weidenfeld & Nicolson, 1965.

Livermore, H. V. *A History of Spain.* New ed. Allen & Unwin, 1966.

Mariéjol, J. H. *The Spain of Ferdinand and Isabella.* Edited by B. Keen. New Brunswick: Rutgers UP, 1961.

Mattingly, G. *Renaissance Diplomacy.* Cape, 1955. *Catherine of Aragon.* Cape, 1942.

Merriman, R. B. *The Rise of the Spanish Empire in the Old World and in the New.* 4 vols. New York: Cooper Square, 1962.

Miller, T. *The Castles and the Crown.* Gollancz, 1963.

Morison, S. E. *Admiral of the Ocean: Christopher Columbus.* OUP, 1942. *Christopher Columbus: Mariner.* Faber, 1956.

Morris, J. and Hofer, E. *The Presence of Spain.* Faber, 1964.

Pita Andrade, J. M. *Treasures of Spain.* Vol. I: *From Altamira to the Catholic Kings.* Translated by I. Quigly. Zwemmer, 1967.

Prescott, W. H. *History of the Reign of Ferdinand and Isabella the Catholic.* Edited by C. H. Gardiner. Allen & Unwin, 1962.

Smith, B. *Spain: a History in Art.* Harrap, 1966.

Walsh, W. T. *Isabella of Spain: the Last Crusader.* Sheed & Ward, 1931.

INDEX

Boldface indicates pages on which maps or illustrations appear

BIBLIOTECA NACIONAL, MADRID: PHOTO, ORONOZ

*In a decorative border to a document authorizing construction of a monastery, the eagle of St. John the Evangelist holds the escutcheon of Castile and Aragón (also shown on page 30), to which has been added (bottom, center) the pomegranate of Granada. Flanking the Catholic Kings' coat of arms are their personal symbols—a yoke (*yugo *in Spanish) for Ysabel and arrows (*flechas*) for Ferdinand. Their motto, Tanto Monta, stood for a phrase citing their equality.*

THE BIG BOOK OF Animals

Editor: Daniela Majerna
Editorial assistant: Tiziana Campana
Scientific Consultant: Marina Mansi
Art Editor: Viviana Cerrato

Illustrations: Cinzia Antinori, Amalia Arosio, Alessandra Micheletti
and in collaboration with Lorenzo Orlandi

Il Grande Libro degli Animali

ISBN 978-0-7097-2223-6

Reprinted 2015, 2018

Printed in China

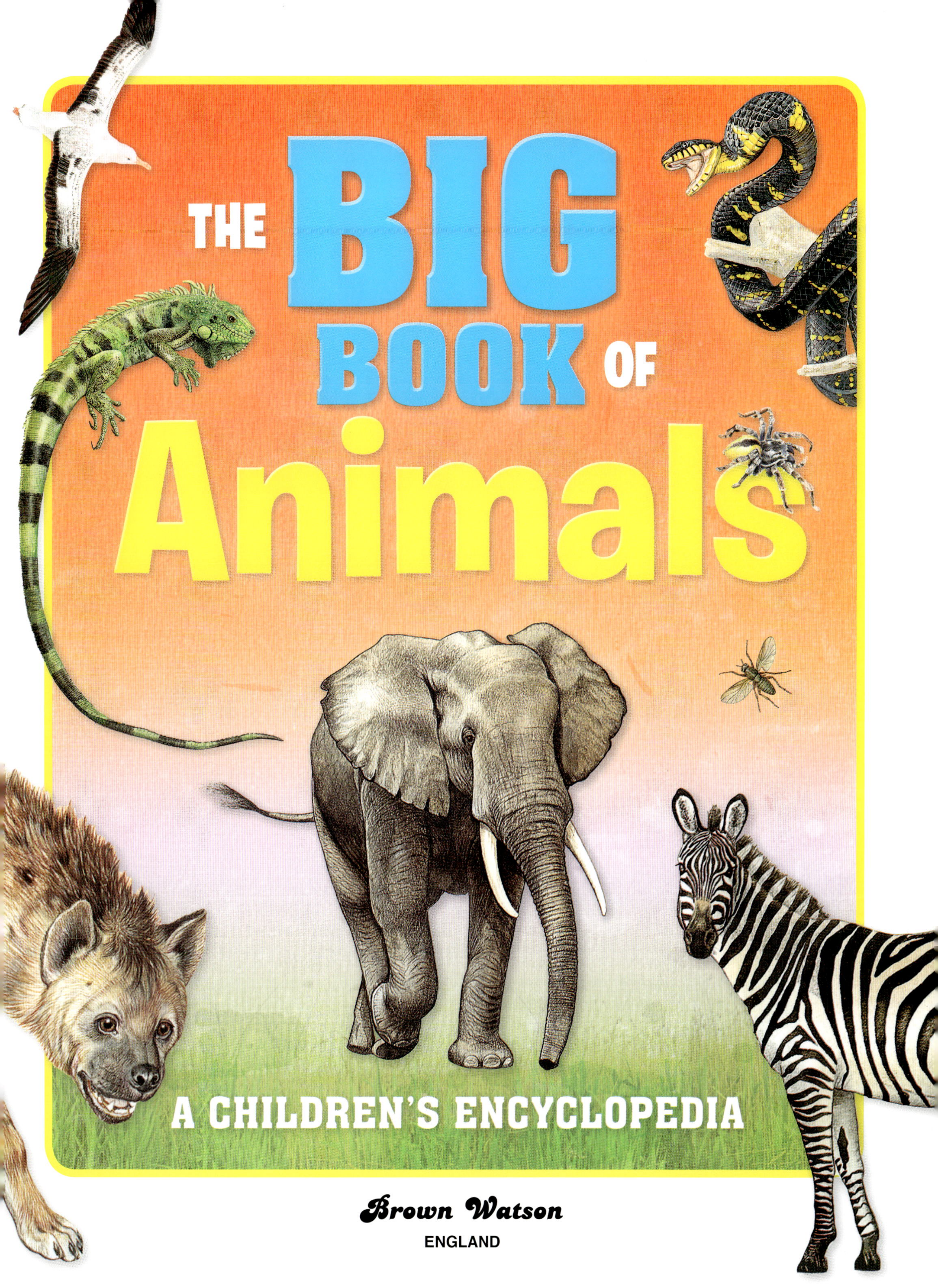

THE BIG BOOK OF Animals

A CHILDREN'S ENCYCLOPEDIA

Brown Watson
ENGLAND

CONTENTS

PORIFERA

EVOLUTION

Animal life on Earth began with invertebrates – creatures without backbones. The oldest phylum (or 'family') of invertebrates dates back over three thousand million years ago. This is Protozoa, one-celled animals, some with simple structures, such as the amoeba, some more complex, like radiolaria. Two thousand million years later, Porifera, commonly known as sponges, probably developed from the one-celled animals of the Protozoa phylum.

NUTRITION

Sponges filter water as this enters through their pores, getting particles of food in this way. Waste products are expelled through larger holes called ***oscula***. Sponges feed on bacteria, protozoa and tiny particles of animal and plant substances. This filtering system is extremely delicate. An excess of food particles can be extremely dangerous.

Right, the ***Axinella polypoides***, a sponge with branches like a little yellow or orange plant, very common on rocky sea beds along the Mediterranean and Atlantic coastlines. The four species of axinella in the Mediterranean are spread in different ways, according to the current and the movement of the water.

The ***spongia officinalis***, the common bath sponge (centre) is widespread throughout the Mediterranean: it has lost its chalky spines, leaving just the spongy fibres of its skeleton. This can absorb a remarkable amount of water, equal to 20-35 times the dry weight of the sponge. Soft and strong, this type of sponge has been used since ancient times, for body hygiene, for medical uses and as a mask to protect against disease.

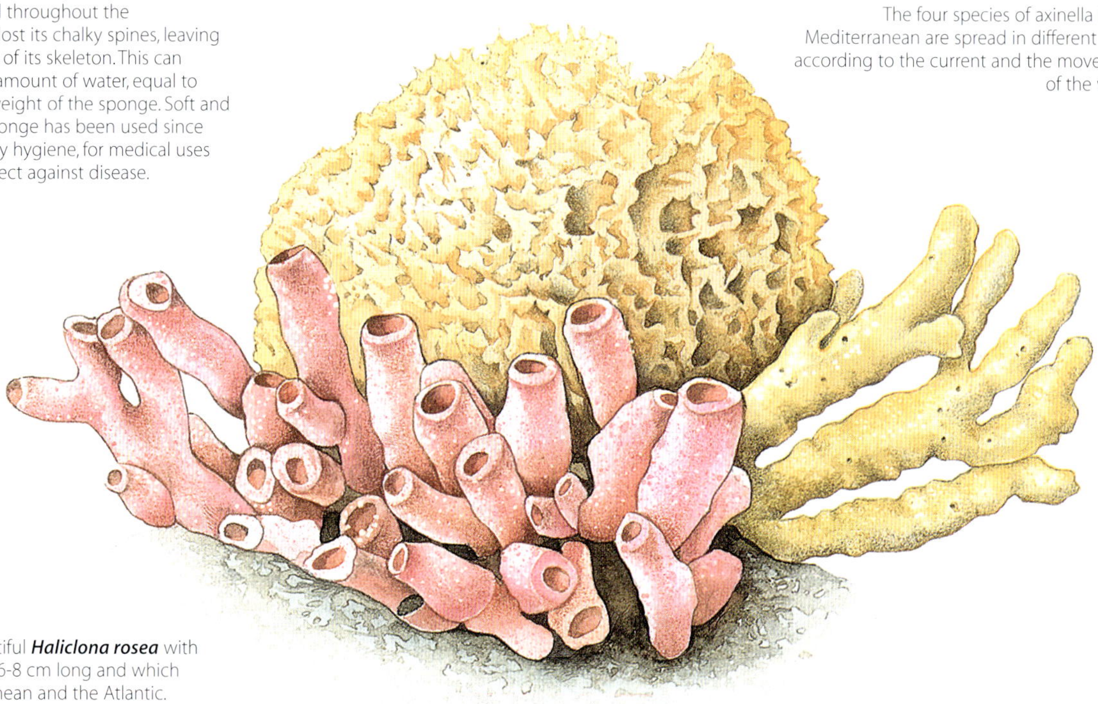

Bottom, left, the beautiful ***Haliclona rosea*** with its delicate red tubes 6-8 cm long and which lives in the Mediterranean and the Atlantic. Clearly visible is the ***osculum*** at the top of each tube and pores on the surface.

COMMON CHARACTERISTICS

All Porifera have a simple structure; a soft, irregular-shaped body, either like a bag or with branches, and supported either by silica or lime spines (or ***spicules***) or by the ***spongin***, a hard, horny-type fibre mesh. Water enters through numerous pores on the surface of the animal and goes out through a larger opening, called the ***osculum***.

HABITAT AND DISTRIBUTION

There are over 3000 species of sponge which live on the sea beds, both in warm and in cold waters. Only some horny-coated species live in fresh water. Some sea sponges live in water which is only slightly deep, others live at depths from between 100 and over 1000 m.

Filtering to Live

Over a three thousand million years ago there were only **single-celled animals**, Protozoa. As animal life began to develop, some species of Protozoa probably **mated** to produce other creatures of two or more cells, which gradually became more complex. Porifera or **sponges** represent the first large group of multi-celled animals. Sponges have an extraordinary ability to **regenerate**. Not only can they renew lost or damaged cells, they can also be reproduced manually, by squeezing sponge cells through gauze or silk, or chemically, by taking out calcium and magnesium from the water in which they live. In both cases, once a group of cells settle on a surface, they begin to develop into a complete, living sponge.

Photograph: Sponges have large openings called oscula, through which water flows.

As well as being able to reproduce by manual and chemical means, sponges can reproduce both sexually, through fertilization of the female egg by the male spermatozoa and asexually, by 'gemmating'. On the surface of a sponge, '***gemmae***' are formed, like little buds. Each ***gemma*** can multiply into a large number, which then detach from the body of the sponge, each one to begin life as an individual animal. The most primitive type of sponge is quite small, with a body in the form of a tube-like bag. But most sponges have irregular shapes, some bushy, others like little trees which can grow up to 2 m in diameter.

The body of a sponge has two layers of cells, criss-crossed with pores which come together in an internal cavity. This cavity is lined with special cells called ***choanocytes***, which have structures like tiny whips which push the water which has entered in through the pores towards the ***oscula***.

Sponges are fixed and unable to move. They feed by filtering useful substances carried along by the current of water.

According to the body structure, sponges are divided into three main classes:

Calcispongiae, the earliest type of sponge, is ***calcareous*** and sponges in this class have lime spicules. They live only in the deepest sea waters and they are small, reaching only 10 cm.

The **Hyalospongiae** have ***silica spicules*** with a body roughly shaped like a cylinder or a funnel, a tubular bag. They are present in all seas, in waters over 100 m deep. One of the most famous in this class is Venus's Flower Basket, with its delicate, white flower-like skeleton.

Sponges in the **Demospongiae** class have a horny, ***spongin*** body. This class is the largest and the most complex. They can be found both in seas, in the deepest waters of the oceans, and in fresh waters – such as the ***Spongilla lacustris*** which lives in lakes and has a characteristic green colour, due to the presence of the algae which live on its body. The ***spongia officinalis*** (common bath sponge) belongs to this class.

PHYLUM	CLASS	SPECIES
PORIFERA OR SPONGES	Calcispongiae	scypha
	Hyalospongiae	Venus's Flower Basket
	Demospongiae	bath sponge

CNIDARIA

EVOLUTION

The Cnidaria (or Coelenterata) phylum developed at the same time as sponges, first appearing in the seas and oceans more than 500 million years ago. Like sponges, Cnidaria are multi-celled creatures with a simple structure, but in many different forms, such as the beautiful **sea anemone**, the **transparent jellyfish** and tree-like **corals**.

NUTRITION

The central cavity of the Cnidaria body begins with an opening on the outside, through which the creature both gets its food and gets rid of waste. This opening is surrounded with numerous tentacles which attract prey, then paralyse it and take it down into the cavity to be digested.

Red coral has been used for centuries in making ornaments and jewellery. Some has been found in a prehistoric tomb dating from the fourth century BC. It was used in Ancient China and Japan, and in Europe, red coral was widely used as a semi-precious stone and in jewellery, especially after the fifteenth century. People in Tibet used coral in place of money during the thirteenth century. At one time, coral was also valued for the power it was thought to have to cure many illnesses.

The coral which we know is in fact an external skeleton of calcium carbonate, hard and brittle and formed by tiny, hollow, tube-like structures called polyps. These form groups of branches, so that corals look like little trees, especially when they move their tentacles like the petals of flowers.

The beautiful red coral, also called the noble coral, is found mainly in the Mediterranean Sea at depths of between 20 and 200 m, forming groups up to 20-25 cm high. Other similar species which live in Japan reach heights of up to 1 m and a weight of over 40 kg!

COMMON CHARACTERISTICS

All creatures in the Cnidaria phylum have a hollow, cylinder-shape body called a polyp. This is generally fixed at the bottom with an opening at the top surrounded by tentacles reaching upward. In some species, such as a jellyfish, a polyp developes into a medusa, where parts of the body are spread evenly around a central point, like the spokes of a wheel.

HABITAT AND DISTRIBUTION

Some Cnidaria can live fixed in water. Others can move about in the sea, especially in the warm waters of tropical seas. The Hydroida class have species both in the form of polyp and medusa. This class is mainly sea-dwelling. The Scyphozoa in medusa form is widespread in all seas. Creatures in the class Anthozoa are in polyp form. They can live alone or in large colonies.

Stinging traps

Cnidaria are multi-celled animals, very simple and with symmetrical shapes (equal on both sides). This type of body means that they can reach out for food and towards enemies with any part of their body and can attack or defend from either side, even when they cannot move.

The body of a Cnidaria animal has a single internal cavity which begins on the outside through a single opening, a mouth. The walls of the body are also used for breathing, digesting food and for getting rid of waste. The mouth is surrounded by tentacles armed with **stinging cells** (**cnidoblasts**). When prey comes in contact with a tentacle, it is stung by a tubular filament containing a poisonous liquid.

Hundreds of these filaments fix themselves like this to the victim and paralyse it. Then, it is taken down into the mouth by the tentacles.

This means that the Cnidaria phylum are **fearsome predators**, often wonderfully hidden among underwater plants and feeding on small crustaceans and tiny fish. The two basic types of Cnidaria, the polyp and the medusa, can change during their development. For instance, the larva can be fixed, in the form of a **polyp**, and the adult mobile, like a **medusa**. Or the adult may be fixed like a polyp, such as the coral.

Photographs above: top, jellyfish, bottom, Sea Anemone.
Facing page: top, Sea Pen, bottom, Tube Anemone.

The walls of the body are composed of two layers of cell tissues. The outer is called the ectoderma, the internal layer is called the endoderma. Between these two tissues there is a **jelly-like layer**, rich in water. The outer fibres make up the muscles lengthways. The fibres inside form the circular muscles. In the **darting movements** which characterise the jellyfish, these two muscular systems work together, relaxing and contracting.

Other cells develop into nerves; these line the internal surface of the body like a network and are extremely sensitive.

The three main classes into which Cnidaria are divided are – Hydroida, Scyphozoa and Anthozoa.

There are over 2700 species of **Hydroida**, all quite small, and in the form of both polyp and medusa. All can reproduce sexually or by 'gemmation' (see Sponges). They can live alone, like the Hydra, or in large colonies, such as the Milleporia corals.

The **Scyphozoa** are always alone, such as the colourful jellyfish with huge 'bells' or canopies. There are around 250 species of scyphozoa. The largest is the ***Cyanea arctica*** with a bell measuring over 2 m in diameter.

Anthozoa, with over 6500 species, is the largest class of Cnidaria. These are only in polyp form. They can be either solitary, such as the sea anemone, or, more often, living in groups, such as the stony corals and the madreporaria.

PHYLUM	CLASS	SUBCLASS	SPECIES
CNIDARIA	Hydroida		Hydra
	Scyphozoa		jellyfish
	Anthozoa	Alcyonaria	corals, gorgonians
		Zoanthinaria	madreporaria, sea anemone

PURPLE JELLYFISH

Diameter: 25-35 cm

Length of tentacles: 45-80 cm

This jellyfish is very common in the Mediterranean and in the warmest parts of the Atlantic Ocean. It has eight tentacles which it can stretch and contract to a remarkable degree and a semi-circular bell. Its colour can range from pale purple to dark red. When disturbed or alarmed, this jellyfish reacts by acquiring a red-mauve luminous shine, easily visible at night-time. The life cycle of the Purple Jellyfish is very unusual. When its larva, called a **planula**, hatches from a fertilized egg, this does not go through the polyp stage. Instead, it develops directly into a baby jellyfish, called an **ephyra**. Because the purple jellyfish does not need to fix itself to a surface, like a polyp, it can live in the open sea. Often, enormous groups of Purple Jellyfish are swept towards a bank or a sea-shore by tidal currents, and then they are a serious danger for bathers who may be stung by their long, stinging filaments.
Some species of jellyfish can be very dangerous to humans, causing shock on contact.

TUBE ANEMONE

Diameter: up to 60 cm, tentacles extended

Length of tube: 20-70 cm

The Tube Anemone belongs to the Anthozoa class and is one of the most beautiful of sea creatures. Its flowery appearance reminds us of the sea anemone, its distant relative.
The Tube Anemone, pictured below, lives in Japanese waters. It lives alone, protected by its tube-shaped skeleton which is rooted to the sea bed. Its tentacles, arranged in two concentric circles, are armed with poisonous glands. When disturbed, the creature withdraws completely inside its tube. The Tube Anemone feeds on small crustaceans and tiny fish, which it paralyses with liquid poison and then captures in its tentacles. Some tropical fish, such as the Clown Fish, are immune to the poison of the Tube Anemone. Instead, it likes to keep the Tube Anemone company, darting in and out of its tentacles. Then, at the least sign of danger, the Clown Fish will hide inside a bunch of the stinging tentacles.

Did you know...

*The **Hydra** lives in fresh water and resembles a fine, delicate seaweed. Its tentacles are about six times the length of its body.*

*The **Xenia** is a most beautiful coral, rather like a little tree flowering with feathery tentacles.*

*The **Bunodactis Verrucosa** is a special type of sea anemone, because its base is covered in lumps. When it senses danger, it will withdraw its tentacles into its mouth, until it becomes a hard ball.*

*Because of its beauty, the **Tube Anemone** is often collected for large aquariums.*

*The **Sea Pen** has an external skeleton that is very soft and flexible, and which looks like a beautiful set of goose feather quills.*

*The **Sea Fan** is becoming very rare. It lives in large groups and its branches reach up to 70 cm in height. It is much sought after by divers because of its beauty.*

*The **Caryophylla Clava**, is a stony type of coral and it is quite common in the Mediterranean. It lives alone.*

*At the base of its tentacles, the **Bell Jellyfish** has cells which are sensitive to light.*

Hydra viridis
HYDRA
Polyorchis
BELL JELLYFISH
Pelagia noctiluca
PURPLE JELLYFISH
Xenia
Eunicella cavolinii
SEA FAN
Cerianthus filiformis
TUBE ANEMONE
Caryophylla
Pennatula phosphorea
SEA PEN
Bunodactis verrucosa

ANNELIDA

EVOLUTION

The name **Annelida** comes from the Latin word *annulus*, meaning ring – because the body is covered with rings from the head to the tail. The segmented body structure of the Annelida phylum signalled the development of the *Arthropods*, such as insects.

HABITAT AND DISTRIBUTION

We know of around 9000 species of Annelida. The earliest form can be found living in the sea, such as the sea worm; the most developed have adapted to fresh waters, such as the leech, or in muddy depths in the earth, such as the well-known earthworm.

In the *Lumbricus Terrestris*, the common earthworm, we can see this marking very well. Each ring has four pairs of short, hairy bristles which enable the animal to make its way along the ground. These bristles are also sensitive to vibrations in the earth, and so enable the earthworm to escape when a mole, its most dangerous enemy, comes near. Only the earthworm's head has no bristles.

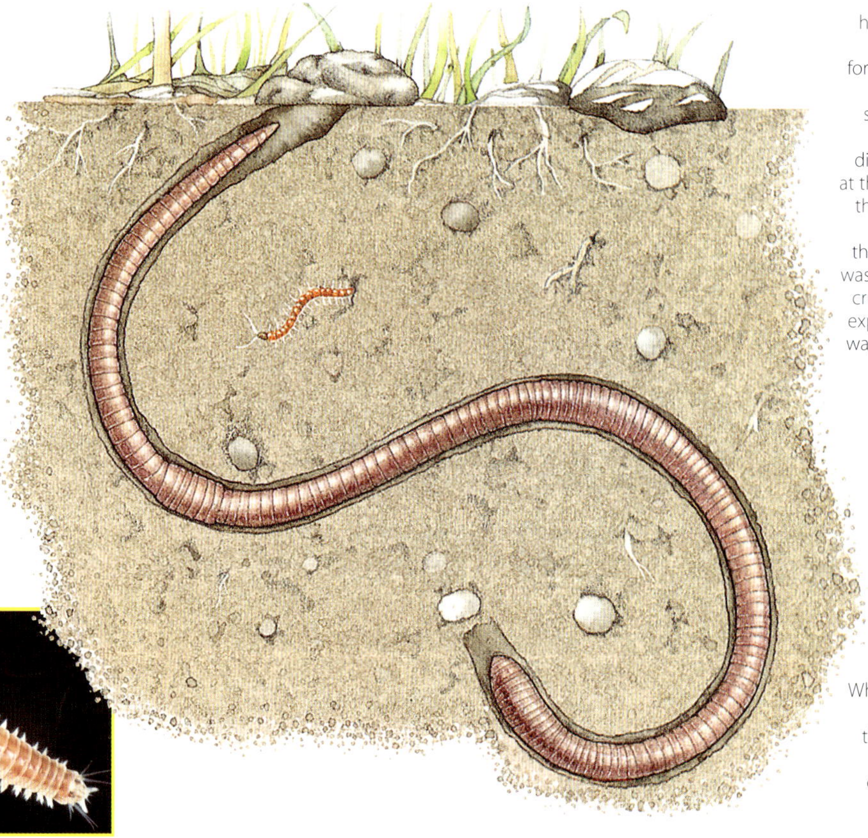

The earthworm seems to have neither a head nor tail, because it can move both forward and back. It opens its mouth in a fraction of a second. All members of the Annelida phylum have a digestive tube which begins at the mouth and opens up at the tail. Earthworms are very useful to farmers, because they feed on both vegetable waste and the soil, which they crumble up, 'digest' and then expel through the tail. In this way, earthworms break down the soil and keep it fertile.

When an earthworm is ready to mate, there appears a thickening in the middle of the body. This is called the clitellum and it is here that eggs develop into young worms, inside cocoons.

COMMON CHARACTERISTICS

The body is segmented. Each segment is called a **metamere**, and is similar to the one next to it. Each **metamere** is practically independent because it has all the organs necessary to survive.

Annelida creatures also have an internal body cavity called a **coelom**, which is full of liquid.

MOVEMENT

A good muscle structure in each metamere means that creatures in the Annelida phylum can wriggle and move in all directions, aided by the **coelom**, which can make a single metamere swell or shrink in many different ways. Some species can expand sideways with **parapodia**, rather like flat feet covered with hair. They use these to lever themselves along the ground.

'The Lord of the Rings'

The bodies of most Annelida are segmented and normally long and rounded. But sometimes, as with some **Polychaetes** (sea worms), it can be broad and flat and the segmentation is not so easily seen. The body is covered by a protective layer, or **cuticle** which is strong and very flexible. In the aquatic form, the body can also have bristles. Beneath the **cuticle** there is a circular, muscular layer, and, inside this, lengthways muscles which are different according to the various species.

This **muscular strength** enables Annelida to move very easily. Animals in the most highly developed classes of Annelida have a **digestive system** with a mouth, intestine and a hole for the waste to pass through.

These also have a circulation system whereby their blood circulates inside through a network of veins.

Reproduction is generally sexual, but some species can reproduce asexually. The **earthworm** is a good example of this. If it loses a piece of its tail or its head, it just grows another!

There are about 9000 species in the Annelida phylum, divided into three classes – **polychaeta**, (marine or sea worms), **oligochaeta** (earthworms) and **hirudinea** or leeches.

There are over 5000 species in the **polychaeta** class and most live in the sea. At the deepest depths, we find the **free-living polychaetes** which move by means of their **strong tails** and **bristles** on the body.

The **sedentary polychaetes** live in tunnels dug out of mud and sand, or they dig tunnels in firm sand. These **polychaetes** cannot move. They get their food by rippling the water with the bristles on their tentacles around their mouths.

The **oligochaeta** comprise around 3000 species which live on the ground or in muddy fresh water. Their bodies are characterized by the presence of the **clitellum**, a swollen group of segments. Here the **cocoons** are formed to contain the eggs that will develop into baby worms. **Oligochaetes** move by contracting and extending the muscular segments. As well as the earthworm, the ***Megascolides Australis*** – a gigantic Australian worm which can reach 3 metres in length – belongs to the same class.

Photographs: top right, a marine worm of the Polychaeta class, above, a spendid Spriographis. Facing page: top right, the Hermodice carunculata (Bearded Fireworm), bottom left, Neried (sea worm).

Hirudinea is the most highly developed class of Annelida, with over 300 species. Some are the predators of the invertebrates, others are **ectoparasites**, such as the leech which attaches itself to the skins of animals and humans and sucks out the blood.

Echiurida constitutes a small group of sea creatures which live mostly along sandy and rocky coastlines. They have a proboscis (a long tube to suck with) at the tip of the front end. They are different to other Annelida because they have the typical metamere segments only in the embryo stage of development.

PHYLUM	CLASS	SUB-CLASS	SPECIES
ANELIDA	Polychaeta	free-moving	
		sedentary	
	Oligochaeta		earthworm
	Hirudinea		leech
ECHIURIDA			spoonworm

MOLLUSCS

EVOLUTION

The origin of molluscs began more than 500 million years ago, probably from the same creatures then living on the sea beds and from which the phylum Annelida also developed. This is confirmed by the fact that the larvae of Molluscs and Annelida are very similar. Some of the oldest fossils ever found have been tiny calcium carbonate mollusc shells.

NUTRITION

Molluscs feed on a variety of food depending on their surroundings. Some feed on algae (seaweed) and Cnidaria, others feed on plankton (tiny plants and sea creatures), algae and small invertebrates. There are some herbivore (plant-eating) species, such as the common snail, others carnivorous (meat-eating) such as squid and octopus.

The sea snail (above) is a beautiful violet or blue. Like the land snail it has a spiral shell produced as its **mantle** (or body cavity). Sea snails are abundant in tropical waters and form part of the oceanic **plankton**. A sea snail spends its life upside down on the surface of tropical waters, attached by a small bubble of air which is so compact, it is impossible to pierce. This bubble is made of mucus produced by the sea snail and which hardens with time.

The pearls found in some bivalve molluscs are formed by **nacre**, a carbonate calcium which lines the shell. The mollusc builds up the nacre around a small foreign body, such as a grain of sand, which gets inside between the nacre and the shell. The most precious pearls are found in oysters such as this ***Margaritifera margaritifera*** (Pearl Oyster).

COMMON CHARACTERISTICS

All molluscs have a soft, compact body without an internal skeleton and which is sub-divided into a head and a trunk. They do not have jointed segments like the Annelida phylum. The particular characteristic of many species is the calcium carbonate shell.

HABITAT AND DISTRIBUTION

There is a large variety of molluscs living in all types of damp conditions. Some live on surfaces below ground or water, others in the mud and the sand, others are swimmers. Some species are able to live in the depths of the sea bed, others can live at an altitude of over 5000 m.

An incredible variety of shapes

All molluscs have soft bodies which are compact and wet because of the liquids made by their glands. There are two main parts – the head, with the sense organs (organs for sight, smell, taste and touch) the nervous system and the mouth, and the main body, with muscular-type feet and an internal bag. On the back of this bag is the mantle, which is wrapped around a cavity where the breathing organs are – **lamella gills** in aquatic molluscs, **lungs** in land molluscs. In the mantle, the mollusc secretes a calcium carbonate substance which forms the **shell** and which is connected to the feet by muscular bands.

Molluscs have a complete digestive system. At the top end of the digestive tube is a rasp-like tongue with lots of little teeth. This is called the **radula**. It is used to gather food in and to break it down.

Molluscs also have a proper heart at the back of the body, which enables the blood to circulate.

The nervous system is made up of a brain and four nerve cords which spread to all parts of the bottom of the body.

In molluscs the muscles are particularly developed. Reproduction of molluscs is sexual, the result of a male mating with a female. Many characteristics are the processes of development from the egg. In many molluscs the development to an adult proceeds through a larva called **veliger**.

There are at least 125,000 living species of molluscs, sub-divided into many classes. The most important classes are the Gastropods, the Bivalves and the Cephalopods.

Photographs: above, the dazzling colours of the Flabellina Lilla octopus, below, a Conus textile or Cone Shell, a marine snail.

The class of **Gastropods** have the most species of molluscs. They are recognized by the **spiral shell**. There are aquatic gastropods, living in fresh waters as well as land gastropods.

Bivalves have a **shell divided into two parts, or two valves** which are joined together by an elastic type of ligament, like a hinge. They do not have a true head or **radula**. Bivalves are mainly sea molluscs.

Cephalopods are almost all **without shells**. They are one of the most developed classes in the mollusc phylum, with a highly developed nervous system and sense organs. The octopus and the squid, have the most complex eyes amongst molluscs. Both live in sea water which has a high salt level.

PHYLUM	CLASS	SUB-CLASS	SPECIES
MOLLUSCS	Gastropods	Prosobranchs	acquatic, with shell
		Opisthobranchs	sea-dwelling without shell
		Pulmonates	land, such as snail and slug
	Bivalves or Lamellabranchs		with shell in two halves
	Cephalopods		with tentacles; nautilus, squid, cuttlefish, octopus

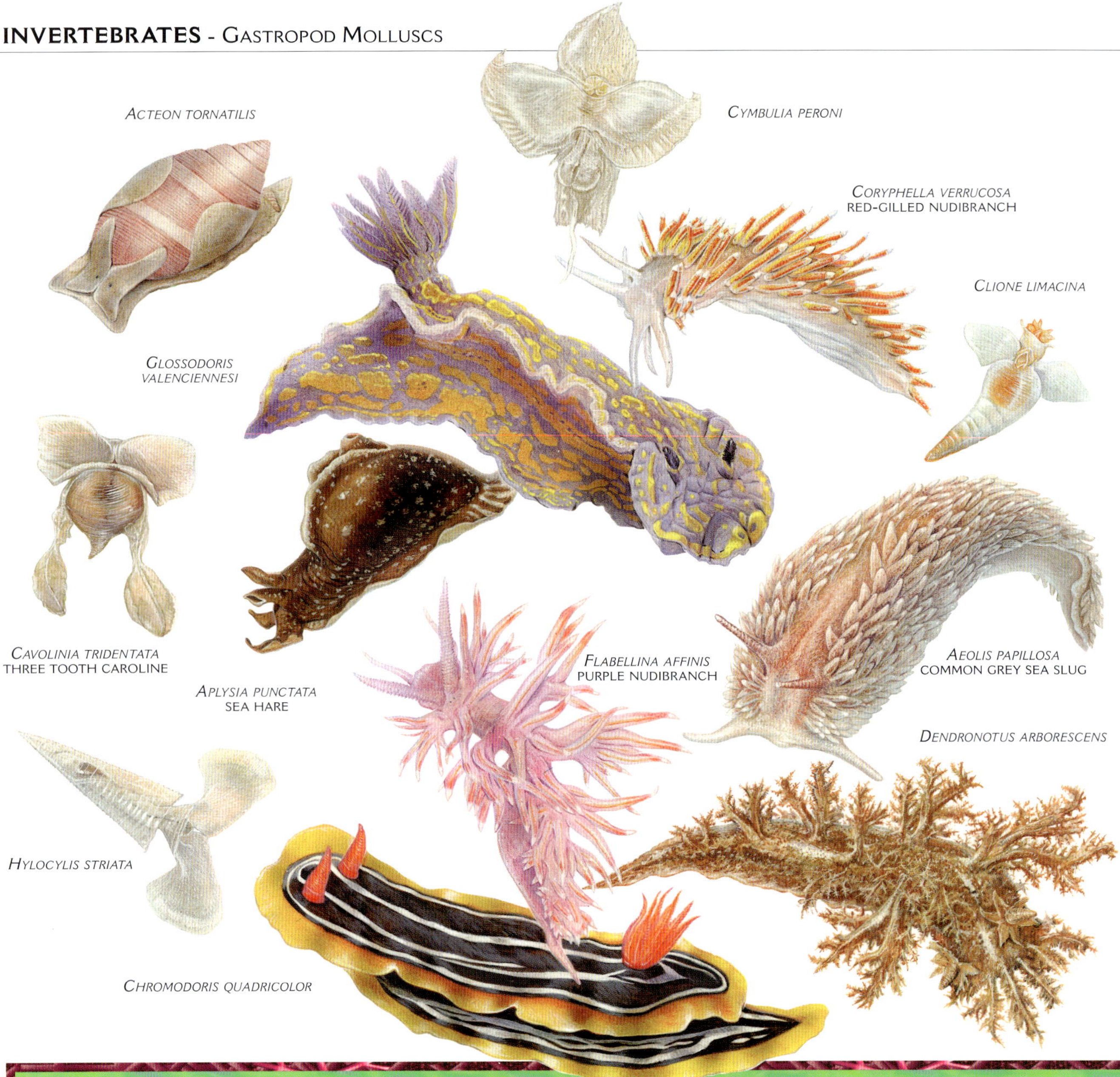

Did you know...

Acteon Tornatilis *usually lives on the sandy beds of European seas. It feeds mostly on a type of seaweed called 'sea lettuce'. This mollusc has a shell rather like a shield, reaching up to its head and which the creature uses as a tool for digging short tunnels beneath the sand.*

Widespread in all seas is the ***Coryphelia Verusca****, or Red-Gilled Nudibranch, one of the most beautiful sea creatures because of its dazzling colours. Instead of a shell, it has numerous long spines around its body.*

The ***Hypselodoris Valenciennesi*** *and the* ***Chromodoris Quadricolor*** *are two gastropods with flat bodies. Each species has a tuft of secondary gills on its back which looks like a beautiful flower.*

Sea butterflies are small plankton gastropods. They are among the most abundant of sea creatures. The ***Cavolina Tridentata*** *(Three Tooth Caroline) is widespread almost everywhere, the* ***Cymbulia Peroni*** *and* ***Hylocylis Striata*** *are found mostly in the Mediterranean; the* ***Clione Limacina*** *is found in the cold waters of the Arctic.*

The ***Flabellina Affinis*** *(Purple Nudibranch) looks rather like a little bush. Feeding on jellyfish, this mollusc has stinging cells in its back spines which it uses to defend itself. Like the* ***Dendronotus Arborescens****, it lives along the European Atlantic coast and has a body with lots of branches. The Dendronotus Arborescens is also a stinging mollusc. It uses its gaudy colours as a signal to the would-be predator to go and find other prey.*

The ***Helix Hortensis*** *(White Lipped Snail) is known for its beautiful colours and the shape of its shell. It is nocturnal, coming out only when it is dark to go in*

search of leaves and nettles.

The ***Delima Itala*** *is a small slug (15-20 mm) which likes hot climates.*

The ***Succinea Putris*** *(Amber Snail) is a European species which lives in very hot places, among the vegetation on the banks of rivers and marshes.*

The ***Limax Maximus*** *(Great Grey Slug) can be up to 15 cm long. It is found in woods and gardens after a storm. Although it seems to be without a shell, it does have a tiny, little one under the mantle.*

The ***Planorbarius Corneus*** *(Great Ramshorn Snail) is a gastropod with lungs. It is widespread throughout Europe and other temperate regions and lives in fresh water. The* ***Lymnaea Stagnalis*** *(Great Pond Snail) is also common in European ponds. Both eat water weeds and rotting plants; some species feed on tadpoles and newts.*

The ***Liguus Dryas*** *lives in hot regions.*

The ***Polymita Picta*** *is a multi-coloured snail which lives in the seas of the Caribbean.*

The ***Helix Pomatia*** *(Roman Snail) lives in vineyards and in zones which are not too damp, in undergrowth and bushes; in some countries, they are bred for food. About 80 fertilized eggs are laid in a hole which is specially dug. They have a calcium carbonate shell and, after 25-27 days, baby snails are born.*

The ***Achatina Fulica*** *(Giant African Land Snail) is a huge land Gastropod which originated in East Africa. It has a remarkable ability to adapt itself according to its surroundings. Every 2-3 months, up to 500 eggs are laid, each as large as a pea. But it needs temperatures above 24°C and very damp conditions.*

Teredo navalis
Shipworm clam
Tridacna gigas
Giant clam
Cyrtopleura costata
Angelwing clam
Cardium aculeatum
Sea nut
Euciroa teramachi
Scallop
Pitar lupanarius
Ensis ensis
Razor clam
Lutraria elliptica
Callanaitis disjecta
Pencillus giganteus
Cuspidaria cuspidata
Glossus humanus
Heart cockle

Did you know...

The ***Teredo Navalis*** *or Sea Worm Clam does considerable damage to the keels of ships by making holes in the wood.*

The ***Cyrtopleura costata*** *(Angelwing Clam) is a large bivalve which digs into clay and rock in the warm waters of the Atlantic.*

The ***Cardium Aculeatum*** *or Sea Nut, is a bivalve with ribs like rays, sometimes spiny. It lives in European seas. It has muscular feet which enables it to stay upside down.*

The ***Cuspidaria Cuspidata*** *is a small bivalve carnivore. It does not have any true gills. The* ***Euciroa Teramachi*** *belongs to the same order.*

The spiny-looking ***Pitar Lupanarius*** *and the* ***Callanaitus Disjecta*** *belong to the same family as the common clam. They live deep in muddy depths.*

The ***Lutraria Elliptica*** *is widespread from Norway to the Mediterranean and on to Senegal.*

The ***Pencillus Giganteus*** *is a small bivalve in a long tube-like shell, rather like an ornamental mace. The embryos of small snails attach themselves to this.*

Glossus Humanus *(Heart Cockle) has a thick, bulging shell in the form of a heart. Large numbers are found in the Mediterranean and in the northern Atlantic.*

GIANT CLAM

Length: up to 1.5 metres or more.

Weight: up to 200 kg. or more.

The Giant Clam is a huge bivalve mollusc which lives on the coral reefs of the Indian and Pacific oceans. It has feet, gills and a downward mouth opening – but the whole mantle is turned 180° degrees, so that the back of the animal faces towards the opening of the shell and all the tummy part is turned towards the inside of the hinges. This strange structure has developed because the weight of the Giant Clam means it cannot move far. So, it needs to twist and turn to catch its prey. It is sometimes thought to be a danger to human beings. In fact, it is harmless and feeds on tiny creatures.

The edges of its mantle houses miniscule one-celled algae which receive food from the Giant Clam; and through the photosynthesis of the algae, the Giant Clam has a supply of oxygen. When the algae multiply too much, the Giant Clam is at risk of suffocation. So, it has to eat the algae to restore the balance. Its enormous hunger is not because of greed. Instead, it is a system of defence against aggressors.

When the Giant Clam senses danger, it slams its heavy valves shut, trapping whatever is in between.

RAZOR CLAM

Measurements: length of shell about 12-15 cm

This bivalve gets its name from the shape of its shell, like an old-fashioned razor.

It is very common in European waters. The **Razor Clam** spends much of its time deep in sand, feeding itself only at high tides. Then, out of the sand comes its short funnel called a siphon. Through this the Razor Clam sucks in water and with it oxygen and particles of food. These food particles become trapped in the mucus of the gills, which are covered with vibrating hair-like structures called cilia, then transferred to the mouth.

In the deepest seas, the Razor Clam retires beneath the sand. Then, every so often there is a tiny spurt of water which it has spat out. The Razor Clam moves by lifting itself off the sea bed with its feet and which it also uses to jump a few centimetres. It can also move by propulsion. First, it fills its body with water. Then it squeezes the water through its siphon, so that the force of the water pushes it forward.

NAUTILUS

This strange-looking cephalopod lives in tropical seas between depths of 50 and 650 m in the Indo-Pacific regions. Throughout its life, it is carried along by the currents.

The female has 90 tentacles and the male 60 which are poisonous. The shell of the Nautilus is a typical spiral shape, divided inside into spaces of air. Its soft little siphon contains an artery and passes through the dividing walls of each space. The siphon enables the Nautilus to vary the quantity of liquid and gas inside, so that it can control how it floats and descend down or rise up. The Nautilus can also move horizontally, by the same sort of propulsion as the Razor Clam; it sucks in water then uses its siphon to squeeze the water out through a cavity in the mantle.

It feeds on molluscs, and small fish which it catches with its tentacles and grasps in its beak-like mouth.

Respiration: 4 gills

Brood: 200-500 eggs, plankton-like larvae and slow development

Diameter of shell: about 30 cm

CUTTLEFISH

The cuttlefish is one of the most interesting molluscs, and one of the most highly developed. It has a remarkable sense of touch and of direction and is very intelligent. Common in the Atlantic and the Mediterranean, the cuttlefish has two gills, two large eyes and ten poisonous tentacles.

Due to the structure of the shell, the cuttlefish can change colour very quickly, according to its surroundings. Despite this, the cuttlefish is the prey of rays, sharks and other fish, as well as human beings. To defend itself against attack, it squirts 'ink', a dark mauve substance which it expels through the funnel of a gland in the intestines. Its shell is internal and is wrongly called 'bone'. The cuttlefish has an system of siphons which enable it to move by means of propulsion, taking in water and expelling it in a jet.

Respiration: 2 gills

Brood: 200-300 eggs: 12 mm long on hatching

Total length: over 60 cm

Did you know...

*The **Spirula Spirula** (Ram's Horn Squid) lives in the depths of tropical seas. It has a luminous organ which glows at the back end of its body.*

*The Californian **Abraliopsis Morisii** captures its prey by catching them on hooks at the ends of its longest two tentacles.*

*The **Chiroteuthis Imperator** is easily recognized by the very long tentacles hanging from its body.*

*The **Nautilus** has good eyesight – but this is no use in the dark depths of the sea where it lives! To find its way around, it uses its sense of smell, taste and touch.*

*The **Sandalops Melancholicus** has tentacles which are either short, or absent altogether.*

*The **Lycoteuthis Diadema** gets its name from about 22 different luminous creatures which cling to its body like a crown of precious stones.*

*Other **cephalopods** pictured on the opposite page have a shell which is reduced to a layer similar to a pen, and a slim, torpedo-shaped body.*

CALLITEUTHIS REVERSA
MASTIGOTEUTHIS GLAUCOPIS
ARROW SQUID
ABRALIOPSIS
MORISII
LYCOTEUTHIS
DIADEMA
BATHYTEUTHIS ABISSICOLA
NAUTILUS POMPILIUS
NAUTILUS
TAONIDIUM SUHMI
SANDALOPS
MELANCHOLICUS
SPIRULA SPIRULA
RAM'S HORN SQUID
CHIROTEUTHIS
IMPERATOR
LOLIGO VULGARIS
COMMON SQUID
SEPIA OFFICINALIS
COMMON CUTTLEFISH

PAPER NAUTILUS

Brood: 500 eggs

Length of adult:
female 30-40 cm; male 1.5-3 cm

Diameter of shell:
up to 25 cm

The Paper Nautilus lives in the Mediterranean and other warm seas. It is similar to an octopus with a bag-shaped body, eight poisonous tentacles, a mouth with a hard ridge and very prominent eyes.

But, unlike the octopus, the female Paper Nautilus makes herself a sort of shell, in the shape of a boat. She does this by using her first pair of tentacles as a framework, and then making a spiral, papery sort of shell.

This 'shell' has two functions. First, it helps the female to control her floating. Second, and more importantly, it is an egg-case, where she carries her eggs to develop.

The male Paper Nautilus is at least twenty times smaller than the female and has no shell. The one purpose of the male's existence is to fertilize the female.

OCTOPUS

Brood: 150,000 eggs

Size on hatching: 3 cm

Diameter of shell: up to 25 cm

The octopus is a cephalopod with eight quick, flexible tentacles each covered with two rows of poisonous cells. There are 150 known species of octopus, some reaching huge dimensions, such as the **Piovra** of Hong Kong, which, with open tentacles, can reach diameters of 10 m. The Common Octopus lives among the rocks in fairly shallow waters of the Mediterranean and in the area from the Canary Isles to the North Sea. It spends a great deal of its life alone in cavities among the rocks or in a shelter which it makes with stones on the sea bed. It moves slowly, clinging with its tentacles. Its presence is only noticed when it shoots out water from its siphon, in the same way as a cuttlefish. When it feels threatened, the octopus will squirt a jet of black ink, and then blend in with its surroundings by changing colour. The octopus is a fearsome predator: with its tentacles it will attack at the rear of its prey, generally crabs, shrimps and bivalve molluscs.

Did you know...

*The **Opisthoteuthis Extensa** is an eight-footed cephalopod with poisonous tentacles, flat body and a tiny shell.*

*The **Eledonella Pygmaea** is also an eight-footed cephalopod but only a few centimetres long, with a jelly-like, transparent body. It is found at depths of 5400 m, or more.*

*The female of the **Common Octopus** watches over her tiny little eggs which she lays in a long, jelly-like row inside grottos and clefts in the sea. She gives them fresh water through her mouth, cleaning and stroking them, and never leaving her babies for at least one month.*

*An octopus which lives in the depths of the Caribbean seas and with a glossy appearance is the **Amphitretus Pelagicus**.*

*The **Vampyroteuthis Infernalis** (Vampire Squid) has a terrifying appearance. About 30 cm long, it lives at depths of between 1500 and 2500 m in tropical seas. Its thickset body is covered with a black membrane, which, with its tentacles forms an umbrella, and it has two dazzling eyes, red like glowing embers.*

*Among the Australian species of octopus is the **Hapalochlaena maculosa** (in the photograph). The saliva glands in the mouth produce a powerful poison which can kill its prey, even without direct contact.*

*The **Octopus cyanea** goes hunting by day in the coral reef, protected by its extraordinary ability to blend into its surroundings.*

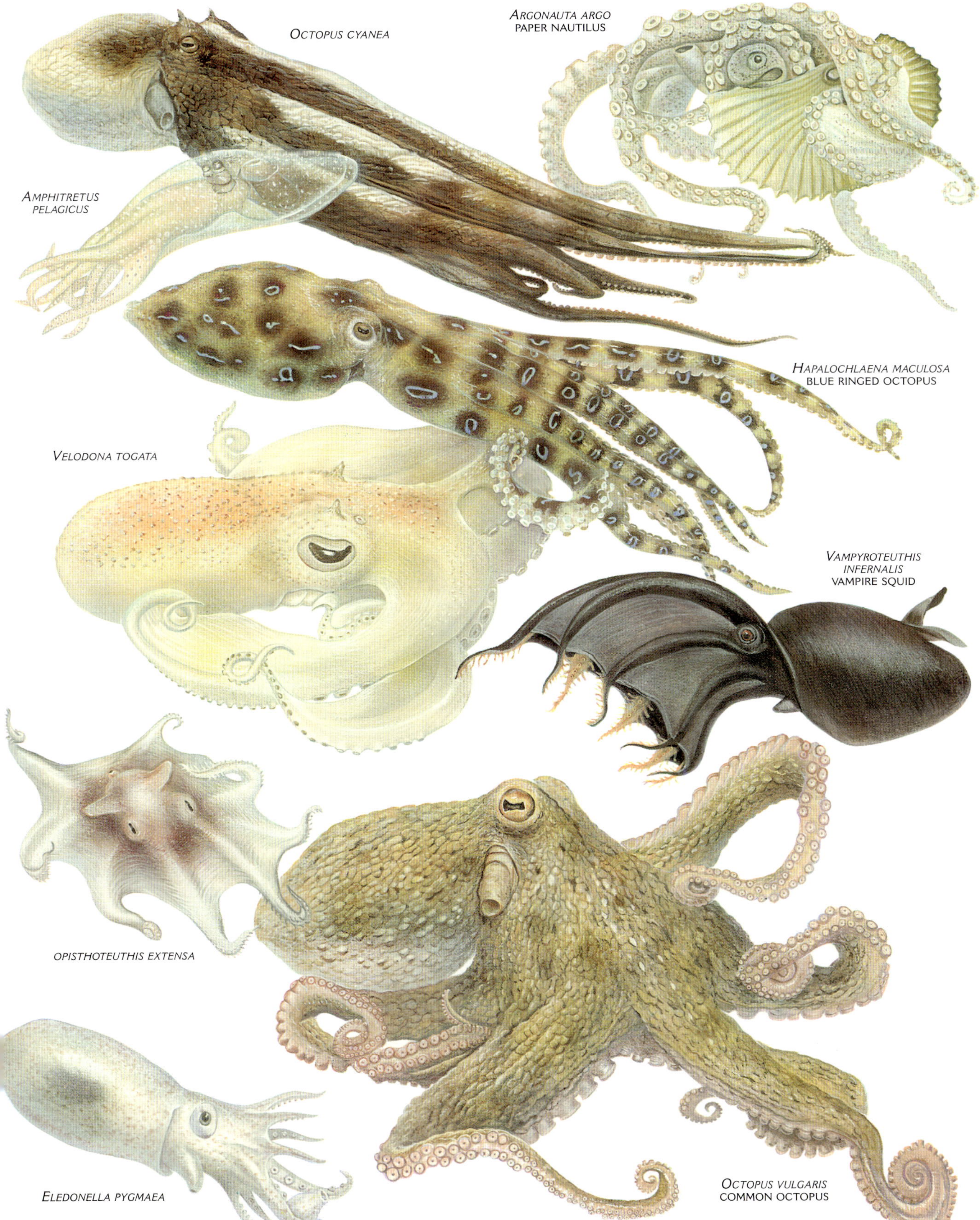
OCTOPUS CYANEA
ARGONAUTA ARGO
PAPER NAUTILUS
AMPHITRETUS PELAGICUS
HAPALOCHLAENA MACULOSA
BLUE RINGED OCTOPUS
VELODONA TOGATA
VAMPYROTEUTHIS INFERNALIS
VAMPIRE SQUID
OPISTHOTEUTHIS EXTENSA
ELEDONELLA PYGMAEA
OCTOPUS VULGARIS
COMMON OCTOPUS

ARTHROPODS

EVOLUTION

Of all the invertebrates, arthropods have been the most successful in development. This phylum originated over 500 million years ago from a branch of Annelida and they have the same segmented bodies.

DEVELOPMENT

All arthropods are covered with a rigid covering called a cuticle. This protects them like armour plating. Little by little, as an arthropod grows, it has to shed its old skin, so that it can grow one which is larger.

The metameric (segmented) body and numerous joints are very noticeable in all arthropods. All chilopods like the centipede have one pair of legs for each segment. Diplopods such as the millipede have two pairs of legs for each segment.

The centipede is a land arthropod. All are poisonous, but most are harmless to humans because they are so small. However, there are some tropical species which can grow to a length of over 280 mm.

Like all arthropods, the centipede has a body protected by a rigid cuticle and which acts as armour plating. It can die without sufficient dampness.

Many arthropods are predators. The centipede feeds on insects and small worms. Large species attack scorpions, lizards and even small birds.

COMMON CHARACTERISTICS

The body is divided into segments, generally in three parts – head, thorax and abdomen.

It has an external skeleton (the cuticle) and articulated (jointed) appendages which enable it to move freely.

HABITAT AND DISTRIBUTION

The Arthropod phylum is the largest and most widespread. There are Arthropods living everywhere – in the seas, in fresh water, on land, on surfaces below the level of the sea and at the highest heights, from the Equator to the North and South Poles.

PHYLUM	SUB-PHYLUM	CLASS	SPECIES
ARTHROPODS	Crustaceans	Branchiopoda	daphnia
		Ostracoda	small water species
		Cirripedes	barnacles
		Copepoda	plankton
		Malacostraca	krill, shrimps, lobster
	Chelicerates	Arachnids	spiders, scorpions
	Antennotrache	Chilopoda	centipedes
		Diplopoda	millipedes
		Insects	bees, beetles

A Successful Development

There are **more species** of Arthropods than any other phylum – about a million and a half species which are known and at least three million species all together, taking into account the ones which have not yet been classified.

Arthropods come in many forms – crabs and shrimps, spiders and scorpions, dragonflies and butterflies, coleoptera (beetles) and bees.

All have **jointed appendages**, legs with movable segments and which enable the animal to go in search of food without any trouble.

The body is **metamerical** (divided into segments) and generally in three parts, the head, thorax and abdomen. In addition, it has a rigid outer covering or skeleton, called the **cuticle.**

This is a substance containing nitrogen and produced within the surface **epithelium** tissue of the body made of chitin.

The cuticle protects the body but allows movement between one segment and the next, not only on the legs, but also various other parts of the animal's body. But because the cuticle itself is so rigid, it does stop the animal growing.

To overcome this problem, Arthropods renew the cuticle regularly in a process called '**sloughing**'. During sloughing, the animal remains weak and helpless.

The evolution (development) of Arthropods has proved very complex. For example – the development of senses in some species is remarkable compared to their intelligence, memory and capacity for learning. Many arthropods have an 'open' circulation system which means that the blood is pumped into body cavities to supply internal tissues. From this point, the blood returns back to where it came from.

Breathing is by gills in aquatic species, by trachea or lungs in land species.

Photograph: a Hermit Crab with a Sea Anemone.

Arthropods are egg-laying; larvae emerge from the eggs and these will develop through successive phases, or **metamorphoses** into adult creatures. The classification system of arthropods is summarized in the table (above). For more details on the types and various species, the following pages will give more information about the three main groups of arthropods – crustaceans, arachnids and insects.

CRUSTACEANS

Common characteristics

All crustaceans have a double pair of antennae, two-part (bifid) joints and one pair of jaws. The body is covered by a skin of varying thickness, called the cuticle. Many species of crustacean have a rigid shell. This can be in the form of a type of shield on the back called a carapace. In some crustaceans, the carapace is closed up and protects the whole body.

Distribution and habitat

Almost all crustaceans are aquatic. Most live in the sea, both in the open sea and near to the coast, in deep seas and on the sea beds, which is why they are often called 'sea insects'.

They are also very common in fresh waters, although a few, such as the Sow Bug, have adapted to life on the land.

The largest crustaceans and the most well-known are those which we see out at sea, in an aquarium, or served as food. Most of these belong to the Decapoda order, so-called because they have ten legs.

The name 'shrimps' actually covers many species of Decapoda which live in both fresh waters and the sea.

Most highly-developed species of Decapoda have a double pair of antennae, two-part legs and an 'armoured plating' (the carapace). Species such as the lobster and the river shrimp also have large, powerful nippers which are used like pincers to tear prey into scraps.

Breathing

The smallest species of crustaceans breathe through a soft, fine film which covers the whole body. Crustaceans with a thick cuticle breathe through gills.

Sense organs

In all crustaceans, the sense organs are highly developed; they only lack the sense of hearing. On the body and legs there are numerous hairs and sensitive bristles for touching. There are also bristles on the antennae, used for taste and smell.

PHYLUM	SUB-PHYLUM	CLASS	ORDER	GENUS	SPECIES
ARTROPODS	Crustacean	Branchiopoda	Cladocera		daphnia
		Ostracoda			mussels
		Cirripedes			barnacles
		Copepoda			plankton
		Malacostraca	Amphipoda		Sand Flea
			Isopoda		Sow Bug
			Mysidacea	*Mysis Relicta*	Opossum Shrimp
			Leptostraca	*Nebalia*	
			Stomatopoda	*Squilla*	Mantis Shrimp
			Cumacea	*Diastylisgoodsin*	Hooded Shrimp
			Euphausia		krill
			Decapoda		shrimps, lobsters, crabs

Plated animals

It is estimated that at least 35,000 species of crustaceans live mostly in fresh water and the seas, and only a few on land. The body is made up of segments. The number of segments can vary. For example, the shrimp is divided into 19 segments, with a head, thorax (covered with a strong plating) and abdomen. Usually, the head and thorax are joined in one piece called the **cephalothorax**.

Each segment has a pair of **appendages** – apart from the first segment, where the eyes are. Crustaceans have two pairs of antennae, which are their tactile (touch) organs. The mouth has jaws, perfect for grinding food, and feet are often used for grinding as well. The joints used for moving sometimes have powerful pincers or **chelae**; these joints often have other special purposes, such as reproduction or carrying eggs.

Male and female sexes are separate in most species, although there are some **hermaphrodite** species (where female and male reproductive organs are present in one individual). Crustaceans can be divided into two large groups – (1) **inferior crustaceans**, with nine classes. The four most important classes in this group are Branchiopoda, Ostracoda, Cirripedes and Copepoda. Then there are (2) the **superior crustaceans** where there are an estimated 18,000 species in just one class, the Malacostraca.

Photgraph: above, a lobster has antennae which are longer than its body.
Facing page: top right, a common crab.

The **Inferior Crustaceans** vary a great deal in size and shape. Many are tiny and live in the ocean, where they make up the major part of **plankton**. In this they play an important part in the food chain of the sea; from one creature comes **huge quantities of larvae**. This constitutes an enormous food reserve for thousands of fish. These fish, in turn, follow shoals of plankton.

Among crustaceans there is a huge range of dimension – some species of daphnia (water flea) are less than 3 mm long, whilst a giant Japanese Sea Crab can be more than 4 m in diameter with its claws open!

DARDANUS AROSOR
PORTUNUS PELAGICUS
BLUE SWIMMER CRAB
EUPAGURUS
BERNHARDUS
HERMIT CRAB
CARCINUS MAENAS
COMMON SHORE CRAB
PORCELLANA
LATYCHELAES
BROAD-CLAWED
PORCELAIN CRAB
HOMOLA BARBATA
INACHUS DORSEMENSIS
INACO
BIRGUS LATRO
COCONUT CRAB

COMMON SHORE CRAB

Habitat: the coast

Region found: Mediterranean

Length of carapace: 4-5 cm

Like all crabs, the Common Shore Crab belongs to the class Decapoda. It can be found along all European coastlines. At the least sign of danger, it runs and hides itself with its typical sideways movement.

The carapace (rigid plating) is black-green, rather like marble to look at, and its pincers are most clearly seen at the tip. At the front of its body, the Common Shore Crab has three triangular teeth, set between its eyes which are on stalks. This crab is well used to ever-changing surroundings, which is why it is so easily found in harbours. It will eat anything, from small fish to shrimps and worms.

It is very aggressive and never lives with smaller crabs, or those which are weak due to the time of sloughing (shedding the carapace plating when this becomes too small). It can often be seen ready to do battle, raising its pincers which are always open; if necessary, it can defend itself by using only one leg if the other is broken off. Then it grows this again by successive sloughing.

COCONUT CRAB

Habitat: coast

Region found: East Pacific

Length: up to 46 mm

Weight of adult: 2-2.5 kg

Coconut Crabs can be found all along the Eastern Pacific coast.

This crab is a large crustacean which gets its name because it is supposed to be able to make coconuts fall to the ground, break them open and then eat them – but this is only a story.

The favourite foods of the Coconut Crab are the remains of fish or rotting molluscs which it finds along the beach. Only rarely does it feed on the flesh of a coconut and then it has to find one which is already open. Despite the strength of its pincers, it would not be able to lift a coconut, let alone open it. But, because of this story, this crab is also called the Robber Crab.

This crab reproduces in the sea and its larvae live in water until they develop into adults; once adult, the Coconut Crab swims towards the land. Because it breathes in oxygen from the air, it does not need to go into water again. The adult grows to about 1 m from head to tail.

Did you know...

*The **Eupagurus bernhardus** or Hermit Crab (see photograph) has a soft, frail body, which is why it lives in empty shells for protection. After each sloughing, its body grows, and so the Hermit Crab goes in search of a larger shell. It lives in symbiosis (each creature helping the other) with the sea anemone. This sea creature feeds on the remains of the Hermit Crab's meals and in return protects the crab from enemies with its stinging appendages.*

*The **Dardanus arosor**, another type of hermit crab also shelters in empty mollusc shells and lives in symbiosis with the sea anemone.*

*The **crab Porcellana platychelaes** (Broad-Clawed Porcelain Crab), is a type of a hermit crab with a flat body – and so it goes in search of flat shells, not spiral.*

*The **Portunus pelagicus** (Blue Swimmer Crab) and the **Inachus dorsemensis** (Inaco) are fairly common in Mediterranean waters. The **Homola barbata** lives in very deep waters.*

*It is easy to distinguish the female of the **Common Shore Crab** from the male by examining the abdomen; in the female, this is large, because it has to carry eggs, whilst the stomach of the male is narrow and triangular.*

ARACHNIDS

DEVELOPMENT, DISTRIBUTION AND HABITAT

The oldest forms of arachnids began a hundred million years ago with creatures similar to today's scorpions. Today, there are arachnids almost all over the world, mostly in very hot, damp regions. But some live in countries with temperate climates, and there are species of spiders which can be found within the Arctic Circle.

NUTRITION

Arachnids are meat-eaters and fearsome predators. For instance – scorpions feed on live prey, such as beetles, cockroaches and other arthropods, which it tears with its claws then sucks out the edible parts. The spider covers its prey with its gastric juices to 'melt' it before eating.

The ***Salticus sanguinolentus*** (Jumping Spider, see bottom left) has eight eyes arranged in three rows; those in the middle are larger than the others and work rather like the telephoto lens of a camera. This gives the spider an enlarged image of its prey, but with a limited range. So, this spider must move in quickly before hurling itself on its victim.

Almost all spiders can spin thread. The common and harmless house spider spins its thread into a web and lives in the thickest part.

The web of the ***Micrathena*** (Orb Web Spider) is much more intricate. With its plated, spiny abdomen, this is a spider typical of those living in hot countries.

COMMON CHARACTERISTICS

The body of an arachnid is divided into a front part and a back part (the cephalothorax and the abdomen). Arachnids do not have antennae; the first pair of appendages are formed either like a hook or as pincers which the arachnid uses to grab, grind up and then put the food in its mouth.

BEHAVIOUR

All arachnids are aggressive creatures, ready to attack and to conquer their prey.

The scorpion uses its poisonous sting to kill. Spiders are experts in making soft webs, transparent and sticky, to trap their prey. An adult Daddy Long-Legs has a pair of glands which secrete an evil-smelling liquid.

Land dwelling creatures

We know of around 36,000 species of arachnid, but many thousands more exist. Most live on the land although a few are aquatic.

All arachnids have the same type of body, in two distinct parts – a **cephalothorax** or a prosoma

Photograph: above, Crab Spider.
Facing page: top right, a Sac Spider.

(head and thorax fused together) protected on the back by a rigid cover called the **carapace**, and six pairs of jointed appendages – a pair of **chelicerae**, one of **pedipalps** and four pairs of jointed claws – and an abdomen or **opisthosoma**, which has no appendages. The pincer-like chelicerae, nearest the mouth, are also used to grip, for the injection of poison, as well as for spinning silk.

The **pedipalps** can have various forms and functions; sometimes they are used for chewing, sometimes as huge, powerful pincers, as organs of reproduction, for taste and for gripping. The fourth pair of claws are for moving, digging, swimming, for **winding silk** and catching prey.

Scorpions, spiders and some mites have **poisonous glands**. Spiders and some other arachnids have spinning glands. Breathing organs are tubular trachea or lungs. In many mites, breathing is through the skin. Males and females are often quite different, as is the case with spiders.

Reproduction is by laying eggs, rarely oviparous (where eggs are hatched in the body of the female before the young emerge).

Arachnids belong to the sub phylum of Chelicerates (animals with pincers) and are divided into many types.

Scorpions have two large pedipalps which end in pincers and a tail with a poisonous sting.

Psuedoscorpions (also called false scorpions) are very small (from 1–8 mm) and without a tail. **Solifugae** (sun spiders) are large in size, covered with a thick fur and with two large claws shaped like pincers. **Uropygi** are nocturnal. They live in damp, tropical and sub-tropical regions. **Amblypygi** have small, flat bodies with very long, spiny pedipalps. **Spiders** are the most numerous of all arachnids. There are over 26,000 known species.

The **Opiliones** (Daddy Long-Legs) have small bodies and very long legs. They live in temperate and tropical regions. Most mites are parasites, living off plant material or other animals. They are widespread on the ground and in both the sea and fresh water.

PHYLUM	SUB-PHYLUM	CLASS	ORDER	SPECIES
ARTHROPODS	Chelicerates	Arachnids	Scorpion	scorpions
			Pseudoscorpions	Book Scorpion
			Solifugae	Sun Spider
			Uropygi	Mastigoproctus giganteus
			Amblypygi	Tail-Less Whip Scorpion
			Spiders	Tarantula, Mygale, Black Widow
			Opiliones	Daddy Long-Legs
			Mites	Castor Bean tick

Did you know...

*The gigantic **Mastigoproctus giganteus** belongs to the **Uropygi** order. It is about 7 cm long and lives in Mexico and in the southern United States. Despite its menacing appearance and its hunger for catching frogs and small toads, it is harmless to humans.*

*The **Prynichus reniformis** moves on three pairs of legs, walking sideways, whilst using its long, spiny **pedipalps** as a trap for catching insects and large cockroaches. The first pair of legs, long and soft, function as organs for touch.*

*The **Phalangim opilio** (Daddy Long-Legs) has only two eyes. It does not spin silk and the male is almost identical to the female. It risks losing at least one of its thread-like legs when it has to flee from danger. Unlike spiders, the lost leg does not grow again.*

*Among other species of **Opiliones** we find the **Ischyropsalis hellwig**, with its huge pincers.*

*The **Galeodes arabs** is a large nocturnal arachnid of the order of **Solifugae** living in Europe and Asia; with its highly-developed pincers it can catch and eat termites, lizards and small mammals.*

*There are thousands of species of **mites**, all very tiny and which live at all latitudes and altitudes (up to 5500m). They are very resistant parasites which feed on plants and animals, such as the Sheep Tick and Castor Bean Tick.*

CASTOR BEAN TICK

Habitat: animal parasite

Length: about 15 mm

The Castor Bean Tick may not seem very interesting. But, this tiny mite can be a very dangerous enemy both for animals and humans.

Like many mites, the Castor Bean Tick can transmit many disease-carrying viruses, some of which can cause death. That is why it can be very dangerous to touch or even to go near certain types of ticks. One bite can have terrible consequences.

The tick feeds on the blood of whatever creature it lives on, sucking through a beak formed by the claws of its pedipalps. It draws out the liquid by pumping movements of the stomach.

The life cycle of the tick is simple. The innocent-looking larva climbs a blade of grass, then waits for an animal to come along on to which it can move. After various moultings, it reaches the adult phase. Then it mates on its host plant or animal and the female falls to the ground to lay her eggs and then die. The new-born ticks will continue the life cycle.

PSEUDOSCORPIONS: BOOK SCORPION

Habitat: fields, libraries

Length: just a few millimetres

There are more than 1500 species of pseudoscorpions which live in all hot and temperate regions throughout the world. They are related to scorpions, but they do not have a poisonous sting and the body is not so long.

All pseudoscorpions are small, no more than a few millimetres. They are lively little creatures which live in cracks, beneath sacks and under the bark of trees, in rotting material, in nests of small mammals and in the honeycomb of bees. One of the most widespread is the Book Scorpion. This is a harmless creature which lives in the pages of books or among powdery paper, in libraries or in old, damp documents. The favourite prey of the **pseudoscorpion** are insects, especially the small, wingless insect, the collembolan. Unlike the spider and the scorpion, the pseudoscorpion is not limited to sucking or eating the soft parts of their prey, they eat everything. Sometimes, they attach themselves to ants or to scorpions to be carried along.

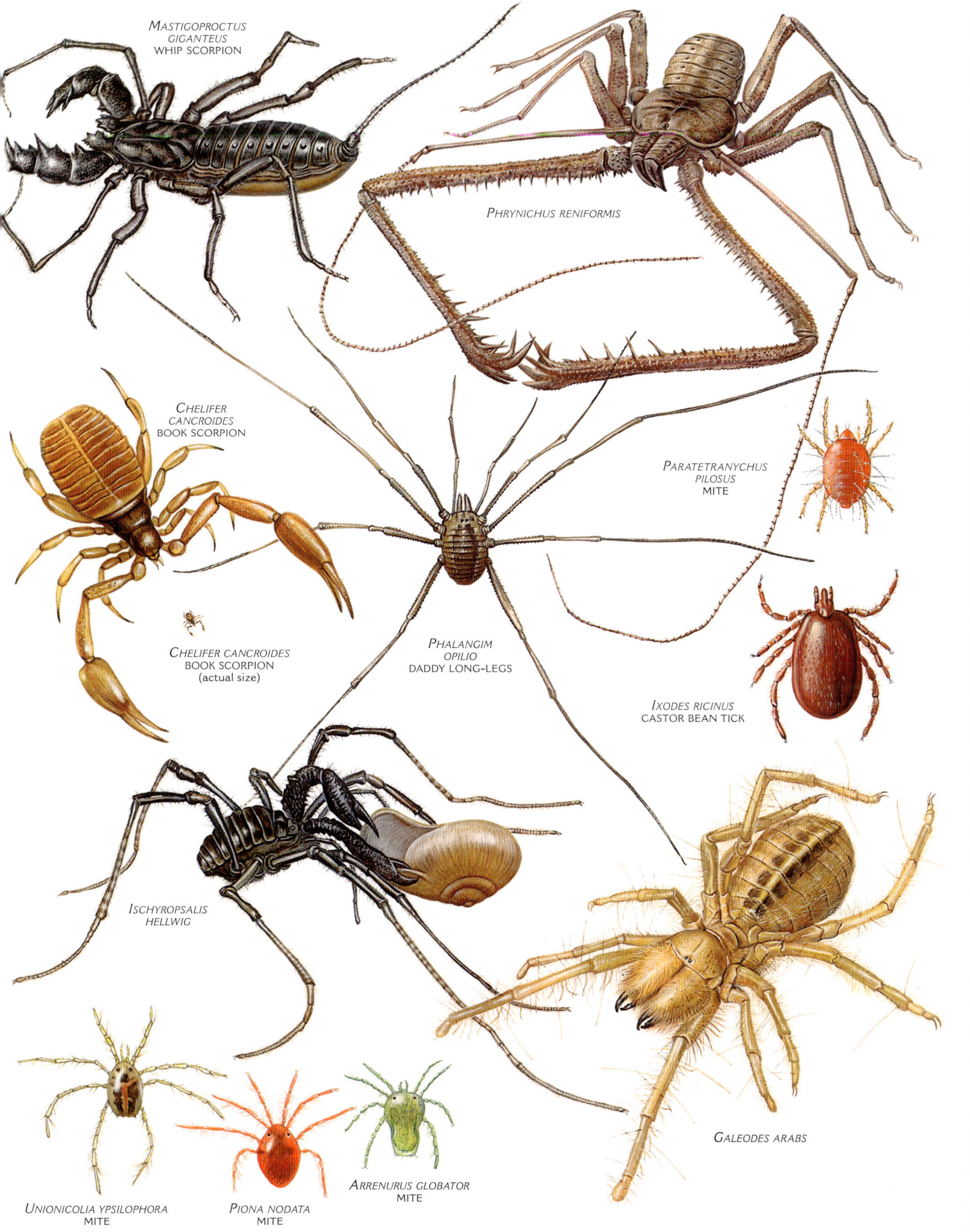
MASTIGOPROCTUS GIGANTEUS
WHIP SCORPION
PHRYNICHUS RENIFORMIS
CHELIFER CANCROIDES
BOOK SCORPION
PARATETRANYCHUS PILOSUS
MITE
CHELIFER CANCROIDES
BOOK SCORPION
(actual size)
PHALANGIM OPILIO
DADDY LONG-LEGS
IXODES RICINUS
CASTOR BEAN TICK
ISCHYROPSALIS HELLWIG
GALEODES ARABS
UNIONICOLIA YPSILOPHORA
MITE
PIONA NODATA
MITE
ARRENURUS GLOBATOR
MITE

LYCOSA RAPTORIA
WOLF SPIDER
LATRODECTES TREDECIMGUTTATUS
BLACK WIDOW SPIDER
THERAPHOSA LEBLONDI
GOLIATH BIRDEATING SPIDER
LATRODECTES MACTANS
AMERICAN BLACK WIDOW
LATRODECTES HASSELTI
REDBACK SPIDER
CTENUS FERUS
CTENUS NIGRIVENTRIS
LOXOSCELES RECLUSA
BROWN RECLUSE or VIOLIN SPIDER
ATRAX ROBUSTUS
FUNNEL-WEB SPIDER

BLACK WIDOW SPIDER

Habitat: low plants, stone, corners or walls

Web: ragged, irregular

Length: about 15 mm

Poison: deadly

Most spiders in the class ***Latrodectes*** are generally small, but poisonous – such as The Black Widow spider. There are six species of Black Widow. The most common lives in the area of the Mediterranean Sea and all very hot zones.

Although it is only 10-15 mm long, this spider is always hungry and is much-feared. It has a poisonous bite which can produce pain, sickness and temporary breathing problems. The bite, however, is rarely fatal to humans.

The female spider looks more frightening than the male, both in size and appearance, with her abdomen usually adorned with blood red spots. She hides among stones and bushes and spins a web made with threads of silk from the top of her body, abdomen turned upwards.

The Black Widow will not hesitate to attack animals larger than itself and grasshoppers are among its favourite prey. The female builds a clear-coloured sphere-shaped cocoon, where she lays her eggs. The spider gets the name of Black Widow due to the female's habit of eating the much smaller male after mating.

Did you know...

*The **spider** usually has 8 eyes, but those which live in caves have none. The average length of the body is 2 cm, but some tropical species, such as the bird-eating Mygale, measure 9 cm and with the legs can even be more than 20 cm!*

*The **tarantula** belongs to the class Lycosa. According to popular belief, its bite can cause a sort of frenzy which can lead to death. In fact, its bite is not very dangerous.*

*The male of the **Black Widow** wraps its sperm in a soft net which it takes to the female to fertilize her eggs.*

*Closely related to the European Black Widow is the famous **American Black Widow**, Latrodectes mactans.*

*The **Theraphosa leblondi** (Goliath Birdeating Spider) is from South America. At the approach of an enemy, it tears into the skin with its rear claws to confuse its victim, causing strong skin irritation.*

*Living in temperate regions are two spiders of the genus **Ctenus**, both very poisonous.*

*Another spider much-feared because of its poisonous bite is the **Atrax** (Funnel-Web Spider) widespread in tropical regions.*

LYCOSA

Habitat: ground

Web: underground covering

Length of body: about 25 mm

Length of legs: 25 mm

Poison: almost harmless

There are many species of **lycosa**, spiders of small or medium size, which are distinguished by their home.

Whilst other spiders make a web, the lycosa species finds a hole in the ground which it lines with a soft web of silk, to make it a comfortable underground sleeve, and which it can then close with a little silk cover. The cover opens or closes the entrance of the hole, making it into a trap. Each hole is inhabited by a solitary female; the male just keeps watch and goes in search of a mate only during the mating period, then flees before being eaten by his mate.

One species of lycosa is called the Ground Spider, or Wolf Spider because of its aggressive nature. The Wolf Spider runs fast across ground in search of prey, especially at night. It identifies its prey with the help of three rows of eyes – two medium-sized eyes in the top row, two large eyes in the middle row, and four small eyes in the bottom row.

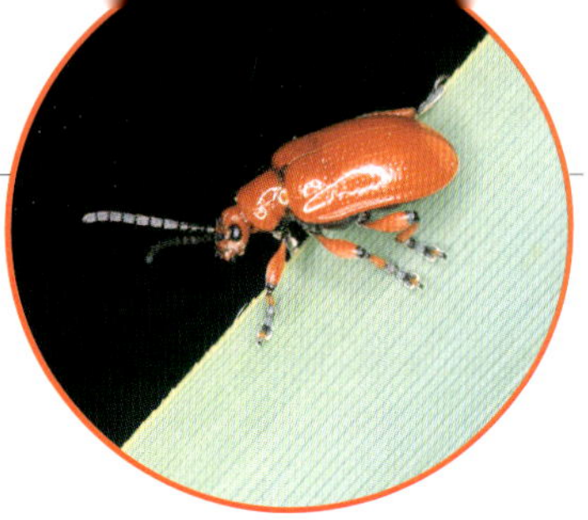

INSECTS

EVOLUTION, DISTRIBUTION AND HABITAT

Even around 400 million years ago, insects were widespread on land, in fresh and salt water and after that, in the air. There are about 800,000 species of insect, more than any other class in the animal kingdom. Some insects, such as bees and silkworms, are useful to humans; others cause damage to agriculture and some are parasites.

DEVELOPMENT

The life cycle of an insect is made up of different stages of development, each one characterized by metamorphosis, or transformation. With each stage of metamorphosis, the insect, once it comes out of the egg, undergoes a transformation of the whole body – from the larva which develops into the chrysalis or pupa which develops into the adult insect.

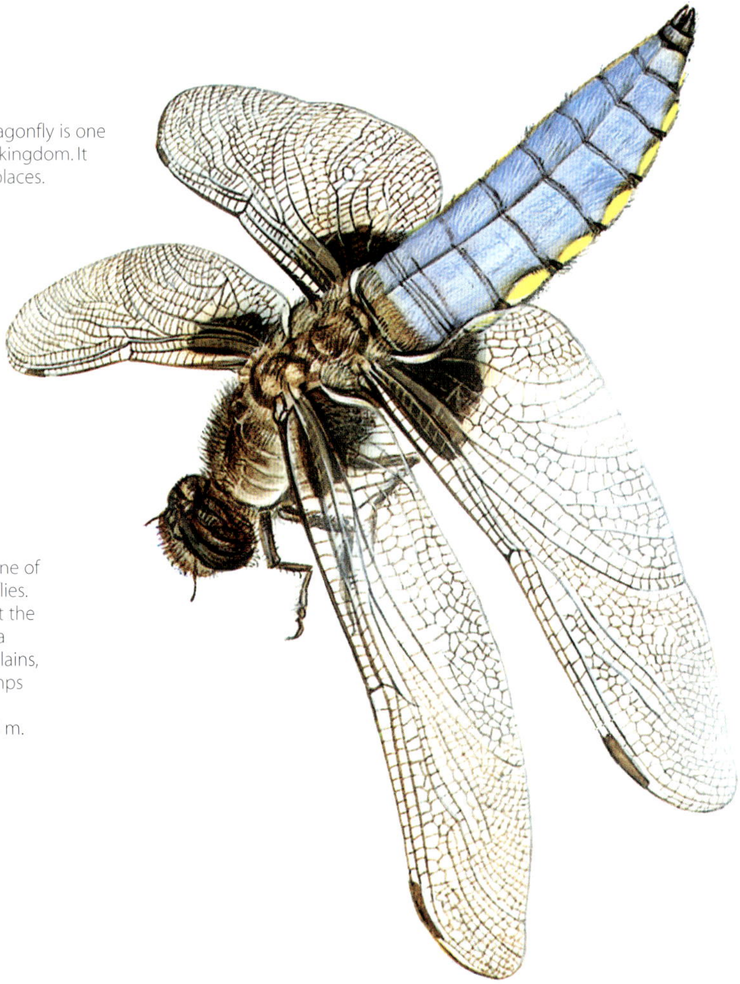

The elegant flight of the dragonfly is one of the fastest in the animal kingdom. It can migrate to far distant places.

This dragonfly prefers the sun and during hot days it stays hidden among the vegetation. It feeds on flies and butterflies and lays its eggs in flight, letting them fall on water plants. Dragonfly larvae are always hungry. They eat tadpoles and other small creatures. The adult dragonfly has a characteristic large, flat abdomen.

The ***libellula depressa*** is one of the most common dragonflies. It is widespread throughout the whole of Europe and in Asia Minor, especially on open plains, in small ponds and in swamps and marshes which do not extend beyond about 1000 m.

COMMON CHARATERISTICS

All insects have just one pair of antennae. These are their organs of touch and smell. Each has a particular form of mouth that can either lick, sting, suck or chew, three pairs of legs and a thorax and abdomen separated by a narrow 'waist'. Like all arthropods, an insect has an exoskeleton (a hard outer body covering).

BEHAVIOUR

Some insects such as ants, bees, wasps and termites, live in communities. In these communities, there are particular groups, each with a particular job to do – such as the feeding and the care of the young, collection of food and the construction of the home. In each group, the insects work together, carrying out their tasks in a very precise, regulated way.

So many different kinds

Nearly all the orders of insects which existed 230 million years ago still exist today, a length of time never achieved by any other animal.

There are many reasons for this success; most important is the conquest of **flight** which has enabled insects to adapt to new surroundings, to gain access to new sources of food and to flee from predators.

Other reasons for this success are – small sizes, modest food needs and a hard body covering called the **exoskeleton** which protects like armour plating. This is solid yet light, made up of chitin and **sclerotin**.

On the head of the insect is a pair of antennae.

Some have two **compound eyes**, made up of many lenses, as well as **simple eyes** which can only make out light, dark and movement.

Most insects have a mouth with many parts, including an upper lip, a pair of jaws and a lower lip. The thorax is made up of three segments, with a pair of legs for each. Generally, the last two segments each have a pair of **wings**. The last segment of the abdomen has appendages which are used for reproduction.

There are some insects which can communicate by using sensitive cells on the legs or abdomen. By rubbing one part of the body against another they can make **sounds** which can be heard at a considerable distance. They detect different smells and explore their surroundings by using their antennae.

Photographs: top, a hornet, bottom, a ground beetle. Facing page: a beetle of the Chrysomelidae family.

The **sight** and the perception of colour in insects is very much less than ours, and this also varies according to the species. Insects breathe by a network of tiny little tubes ('trachea') which distribute oxygen to the whole body. Almost all insects are born from the egg; only aphids and a few others give birth to live young.

In some insects, the **larvae** which emerge from the egg develop directly into an adult. This is known as incomplete metamorphosis. In others, the larvae pass through the phase of pupa and then into an adult insect. This is known as complete metamorphosis.

In insects, there is an enormous variety in the development of life cycles, which we often refer to as different **phases**.

PHYLUM	CLASS	SUB-CLASS	ORDER	SPECIES
ARTHROPODS	Insects	Pterygota	Odonata	dragonfly
			Orthoptera	cricket, locust
			Homoptera	aphids, cicadas
			Isoptera	termites
			Ephemeroptera	mayflies
			Heteroptera	bugs
			Thysanoptera	thrips
			Psocoptera	bookworms, silverfish
			Coleoptera	beetles, weevils
			Hymenoptera	bees, wasps, ants
			Raphidoptera	snakefly
			Neuroptera	lacewings
			Lepidoptera	butterflies, moths
			Diptera	flies, mosquitoes

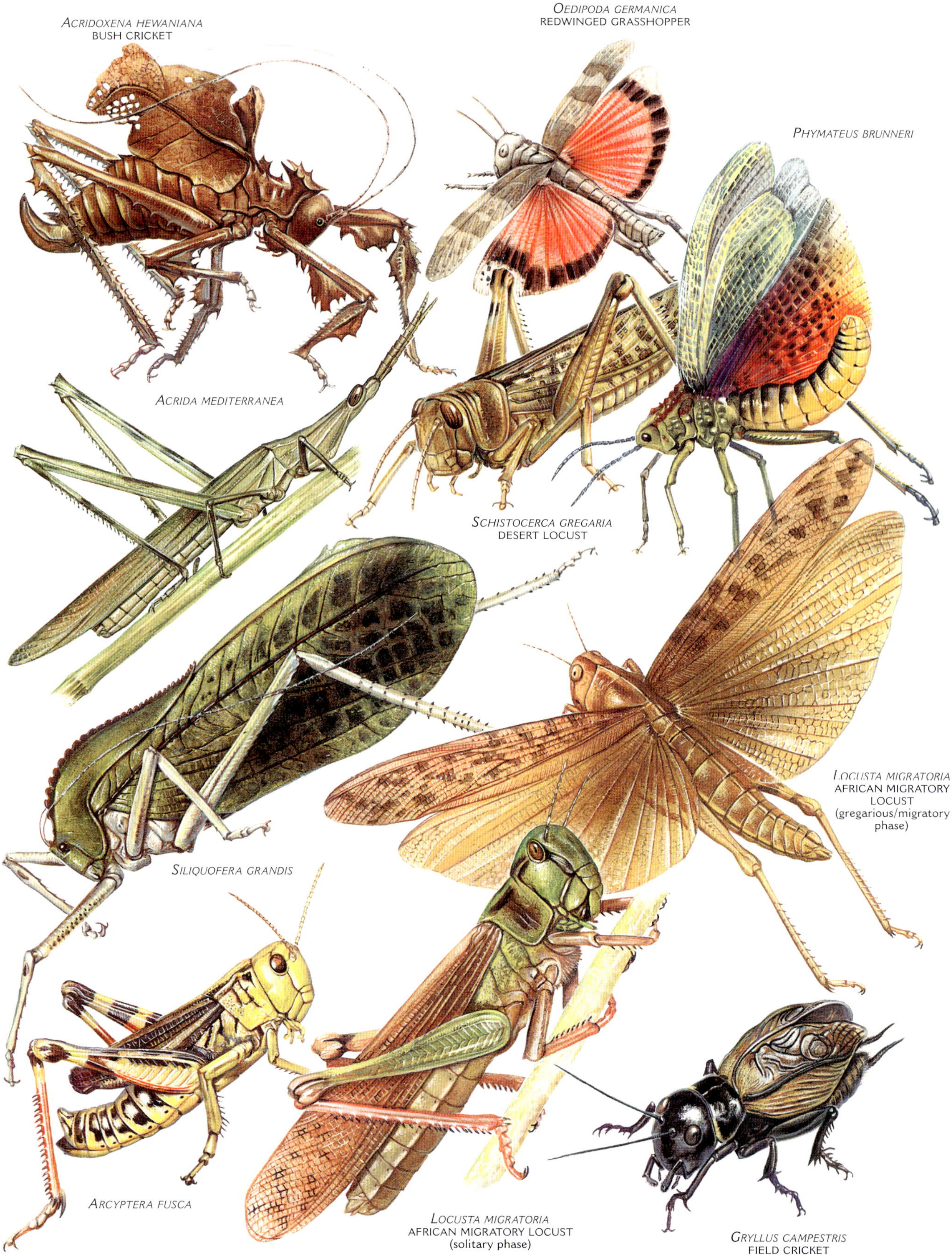
ACRIDOXENA HEWANIANA
BUSH CRICKET
OEDIPODA GERMANICA
REDWINGED GRASSHOPPER
PHYMATEUS BRUNNERI
ACRIDA MEDITERRANEA
SCHISTOCERCA GREGARIA
DESERT LOCUST
LOCUSTA MIGRATORIA
AFRICAN MIGRATORY
LOCUST
(gregarious/migratory
phase)
SILIQUOFERA GRANDIS
ARCYPTERA FUSCA
LOCUSTA MIGRATORIA
AFRICAN MIGRATORY LOCUST
(solitary phase)
GRYLLUS CAMPESTRIS
FIELD CRICKET

Did you know...

*The ability of the **Desert Locust** to destroy a whole crop of corn with astonishing speed has been known since Biblical times – a plague of locusts was one of the seven plagues to hit Ancient Egypt. The Desert Locust is gigantic, about 75 mm long.*

*The **Acrida mediterranea** is a grasshopper with a cone-shaped head and tapering legs. The female is up to 75 mm long. It does not 'sing' and lives in the warm, damp plains of southern Europe.*

*The **Redwinged Grasshopper** is a species with red rear wings. It prefers dry places.*

*The sound of the **Arcyptera Fusca** is high-pitched and consists of many short notes followed by a burst of three seconds. It lives in the mountain pastures of the Alps and the Pyrenees. The male insect flies, but the female does not.*

*The **Acridoxena Hewaniana**, (Bush Cricket) has wings looking like small, curled-up leaves. It is an excellent example of an insect which can mimic its surroundings.*

*In some parts of Africa, people eat **locusts**. As a food, they are a good source of protein.*

*A swarm of **locusts** can cover an area of over 1000 square kilometres and eat up to 20,000 tonnes of food per day.*

FIELD CRICKET

Mouth: chewing

Movements: rarely jumps, prefers to fly and to run

Metamorphosis: incomplete

Length of adult: 3 cm

The Field Cricket belongs to the order of **Orthoptera** of insects and is found all over the world.

It is sometimes called the 'singing cricket' because of the loud, continuous sound it makes by lifting its front wings and rubbing one against the other. The sound is clearly heard, especially on warm, summer nights, but it is difficult to see the Field Cricket at close range.

It is very timid and at the slightest sign of anything strange, it will quickly flee into its underground lair. This can be up to half a metre deep, ending in a round chamber which is the actual nest. In front of the entrance to the lair, the cricket prepares a clean area where it sings to attract the female and to mark its territory. The female Field Cricket lays between 20 and 60 eggs during the autumn months. Young crickets are called 'nymphs' and they hatch in the spring. They shed their skins up to twelve times as they grow into adult crickets. Not all crickets sing. Some spend almost their whole lives underground. Others live in ant-hills, in symbiosis (helping and receiving help) with ants.

Although the cricket is related to the grasshopper and has hind legs ideal for jumping, it is not a great jumper and prefers to fly or to run rapidly over the ground. It eats leaves and small insects.

AFRICAN MIGRATORY LOCUST

Mouth: chewing

Movements: jumps, flies; migrates long distances

Speed: 10-25 km per hour

Metamorphosis: incomplete

Length of adult: 5 cm

The locust belongs to the **Orthoptera** order of insects. It is a species of grasshopper, and it can be found almost all over the world. From Biblical times, the African Migratory Locust has been known to cause serious damage to crops. But not all types of locusts are harmful. They are insects which usually live alone in what is called their 'solitary phase'. It is only when they congregate in thousands that they pose a danger. Then, under the influence of so many others, each locust changes its food, its size growing up to 70 mm long, and also its colour (yellow or mauve), becomes less bright. This is known as the 'gregarious phase'.

The locust eats mostly grasses; so, when thousands get together, no crops can stay undamaged. Also, eggs laid by a solitary female locust will spread about as they fall. But with many locusts all together, eggs will fall in a compact mass.

When the larvae hatch, they begin to group together in search of food - and once they reach the adult stage, they take to flight in swarms, swooping down on a field of crops and destroying it in a few hours.

PRAYING MANTIS

Mouth: chewing

Development: Metamorphosis: incomplete; larvae moult 6-12 times before reaching adult stage

Length of adult: 7-8 cm

Mantids belong to one of the higher order of insects. There are more than 200 species of mantids, all related to the grasshopper, to the leaf-insect and the stick-insect.

The Praying Mantis gets its name because of its pose as if in prayer, with 'hands together', when it is still, waiting to trap another insect. Then, with a sudden stretch of its front legs, the Praying Mantis catches its victim, holding it between two rows of sharp points then taking it to the mouth to be eaten.

The reproduction of the Praying Mantis is also very strange. Whilst male and female are in the act of mating, the female grasps the head of the male and begins to eat him, even before mating is finished.

As well as insects, a large Praying Mantis will even eat lizards, small frogs and birds. By doing this, the female will take in a sufficient amount of protein to hatch a lot of eggs – up to 200 in a sort of cocoon called an ootheca. The female makes the ootheca from a sort of foam and this soon hardens, becoming like parchment.

This opens in the spring-time, having survived even very low winter temperatures. Most species of mantid spend the winter in the egg phase. The first moulting of the larva happens with the splitting of the membrane in which it has been enclosed from birth. Even then, although it is still immature, the larva already has front legs with which it can catch its first tiny, little prey.

There are over 2,000 species of mantids. Most are found in tropical regions throughout the world, but some species are found further north, in Mediterranean countries.

Some mantids from Africa and South-east Asia are highly coloured, looking exactly like a flower or a leaf. European mantids can also blend in perfectly with the leaves among which they hide to attack their victims.

The Chinese Mantid originates in eastern Asia, but it is also the largest mantid found in North America, measuring up to 10 cm long.

Díd you know...

*The **Hymenopus coronatus** (Orchid Mantis) is a very beautiful species which comes from Java. Its vivid colouring simulates the shadow and the streaks of the flowers among which it lives. During moulting it can change colour, according to the flowers surrounding it.*

*The **Empusa pennata** (Horned Praying Mantis) is easily recognized by its long neck. It can be green or mauve and spends all winter in the larva phase. Generally, this is recognisable, because, only at this stage of development, the mantid keeps the abdomen folded up above the thorax.*

*Most **mantids** (see photograph) are generally bad fliers. Desert species have no wings, and so they do not fly. But they are excellent runners.*

*The **Idolum Diabolicum** (African Devil Mantis) is one of the largest in the world. It blends in so perfectly with flowers that it is very difficult to see. It hangs on a plant with its head down and the bumps which cover its front part become easily mistaken for the tiny insects intent on eating the 'flower'.*

*Among the most curious of mantids are the **Acanthus falcate** found in Brazil, the **Pseudocreobotra Wahlbergi** (African Flower Mantid) and the **Gonylus Gongyloides** (Wandering Violin) from South-east Asia.*

CHOERADODES LATICOLLIS
IDOLUM DIABOLICUM
AFRICAN DEVIL MANTIS
PSEUDOCREOBOTRA WAHLBERGI
AFRICAN FLOWER MANTID
GONYLUS GONGYLOIDES
WANDERING VIOLIN
ACANTHUS FALCATA
EMPUSA PENNATA
HORNED PRAYING MANTIS
HYMENOPUS CORONATUS
ORCHID MANTIS
MANTIS RELIGIOSA
PRAYING MANTIS

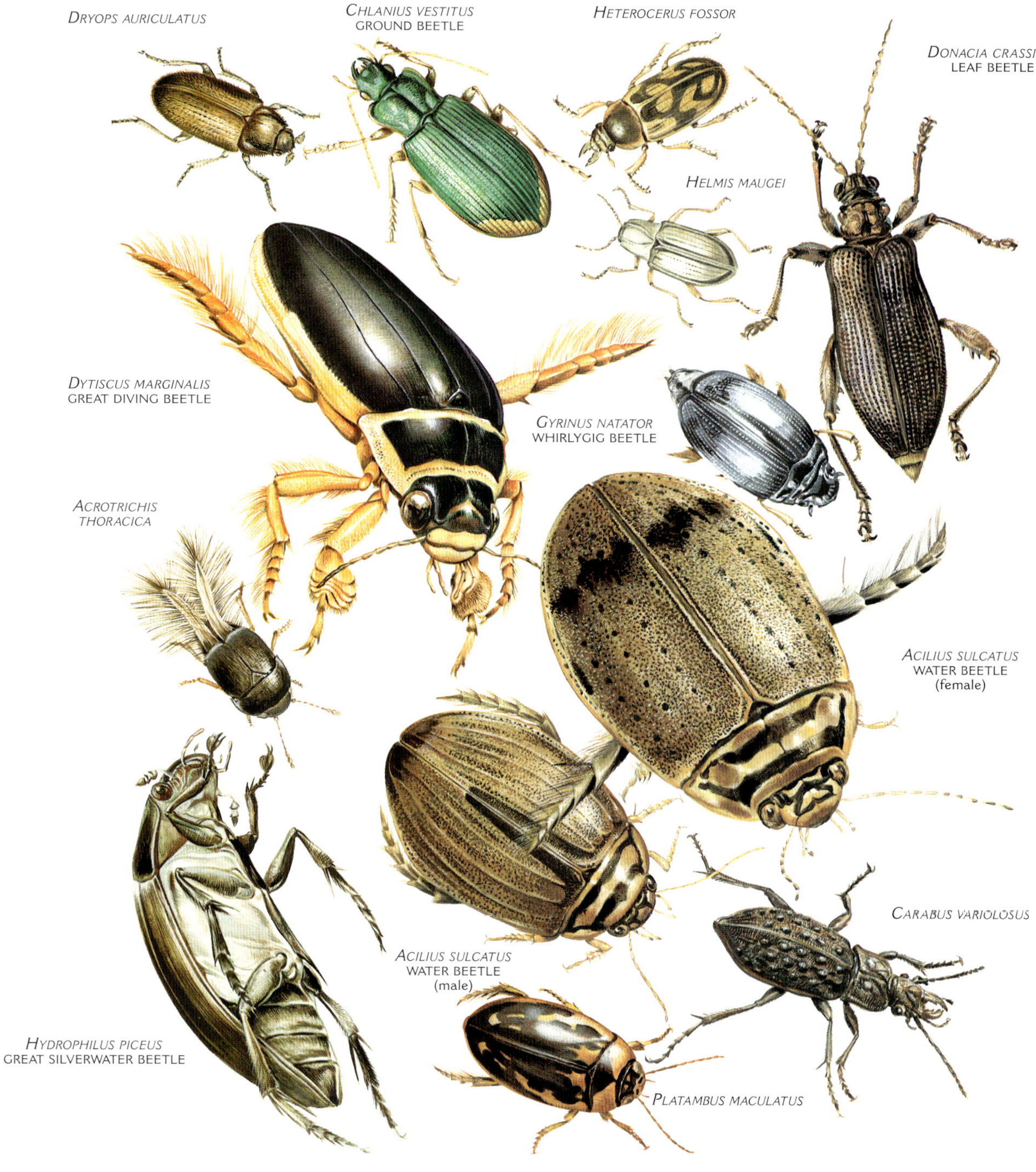

Did you know...

There are many thousands of species of water beetles, all belonging to the insect order Coleoptera. The beetle, ***Platambus maculates****, is a predator which lives mostly in flowing water.*

The larvae of the ***Dytiscus Marginalis*** *('water tiger') pierces and sucks its prey until it is dry. The slow-moving adult is a water carnivore.*

The ***donacia*** *(leaf beetle) lays its eggs in the submerged stems of water plants. Their larvae never leave the bottom of ponds.*

The female ***Water Cockroach*** *fixes her cocoon containing her eggs beneath a leaf near the surface of the water.*

The ***Elater sanguineus*** *is one of the most brilliantly coloured coleoptera. It is a very rare species which feeds on dead and decomposing wood, especially the wood of the conifer.*

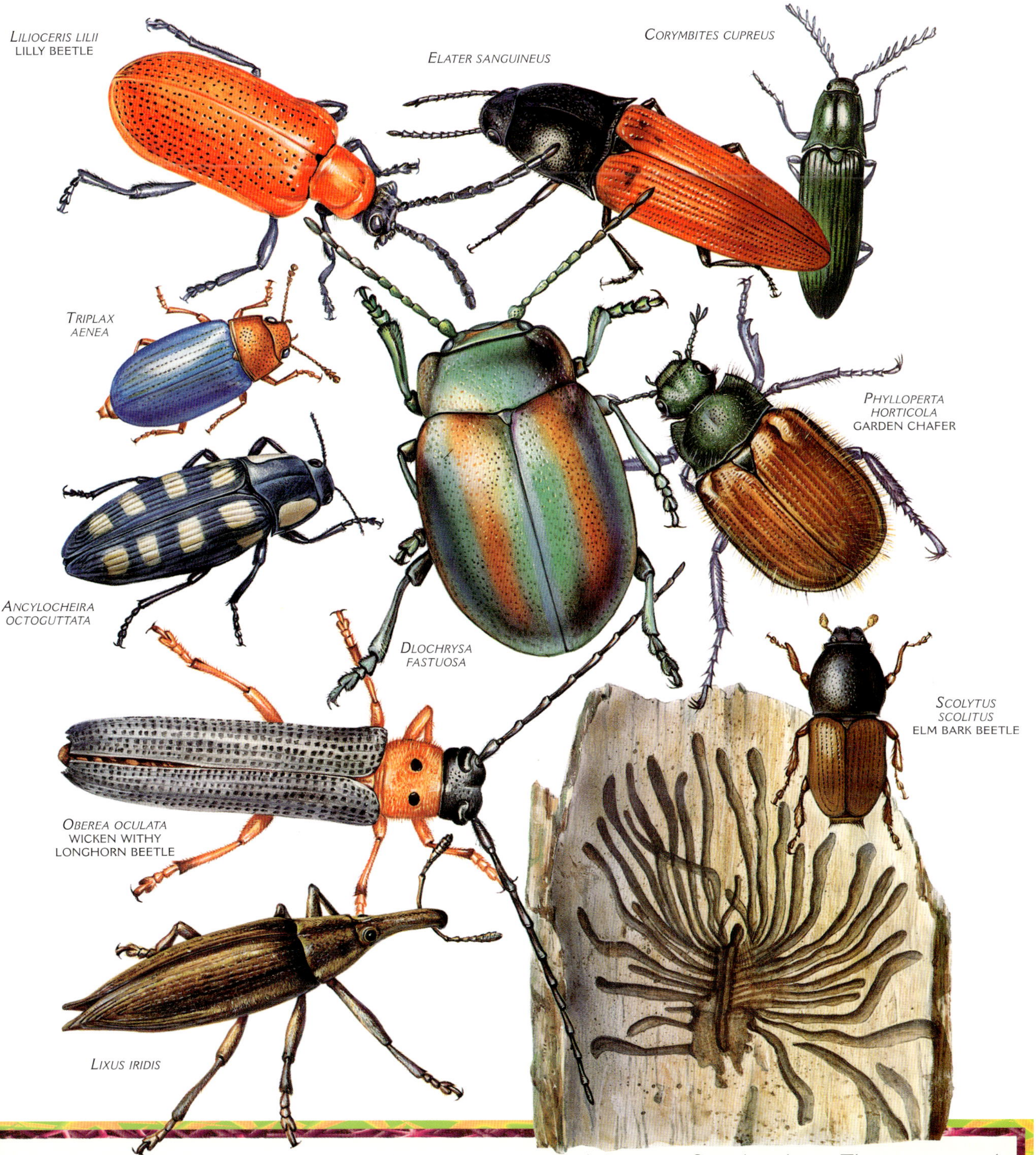

*The **Lilly Beetle** belongs to the **Chrysomelidea** family, coleoptera characterized by bright colours and metal-like effects.*

*The **Elm Bark Beetle** digs long galleries beneath the bark of elm trees, seriously weakening the structure. Many species of coleoptera, commonly called **woodworm**, attack wood in our homes and in buildings.*

*To the Ancient Egyptians, the **Scarab Beetle** was a holy symbol of the Sun.*

*Coleoptera belonging to the genus **Lixus**, are wood-boring weevils. All are members of the family of Curculionidae. They are easily recognized by the rigid proboscis (nose tube) on which angular antennae are attached.*

*The **Phyllopertha orticola** belongs to the important family of Scarabaeidae of which there are at least 30,000 species, all feeding on plants.*

OAK LEAF BUTTERFLY

Mouth: sucking

Metamorphosis: complete

Length of caterpillar: up to 8 cm

Wing span: 8-9 cm

The scientific name for this butterfly is ***Kallima***, which in Greek means 'magnificent'. It is a name given to a group of tropical daytime butterflies which live mostly in Asia, but which can also be found in Africa and in Madagascar. The most notable feature of this butterfly is the way it can imitate perfectly the shape and the colour of narrow leaves when settling on trunks and branches of trees. In a resting position, the Oak Leaf Butterfly shows a central vein on the underside of its wings, which looks like a leaf. This 'disguise' protects the butterfly very well. Without ever showing itself too soon, it can fly away at the very last minute, when danger is almost upon it. Then it shows the coloured side of its wings, frightening away the unsuspecting predator and giving itself time to escape.

The Oak Leaf Butterfly prefers shady places and thick vegetation. It does not stay still for long and so it is difficult to pick out. It does not go in search of nectar from flowers, but feeds on the sap which oozes from the bark of the trees.

The larva (caterpillar) which feeds on nettles has two long 'horns' on its head, with rows of little spines on its back.

OWL BUTTERFLY

Mouth: sucking

Metamorphosis: complete

Length of caterpillar: up to 16 cm

Wing span: 12-14 cm

The butterflies of this genus are among the most beautiful in tropical America. They have brown wings, streaked on the upper side with a rich shade of shiny violet, like velvet. The underside is always brown with clear, black streaks.

The main feature of the Owl Butterfly are the two 'eyes' at the centre of the underside of the rear wings. In the past, some naturalists thought that these were 'false' eyes, similar to those of an owl. According to them, when predatory birds came near, the Owl Butterfly would bend its head down, and show its 'eyes' on the underside part, looking something like a bird of prey and scare enemies away.

In fact, the Owl Butterfly poses just like other butterflies; its head up, and resting its wings one on top of the other, with its strange 'eyes' no more than a beautiful decoration. It can look rather like a bat as it flies around at twilight and by night.

Did you know...

*The order of **Lepidoptera** is one of the most numerous and most colourful groups of insects, with 165,000 known species. The word Lepidoptera means 'with scaly wings'. The tiny scales, however, fall off, even when they are not touched.*

The mouth is in the form of a long proboscis, a tubular structure of two horny filaments which the insect uses for sucking. As the insect rests, the proboscis is sometimes curled up on top of the head. To suck the nectar from flowers, the butterfly extends its proboscis, thanks to a combined action of the muscles and pressure of the blood.

*The **Lichen butterfly** belongs to a group of nocturnal butterflies. It has bright colours on the back of its body.*

*The **Death's Head Sphinx** butterfly gets its name because of the skull shape on the back of the thorax. It comes from Africa, but migrates every spring to Europe. Its back wings are yellow, with black streaks.*

*When the caterpillar of the **Cerura vinula** (Puss Moth) is disturbed, it withdraws its head and sticks out two long 'tails', looking like one, at the back.*

*The **cocoons** of Lepidoptera can last for years in dry conditions. The silk of which these are made is very strong and does not rot easily.*

AUTOMERIS IO
THECLA LINUS
CALIGO EURILOCHUS OWL BUTTERFLY
CATOCALA NUPTA
KALLIMA INACHUS OAK LEAF BUTTERFLY (upper side)
THYSANIA AGRIPPINA
KALLIMA INACHUS OAK LEAF BUTTERFLY (underside)
ANOPHYLLA MAGNIFICA
ACHERONTIA ATROPOS DEATH'S HEAD SPHINX
CERURA VINULA

Did you know...

The ***Garden Tiger Moth*** *may have markings similar to that of a tiger – but it avoids birds! However, it does make poisonous substances.*

The rear wings of the ***Comet Moth*** *end in two long 'tails'. In the male these can reach up to 20 cm. The female accumulates all the food she needs when in the caterpillar phase. As an adult, she neither eats nor drinks.*

The Madagascar Giant Urania *is one of the most colourful butterflies. Its colours take a long time to fade, even when exposed to strong light.*

The bushy wings of the tiny ***Alucita Pentadactyla*** *(Plume Moth) is split into five 'feathers'. It is so light, that it can easily be blown about by the wind.*

When the ***Zigaena Filipendulae*** *(Six-Spot Burnet Moth) is resting it almost seems to lose consciousness. Like this, it can be caught easily – however, the moment it is trapped, it releases a poisonous, yellowish liquid.*

The ***Owlet Moth*** *lives in all the temperate regions of the northern hemisphere. It belongs to the* ***Noctuidae*** *which is the largest group of Lepidoptera.*

***Moths** and **butterflies** both belong to the Lepidoptera family. But, although some moths can be seen during the day, most moths are active by night. They also tend to be duller in colouring than butterflies, with smaller wings, fatter bodies, and distinctive bushy-like antennae.*

*The larvae of the **Tortrix viridiana** (European Pea-Green Moth) are a serious pest. They will attack any tree, particularly the oak, often stripping its leaves completely.*

*The **White Lined Sphinx** belongs to the **Sphingidae** (hawk moth) family. In tropical countries the wing span of a hawk moth can reach up to 20 cm.*

*The **Currant Moth** is one of the most typical of **geometrid moth**. The caterpillars of all moths in the geometrid family move with a strange 'looping' movement.*

***Brush-footed butterflies** get their name because their forelegs are short and look like tiny brushes. Instead of proper front legs, this butterfly has a special organ which it uses to distinguish taste.*

*In the Amazon rainforest there are many examples of **nymphalaedea** butterflies. Their rare beauty in species such as the deep metallic blue **Morpho didius** makes them a sight to see.*

COMMON MOSQUITO

Mouth: sucking

Metamorphosis: complete

Length: 8-13 cm

Life of Adult: 3-5 days

The common mosquito can be found almost anywhere in the world. It is harmless but very irritating. Normally, it sucks the blood only of birds and humans. Its abdomen is rather wide and covered in scales. The female's antennae are short; in the male, the antennae are long and feathery, so that the insect can collect the nectar on which it feeds. The male's antennae are also used to sense the high-pitched humming sounds made by the female and to recognize the presence of other mosquitoes. Both male and female have a long, hollow proboscis through which they suck their food. The common mosquito lays its eggs in all types of stagnant water.

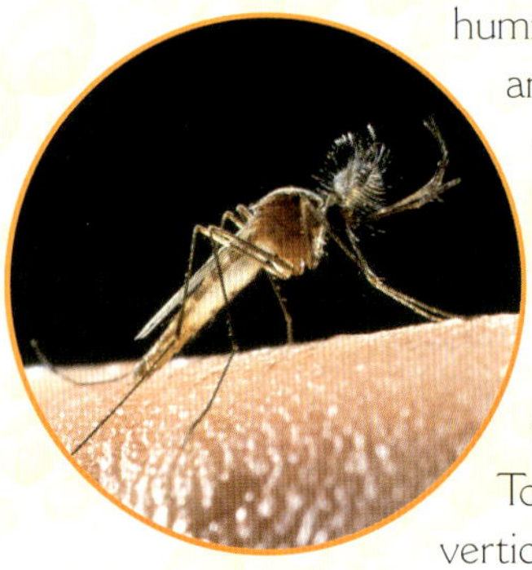

To breathe, the larva positions itself vertically below the skin of the water, so that the bottom part of its body, where the respiratory organs are, is above the surface. The larvae feed on animal and vegetable matter, but the males of many species of mosquito 'fast' for the whole of their brief lives and survive by using up the reserves of food taken in during the larva stage. The female lives much longer than the male.

GARDEN MOSQUITO

Mouth: sucking

Metamorphosis: complete

Wingspan: 10 cm in the largest examples

The Garden Mosquito can be found almost all over the world. It is mostly noctural and looks very much like a giant mosquito, although it does not bite or sting.

The Garden Mosquito lives almost wholly in fields. It is quite lazy and, right from the start of its brief life, prefers to spend its time among the grass, always flying just above the surface of the ground in a clumsy, heavy sort of flight.

The adult feeds on nectar and other vegetable juices. But the long, soft larvae are very harmful because they feed only on the roots of vegetables, especially carrots. They are very resistant to pesticides and farmers see them as a pest and a danger.

Like all mosquitoes, the Garden Mosquito needs water. The female lays her eggs on muddy or moist surfaces, and as rain falls, so the larvae hatch. When the land is uncultivated, the larvae satisfy themselves with the roots of grasses.

Did you know...

*Unlike all other flying insects, those which belong to the **Diptera** class have only one pair of true wings. The hind wings are in the form of 'halteres' – tiny, little fin-like organs which enable the insect to keep its balance during flight. In many of the larger insects in this class, the halteres are covered with scales.*

*The male and female of **Bibio Hortulanus** (Mark's Fly) are quite different to each other; the female has a more narrow head and smaller eyes. They live particularly near trees, and close to farms and gardens where they attack the roots of plants.*

*The adult male of the **Anopheles Mosquito** (see larvae in photograph) is easily recognized by its feelers, which look like little golf clubs. It is the female which carries malaria, and she, too, has long feelers. In a resting position, the adult keeps the abdomen leaning towards the surface on which it has landed, whereas the mosquito larvae lie horizontally on the surface of the water.*

*The female of the **Black Fly** sucks the blood of many mammals, injecting them with a poisonous substance with her saliva.*

*The female **Sand Fly** carries the germs which can cause serious illnesses in humans, for example, a tropical disease called Pappataci Fever.*

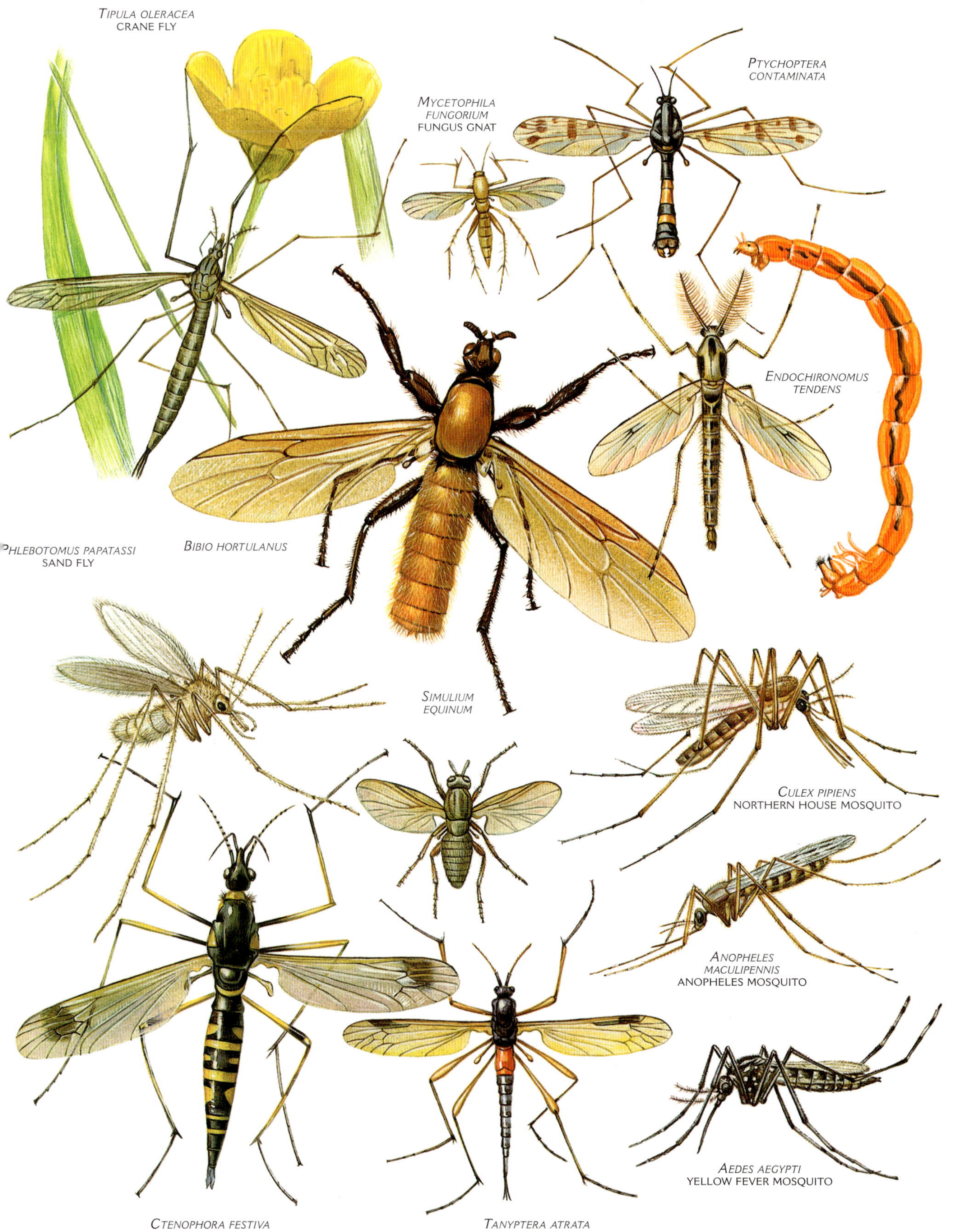
TIPULA OLERACEA
CRANE FLY
MYCETOPHILA
FUNGORIUM
FUNGUS GNAT
PTYCHOPTERA
CONTAMINATA
ENDOCHIRONOMUS
TENDENS
PHLEBOTOMUS PAPATASSI
SAND FLY
BIBIO HORTULANUS
SIMULIUM
EQUINUM
CULEX PIPIENS
NORTHERN HOUSE MOSQUITO
ANOPHELES
MACULIPENNIS
ANOPHELES MOSQUITO
AEDES AEGYPTI
YELLOW FEVER MOSQUITO
CTENOPHORA FESTIVA
TANYPTERA ATRATA

TSETSE FLY

Mouth: sucking

Metamorphosis: complete

Length of adult: 10-13 mm

The Tsetse Fly may look harmless, but it is the carrier of terrible disease which can strike many species of animals, including all domestic animals, as well as humans. It lives mainly in Africa in the southern region of the Sahara Desert.

It feeds by breaking the skin of its victim with its beak-like mouth then sucking the blood. Unlike the horsefly and the mosquito, where only the female feeds on blood, in the Tsetse Fly, both sexes are 'blood-suckers'.

So far, twenty similar species have been discovered, some living in the open spaces of desert and jungle regions, others among thick vegetation.

When it comes to reproduction, instead of laying a certain quantity of eggs, the female Tsetse Fly produces only one which she carries inside her body. Here, the larva develops, nourished by a food which the female produces by special glands, having sucked enough blood for the development of the egg. As the larva develops, it breathes through two ducts which develop from a swelling on the female's body.

FRUIT FLY

Mouth: sucking

Metamorphosis: complete

Length: 4.5-5 mm

There are many species of Fruit Fly, all seen as serious pests to fruit farmers. The Mediterranean Fruit Fly is common in the USA and most of Europe. It has grey-red wings, spotted with black at the bottom. The thorax is black with spots and covered with a fine silver down. After mating, the female will use a type of sting to break open the ripened skin of citrus fruits such as oranges and grapefruit, then lay her eggs inside. Some are killed by the oils in the skin of certain types of citrus fruits, but with up to 500 eggs laid, some are bound to survive. Once hatched, the larvae feed on the flesh of the fruit. In the case of the Apple Fruit Fly, the larva will eat the flesh to the core, so the fruit goes rotten and then falls. The Fruit Fly was identified in Malaysia at the beginning of the 18th century. After that, it invaded all hot and temperate countries in the world. In the USA, a 'war' is constantly being waged against all types of fruit flies, especially in Florida and in Hawaii. Crops are sprayed with pesticides during the mating season of the Fruit Fly, and there are strict quarantine laws to regulate all fruit imported from other countries. But, the Fruit Fly continues to survive.

Did you know...

*The female of the **Tabanus Atratus** (Black Horse Fly) can suck up to 55 mg of blood with just one bite, attacking its victim near the eyes. But the male Black Horse Fly feeds only on nectar and dew.*

*One of the names for the Devil is Beelzebub, which means 'Lord of the Flies' – and the ugly appearance of the **fly** does make many of us think of this insect as a 'devilish' creature, especially when we know the fly to be a tireless pursuer which is able to feed on rubbish. In addition, a flies can cause illness and disease, and it is virtually indestructible.*

*The **Killer Shore Fly** from North America has front legs similar to the Praying Mantis.*

*The larvae of both the **Warble Fly** and the **Cattle Fly** develop beneath the skin of cattle. This causes painful lumps which have to be lanced.*

*Flies belonging to the **Diosophilidae** family and especially the **Vinegar Fly** were first studied by an American biologist in 1900. It was soon discovered that they can be easily bred and reproduce very quickly, which is why these insects have since become important in the study of the heredity of characteristics in human beings. In addition, the Vinegar Fly has only four pairs of chromosomes in the protein molecule nucleus of a cell, and this feature makes research into the DNA which carries hereditary information much easier.*

BRAULA COECA
BEE LOUSE
OCYPTERA BRASSICARIA
ASILUS
CRABRONIFORMIS
ROBBER FLY
OCHTERA MANTIS
KILLER SHORE FLY
CERATITIS CAPITATA
MEDITERRANEAN FRUIT
FLY
LUCILIA
CAESAR
GREEN BOTTLE
CHRYSOPS
CAECUTIENS
HORSE FLY
CONOPS
QUADRIFASCIATUS
WASP FLY
GLOSSINA
PALPALIS
TSETSE FLY
HYPODERMA BOVIS
WARBLE FLY
CYRTUS GIBBUS
CELYPHUS

ECHINODERMS

Evolution

The earliest fossils of Echinoderms are the remains of species fixed to levels of ground beneath the sea. These fossils date back 500 million years ago – although these sea creatures probably originated much earlier than that. The reason that Echinoderms have survived to the present day is largely due to their strong, protective covering.

Main characteristics

All Echinoderms, although in different shapes (star, spherical or elongated) have the same type of symmetrical shape (equal on all sides). This means that the body can be divided into five equal parts branching off from a central disc or axis, like the spokes of a wheel.

The ***Marthasterias glacialis*** *(Spiny Starfish)* is very large, with a strong body. Its five arms are covered with six or seven rows of sharp, cone-shaped prickles. It is dark green in colour and is usually found in deep waters.

Like all starfish, it is a hungry predator and attacks its victims oysters, mussels and other bivalves, grasping it with the strong pressure of its arms. It uses the suction of its tube-feet to force open the shell of the bivalve, then it takes the prey into part of its stomach where it is digested before the starfish actually swallows it. When it is very hungry, it will not hesitate to attack also the well-armoured sea urchin. Starfish will generally avoid the light, but the *Spiny Starfish* is attracted by it

Other larger species which live at depths of over 100 m have brighter colours, from red to red-brown with white blotches.

Movement

Except for the feather stars and sea lilies which are fixed on the sea bed, Echinoderms can move about on tube-feet on the lower part of the body. As these tube-feet draw in water and pump it out, they contract and expand, enabling the animal to make slow movements.

Habitat and distribution

Echinoderms live mostly in the sea. Only those belonging to the Crinoidea family (feather stars and sea lilies) are fixed to underwater surfaces, many of these species in the deepest depths of the ocean. All other species live and move about close to the sea bed, in shallow and deep depths.

Rays in a Pattern

Even by looking closely at a starfish or a sea urchin it is not easy to distinguish the front part from the back. The sea urchin looks like a ball covered with prickles, the starfish has two flat surfaces with five points, apparently all the same. In fact, all Echinoderms are without a head and almost all have the mouth and the anus in line, on an upright axis through the centre of the body. The mouth is on the bottom surface whilst the anus, if there is one, is on the upper surface. The structure leads out from a central disc, spreading into **five equal parts**. This structure is very clear in the starfish, but it is more difficult to make out in the sea urchin. The walls of the body are divided into three layers – external, intermediate and internal. The external is the most delicate and covers the entire body. The intermediate layer is the connective tissue and forms the internal skeleton. In sea urchins, this skeleton is a compact shell covered with prickles; in the starfish and other species it is made of plates fused together. The internal layer covers the whole of the inner cavity which contains organs for digestion and for reproduction. It is inside this cavity that the five double rows of **tube-feet** begin, appearing on the body like rays. These tube-feet can move by a system of little canals which fill up with water and then empty, making the tube-feet contract and expand so that the animal can crawl along the sea bed. Breathing is done in the same way, taking in oxygen from the outside and expelling waste air.

The circulation system is made up of blood vessels called lacunas. These penetrate between the organs and the tissues, and the muscles in the lacunas contract to make the blood circulate.

Photograph: a colourful example of a Starfish

The nervous system also follows the same radial (ray-like) pattern.

The sex of echinoderms is almost always separate, but there are some species which are hermaphrodite; as a rule, the egg and sperm come together in the water. Swimming larvae develop from the egg, and these have a bilateral (two-sided) symmetry. Only after a complex metamorphosis (transformation) do the adults develop the five-point symmetry. Echinoderms are divided into six classes:

Crinoidea, comprising about eighty known species. These have one foot which fixes the animal to the sea bed. The body is cup-shaped with five or more coloured, feathery arms. After the early stages, some species detach themselves from the feet and move around freely.

Echinoidea, the family of sea urchins, with more than 850 species.

Holothurioidea, commonly known as sea cucumbers, of which there are 500 species, all with the typical long shape. The body has calcite little plates and moving tube-feet.

Asteroidea, the starfishes with over 2000 species.

Ophiuroidea (basket stars and brittle stars), with a little less than 2000 known species. These are rather like starfish, but with softer arms, long and flexible.

Concentricycloidea (sea daisies).

PHYLUM	CLASS	SPECIES
ECHINODERMS	Crinoidea	feather stars, sea lilies
	Echinoidea	sea urchins
	Holothurioidea	sea cucumbers
	Asteroidea	starfish
	Ophiuroidea	brittle stars
	Concentricycloidea	sea daisies

EUCIDARIS TRIBULOIDES
PENCIL URCHIN
RHIZOCRINUS LOFOTENSIS
CENOCRINUS ASTERIA
ANTEDON MEDITERRANEA
FEATHER STAR
ECHINUS MELO
SEA MELON
SPHAERECHINUS GRANULARIS
PURPLE SEA URCHIN
PELAGOTHURIA LUDWIGI
ECHINARACHNIUS PARMA
SAND DOLLAR
CUCUMARIA PLANCI
SEA CUCUMBER
RHOPALODINA LAGENFORMIS

PURPLE SEA URCHIN

Shape: spherical

Diameter: from 6-7 cm to 15 cm

Appendages: numerous short prickles

This sea urchin is one of the largest. It often reaches a size of 6 or 7 cm in diameter, but sometimes the largest reach a diameter of up to 15 cm.
Its surface is thickly covered with short, violet-coloured prickles, fading to white at the ends.
It is very common in the Mediterranean and in the Atlantic, and from the English Channel to the Gulf of Guinea, at depths of between 30 and 100 m.
It lies on its lower side, the same side as its mouth opening. The anus is on the opposite side, in the centre.
As in all sea urchins, the internal skeleton is made up of 10 double rows of calcite plates which form a rigid covering. Some of these plates have the larger nodules (swollen bases) for the roots of the prickles; other plates have the smaller nodules where the pedicellariae (small pincers) begin. The sea urchin uses these pedicellariae to defend itself.
From the actual body of the sea urchin emerge the five double rows of tube-feet which enable it to move.
The sea urchin is egg-laying.

Did you know...

*The **Rhizocrinus Lofotensis** is a 'dancer' of the sea, with a delicate hanging stem, about 7 cm long, and short grey or yellowish feathery arms. It is found in the northern Atlantic at depths of between 140 and 480 m. The **Cenocrinus Asteria** is another 'dancer'; it lives in the western Indian Ocean at depths of between 200-600 m; it is yellowish-brown, with a body 1.7 cm diameter and arms 10 cm long.*

*The feather star **Antedon Mediterranea** is very common in the Mediterranean. It lives among the algae on the sea bed, attached to the rocks and on the deep coral beds, at depths of up to 200 m.*

*Sea cucumbers in European seas are purple, grey or yellow, whilst those on or near coral reefs are brightly coloured. On the **Pelagothuria Ludwigi**, the front part of the trunk is encircled an umbrella-shaped membrane, with 13-16 tentacles which enable it to swim like a jellyfish.*

*The **Rhopalodina Lagenformis** keeps the lower part of its body sunken, whilst the upper part comes out of the sand to take in the water.*

*The **Sea Urchin** feeds on dead fish and carcasses of other sea creatures, seaweed and even wood. There are three main examples: the **Eucidaris Tribuloides** (Pencil Urchin), which lives in the Caribbean seas and the East Atlantic coast; the **Echinus Melo** (Sea Melon) in the Mediterranean Sea and the Atlantic Ocean, and the **Echinarachnius Parma** (Sand Dollar) which lives along the Atlantic coast of North America.*

SEA CUCUMBER

Shape: long

Length: 20-35 cm

Appendages: 20-39 tentacles around the mouth

The Sea Cucumber has a cylinder, worm-like shape, which makes it look rather like a vegetable cucumber. Unlike the starfish and the sea urchin, it does not rest on the ground on its 'face' but on one side of its body. This is because its mouth is at the end of its front and its anus at the end of its rear. It moves on the sea bed by moving tube-feet on the abdominal surface of the body; only a very few species swim by means of moving appendages. About twenty tentacles encircle the mouth. Some tentacles are branching. Others are simple, especially those which retract most quickly and which are used to catch small prey.
The Sea Cucumber can also take in the sand and the mud on the sea bed, from which it can absorb organic particles for nourishment. Its eggs are laid in the sea and incubated in the body of the female. Some species live in symbiosis with the small pearlfish, which shelter inside the body of the Sea Cucumber.

ASTROPECTEN AURANTIACUS

Diameter: from 10 to 28 cm

Shape: arms branching out from a central disc; plated edging and prickles

Habitat: sandy surfaces from depths of 1 to 20 m.

This starfish belongs to the family of Asteroidea. The Echinoderma in this family have broad plates on the edges of the body, which is covered by prickles and bumps. It is easy to see this beautiful red-orange starfish, with its five arms edged with prickles, on the fine, sandy beaches of the Mediterranean and the Atlantic.

It lives very near the coast, and at low tide, it buries itself in the sand, showing only the centre of its upper side which is swollen in the form of a cone. This cone acts like a sense organ. When it is touched, the cone contracts and the starfish buries itself in the sand once again. Then at high tide, the starfish reappears on the surface of the water. It buries itself in the sand to shield parts of its body from the rays of the Sun, especially the bottom surface which has no coloured pigment and is therefore particularly sensitive.

This starfish is a carnivore and feeds on the molluscs which it catches with its arms and then takes to the mouth. The prey is then trapped by the long, moving prickles around the mouth cavity.

BRITTLE STAR

Diameter: from 5 to 15 cm, arms extended

Shape: arms clearly separated from the central disc; presence of spines

Habitat: from depths of 0 to 475 m on rocky or slimy sea beds

This highly-coloured species is sometimes called the Serpent Star. Its arms are flexible, but so fragile that they can be broken off and re-grown easily.

The Brittle Star is found on the rocky coasts of Europe, in the Mediterranean and in the Atlantic, where it lives at the flattest part possible on the rough surfaces there. Sometimes it likes to be with other Brittle Stars, when a large group will gather together.

If the current is calm enough, the Brittle Star will extend its arms upwards to catch tiny food particles and fragments of debris floating in the water. The food reaches the mouth by quick movements of the tube-feet along the arms.

If the current is strong, the Brittle Star will squash itself against the sea bed, so as not to be carried away.

When a Brittle Star finds itself on its wrong side, it can turn itself back again within about 40 seconds. It can do this by making all the muscles in its body work together.

Did you know...

*The **Linckia laevigata** (Blue Linkia) is a blue starfish which lives in the seas of Australia. It is remarkable because of its ability to re-grow; from a single arm, it can grow its whole body again. During this process, it presents the so-called 'comet shape' with one large and four small arms.*

*Unlike many other starfishes, the **Astropecten Aurantiacus** moves very quickly, between 30-60 cm in one minute.*

*The **Acanthaster planci** (Crown of Thorns Starfish) is a poisonous starfish. With its 11-21 short arms and its long prickles, it can attack and destroy the coral reefs.*

*The **Crossaster Papposus** (Spiny Sun Star) lives in the northern Atlantic. It has 8-14 arms, and a large disc with a diameter of 34 cm.*

*The **Fromia ghardaqana** (Hurghada Star) with its gaudy red colouring and white spots, is found in the Red Sea.*

*In the **Amphilycus androphorus**, the female is large-sized. The male is much smaller and stays with the female all his life.*

*The **Ophioderma longicauda** lives in the Mediterranean and in the Atlantic Ocean. The **Ophiura albida** lives in the Mediterranean in the Atlantic and in the eastern Baltic Sea.*

Linckia laevigata
Blue linkia
Acanthaster planci
Crown-of thorns starfish
Ophiothrix fragilis
Brittle star
Astropecten aurantiacus
Ophioderma longicauda
Crossaster papposus
Spiny sun star
Amphilycus androphorus
(with the male on the disc)
Ophiura albida
Fromia ghardaqana
Hurghada star

CHORDATA

Evolution

The phylum Chordata represents the link between the most primitive animal life through to the most highly developed, and on to human life. The earliest forms of Chordata were small animals, thin and sometimes transparent.

Main characteristics

The body of a chordate is always symmetrical. This means that is is equal on both sides. In the embryo or larva state, it is supported by the **notochord**, which is like a flexible rod and runs through the length of the body.

The Ascidian or Sea Squirt (right) is a very primitive Urochordate. It has a distinctive pouch-shaped body covered by a rigid 'tunic' of live tissue, with a siphon at the mouth through which water enters (and from which the animal extracts food particles and gets the oxygen for breathing), and another siphon through which the water is expelled.

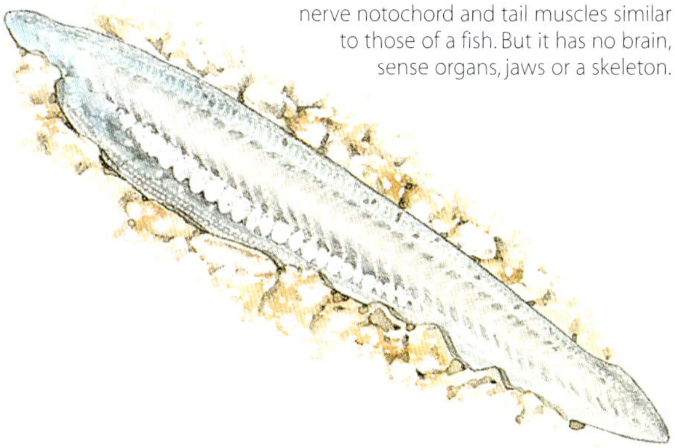

The Amphioxus or Lancelet is a Cephalochordate. It is a tiny, semi-transparent sea creature which lives in the bottom of the sea. In the Lancelet, we can see evidence of the transition from invertebrates to vertebrates. It has a nerve notochord and tail muscles similar to those of a fish. But it has no brain, sense organs, jaws or a skeleton.

The Lamprey is one of the most primitive vertebrates. It has a long body but without jaws or limbs. Its system of feeding is very strange; the mouth is a sort of funnel covered with numerous horny teeth, set out in concentric circles at the centre of which is the tongue. The lamprey seizes its prey in its mouth, sticking on to it like a sucker and sucking out the blood. From pictures, it is possible to make out the gill slits through which the water which the Lamprey has ingested is expelled.

Nervous system

Another characteristic which distinguishes Chordates from invertebrates is the presence of a central nervous system, which is always controlled from the top of the **notochord**. In the earliest Chordates, the nervous system was very simple. In the course of evolution, this has developed into various degrees of complexity.

Habitat and distribution

Among Chordates, only the vertebrates have been able to adapt to every type of surroundings, from water to the land and at all latitudes. The other two groups – the Urochordates and the Cephalochordates – have never succeeded in abandoning the waters of the seas and the oceans.

A strong and flexible support

All Chordates have a **notochord** – a solid, yet flexible rod of cells to support the body and the muscle structure. They also have organs necessary for the capture and filtering of tiny organisms, the 'gill slits'. Water is inhaled through the mouth and expelled through the **gill slits**, whilst the food particles present in the water are trapped.
The gill slits are also used for breathing.
The circulation system comprises a pulsating organ, the **heart**, and veins through which blood circulates. Over the centuries, this system has developed more and more, also becoming an important part of the process of respiration (breathing). In the majority of cases, reproduction is by the meeting of two animals of the opposite sex, although some Urochordates are hermaphrodite.

Chordates are divided into three subphylums – **Urochordates** or **Tunicates**, Cephalochordates and Vertebrates. Urochordates or Tunicates are sea creatures which can vary in size from a few millimetres to tens of centimetres. Some are fixed, like the Sea Squirt, others move freely. They are formed in one 'piece' and the body is similar to a pouch or bag. The spinal chord (or notochord) is present only in the tail and in some species only in the larva stage. Also, both the circulation system and the nervous system are reduced.
Cephalochordates are the closest to Vertebrates. The body is compressed sideways and covered with an epidermis (top layer of skin). The notochord runs through the whole body and remains in adults. Instead of limbs, **Cephalochordates** have fins for swimming, with various segmented organs repeated along the whole body. The mouth is at the front of the stomach and this has many filaments which trap food filtered through the water. The circulation system has only veins. Sexes are separate and fertilization takes place in water.

Photograph: above, a Sea Squirt, below, a colony of salps, creatures belonging to Thaliacea, another class of Urochordates.

Vertebrates are the most complex of Chordates. The notochord is present only in the embryo; in the adult this is replaced by the **spinal column** which supports the whole skeleton. Two pairs of limbs are joined to the spinal column and these are used for movement. The skin protects the body and this has two layers, the dermis and the epidermis.
The **nervous system** consists of the brain, the spinal cord and various nerves. Blood is pumped around by the heart, circulating through a network of arteries, veins, and capillaries,. The respiratory (breathing) system consists of gills in aquatic vertebrates and lungs in land-dwelling vertebrates.

PHYLUM	SUB-PHYLUM	CLASS	SPECIES
CHORDATES	Urochordates		Sea Squirt
	Cephalochordates		Lancelet
	Vertebrates	cyclostomes	Lamprey
		fish	
		amphibians	
		reptiles	
		birds	
		mammals	

All vertebrates have separate sexes and can be oviparous (egg-laying), ovoviviparous (the egg is carried inside the female's body until it comes out) or viviparous (the embryo develops in the uterus of the female).

FISH

BEHAVIOUR

Most fish take little care or no care at all of their young. However, there are some species that build nests and keep guard over both the eggs and the young.
There are fish which 'walk' on the sea bed, fish that can fly, change colour or look like rocks and there are fish which can inject poison.

COMMON CHARATERISTICS

A fish has a tapering body which, in most cases is protected by scales, and dorsal (back), ventral (stomach), pectoral (breast) caudal (tail) and anal (end) fins. They breathe through gills and most species have a swim bladder which prevents them from sinking when they are not swimming.

The Coelacanth ***(Latimeria Chalumnae)*** shown below was discovered in the1930s. Before then, it was believed that this strange fish had been extinct for 70 million years. The coelacanth is a 'living fossil', one of the few survivors of a group of prehistoric fish, unaltered since the time it was widespread on Earth. It has a swim bladder, but this is full of fat and does not carry out any function in keeping the fish afloat. It has long pectoral fins which enables it to 'walk' on the sea bed, supported by 32 feelers.

The Siamese Fighting Fish ***(Betta Splendens)*** in the photograph above, originated in Thailand. Little by little, as they were inbred by humans, they became more warlike, hence the name. It is not a sociable fish. A male lives with a female, defending its territory from other males. Unlike most fishes, it takes care of the young along with the female. She lays her eggs in a nest of bubbles of air, covered by saliva. The nest floats to the surface of the water.

FEEDING

Fish can be herbivore or carnivore. In both cases, there are some species with teeth strong enough for them to feed on large plants and sea creatures and some which can eat only the microscopic organisms in plankton which they filter through the water.

EVOLUTION, DISTRIBUTION AND HABITAT

The first vertebrates lived on earth about 400 million years ago. There are believed to be around 27,000 species, most living in seas and oceans all around the world, up to 9000 m deep. An estimated 5000 live in fresh water. Some can migrate from the seas to rivers and vice versa.

Gills, fins and a spinal column

Fish are divided into two main classes – cartilaginous fish and bony fish. Among the **Cartilaginous Fish** are the sharks, rays and chimaera. All sharks have a body which is perfectly hydrodynamic – ideal for moving through the water – but they do not have a **swim bladder**. Therefore, so as not to sink, the shark has to swim continuously. The ray has a very flat body which is made larger by the pectoral fins being fused together. Sometimes, the tail is sharp and has a very poisonous sting. Some rays give off strong electrical charges. The Chimaera, which lives in the deepest depths of the sea, is of medium size and has a very odd, somewhat monster-like form. Its appearance is somewhere between a cartilaginous and a bony fish. Like the Coelacanth, the Chimaera is an ancient species and a 'living fossil'.

Photograph: an example of a ray.

All the other known species of fish, of many different shapes and colours, belong to the class of **bony fish**, with a bony skeleton. Bony fish have separate sexes and often the male and the female are very different. The eggs can be pelagic (floating in water) and may contain a **drop of oil** which enables them to be carried along by the currents, drifting some way down from the surface and getting mixed with the plankton. The tuna lays this type of egg. Alternatively, the egg may have a heavier density than that of the water and so sinks to the bottom. This is what happens with the eggs of the coast-dwelling, and fresh water fish. When fertilization is 'external' the male bathes with his sperm the mass of eggs which has been laid; when fertilization is internal, the male fertilizes the female by way of a special organ called a **gonad**. In fish, the number of eggs laid can reach many millions; but then the larvae become easy prey for other animals and so few reach the adult stage.

PHYLUM	SUB-PHYLUM	CLASS	ORDER	SPECIES
CHORDATA	Vertebrate	Cartilaginous Fish	Selachii	shark
			Batoidei	ray
			Chimaeriformes	Chimaera
		Bony Fish	Acipenseriformes	paddlefish
			Coelacantthicae	Coleacanth
			Clupeiformes	herring
			Mormyriformes	Nile Mormyrus
			Anguilliformes	eel
			Cypriniformes	carp, tench
			Bolonidae	garfish
			Gadiformes	cod
			Gasterosteiformes	Sea Horse
			Lampridiformes	Moonfish
			Zeiformes	Boarfish
			Perciformes	tuna
			Pleuronectiformes	sole, flatfish
			Tetraodontiformes	Puffer Fish
			Lophiiformes	Batfish

WHITE SHARK

Length: 6-7.5 m

Reproduction: viviparous

Distribution: worldwide

The White Shark belongs to the class of cartilaginous fish. It can be found in the open seas, especially in areas up to 1000 m deep, but also near to the coast. It lives all over the world, but is rare both in the tropics and in the Mediterranean. Its huge, solid body is silver grey or brown and the stomach white. It is recognized mainly by its cone-shaped, pointed nose, its tail shaped like a crescent moon and its terrifying mouth, with its grooved lips and rows of pointed teeth, all razor-sharp. The White Shark's largest teeth can reach a length of 8 cm. So, we can understand how easily it can tear its prey apart. The White Shark feeds on small sea creatures and occasionally larger prey such as dolphins, seals and other sharks. The stomachs of White Sharks often contain rubbish thrown overboard from ships, even tin cans. The shape of its body may lead us to think that it can swim very fast, but hunters using harpoons would seem to contradict this. Although the White Shark is so easily recognized, there are still many aspects of its life about which we do still do not know very much.

CARCHARODON CARCHARIAS
WHITE SHARK

Did you know...

The largest shark – and the largest of all fish – is the ***Whale Shark*** *with a length of 20 m and a weight of about 20,000 kg.*
It swims slowly, keeping its mouth open. This alone can reach 2 m in length, but the Whale Shark eats only plankton and small fish.

The tail of the ***Thresher Shark*** *is as long as the head and the body put together and it weighs up to 450 kg. But despite its appearance, it is quite harmless to humans. It is always hungry and can swallow in one gulp almost all the fish it can hold in its mouth at one time. It is unpopular with fishermen because it causes serious damage to their nets.*

The young of the ***Basking Shark*** *have a very long nose, like the proboscis of an insect. In the adult, this stays relatively thin.*

The skin of the ***Small Spotted Catshark*** *is covered with scales like dense teeth – no wonder this was once used like sandpaper! This fish lays its eggs in hard, rigid cases 10 cm long.*

The mouth of the ***Spotted Wobbegong*** *shark is surrounded by numerous curling skin appendages rather like the form of some invertebrates or algae. These helps them to disguise themselves as creatures on the deepest sea beds and to identify food.*

The ***Sand Shark*** *lives in the shallow water along all ocean coastlines except those which are too cold. Only two species of sand shark are considered to be dangerous.*

Although it is always very hungry, the ***Blue Shark*** *is not one of the most dangerous sharks. It also swallows non-edible material and often follows the whaling ships to eat the rubbish thrown overboard.*

CETORHINUS MAXIMUS
BASKING SHARK
CHLAMYDOSELACHUS ANGUINEUS
FRILLED SHARK
SCYLIORHINUS CANICULA
SMALL SPOTTED CATSHARK
ALOPIAS VULPINUS
THRESHER SHARK
RHINCHODON TYPUS
WHALE SHARK
CARCHARODON CARCHARIAS
GREAT WHITE SHARK
CARCHARIAS FEROX
SAND SHARK
PRIONACE GLAUCA
BLUE SHARK
ORECTOLOBUS MACULATUS
SPOTTED WOBBEGONG

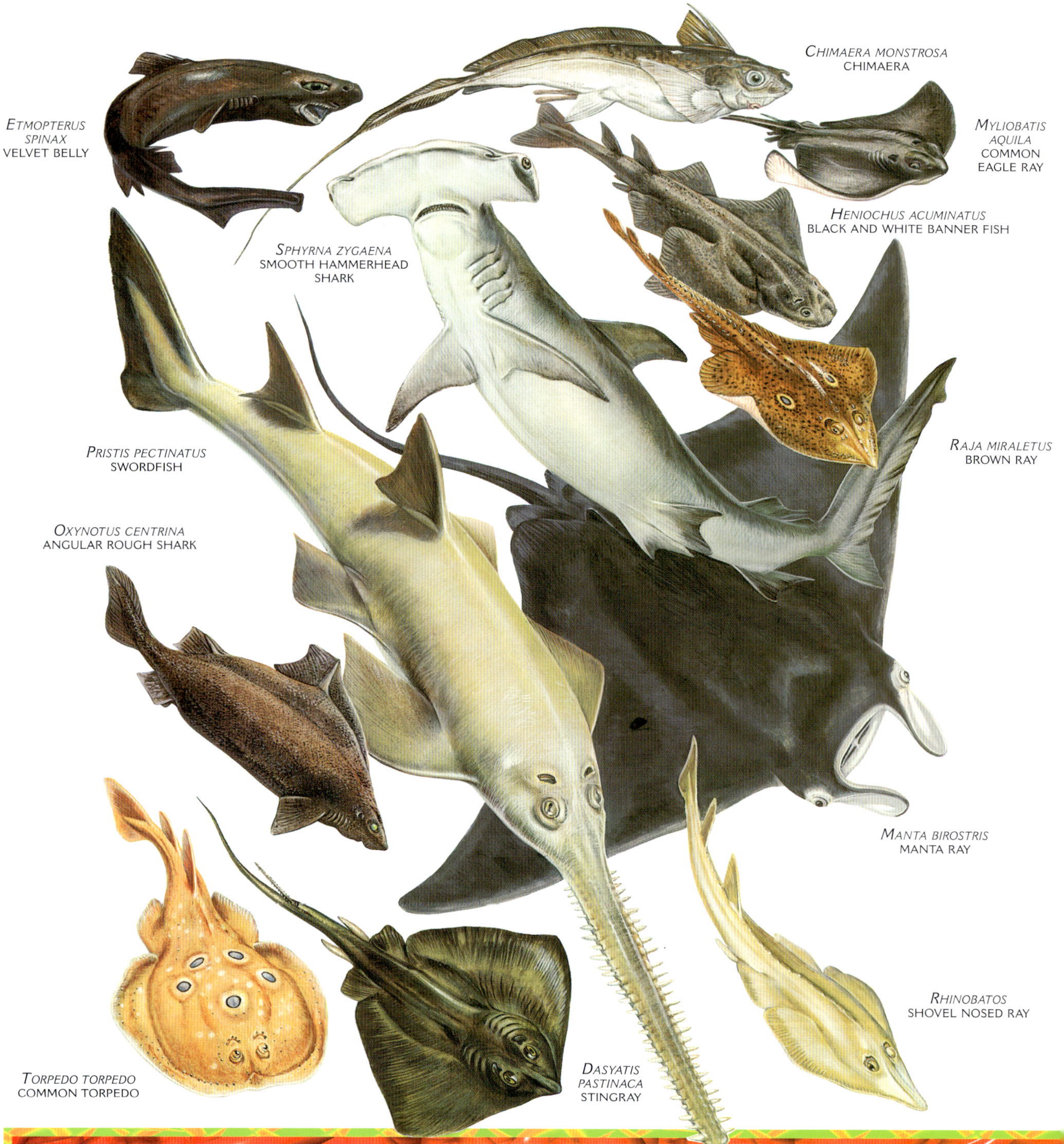

Díd you know...

Because of its appearance and its position in the table of evolution, the ***Black and White Bannerfish*** *is widely regarded as the link between sharks and rays.*

The weapon of the ***Swordfish*** *is a very long armed beak equal to a quarter or a third of its total length. As well as stunning other fish, the Swordfish also uses its beak to search in the mud of the sea bed for molluscs and other sea creatures to eat.*

The ***Common Eagle Ray*** *often appears as the 'monster' in sea legends. It has a tail like a dagger, 'armed' with spines and both tail and spines are poisonous. By moving its pectoral fins very rapidly, it can suddenly rise up from the deep and this is when it seems as if it is really flying.*

The dorsal fins of the ***Angular Rough Shark*** *each have a large spine; the front fins lean forward.*

STINGRAY

Reproduction: viviparous

Number of young: 2-4

Length: up to 1.5 m

Weight: up to 20 kg

Where found: sea beds of the Mediterranean and the Eastern Atlantic

Feeds on: molluscs and crustaceans

The Stingray (in photograph) is one of the most common species of poisonous rays. It can be reddish in colour, or olive green or grey and it has a triangular snout and a long tail, one and a half times the length of the body, and this has at least one sharp spine. To defend itself, the Stingray uses its tail like a whip. This can pierce the bottom of a boat and cause serious injury to humans. In some regions of the Pacific, the caudal spines of the Stingray are used by inhabitants to make poisonous daggers.

The Stingray lives half-hidden in the sand, which it leaves only to go in search of food – molluscs, crustaceans and other invertebrates. Its enormous pectoral fins act as huge fans which, when waved about near muddy depths, uncovers the sea creatures which live hidden in the sediments of the sea bed.

The Stingray chews its prey with lots of strong teeth which are set out in rows. Some species of stingray have been known to swim up into the estuaries of rivers.

HAMMERHEAD SHARK

Reproduction: viviparous

Number of young: 30-40

Length of adult: up to 6 m

Width of head: up to 80 cm

Where found: worldwide

The Hammerhead Shark is one of the strangest members of the shark family. The reason for its curiously-shaped head, which can be up to 80 cm wide, is not known. Its side lobes, with the eyes at the end of two bulges in the skin and which are very large compared to the tiny pectoral fins, wave about to enable the Shark to maintain balance. Hammerhead Sharks can be up to 6 m long. They live in open, deep waters, in tropical and in temperate seas. In the Caribbean seas and along the coast of Florida, gatherings of hundreds of young Hammerhead Sharks can often be seen.

The Hammerhead Shark feeds mainly on cartilaginous fish, crustaceans and crabs which it finds on the sea beds. When it is hungry, there is nothing safe from its poisonous rays – the unmistakeable sting of the Stingray has been found in the flesh of some Hammerhead Sharks which have been caught.

The female gives birth to lots of living young many times in her lifetime.

*The **Brown Ray** is egg-laying and fertilization is internal. The egg comes contained in a horny, rectangular shell with an appendage of variable length at any angle.*

*The **Manta Ray** is a real giant; it can measure over 8 m wide and weigh 3000 kg. Its square-shaped head has two small fins turning inward.*

*In the times of the Ancient Romans, doctors would use the **Common Torpedo** to cure some illnesses. The Torpedo is one of the few fish which can give an electric shock. This can be equal to approximately 200 volts. The organs which generate electricity are luminous and situated at the each side of the head, within the folds of the body.*

*The **Chimaera** is sometimes known as the Ghost Shark and is a fascinating creature. It lives in the depths of the sea bed and has a tail similar to a thin whip. As the last representative of an extinct animal family, it is regarded as a 'living fossil'.*

SALMON

The Common Salmon lives in fresh water and also the sea. It is a tireless traveller. This fish leaves the sea when, having reached sexual maturity and accumulated remarkable reserves of fat, it feels the moment of reproduction has arrived. Then it swims towards the mouth of the river, which leads it to the source. Here, the very cold waters are rich in food. Overcoming many obstacles, waterfalls, rapids and barriers, the salmon covers hundreds of kilometres. After losing one third of its weight, it reaches the place where it will reproduce.

Because of serious organic deterioration, some salmon die after laying their eggs, or become easy prey for fish, birds or mammals. It is calculated that only 5% of salmon have carried out a first migration successfully and returned to the sea. The young salmon, born in the highest reaches of the rivers, migrate towards the sea when they reach a length of 10-12 cm, at 1 - 2 years of age. When they reach the time to reproduce, they go back to exactly the same river in which they were born. It seems that salmon have a highly developed sense of recollection, even after 5 - 6 years, for the smell, the temperature and the chemical features of the water in which they were born.

Reproduction: egg-laying

Length: 1.20-1.50 m

Lives in: the sea and fresh water

PIKE

Reproduction: egg-laying

Length: up to 1.35 m

Weight: up to 23 kg

Lives in: fresh water

The Pike is a freshwater fish with a dreaded reputation; it is the most fearsome predator of European rivers, if not the whole northern hemisphere. The Pike does not like strong currents of water because these prevent it from staying still so that it can surprise and then attack its prey. Instead, it prefers the calm waters of lakes and non-raging rivers.

The clear colour of the skin with dark stripes enables it to stay hidden among the water plants. The strong tail and the backward-bending fins give the Pike enormous spurts of speed, so that it can dive down on a fish from a distance of 10 m before the prey is aware of the danger.

The adult Pike even attacks ducks, coots and other water fowl. Because of its appetite and its mouth with its sharp teeth, curving inside like hooks, the pike is called 'the freshwater shark'.

The female lays her eggs on submerged water plants.

Did you know...

Freshwater fish *live at different levels of rivers according to the temperature of water and the amount of oxygen that they need.*

When the time comes for them to spawn, all species of ***Shad*** *swim from the sea to rivers ready to lay their eggs, like the salmon. The Shad belongs to the herring family.*

The ***Grayling*** *has a huge dorsal fin, rather like a sail. Because it needs lots of oxygen, this fish is sensitive to the features of water and also pollution.*

The ***Red Salmon*** *lives in the Pacific and returns to Asiatic and African rivers.*

The ***Schelly*** *belongs to the group of white fish which in European waters are almost extinct.*

The ***Gizzard Shad*** *is found in the rivers of Central and Northern America and in Atlantic waters. It belongs to an order of fish which is very popular as a food in the United States.*

The ***Alaska Blackfish*** *is found in North America and Siberia. It also lives in the stagnant waters of the Arctic tundra.*

ALOSA FALLAX
THWAITE SHAD
HIODON TERGISUS
MOONEYE
THYMALLUS THYMALLUS
GRAYLING
COREGONUS LAVARETUS
SCHELLY
ONCORHYNCHUS NERKA
RED SALMON
ESOX LUCIUS
PIKE
DALLIA PECTORALIS
ALASKA BLACKFISH
SALMO SALAR
ATLANTIC SALMON
DOROSOMA CEPEDIANUM
GIZZARD SHAD

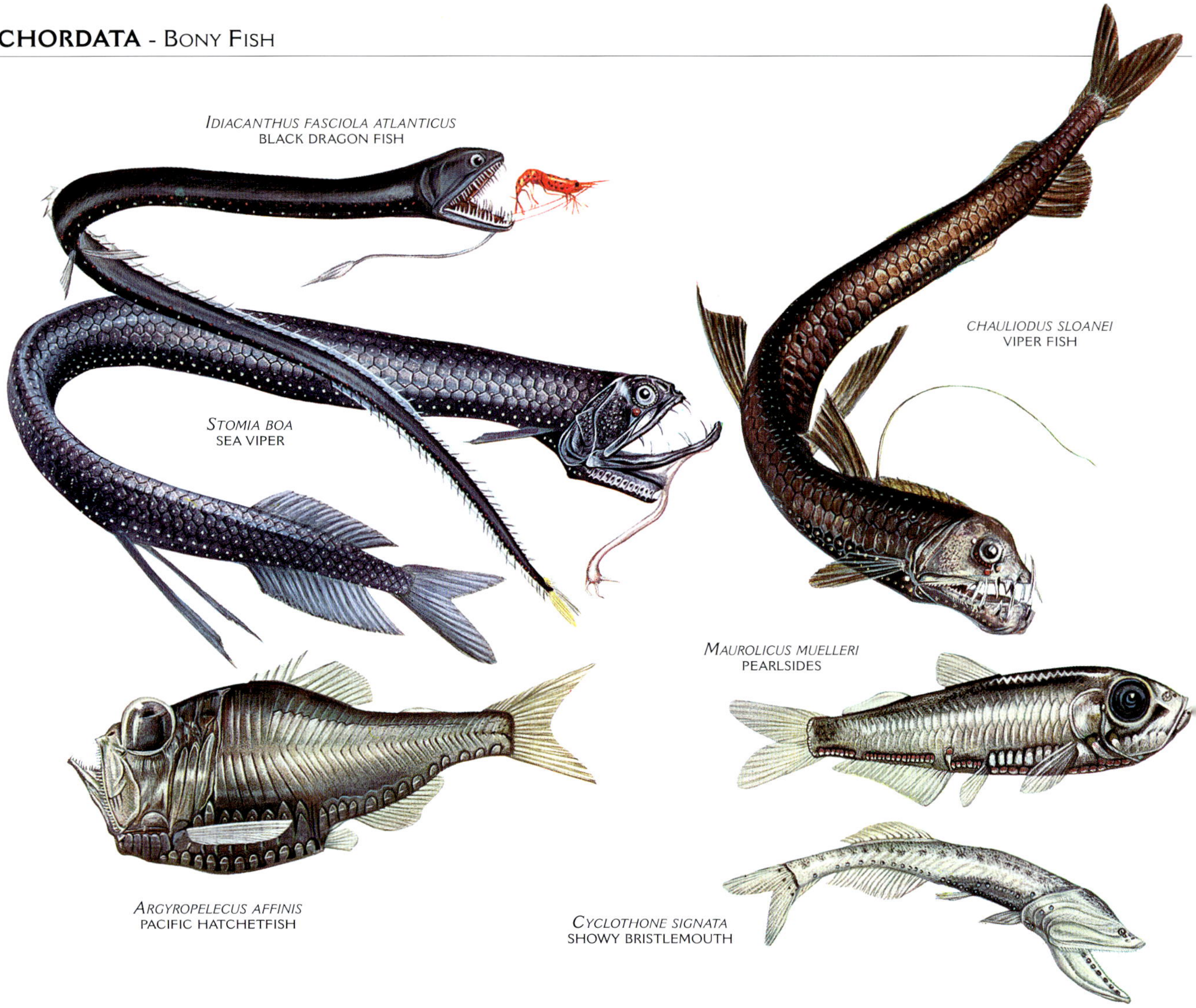

FISH OF THE OCEAN DEPTHS

The Black Dragon Fish lives at depths of between 300 and 2000 m. This fish has a monster-like appearance, with its long body, coloured spots, hexagonal scales, and dorsal and anal fins near the tail. It also has huge jaws which it can open very wide, showing its sharp teeth, some long, others short and which are typical of carnivores. Its 'squashed' looking sides and the snout have numerous luminous organs called **photophores**, one under each eye.
The jaw has a filament, or thread, which is often very long. This, too, is luminous and attracts prey like a bait. Black Dragon Fish probably live in groups.

The **Sea Viper** is grey and has many luminous organs on its underside.

The **Viper Fish** has a very mobile head and the jaw can open up wide enough to form an angle of 90 degrees.

The **Pearlsides** and the **Showy Bristlemouth** are mainly found in the Mediterranean, but also everywhere else except the Arctic. Both have a slightly bony skeleton, a silver body and **photophores**. Both are an important part in the food chain of sea creatures, because they are the food of other predators which live at the same depth. The **Pearlsides** has very large eyes and goes up near the surface of the water only at night-time.

The **Pacific Hatchet Fish** is widespread in the Mediterranean. It has large photophores which glow with a pinkish light.

The larvae of the **Black Dragon Fish** have long stalks. Their eyes are located at the end of these stalks.

SOME TROPICAL FRESHWATER FISH

Some of the largest tropical freshwater fish belong to the order of Clupeiformes.

The **Knifefish** has no dorsal fins and has just one very long fin, because the caudal, the ventral and anal fins are all joined up together.

The **Piracucu** is a terrible freshwater marauder which lives in the large rivers of Brazil and Guayana. It can measure up to 5 m long and weigh over 200 kgs. The male builds a nest where there is no vegetation, ready for the female to lay up to 11,000 eggs. The male always takes care of the young, and when it has to move, he carries them in his mouth for protection. The North American Indians catch them for meat, using the tongue for a nail-file.

The **Arawana** hatches its eggs in its mouth, in a sort of pouch between the bones of the lower jaw. It has a very flat, metallic-looking long body and beneath the lower jaw are the long whiskers, or sensory appendages, with which it identifies its prey, especially shrimps. It swims just below the surface of the water and shelters among the plants which grow on the bank.

The **Butterfly Fish**, which lives in the rivers Niger and the Congo, is also very flat. Its pectoral fins are similar to large wings, whilst the ventral fins have large thread-like rays which it uses as organs of touch.
The Butterfly Fish feeds on small fish, and, by staying quite still just below the skin of the water, it can also catch the insects bouncing down outside on the surface and making the most of those which fall into the water. The name 'Butterfly Fish' is misleading. Rather than being a true 'flying fish' it can only jump up in the air and cover a distance of about 2 m.

Xenomistus nigri
KNIFEFISH

Pantodon bucholzi
BUTTERFLY FISH

Osteoglossum bicirrhosum
ARAWANA

Arapaima gigas
PIRARUCU

EEL

Reproduction: oviparous

Length: up to 1.5 m

Habitat: sea and fresh water

The freshwater eel is a night-time predator, long-lived and a remarkable migrator. The eel looks very much like a snake. It has no ventral fins. Instead it has a dorsal fin more or less the length of its body, which is supported by about 260 vertebrae. The male is about 50 cm long; the female can be up to 1.5 m. Outside the water, the eel can crawl through damp and wet grassy land for hours, sometimes days. Its thick, slimy skin may seem bare, but it is actually embedded with tiny scales.

The most notable feature of this 'sea snake' is its capacity for covering very long distances; the spread and the migration of eels has been one of the most long-standing mysteries which has puzzled naturalists for hundreds of years. Only in the twentieth century did some aspects of its life become clearer, when in particular zones of the Atlantic Ocean between the Sargasso Sea and the European coast, groups of eel larvae were identified at different stages of development and which gradually moved eastward. From this, it was discovered that the female eel lays her eggs only in the depths of the Sargasso Sea.

The long journey of the larvae begins at a depth of 400 m. It lasts around four years, mostly at more shallow depths, carried along by the Gulf Current. By the time it nears the European coast, the larva has been transformed into an eel with a short, cylindrical body, between 6 and 8 cm long, after which time it follows the course of rivers until it reaches a calm location.

The eel does not abandon fresh water for many years, until it reaches sexual maturity. Then, from the end of the summer until the beginning of autumn it begins to swim along the course of European rivers ready to swim into the sea, across the Atlantic Ocean and reach the Sargasso Sea. After spawning, the eels die. But the larvae, as if attracted by an irresistible call, undertake their first journey towards the places where their parents came from.

ANGUILLA ANGUILLA
EEL

MURAENA HELENA
MEDITERRANEAN MORAY EEL

MORAY EEL

Reproduction: oviparous

Length: up to 1.3 m

Habitat: sea water

The Moray Eel belongs to the same family as the common eel. It looks very much like the eel, except that it is much more muscular and compressed at the sides. It has a more aggressive appearance, with pointed teeth, strong jaws and small eyes, its long snout thrusting forward. Its head is small and it has no tongue or fins.

The skin of the Moray Eel is very thick, without scales but with glands which secrete mucus. In almost all species the skin is coloured and patterned; some moray eels – for example, the Mediterranean Moray Eel – have a marble-like yellow and black pattern; others have a pattern like the coat of a leopard. It has a very distinctive snout, with nostrils at the front which are like tiny little tubes, and rear nostrils in the form of simple, round holes situated above or near to the eyes. Also, the mouth of the Moray Eel can open up wide enough to reach its eyes. This fish is very strong in any kind of battle and can survive for a long time outside water.

The Moray Eel is found in all tropical and sub-tropical seas. It lives along rocky coasts and in coral reefs where it shelters in cracks and cavities in wait for prey. Normally, it spends the daylight hours in this way, going in search of food at night.

It feeds mainly on fish, molluscs and crustaceans which it seizes and then grinds up between rows of strong teeth. For humans disturbing the Moray Eel, a bite from those teeth can cause serious injury.

The Moray Eel is eaten in some countries, but the flesh of some species can be poisonous. In 92 BC, Ancient Romans built reservoirs to breed the Mediterranean Moray Eel for this purpose, considering it to be a great delicacy.

SEA HORSE

Reproduction: oviparous

Length when born: about 1 cm

Length of adult: 15-20 cm

Habitat: sea

The Sea Horse is one of the most eccentric and delightful of sea creatures. Its shape is well-known, with its head bent down towards its body and its tube-shaped mouth, making it look rather like the 'knight' piece in a game of chess. Its way of reproduction is interesting. After a refined courtship, which consists of little more than the male parading in front of the female, the male empties its ventral pouch of water so that the female can insert her ovipostor (egg-laying organ) and lay her eggs inside. The male then fertilizes the eggs and incubates them for some weeks.

When the eggs are ready to come out, the male pouch opens and the moment the little Sea Horses touch the water, they swim towards the surface to fill their swim bladders with air. Sea Horses can move upright because of their dorsal fin, which vibrates very quickly. This also draws in plankton and small prey. When the sea horse is still, it can 'anchor' itself to algae and coral with its prehensile (able to grasp) tail.

GARFISH

Reproduction: oviparous

Length when born: 13 mm

Length of adult: 90 cm

Habitat: sea

The common Garfish with its brilliant green-blue colouring, belongs to the group of fish which are able to jump out of the water when they are being chased or simply for fun.

An elegant and fast predator, the Garfish has a thin body, with jaws in the form of a long beak, thin and flat. The beak's small, sharp teeth enables the Garfish to skewer small prey such as herring, squid and crustaceans as it swims.

The Garfish reproduces along the coast in spring-time, the female laying round eggs which she 'anchors' to algae and submerged plants with a thin adhesive thread from her body. In winter, the Garfish goes out into the open sea. It lives in banks in the Mediterranean and the Atlantic Ocean. In the Black Sea and in other seas the Garfish is widely fished.

Did you know...

*The **Stickleback** builds its nest by binding the fronds of plants together with its mucus; Sticklebacks mate for life and both take care of their young.*

*The **Blackspot Grenadier** belongs to the family of **Macrouridae** which are also called the rat-tails or rat fish.*

*The **Trout Perch**, otherwise known as the **Bull Trout**, is recognized by its upper 'lip', which is significantly larger than the lower jaw. Its scales are embedded with tiny little spines.*

*The **cod** is one of the most numerous fish in the world. Cod are excellent migrators and because they live in enormous shoals, they are widely fished.*

*Because of its rigid, flat body, the **John Dory** has had to develop a very careful plan of attack. By making barely noticeable movements only of the fins and the tail, it succeeds in nearing its prey slowly and without being seen.*

*The **Cypselurus Heterurus** belongs to the family of flying fish. A vigorous push from the tail gives it a short burst of speed outside the water and then it can jump up to a metre above the surface of the water and float for up to 10 metres with its wide pectoral fins open like the wings of a bird.*

*The **Opah**, also called the **Moon Fish**, can reach a length of 2 m and a weight of 100 kg.*

*The pipefish **Syngnathus Typhle** has a long, thin body and can stay immobile vertically, hidden among water plants.*

*The **Tetra Black Ulrrey** is an eel that can withstand periods of drought by immersing itself in mud and breathing through its skin.*

COELORHYNCHUS COELORHYNCHUS
BLACKSPOT GRENADIER
BELONE BELONE
GARFISH
TRACHIPTERUS TRACHIPTERUS
RIBBONFISH
MACRORAMPHOSUS SCOLOPAX
COMMON SNIPEFISH
ZEUS FABER
JOHN DORY
HIPPOCAMPUS GUTTULATUS
SEA HORSE
PSELURUS HETERURUS
SYMBRANCHUS MARMORATUS
TETRA BLACK ULRREY
PERCOPSIS OMISCOMAYCUS
TROUT PERCH
LAMPRIS REGIUS
OPAH
SPINACHIA SPINACHIA
15-SPINED STICKLEBACK
GADUS MORHUA
COD
SYNGNATHUS TYPHLE

SPHYRAENA BARRACUDA
BARRACUDA
AMMODYTES LANCEOLATUS
SAND LANCE
ATHERINA HEPSETUS
MEDITERRANEAN SAND SMELT
THUNNUS THYNNUS
TUNA
MUGIL CEPHALUS
MULLET
CYCLOPTERUS LUMPUS
LUMPSUCKER
XIPHIAS GLADIUS
SWORDFISH
ANARHICHAS LUPUS
WOLF FISH
PTEROIS VOLITANS
LION FISH
TRIGLA LUCERNA
TUB GURNARD
DACTYLOPTERUS VOLITANS
FLYING GURNARD

TUB GURNARD

Reproduction: oviparous

Length: 40-70 cm

Habitat: sandy sea beds of the Mediterranean

The Tub Gurnard (also known as the 'Sea Robin') has very long pectoral fins each with three lower rays, rather like thin fingers, which sink into the sand like the runners on skis. By beating its fins energetically, the Tub Gurnard stays on the sea bed with the same ease as with which it swims. At the point of these 'fingers', which are its organs of touch, the fish has taste buds by which it can find a feed of crustaceans and molluscs in the sand.

Its enormous pyramid-shaped head is not the only reason for this fish being considered rather odd. By using a group of special muscles to make the walls of its swim bladder vibrate, it can make loud, rumbling noises of various kinds – moans, grunts, rumblings. Experts believe that this is a way of communicating with other fish of its own kind.

TUNA

Length: up to 3 m

Weight: up to 400 kg

Habitat: Mediterranean Sea and Pacific Ocean

The Blue Fin Tuna with its characteristic long, scythe-shaped fins, is one of the largest fish in the sea. The body temperature of all tuna fish is 10°C higher than that of the water and this helps to increase its power – a warm muscle contracts and relaxes more quickly than a cold muscle.

No other bony fish carries out a longer migration in the oceans at such high a speed: the tuna swims at 15 - 30 km/h, but, when it sees prey it can reach a speed of 70 km/h in just a few seconds.

As with all cold-blooded animals, all this movement uses up great quantities of oxygen. But the tuna has a small mouth, and to draw in lots of water, it must swim always holding this open wide.

The favourite prey of the tuna are herring, mackerel and sardine.

Did you know...

*The **Wolf Fish** feeds on shrimps and echinoderms. It breaks the shells of crustaceans and molluscs before eating them, using its teeth which it re-grows each year.*

*There is no other fish which is more hungry or more 'nosey' than the **Barracuda**. It can grow up to 2 m long, and when it hunts, it proves that it has very sharp eyesight.*

*Unlike most other fish, the **Swordfish**, up to 6 m long, has an active circulation, and, like the tuna, has a body temperature higher than that of the water which surrounds it. So it can move very quickly and do spectacular jumps.*

*During winter, the **Mullet** stops feeding and stays almost quite still, in a state of torpor, near the sea bed. It carries out seasonal migrations in search of cooler water in summer and hotter water in autumn.*

*The **Sand Lance** is also called the sand eel – although it is not an eel at all. It inhabits the coastal waters of the Mediterranean, and, like the **Sand Eel**, it hides itself among the sand on the sea bed.*

*The **Mediterranean Sand Smelt** is sometimes called 'silverside' because of its colour. It lives in shoals in coastal waters and swims very near the surface with its snout barely submerged.*

*During the period of reproduction, the **Lumpsucker** acquires a bright red-orange colour.*

*The **Lion Fish** (in the photograph) with a length of 20 - 30 cm has a majestic ornamental skin and looks like a Chinese dragon. For this reason it is much loved by aquarium owners, even though it is poisonous; the stings of the dorsal fins are as poisonous as the teeth of a cobra.*

AMPHIBIANS

COMMON CHARACTERISTICS

All amphibians have two pairs of limbs with claws. In the larva phase, breathing is by gills. By the time the amphibian is adult, its body has adapted from living in water to living on land, and so it is has lungs for breathing. The skin is bare and with glands which produce mucus to avoid dehydration.

EVOLUTION AND HABITAT

Over 350 million years ago, some of the animals living in water began to live on land, some partially, some completely. These animals were the first amphibians, a name meaning 'of double life'. All four-legged vertebrates (tetrapods) evolved from these creatures.

The Caecilian is very different to other amphibians. Because of its long-shaped body, it is impossible to disguise its head from its tail. The skin is covered with bony scales and it has a skull with jaws – very much like prehistoric amphibians. In fact, it is generally believed that the Caecilian is a direct descendent of the first amphibians.

The Caecilian lives hidden in warm, damp earth near water. It rarely comes to the surface. After hatching from the egg, the larva becomes adult in the same way as other amphibians. When its metamorphosis is complete, the adult Caecilian comes out of the water of a pond and crawls on to the bank, where it digs a hole for shelter.

The length of the Caecilian can vary from a few centimetres to over one metre. It is a carnivore feeding on earthworms and arthropods. The largest species also eat small earth-digging snakes. It lives only in America

BEHAVIOUR

Amphibians are very intelligent. Each one lives within a precise territory, which it can recognize on its return, even after it has travelled a considerable distance. Amphibians can defend themselves against enemies, either by changing colour to blend in with their surroundings, or by poisons which they secrete in special glands in their skin, as happens with some toads.

FEEDING

As amphibians adapted to life on land, they developed a proper tongue, with its own muscles and glands. An amphibian uses its tongue to grasp, to moisten and to swallow its food. At the larva stage, it eats only plant matter. Adult amphibians also eat arthropods, molluscs, worms and also small vertebrates.

The double life of Amphibians

There are about 25,000 species of amphibians, with a great variety of shapes and sizes. Most are quite small, usually with four limbs which have claws. They are cold-blooded animals, which means that their body temperature depends on the temperature of their surroundings. Because amphibians cannot stand very low winter temperatures, they go into hibernation. In summer, if the temperature is too high, they remain inactive and immobile.

Their **bare skin** is rather thick, warm and damp, but offers little protection against dehydration. However, many amphibians do breathe through the skin. Due to the risk of dehydration, all amphibians live near lakes, rivers and fresh water or in very damp conditions. As well as the skin, which is shed and renewed at regular intervals, it has numerous **glands** in layers of skin below the surface. These glands secrete irritating and toxic substances, and also **chromatophores**. These are the pigment-bearing cells. By contracting and expanding, they determine the colouring of the body.

With most amphibians, the eggs are fertilized outside the body; the female lays the eggs in water, where the male fertilizes them. The eggs are round and non-transparent, often with gelatinous layers which absorb the water and swell up until each egg reaches a diameter five times larger than when it was laid. The larvae are born in water, breathing through gills. Later, they will undergo a **metamorphosis** and be transformed into adults. Some larvae grow rapidly and leave the water after one week. Others stay there for up to five years. In adults, which are capable of breathing on land, lungs are formed, limbs sprout, the skin thickens, the tail shrinks, muscles and the skeleton changes and the digestive tube shortens until it reaches the same measurement as the body. Some amphibians lay their eggs on land. Others give birth to perfectly-formed young, ready to breathe with their lungs.

Photographs: top, a Fire Salamander with its gleaming black background and yellow patches, which can vary so much in number and size that no two individuals are the same.
Bottom, a Tree Frog.

Amphibians are divided into three orders.

Apoda are amphibians similar to earthworms. They live in the layer of humus and rich soil which covers the ground of tropical forests. Fertilization is internal and the eggs are laid in the ground; the larvae complete their metamorphosis in water. **Caudata**, such as the salamander and the newt, are characterized by a **tail** which the animal has all its life. **Anura**, such as the frog, the toad and the tree frog have a bulging body, hind legs which are more developed than the front legs, and bulging eyes. The adult has no tail. The embryo hatches from an egg, becomes a **tadpole** and undergoes a slow metamorphosis to the adult stage. By that time, it will have a strong skeleton, a bony skull, jaws with teeth and a spinal column composed of a varying number of elements (6 - 9 in Anura, up to 300 in some Caudata).

PHYLUM	SUBPHYLUM	CLASS	ORDER	SPECIES
CHORDATA	Vertebrates	Amphibians	Apoda	Caecilian
			Caudata	salamander, newt
			Anura	frog, toad

Did you know...

*The **Giant Salamander** is a land-dwelling species distinguished by its large head and small eyes. Despite its 30 cm length and its slightly comical appearance, it can climb trees easily. It feeds on invertebrates, small snakes, amphibians and small mammals.*

*The **Hellbender**, sometimes called the **Alligator Salamander** can grow up to 70 cm long. It only undergoes partial metamorphosis from larva to adult, and so the adult has withered gills. It lives in the rivers of the eastern United States. In summer, the female lays a gelatinous chain of hundreds of eggs, ready to be fertilized by the male.*

*Commonly found in North America is the **Tiger Salamander**. It has a strong body with a length which can vary from 19 to 33 cm. This salamander can withstand periods of drought by seeking refuge in the lairs of other animals. At the larvae stage, it is frequently known as an **axolotl**.*

*The **Red Salamander** breathes*

CALIFORNIA SLENDER SALAMANDER

Habitat: underground

Respiration: through the skin

Length: up to 10 cm

The California Slender Salamander has a very tapering body and tiny limbs. Along with another 180 species, it belongs to a family of Caudata living mainly in North and Central America and is characterized by having no lungs. Instead, this animal breathes through its skin and by means of the mucus which covers the mouth cavity.

The California Slender Salamander has adapted to life on the land, where it spends the major part of the year. When it leaves its lair, it crawls along in the same way as a snake. If the ground is wet, it will dig a proper lair for itself; but if the ground is too compact, then it will find a hole which another animal has left empty.

It does not dig very deep and it can often be found beneath the trunk of a tree or under large stones, rolled up in a ball. It is very important that the surroundings are damp.

It feeds on small insects, their larvae and those of other invertebrates which it finds in the ground. It lays its eggs in water in the autumn.

Often a few females lay their eggs (up to 21) in the same place. Then, for most of the time, one or more females stay near the place where the eggs are. The tadpoles are born the following spring,

GREATER SIREN

Habitat: water

Breathing: gills and lungs

Number of eggs: up to 300 laid in bunches

Length of larvae: 5-10 mm

Length of adult: up to 1 m

The Greater Siren is one of only three species belonging to the Sirenidae family. From the time that metamorphosis is complete, it has an appearance which is halfway between a larva and an adult form. The eyes are not covered by eyelids, and although it has lungs, it also has the remaining tufts of external gills and only the skin is different.

The Greater Siren has a long body, but it has no back legs and the front legs are very short and weak, with four thin claws. It lives in water and rises to the surface to breathe, because it mostly uses its lungs. The Greater Siren lives in fairly shallow streams, stagnant waters and swamps, hiding itself by day and coming out at night to get food. It has no teeth, but its jaws have horny plates which it uses to catch worms, water snails, crustaceans and plants. In times of drought, it hides itself in mud and survives within a sort of cocoon for about two months. During this time, its skin dries up and the gills become numb. But when the rain returns, it regains its normal size and the body becomes moist once more.

through its skin and it is very susceptible to dehydration. It can die within a few minutes if exposed to the Sun.

*The **Texas Blind Salamander** lives mostly in water below the surface of the ground. It is in danger of extinction because so many have been hunted down and captured.*

*The **Mudpuppy** is a North American salamander which keeps its larva form when it is adult, with colourful tufts at the sides of its neck, corresponding to external gills. It is nocturnal and lives in the waters of ponds, lakes and rivers.*

*The tiny **Common Red Spotted Newt** lives in the mountainous zones of United States. It looks rather like a primitive amphibian.*

*The **Siberian Salamander** lives within the Arctic Circle. Before hibernating in icy mud, the water in its body is reduced. This is so that its tissues become less dense to avoid the danger of the animal becoming stiff.*

CAVE SALAMANDER

Habitat: caves

Breathing: gills and lungs

Reproduction: internal

Length of larvae at birth: 2 cm

Length of adult: 25-30 cm

The Cave Salamander belongs to the order of Caudata. It is one of the strangest vertebrate animals in Europe. It lives all the time in caves and never goes out in daylight. Its body has no pigment at all and so the blood inside makes it pale pink in colour – only the external gills are a vivid red. The Cave Salamander is a neotenic – becoming adult without undergoing a complete metamorphosis – and so it does not lose its gills before getting lungs.

The eyes are fixed and hidden under the skin. It is not very active, mostly staying hidden underneath stones or in the cold waters inside grottos. But every so often it gets the urge to swim up to the surface, to take a mouthful of air. On these occasions, its skin can dehydrate, but this soon becomes damp and rich in mucus once again, the moment the Cave Salamander gets back into the water.

It feeds mainly on small crustaceans and cave-dwelling creatures and tiny fish. It is very sensitive to the vibrations of water and these help it to find its prey more easily. The female, after having laid her eggs, strongly defends them against any intruders.

FIRE SALAMANDER

Habitat: damp, dark places

Breathing: lungs

Reproduction: ovoviviparous

Length of adult: 12-28 cm

The Fire Salamander is the best-known salamander in Europe, especially for its bright colouring, with yellow spots on a black background. Its skin is impregnated with a highly poisonous substance produced by glands directly behind its eyes and nothing can attack it without feeling the effects of the irritant milky liquid which it produces in times of danger.

Slow and lazy, this salamander looks for dark, damp places where it can live in peace. It prefers to hide among layers of moss or rotten wood, underneath stones or in caves. It is very sensitive to heat and dies as soon as the temperature goes above 20°C. In the autumn hundreds of salamanders arrive to settle in particular places to lie underground and spend the winter in hibernation.

Mating takes place on land. The female goes to the water only to give birth to the larvae, which develop inside her body.

Did you know...

*The **Crested Newt** gets its name because of the crest which the male gets on his back during mating time. It is found in western Europe and central Asia.*

*The **Smooth Newt** lives in almost every part of Europe, especially on the plains. It likes the water of puddles, lakes and reservoirs.*

*The bright colours of the **Alpine Newt** are only present at mating-time.*

*During the courting stage, the male of the **Marbled Newt** shows the female its high dorsal crest and vibrates his tail, with which he can hit his own head.*

*The **Corsican Brook Salamander** is very rare. It is completely nocturnal, never coming out during the day.*

*The **Italian Cave Salamander** is a small meat-eating salamander. It is not poisonous, but defends itself by changing colour.*

TRITURUS CRISTATUS
CRESTED NEWT
TRITURUS VULGARIS
SMOOTH NEWT
PROTEUS ANGUINUS
CAVE SALAMANDER
TRITURUS ALPESTRIS
ALPINE NEWT
TRITURUS MARMORATUS
MARBLED NEWT
EUPROCTUS MONTANUS
CORSICAN BROOK
SALAMANDER
HYDROMANTES
ITALICUS
ITALIAN CAVE
SALAMANDER
SALAMANDRINA TERDIGITATA
SPECTACLED SALAMANDER
SALAMANDRA SALAMANDRA
FIRE SALAMANDER

TREE FROG

Habitat: tree-dwelling

Breathing: lungs and through skin

Length of tadpole: about 4 cm

Length of adult: 7-12 cm

The Tree Frog is one of the most popular amphibians in Europe, characterized by its brilliant, shiny green skin and friendly appearance.

With the claws on its legs working like suckers, the Tree Frog can climb trees and bushes easily, staying attached to the leaves, where it finds insects to eat. It skin secretes a mucus and this gives good protection against the danger of dehydration.

It lives near swamps and marshes only until the age of two years. Afterwards it returns to these places only at mating-time, from April to June. Then, after sunset, it is easy to hear the high cry coming from the courting males. The females ready for reproduction are attracted by the croaking and they lay their eggs in late evening. These are attached in small masses to the plants near water and after they become fertilized by the male, the eggs fall into the pond, ready for larvae to develop.

COMMON TOAD

Habitat: land

Breathing: lungs and through the skin

Length of adult: 7-18 cm

The Common Toad has a squat, solid body, large, bulging eyes and a dry skin covered with warts.

It is a very useful animal, because it eats snails, insects and other agricultural pests. Its skin is non-irritant, but in times of danger it can secrete poison from two glands, one at each side of its head.

The toad is land-dwelling and nocturnal. In March, after its winter hibernation, it leaves the woods where it has slept and returns to the ponds to reproduce.

When the male meets a female, it grabs hold of her and takes her to the nearest stretch of water. After some days, the female lays a long chain of eggs which the male then fertilizes in the course of five to ten hours. After reproduction, the male and female separate and return to the woods, fields and gardens.

In autumn, the Common Toad goes back to the place where it spends its winter hibernation.

Did you know...

*The male **Common Green Frog** gives out a loud series of sounds to attract a female. Vocal sacs emerge from a split at either side of the mouth and these swell in proportion to the pitch and the duration of the song.*

*The **Green Toad** has customs which are more aquatic than the Common Toad, and it is more agile. In mountain areas they can reach altitudes of about 1500 m.*

*The **Yellow Bellied Toad** is found throughout Europe. It has a bright yellow speckled abdomen which it shows the moment it feels threatened, turning over and arching its back to scare away an enemy and sending out a shower of poisonous foam at the same time.*

*The **Midwife Toad** can be found all over central and southern Europe. It is the male which takes particular care of the eggs. After having fertilized the chain of eggs, it then attaches them around its hind legs, carrying them around with him until they hatch.*

*The **Common Spadefoot Toad** lives in Europe and western Asia. It has adapted to life on dry, sandy ground. It comes out mainly at night and at dawn buries itself, digging a shelter with the horny growth on its hind claws. At evening, it comes to the surface to catch insects and worms.*

*The **Common Frog** is the brown frog most widespread throughout Europe. It is found at altitudes up to 2000 m. It lives in woodland during Spring and Summer, returning to the water only to reproduce or to spend the winter.*

RANA ESCULENTA
COMMON GREEN FROG
HYLA ARBOREA
EUROPEAN TREE FROG
BUFO VIRIDIS
GREEN TOAD
BUFO BUFO
COMMON EUROPEAN TOAD
PELOBATES FUSCUS
COMMON SPADEFOOT TOAD
ALYTES OBSTETRICANS
MIDWIFE TOAD
BOMBINA VARIEGATA
YELLOW BELLIED TOAD
RANA TEMPORARIA
COMMON FROG

REPTILES

COMMON CHARACTERISTICS

Reptiles are cold-blooded vertebrates, with a skin covered with scales or horny plates, or – as in the case of turtles, tortoises and crocodiles – a bony shell which protects the animal against dehydration. Reptiles do not go through a larva stage, because the egg is enclosed in a shell which is full of liquid containing all that the embryo needs to develop.

EVOLUTION, DISTRIBUTION AND HABITAT

The first reptiles appeared on Earth over 260 million years ago. These were the first vertebrates to live completely free of the water. Reptiles are found all over the world, except for the Antarctic, in places where the temperature is either hot or temperate. Some live in water, others on land.

The Tuatara is a small reptile, about 60 cm long. It belongs to the order of Rhynchocephalia, a group which began 200 million years ago and which were believed to have been extinct. Then at the beginning of the 18th century, some Tuatara were discovered on islands of New Zealand. The remote geographical position of these islands made access difficult and this protected the Tuatara and their home and helped them to survive. It is a species which is still at the risk of extinction and so it is strongly protected.

The Tuatara has a squat body an a clumsy walk, a strong head and a jagged cre on its back. It likes lower temperatures and calr surroundings, feeding on worms, insects, plants an sometimes the egg of the Puffin, a bird with which sometimes shares its lair. The Tuatara takes a lon time to digest its food and its breathing is very slov It can stop breathing for up to an hour. It is ver long-lived, up to a hundred year

BEHAVIOUR

Reptiles have very keen eyesight, which they use to pick out enemies, prey and companions. They can detect smells through the tongue as well as the nose. Some species have skins with bright colours and patterns. The ability to change colour is not only a means of blending in with its background, but also depends on their activity or their mood, such as aggression or fear.

NUTRITION

Most reptiles are carnivores, but there are some herbivore species. The activity of their well-developed saliva glands not only aids ingestion of food but also digestion. In many reptiles, the digestive system is very roomy and 'stretchable', so that many are able to ingest prey considerably larger than them – for example, some snakes can devour a whole deer.

Relatives of dinosaurs

Reptiles were the first animals which were completely land-dwelling. The body is covered with **scales** or horny shields which reduce the evaporation of the skin; this is almost always without glands, so that keeping a constant body temperature depends entirely on the skin taking in heat and not losing it.

Once the scales and the horny plates are completely formed, these do not grow any more, because they are made of dead cells. This means that the skin must be renewed at regular intervals (moulting) especially in the young and during the stages of development.

Breathing is by lungs.

Except for snakes, reptiles have two pairs of limbs, each limb having five claws with talons. The jaw is joined to the skull by a bone which enables the mouth to open very wide. The mouth has strong teeth which the reptile uses to grab and to hold its food .

The skull and the skeleton form a rigid bony structure, which gives the reptile the power to attack as well as muscular development.

Reproduction is almost always **oviparous**. The presence of a solid shell indicates internal fertilization, where the egg becomes fertilized inside the female's body and the albumen and shell are formed before it is laid. Some species of reptile are **viviparous** where the fertilized egg develops inside the female body. At birth, the membrane of the egg breaks and the young come out already perfectly formed.

Photographs: top right, a sea turtle, above, the ***Furcifer pardalis***, a chameleon from Madagascar. Facing page: an Iguana.

Unlike their dinosaur ancestors, reptiles today are of modest size, except for crocodiles which can grow up to 7 m in length. Present-day species are less numerous than dinosaur species and are divided into four main orders – **Chelonia**, the order to which the tortoise belongs, with its short, squat body, protected with a hard shell; **Rhynchocephalia** an order which today comprises just one special species, the tuatara; **Squamata** characterized by a body protected by horny-type scales; lizards and snakes also belong to this order; and **Crocodilia**, to which crocodiles belong.

PHYLUM	SUB-PHYLUM	CLASS	ORDER	SUB-ORDER	FAMILY	SPECIESS
CHORDATA	Vertebrates	Reptiles	Chelonia			tortoise
	Rhynchocephalia				Tuatara	
				Sauria		lizard
			Squamata	Amphisbaeria		Spiny-Tailed Lizard
				Serpentes		snakes
					Gavialidae	Gavial
			Crocodilia		Alligatoridae	crocodile alligator Caiman

Chelodina longicollis
Long necked turtle
Gopherus polyphemus
Gopher tortoise
Dermochelys coriacea
Leatherback turtle
Hydromedusa tectifera
South American terrapin
Pelusios niger
West African black turtle
Geochelone pardalis
Leopard turtle
Terrapene carolina
Common box turtle
Platysternon megacephalum
Big headed turtle
Tryonix spiniferus
Eastern spiny soft shell turtle

LEATHERBACK TURTLE

Habitat: the sea

Reproduction: oviparous

Diameter of eggs: 5-6 cm

Length of adult: over 2 m

Weight: up to 600 kg

The Leatherback Turtle is the largest turtle in the world. Its plating consists of hundreds of irregular, bony plates, joined together like a mosaic pattern and covered with a smooth, strong, skin.

Its huge, rounded head, with its hooked jaw cannot withdraw back inside its shell, neither can the limbs, which are like fins.

The Leatherback Turtle is the rarest of sea turtles. It lives by itself in the warmest of seas in Malaysia, Indonesia and Central America – but during the breeding season, the female swims to the land where she digs a nest in the sand, ready to lay her eggs. The Leatherback Turtle is a tireless swimmer, feeding on fish, molluscs, echinoderms and jellyfish, even the poisonous ones. The inside of the mouth and the throat is covered with horny spines which help with the ingestion of slippery prey, like the jellyfish.

COMMON BOX TURTLE

Habitat: land

Reproduction: oviparous

Length of shell at birth: 26-30 mm

Length of shell when adult: 12-16 cm

The Common Box Turtle lives in Mexico and North America. It has a shell with a type of hinge, so that when the turtle withdraws its claws and legs, the front section of the lower shell lifts up. Then the opening of the 'box' closes up, protecting the turtle from its many carnivore enemies.

The Common Box Turtle is an omnivore and so feeds on fruit, berries and other vegetable matter, as well as insects and various tiny animals.

It lives in central areas for most of the year, but in autumn it travels further north, becoming slower and slower and then digging a deep shelter. It likes to stay in one place, and generally spends the whole winter in the same lair. When it wakes, it reproduces; the female can lay fertile eggs up to four years after her last mating.

Did you know...

*The **Gopher Tortoise** has a shell about 25 cm long and lives in the central strip of the United States and Mexico. The heat of the deserts and the steppes prevent it from being active during the day, and so it digs an underground tunnel, where it spends the hottest hours. It comes out only in the evening to go in search of grassy plants to eat.*

*The **Long Necked Turtle** is an Australian species which has highly developed back vertebrae, to which is attached powerful muscles which enable the turtle to bend its neck, like a snake.*

*Another turtle with a long neck is the **South American Terrapin** which lives in lakes in Argentina. It is mainly active in the evening and the night-time, feeding mainly on water insects.*

*The **West African Black Turtle** can be found in most parts of Africa. It has a shell about 35 cm long. This species has an extra cover of bony plates which are not visible on the outside.*

*The **Leopard Tortoise** is a tortoise which lives in the savannah of the African high plains. To go in search of food it will even cross desert zones.*

*The **Eastern Spiny Soft Shell Turtle** from North America is both agile and aggressive. It has a long neck which it pushes out when it is about to bite.*

*The **Big Headed Turtle** has an over-sized head which it cannot hide inside its shell. This is, however, covered in hard, bony plates which protect it very well. It lives in the flowing waters of mountains in tropical Asia.*

COMMON GECKO

Habitat: land, trees

Reproduction: oviparous

Length of adult: 35-40 cm

The Common Gecko belongs to the family Geckonidae. These reptiles have highly developed eyes, covered with eyelids joined together to form a transparent horny covering. In the Gecko, the eyes have a vertical pupil adapted to night-time vision.
Its legs have five broad claws with adhesive pads and tiny hooks which are able to grasp surfaces. When it climbs, the Gecko moves in the same way as a cat, alternately stretching out and sinking down, then contracting and lifting up each claw. The Common Gecko lives in South-East Asia as far as western Indonesia. It can rapidly darken or lighten the colour of its skin. It lives in large trees and also in houses, where it is always welcome, because it is such an excellent predator of insects and even mice. Perhaps this is why it is considered to be an animal which brings good fortune. Only the male emits the cries which ring out to defend its territory, especially in spring during the mating season.

IGUANA

Habitat: land, trees

Reproduction: oviparous

Lenght at birth: 20-25 cm

Length of adult: 1.5-2 m

The Iguana is the most important member of the family Iguanidae. It has a squat body and a tail which takes up two thirds of its total body length. It has four strong legs with long claws and strong talons. Both sexes have a large sac under the throat. The head has a series of round scales and plates beneath the eardrum and on the back is a crest which is up to 8 cm high in the male. In young Iguanas, this is a deep green colour, in adults it is crossed with wide crossway stripes.
The Iguana is commonly found in Central and Southern America. It lives along the banks of rivers and among the leafy branches of trees. When it senses danger, the Iguana can jump to a height of 5 or 6 m, before diving into water. Its feeding habits vary with age – the young go in search of insects, whilst the adults prefer eating plants. They eat heartily during the rainy season, when vegetation is most plentiful, accumulating reserves of fat at the corners of the jaws which enable an Iguana to survive during periods of drought and shortage of food.

Díd you know...

*The Australian **Jumping Snake** has a body up to 70 cm long, which is covered in scales. It has a highly-developed tail but no front claws – and its back claws are no more than two tiny, little appendages.*

*The **Agama** (in the photograph) lives in Africa. Like the chameleon, it can easily change colour if it is disturbed or frightened.*

***Fischer's Chameleon** is also found in Africa. The male uses the horn on its head as a weapon. When it arrives in new surroundings, it settles down by licking the roots of trees.*
As well as being able to change colour,

chameleons have a very long and very flexible tongue, which is always moist and sticky. This is an effective weapon when it wants to catch prey.

Common in the Indonesian islands of Moluccas and the Philippines, the **Asian Water Dragon** looks like a dragon, but it is only 1 m long. It spends much of its life in flowing water, where it swims with great ability, using its flat tail.

The **Yellow Headed Gecko** is from Central America. It has a yellow head and a black body. It is active by day, sometimes even jumping into people's homes.

The **Island Night Lizard** is found only on three islands off the coast of California. It feeds on flowers and on fruit.

Mellers Chameleon is the largest of all those found in Africa. Including its tail, it can reach a length of around 32 cm. Like all other chameleons, it can change colour by increasing or decreasing the concentration of pigment in certain body cells which determine its colouring. It changes colour according to light, temperature, as well as its mood – not because of its surroundings.

ELEPHANT TRUNK SNAKE

Habitat: water

Reproduction: viviparous

Length of adult: 1.10-1.85 cm

The Elephant Trunk Snake is very unusual and easy to identify because of its large body and square-shaped head.

It lives its whole life in the salt water at the mouths of rivers, also going into the sea for short spells. It is a lazy, apathetic creature, which moves extremely slowly and can also remain for a long time on the river bed. When it needs oxygen, it only has to raise its nostrils on the upper part of its nose. The few times it goes on land, it crawls along the river banks with clumsy movements. The body seems to advance independently of its skin, which seems to be too roomy and too pleated – hence the name, the Elephant Trunk Snake.

It eats only fish of various sizes which it catches by remaining perfectly still, trapping them in much the same way as a crocodile. It is always wise to be wary of its poisonous bites, which can cause death, even to humans. Even when it seems to be asleep, it can bite with extreme speed. The female, which gives birth to perfectly-formed young, is more powerful than the male and can reach lengths of over 2 m.

Did you know...

*The **Boa Constrictor** lives in Mexico and throughout the whole of South America. It can reach a length of 4.5 m and lives in forests near to swamps. It feeds on small mammals and birds but shows no aggression when confronted by humans.*

*The **Madagasca Tree Boa** (in photograph) is similar to the Boa Constrictor.*

*The **Emerald Tree Boa** lives solely in trees in the north of South America and in Brazil. It can grow up to 3 m long and gets its name from its beautiful green colour, criss-crossed with white. Its colouring enables this snake to hide itself perfectly in trees, where it rolls itself up, remaining quite still for days, sometimes for weeks.*

*The **Sunbeam Snake** is common in southern India and Indo-China. Its body is about 1 m long, covered with brown scales. When these catch the light, they give off the most beautiful shimmering reflections, which is why it is also called 'the rainbow snake'. When excited, it will agitate its tail. It eats small snakes, rodents and birds.*

*The American **Desert Blind Snake** and the **Worm Snake** which comes from Greece and South-West Asia, are medium-sized snakes. They both like dry places, living hidden in the sub-soil or sheltered by rocks underneath stones. They mostly eat insects, ants and termites.*

*The **Pipe Snake** is also medium-sized, about 75 - 85 cm in length and scattered throughout South America. It has a vivid red colouring with black rings around the body which makes it look similar to the poisonous **Coral Snake** which has yellow rings either side of black. The Pipe Snake feeds on lizards and smaller snakes.*

*The **Anaconda** (photograph below) with its 9 m maximum length, is an almost mythical snake. It has amphibian habits, because it spends most of its life in the waters of the basin of the River Amazon and the Orinoco River. It is an excellent swimmer and can remain under water for a long time. Its extremely stretchable muscles enable it to swallow even large mammals. The female gives birth to 30 - 80 young which, when they are born, are about 70 cm long.*

NDIAN PYTHON

Habitat: tropical forests

Reproduction: oviparous

Length of adult: up to 8 m

Weight of adult: up to 110 kg

The Indian Python is one of the longest snakes in the world.
t lives in India, preferring the areas with thick vegetation, such as the ungle. It particularly likes the banks of running water, in which it can ometimes stay immersed for a long time. Although it stays mainly at ground level, the Indian Python can also climb and move among the branches of trees easily.
t has a varied diet which consists of monkeys and different rodents, although it will also attack small deer and birds. Like other members of the python family to which it belongs, the Indian Python has exceptional strength. Its body muscles are o strong that it can wrap its spirals around its prey, strangling it to death. The Indian Python is a nocturnal snake and pends most of its time quite still. It rouses itself and goes into action only when it is hungry. In captivity, it can fast for ome months without food and then suddenly decide that it is time to eat. After laying her eggs, the female wraps a spiral of her body around them and does not abandon them until the time that they hatch. When the young are born, they are bout 50 cm long.
The Indian Python is a species which is at risk of extinction.

Did you know...

The favourite home of the ***African Boomslang*** *are the branches of trees in tropical Africa.*

The ***Mangrove Snake*** *lives in Malaysia and the neighbouring islands. Its diet is comprised of birds, lizards and smaller snakes.*

The ***Sidewinder*** *is a water snake from Thailand. It likes salt water with a lot of vegetation.*

The ***Corn Snake*** *is a North American species with a most beautiful red-brown pattern.*

The ***Ribbon Snake*** *is a thin, medium-size snake, most common in North America.*

The ***African Egg-Eating Snake*** *has a diameter equal to the finger of a grown man, yet it can swallow a hen's egg whole.*

The ***Horseshoe Whip Snake*** *is very aggressive. At 2 m long, it is the largest snake in Europe.*

In Europe we also find the nocturna ***Cat Snake****. This snake is no dangerous to humans.*

The ***King Snake*** *is most widespreac in the forests of the southern Unitec States and in Mexico, and is often brec in a terrarium.*

Like all rattlesnakes, the ***Diamonc Back Rattlesnake*** *really does have c tail that makes a rattling sound. It lives in the USA and South America. The female can give birth to up to 6C young.*

*The **Red-Headed Krait** is a nocturnal snake. It is poisonous and lives in the forests of South-East Asia.*

*The **Sidewinder** can grow up to 60-70 cm long. This is also a nocturnal snake, found in the deserts of the USA.*

*The **Common Sea Snake** lives along the coasts of South-East Asia. Its poison is twice or three times more toxic than land-dwelling poisonous snakes.*

*The **Common Death Adder** is an Australian snake, so-called because of its highly toxic poison.*

*The **Spectacled Cobra** lives in Asia. It gets its name from the pattern which appears on its back when it is excited. From a half-open mouth it can spit its highly toxic poison up to a distance of 1 m.*

*The **Gabon Viper** lives in the rainforests of tropical Africa. The damage it can do with its teeth, up to 38 mm long, adds to the danger of its poison, which is one of the most dangerous of all vipers.*

*The **Pit Viper** is widespread in the tropical forests of Central America. The female gives birth to up to 70 young at a time.*

*The **Coral Snake** of North America is a brightly-coloured, beautiful looking snake. It is not aggressive, but its bite is very dangerous, because its poison attacks the nervous system of its victim.*

MELANOSUCHUS NIGER
CAIMAN CROCODILE

ALLIGATOR MISSIPPIENSIS
AMERICAN ALLIGATOR

PALEOSUCHUS PALPEBROSUS
DWARF CAIMAN

CAIMAN CROCODYLUS
SPECTACLED CAIMAN

CAIMAN CROCODILE

Reproduction: oviparous

Brood: 35-50 eggs

Length: up to about 5 m

The Caiman Crocodile is the most common crocodile in South America, widespread throughout the Amazon Basin. It resembles its close relative, the alligator, but, unlike the alligator, the skin on its abdomen is reinforced with bony plates. Its favourite habitat is within the calm areas of running water. Often, however, it is also found in the swampy stretches and the forests. These remain flooded for some months in the year and are known by the Brazilians as the 'igapos'.

Very aggressive and hungrier than any other crocodile, the Caiman Crocodile will attack and eat many types of animal. It is a reptile which does not always have a particular liking for certain food, eating anything and everything, and nothing is too big or too small – fish, small mammals, grasshoppers and locusts.

The Caiman Crocodile migrates from the Amazon Basin in the summer to spend the hottest months in lagoons. It will also drink from lagoons at other times, during spells of dry weather.

AMERICAN ALLIGATOR

Reproduction: oviparous

Brood: 20-80 eggs

Length of adult: up to 6 m

More than any other species of Crocodilia, the alligator has been hunted and persecuted for years, because its skin is so much valued in the leather industry. Anti-hunting laws have now been passed to protect crocodiles and the Caiman as well as the alligator.

The geographical distribution of the alligator is at opposite ends of the globe – one species is found in China, another in the United States. They live in rivers and swamps and spend most of their time in water, where they move with surprising rapidity. The alligator has a type of 'radar' on its cheeks. Here is a body zone sensitive enough to sense the least ripple of water and which signals to the alligator the presence of prey.

In winter, the American Alligator hibernates, burying itself in mud.

Did you know...

*The **Dwarf Caiman** is one of the smallest crocodiles living, reaching a length of just 145 cm. The shape of its head is very convex (curved), with noticeable bony plates above its eyes and its nose turned up instead of being flat. Despite its fierce appearance, it is not very dangerous. It lives in small groups in the flowing waters of the Amazon Basin, preferring moving water rich with the oxygen of rapids and waterfalls. It eats fish. During the period of drought, it finds a deep well and stays down there, lying for two or three days without eating.*

*The expression 'crocodile tears' comes from the fact that the **alligator** (in photograph) secretes tears with bubbles of air to protect its highly sensitive eyes.*

*A close relative of the American Alligator is the **Chinese Alligator** (not illustrated). There are only a few examples of these, in the waters of the Yang-tze Kiang river. It differs from the American Crocodile because it has a shorter nose and its front claws are not webbed.*

*The **Spectacled Caiman** is the most common caiman in South-Central America. It has quite a short head, its eyes are large with green irises and the bony crest which connects the two sockets gives the Caiman the appearance of wearing spectacles, which is how it gets its name. It lives in slow-flowing water with a muddy bed in the Amazon rainforests. It shares the same water surroundings with the Anaconda, which means that young caiman are often caught by this large snake.*

BIRDS

COMMON CHARACTERISTICS

The body is covered with feathers. The mouth is in the form of a beak. Body temperature is constant, as in mammals. The body is built for flight – birds have lungs and very strong pectoral (chest) muscles, a light skeleton with hollow bones. The wings have longer feathers than on the body.

EVOLUTION, DISTRIBUTION AND HABITAT

150 million years ago a group of reptiles similar to the offspring of dinosaurs began to adapt to different surroundings and climates. The front limbs became transformed into wings, so that a bird could maintain flight. After that, birds became creatures of the air – although there are some species which cannot or do not fly. Now, there are birds in all parts of the world, including cities.

Unlike most other birds, the ostrich has a very long neck and long legs covered with tiny feathers. Its legs are very strong, not only to support the weight of the body, but also to sustain the high speed which the bird reaches, with strides up to 4 metres long.

Ostriches mate for life. The female lays her one egg in a depression in the ground, together with the eggs of other female ostriches. Both male and female sit on the egg, but it is the male which takes the most care of the young. An ostrich can live at different temperatures in climate. In winter its body has more feathers which are also thicker. The ostrich is famous for eating strange things. Although it feeds mainly on vegetable matter, it will swallow other objects especially if these are shining.

This male Ostrich can reach a height of 2 metres, the female is smaller. It lives in groups, both small and large, mixing with zebras, antelopes and gazelles.

BEHAVIOUR

Birds react to their surroundings by singing, as well as singing at special times, such as the mating season. Many birds parade and some 'dance' to attract a mate. Birds build nests of different shapes and sizes. Some care for their young, a few leave the rearing to other birds. Only some birds mate for life. Some birds stay in one place, others migrate, using the Sun and stars to find their way.

FEEDING

The type of beak a bird has depends on the food it has. Its diet can include vegetable matter, insects and grubs or food which other creatures, even mammals, cannot get or eat. Many are omnivore and without any particular preference in food. Others have developed a need for a particular diet.

Made to fly

On the Earth there exists at least 8580 species of birds living in all habitats, from the icy wastes of ne Antarctic to the torrid climates of the deserts. ome live their lives almost completely at sea. The urvival of birds has been mainly due to different oods becoming available and their ability to find and at them. As well as food being easily found, there is lso the dispersal of seeds of different species of plant, s well as some birds being able to eat the remains of lead plants and animals.

The bird's light skeleton has **hollow bones**, a very lexible neck and a wide, smooth breast-bone to upport the powerful muscles which are necessary to eat the wings in flight. The wing **feathers** make the ight of a bird the best in the animal kingdom. Each eather is flexible and light, yet very strong. It has a entral shaft and from this a series of parallel barbs ranch off, each barb slanted down towards the next nd with microscopic filaments called **barbules**, losely fitting together to form a surface smooth nough to overcome the resistance of the air. A bird's eathers, and especially he layer of fine down next to the body, insulate he body. In an adult bird, he top feathers are enewed each year, in the process of moulting during ate summer. To slow down the fall of these eathers and also to keep ts body waterproof, a bird vill then spread itself with a natural grease which it secretes in special glands at the base of the neck.

With birds perching on branches and slowly falling asleep, an automatic reflex of the talons on the claws stops them from falling.

The shape and the strength of the **beak** depends on the diet; strong and hooked in birds of prey, short and cone-shaped in seed-eaters, slender beaks in birds which feed on insects, long and sensitive in those which search for food in mud and slime.

During courtship, many birds dance and carry out particular ritual, showing off the colours of their feathers and how well they can fly and sing. Most birds are particularly dedicated to the care of their eggs, the nest and their young.

Photographs: left, a Buzzard, a bird of the Falciforme order. Facing page: a swan followed by her cygnets.

PHYLUM	SUB-PHYLUM	CLASS	MAIN ORDERS	SPECIES
CHORDATA	Vertebrates	Birds	Sphenisciformes	penguin
			Procellariiformes	albatross
			Ciconiformes	heron, stork
			Anseriformes	goose, duck
			Falconiformes	falcon
			Galliformes	Pheasant
			Gruiformes	Crane
			Charadriformes	Tern
			Columbiformes	pigeon
			Psittaciformes	parrot
			Cuculiformes	Cuckoo
			Strigiformes	owl
			Caprimulgiformes	Nightjar
			Apodiformes	hummingbird
			Trogoniformes	Trogon
			Coliiformes	Mousebird
			Coraciiformes	Kingfisher
			Piciformes	Woodpecker
			Passeriformes	sparrow, swallow

Did you know...

The body of a ***penguin*** *is entirely covered with feathers so smooth as to be like a skin. The wings have been transformed into strong flippers – a penguin is a fine swimmer but can no longer fly. The funny walk of the penguin is because the turned-out position of the feet means that the bird cannot keep upright. The short, rigid tail is used as a support.*

Penguins belonging to the order of ***Sphenisciformes****. Other species are the Galapagos Penguin, the Magellan Penguin and the Humbolt Penguin which lives in the freezing waters along the cost of Peru Numbers of this species have faller dramatically due to hunting Deposits of their guano (excrement is used as fuel.*

Although different species o penguin live near the Antarctic some can be found further north nesting on a few sub-Antarcti

EMPEROR PENGUIN

Number of eggs: only one, weighing 450 g

Height: 1.35 m

Weight of adult: 45 kg

Home: the Antarctic

The Emperor Penguin is the largest of all living penguins. The only penguin larger than this was a fossilized species discovered in Patagonia at a height of around 1.75 m. All penguins have a body structure which is perfectly adapted to movements under water.

The Emperor Penguin does not swim in the true sense, but 'flies' beneath the water. Athough water is much more dense than air, the bird can reach a remarkable speed by batting its wings like flippers, using its legs as rudders.

To eat, the Emperor Penguin goes hunting for fish and molluscs. It is also an expert diver, reaching depths of up to 250 m as well as being able to stay up to 18 minutes under water.

The Emperor Penguin reproduces in autumn, during the long polar nights. After laying her one egg, the female goes off and the male hatches it. He does not build a nest, but pushes the egg between his feet with his beak. It is wrapped in a pouch-like fold, which is criss-crossed with many blood vessels. These blood vessels keep the egg very warm.

The male fasts for two months, until the egg hatches. The female then returns and begins to provide food for her young. At this point the male, now much, much thinner, goes to the sea to start feeding again.

JACKASS PENGUIN

Number of eggs: 2-3

Height: 60 cm

Speed swimming under water: 20-33 km/h

Home: South Africa

The Jackass Penguin is the only penguin which lives in Africa, on the islands of Dassen and Dyer near to the Cape of Good Hope.

It does not undertake long migrations like other penguins in its family. When it finishes raising its young, the Jackass Penguin is content to gorge itself on the fish in the fresh waters which extend between Africa and the Antarctic Ocean. The Jackass Penguin digs a nest in sandy ground and often builds a proper shelter, which is shielded at the bottom from the rays of the Sun.

The female lays two or three eggs at a time, then she and the male take it in turns to sit on them, and share in the rearing of the young, which lasts for more than eight weeks. On the islands of Dassen and Dyer, the Jackass Penguin form numerous colonies which are always very noisy. Yet, like other large assemblies of penguins, the young penguins and their parents always succeed in finding each other – it seems that it is all a question of the right 'ear' to recognize the 'right' voice. Then they greet each other by stroking each other's neck with their beaks.

islands. One of these is the small ***Galapagos Penguin*** *which lives on the Galapagos Islands, at the same latitude as the Equator.*

The smallest penguin, the ***Little Penguin****, is only 40 cm long and so is often caught by seals. Snakes, lizards and mice eat its eggs. However, the winter weather and human beings are its enemies too. When it goes out at night to catch fish, it gives a cry rather like a howl.*

The male and the female ***Adelie Penguin*** *both stay close to the nest, which is built with flat stones and pebbles placed perfectly together. They return to the same nest each year to hatch their eggs.*

The ***Rockhopper Penguin*** *is easily recognized by the tufts of yellow feathers at either side of its head. Its limbs enable it to climb on rocks. This is the reason it got its name.*

The ***Gentoo Penguin*** *is about 75 cm tall. It lives on the islands around the Antarctic.*

Did you know...

*The **Northern Fulmar**, white as the snow, is found in the seas of the Arctic, whilst the **Cape Petrel** lives in the islands in the Antarctic and also in New Zealand.*

*The **Black Footed Albatross** lives in the Northern Hemisphere of the world.*

*The **Wandering Albatross**, a massive and strong flier of the southern oceans. With a length of 140 cm and a wingspan of more than 3 metres, this is the largest sea bird. It builds its nest on the ground and the female lays one egg. From that moment, the hatching lasts 2-3 months and the rearing of the young, in the care of both male and female, lasts between 5-9 months. Albatrosses reproduce only every two years.*

***Wilsons Storm Petrel** is found in the Southern Hemisphere. This bird takes off from the water with its wings open and feet on the surface of the sea, as if it is walking in search of food.*

*The **Great Shearwater**, recognizable by its characteristic nocturnal calls similar to a lament, nests in the south Atlantic but winters in European seas.*

*The **Pelican** likes swamps and marshes in summer, and sheltered bays and coastlines in winter. It nests in large colonies and, despite its strange appearance, it is a fine swimmer and flies at speed. The young eat pre-*

digested food which is collected in a sac under the beak of each parent bird.

The ***cormorant*** *(in the photograph) lives along sea coasts, besides lakes and some rivers in most parts of the world. It swims perfectly, but walks with difficulty. Its talent for catching fish is used by the Chinese who have tamed them to 'fish'.*

The ***Lesser Frigate Bird*** *prefers to steal prey from other sea birds. During the mating season the male shows himself off, a vivid red sac swelling out beneath his throat.*

The ***American Darter****, sometimes called the 'serpent bird', due to the shape of its long neck, lives near lakes and rivers.*

The ***Brown Booby*** *catches fish by diving underwater.*

The ***Double-Crested Cormorant*** *lines its nest with vegetable matter and algae, held together with mud and remains of food.*

The ***Red-Tailed Tropic Bird*** *has an elegant flight. It nests on the most inaccessible rocks in the tropical islands where it lives.*

WHITE STORK

Habitat: swampy areas, farmland rich with water, fields

Wingspan: 170 cm

Length: 100 cm

Food: reptiles, amphibians, fish, chicks and small mammals, insects, vegetable matter

The White Stork spends the summer in Europe and the winter in South Africa. It is an expert at soaring and being carried along by currents of air, usually flying above water only if land is in sight. Yet, when it migrates, it has to cross large expanses of open sea without currents of air to help it along. Many storks from Northern Europe cross the Mediterranean then fly over eastern Asia and on to South Africa. It is estimated that a Danish stork undertakes a journey of 13,000 km! Groups from Holland and Spain follow a south-east route towards the Straits of Gibraltar, where they take advantage of the warm rising currents before plunging down 16 km towards the sea, flying to Africa. However, the numbers of White Storks in South Africa has diminished, perhaps because the horse-flies and other insects on which it feeds have been killed by insecticides. But in some cities in northern Europe, people have built artificial nests for White Storks to rear their young, which are then protected. From March to September the female lays 3 to 5 white eggs in the nest made of feathers, hairs, paper and moss. The stork makes no noise but communicates by batting its beak in a particular way, reinforcing the message with ritual parades and movements.

CICONIA CICONIA
WHITE STORK

Did you know...

The ***Spoonbill*** *does not have its wide-ended beak from birth. It gets likes this little by little, as the bird grows. The reason for this unusual shape is not yet known.*

The ***Yellow-Crowned Night Heron*** *builds large nests with branches woven together in trees, in swamps or on floating vegetation. By day, it remains well-hidden, whilst at night it hunts for fish, reptiles, frogs, crustaceans, insects and vegetables.*

The ***Marabou Stork*** *lives in Asia and in the African savannah. Usually it feeds on the rotting meat which is easier to find in dry areas. This bird looks similar to the vulture, with which it often has to compete for the same food.*

Probably because of its splendid plumage, the ***Scarlet Ibis*** *(in photograph) with its long, thin, downward curving beak attracts hunters which capture them to sell to zoos. However, in captivity, the Scarlet Ibis pales in colour and becomes a clear pink. For this reason, the owners take care of its diet, enriching it with carotene. It gathers in colonies, together with the White Ibis.*

The ***Shoebill*** *or* ***Whale-Headed Stork*** *grows up to 120 cm tall. It is lazy, very shy and silent. It flies with great agility keeping its wings firm and taking advantage of the currents of air. It lives in swampy areas of tropical Africa. Special laws protect it from killing because it is at risk of extinction.*

There are about 100 species of ***bittern****, many living in Europe. It flies very slowly, keeping its neck folded in the form of a letter 'S'. In some species, the head has long feathers, a feature of great importance during the mating season and communication with other bitterns. The thin legs enable the bittern to walk in the water of marshes, swamps, lakes or along sandy coastlines.*

Tigriornis leucolophus
WHITE CRESTED BITTERN
Egretta gularis
WESTERN REEF HERON
Nycticorax violacea
YELLOW-CROWNED
NIGHT HERON
Eudocinus ruber
SCARLET IBIS
Ajaia ajaia
SPOONBILL
Balaeniceps rex
SHOEBILL OR
WHALE-HEADED STORK
Leptoptilos
crumeniferus
MARABOU STORK

BRANTA SANDVICENSIS
HAWAIIAN GOOSE
ANSERANAS SEMIPALMATA
MAGPIE GOOSE
ANHIMA CORNUTA
HORNED SCREAMER
ANSER ALBIFRONS
WHITE-FRONTED GOOSE
ANAS CRECCA
TEAL
DENDROCYGNA ARCUATA
WANDERING WHISTLING DUCK
CYGNUS MELANOCORYPHA
BLACK-NECKED SWAN
MERGUS CUCULLATUS
HOODED MERGANSER

MUTE SWAN

Weight of egg: 350 gr

Weight of adult: 12-18 kg (up to 23)

Wingspan: 2.20-2.50 m

The Mute Swan is the largest bird in Europe. In the wild state, it lives in the area which encircles in the Baltic Sea and in the swampy expanses of northern Germany. It can also be found in northern Asia.

The Mute Swan is recognized by a long neck which is almost serpent-like and a fleshy lump (tubercle) at the base of the beak. Despite its heavy appearance and the comical way in which it walks, there is no bird more elegant in flight. It rises swiftly with majestic batting of wings, keeping its neck straight in front like the stork and the crane. Its wingspan can reach 2.50 m and it is one of the heaviest birds able to fly, reaching speeds of 60-65 km/h. Unlike many other birds which live in pairs only during the mating season, Mute Swans make partners which last all their lives.

This bird builds a large nest with straw and vegetable fibres, slightly raised up above any surrounding water. Once a year, the female lays between 3 to 12 grey-green eggs which she sits on for about 5 weeks. The number of eggs depends on the age of the female – younger females tend to lay more eggs. The chicks are grey, becoming white by about 2 years old. This swan, also called the Royal Swan, makes a cry similar to the sound of a small trumpet. It eats aquatic insects, as well as aquatic vegetable matter. In Europe, they appear in late autumn or in winter and nest in the north of the Alps.

Did you know...

*The **Hawaiin Goose** has completely adapted to life on the land, living between the blocks of lava of the Hawaiin volcanoes. As well as being hunted by humans for its meat, it is also in danger from cats, pigs and mice which eat its eggs and its young. This bird was at risk of extinction until the inauguration of a programme of breeding in zoos.*

*The **White-Fronted Goose** lives in the Arctic tundra in summer, returning to southern Europe in the autumn. Its cry is similar to a laugh. It builds its nest in a simple depression in the ground with twigs and dried grass and lining with feathers.*

*The small **Teal** is only 35 cm long. It lives in groups near fresh water in summer and coastal waters in winter. It is very common in Europe, and in May the female lays from 8 to 10 white-yellow eggs.*

*The **Black-Necked Swan** is not very common. Smaller than the Mute Swan, it has shorter legs and a longer body, so that it has to toss itself about with every step. To take off is very tiring for this bird and it needs a long run. It flies rapidly making a characteristic rustling noise. It lives in large groups, in swamps, in ponds and in the marshes of Patagonia.*

*The **Magpie Goose** lives in Australia, whilst the **Wandering Whistling Duck** is most common in America, Asia and Africa.*

*Among the tufted ducks is the **Hooded Merganser** with the crest on its head similar to a horse's mane. During summer, this bird can be seen along the coastlines of many Mediterranean countries.*

*The **Horned Screamer** from Central America has two pointed spurs, about 5 cm long, at the front edges of its wings. Its legs are relatively long and the claws very spread out.*

KING VULTURE

Length: 1.10 m

Weight: 3-3.7 kg

Habitat: tropical forests and savannahs with trees

Home: from southern Mexico to northern Argentina

This bird 'looks the part' as the king of the vultures. It acquires its distinctive plumage and the vivid colours of the fleshy growth (caruncle) on its head, by the age of 3-4 years. Until then, the young are completely brown.
Although it can kill animals to eat, the King Vulture feeds mostly on the rotting flesh of dead animals, which it can pick out, even when these are covered by vegetation, by a highly-developed sense of smell which is rare in birds. When it makes its appearance at a 'feast' the other vultures give way, scared off by the strength of its beak. It has the habit of flying over the territory.
It does not build a nest. The female lays one egg in the cavity of a stunted tree or a trunk hit by lightning or by old age. Both male and female take it in turns to look after the young.

HAWK

Length of adult: 50-54 cm

Wingspan: 1.15-1.40 m

Weight of adult: 500-1000 g

Lives in: woodland regions

A hawk is easily recognized by its heavy flight, large wings and short, rounded tail, all of which distinguish it from other daytime birds of prey. There are many species (right, the Bat Hawk of Africa and Asia) and the largest population is found in the hottest regions. At the onset of winter it migrates towards the sunshine of southern countries, where it can continue hunting its prey.
By autumn, many hawks arrive in central and southern Europe. The hawk hunts mostly in open farmland, which it explores from on high with a gliding flight, sometimes staying quite still in the air. Its style of hunting is to trap – it stays at the top of a pole or on a branch where it is easily seen, surveying a wide part of the territory. With its keen eyesight, it can pick out field mice or a vole as soon as the animal puts its nose out of the lair. At the right moment, the hawk throws itself down and grabs the prey in its claws. The hawk nests in forests, among the leaves of trees. The nest is made of twigs lined with leaves and paper.

Did you know...

*In **birds of prey** the long fringes of feathers on the wings and the tail act as 'silencers' in flight.*

In ancient civilizations, birds of prey were regarded in the same way as gods. One of the most ancient was Garuda, an eagle in Hindu mythology with wings of gold. The Babylonians and the Hittites built temples for the eagle and the most popular Egyptian divinity was Horus, a falcon.

*Falcons, eagles, hawks and vultures are **daytime birds of prey** (active from dawn to sunset). These represent two thirds of birds of prey.*

*The African **Bateleur Eagle** and the **Serpent Eagle** from Asia are both predators of reptiles.*

***Hawks** can see better than humans because the retinas of their eyes are richer in photo-sensitive cells. At the point most dense in photo-sensitive cells, there are a million and a half in the hawk, opposed to the 200,000 in the retina of the human eye.*

*Like many other falcons, the **Gyrfalcon** is used in the sport of falconry. In this pursuit, a falcon is trained to catch prey and bring it back to a particular spot.*

*Eagles are very widespread. In Europe, there is the **Royal Eagle** (see photograph), whilst the **Wedge-Tailed Eagle** and the **White-Tailed Eagle** live in Australia, the latter also in Asia.*

Machaermamphus alcinos
BAT HAWK
Rosthramus sociabilis
EVERGLADE KITE
Terathopius ecaudatus
BATELEUR EAGLE
Haliaëtus leucogaster
WHITE-TAILED EAGLE
Aquila audax
WEDGE-TAILED EAGLE
Sarcoramphus papa
KING VULTURE
Microhierax caerulescens
RED THIGHED FALCON
Falco rusticolus
GYRFALCON
Spilornis cheela
CRESTED SEA EAGLE

GROUSE

Lives: mountain woodland and clearings up to 3000 m altitude

Length: male 85 cm; female 60 cm

Food: shoots, fir cones and pine needles

There are many types of grouse and they all belong to the Tetraoyndae family in the order of Galliformes. This family also includes the Partridge, the Mountain Pheasant, the Ptarmigan and the Prairie Chicken. The largest member of the grouse family is the Capercaillie (illustrated right). The male Capercaillie is as big as a turkey, with a large head, pale yellow beak, metallic green breast and a large rounded tail. The female is smaller and is sometimes confused with the Mountain Pheasant, from which it is distinguished only by a large reddish patch on the breast and the wider rounded tail. The Capercaillie also has a scarlet caruncle (fleshy growth) above each eye and a bristly 'beard'. It can be found in most mountainous regions of Europe, especially in the east-central chain of the Alps. Because the Capercaillie is still hunted, there are not so many areas in which it lives in the wild. It prefers forests of conifer trees. Here, the Capercaillie moves along the ground in summer and up in the trees in winter. Its nest is a depression in the ground hidden with leaves, feathers and moss, where the female lays her yellow-red eggs spotted with brown in May. These can only be hatched once a year. The chicks are fed with worms and insects, whilst the adults eat vegetable matter.

Did you know...

The order of Galliformes comprises many domesticated birds which have more or less accompanied the development of the human race from ancient times. Birds belonging to this order are not very good fliers. They have a slightly elevated first digit on each claw, a strong, curved beak and very distinctive nostrils to take the position of a bony septum.

A full-grown ***Hoatzin*** *is quite a large bird – about 65 cm long – yet it weighs less than 1 kg. It lives in the swamps of South America.*

The ***Ocellated Turkey*** *hides its showy feathers hidden in thick vegetation in small groups of 5-6 individuals. But it seems that the males live alone for some months returning to the females at mating time. It runs quickly and only takes to flight when it is in danger.*

The ***Congo Peafowl*** *has sumptuously-coloured feathers.*

The ***Plumed Quail*** *is a small member of the pheasant family, very similar to the Common Quail, but with a plume on its head and a brown patch at the throat.*

In the United States and the north of Argentina lives the ***Great Curassow****, with its backward-sweeping feathers on its crest and a large protruberance (swelling) at the top of its strong beak.*

The colourful ***Tragopan Satyra*** *is a species of quail from central Asia. It is in danger of extinction.*

The ***Mallee Fowl*** *from Australia lays its eggs in holes in the ground which it then covers with sand or vegetable matter, so that the eggs can be incubated by the warmth of the Sun. The parent birds can attend to the eggs only when the temperature suits them.*

The ***Spruce Goose*** *belongs to the same family as the Capercaillie. It has a massive form, sometimes reaching a weight of up to 7 kg. Members of the Tetraoyndae family live in temperate regions and in the Arctic in the Northern Hemisphere.*

AGRIOCHARIS OCELLATA
OCELLATED TURKEY
OPISTHOCOMUS HOAZIN
HOATZIN
AFROPAVO CONGENSIS
CONGO PEAFOWL
LEIPOA OCELLATA
MALLEE FOWL
OREOTYX PICTUS
PLUMED QUAIL
SATYRA TRAGOPAN
TRAGOPAN SATYRA
CRAX RUBRA
GREAT CURASSOW
CANACHITES CANADENSIS
SPRUCE GOOSE

Did you know...

*The **Pale-Winged Trumpeter** does not like flying. Even when it has to flee from predators, it prefers to keep its 'feet on the ground' or sometimes to swim in currents of water.*

*The **Limpkin** is very similar to the heron. It lives in the swamps of the United States and Argentina and feeds on water snails, which it swallows without breaking the shells.*

*The **Red-Legged Seriema** is found in Brazil, Argentina and Uruguay.*

*One of the largest examples of dove is the splendid **Victoria Crowned Pigeon** from New Guinea. It is about 65 cm tall and its tuft of feathers of different lengths give it a majestic air.*

*The **Kagu** exhibits complex parades during which it will use its beak to grasp the feathers on its tail and wings*

*The **Takahe** is a species native to New Zealand. It has lost its ability to fly and uses its wings only to swim during the mating season.*

*The **African Finfoot** can be found throughout almost the whole of Africa It is an excellent swimmer, with the*

habit of carrying its young on its back, even when going under water.

The ***Sun Bittern*** *lives in the rainforests of Central America and Brazil.*

The ***Lichenstein Sand Grouse*** *is a bird of the African savannahs, living in groups of up to 400 individuals. Its plumage blends in with its surroundings.*

The female ***Button Quail*** *is larger than the male. She is the one who approaches the male at mating time.*

The ***Pin-Tailed Sand Grouse*** *nests in Spain, Portugal and parts of northern France.*

The ***Laughing Dove*** *also belongs to the dove family, as well as the beautifully-coloured* ***Pheasant Pigeon****, the* ***Plumed Guineafowl*** *with the plume on its head and the* ***Common Ground Dove*** *which has a more rounded shape than the others.*

The ***Large Green Pigeon*** *can be found in the tropical zones of Africa and Asia. It lives in groups and is guided along by one adult bird. They travel in search of fruits and berries, which is the only food they can eat.*

Did you know...

*The **Kea** is a parrot from New Zealand which feeds on fruit and the larvae of insects.*

*Other members of the parrot family are the **Crimson Rosella**, a very able flier, and the 'hanging parrots' with short, blunt tails, such as the **Blue-Crowned Hanging Parrot** which lives on the ground.*

*The **Palm Cockatoo** has dark feathers and a strong beak which is large compared to the size of its head.*

*The **Rainbow Lorikeet** lives in Australia, New Zealand and the islands of Oceania. It eats only nectar, which means that it has to move around according to the flowering of plants.*

*The **Kakapo** also lives in New Zealand. It spends the day among the roots of trees and goes hunting at night, sometimes throwing itself down from trees.*

*In the wild, the little **Golden Parakeet** is found in India, Sri Lanka, South-East Asia and tropical America.*

*Among the lovebirds of Africa and Madagascar is the **Masked Lovebird**, so-called because of its black 'mask' of feathers which extends from below its nose to the middle of its head.*

*The Indian **Red Wing Crested Cuckoo** and the **African Emerald Cuckoo** belong to the cuckoo family. These are typical of birds living in tropical zones throughout the world.*

*The **Violet Tauraco** lives in Africa, in Gambia and Nigeria.*

*The **Black-Faced Coucal** lives in the Philippines. It is a poor flier and flees from danger by running along the ground.*

*In China, India and New Guinea the **Koel** is found. Like many other members of the cuckoo family, the female lays her eggs in the nests of other birds, which then incubate the young.*

*Another parasitic cuckoo is the **African Emerald Cuckoo** with its head coloured in a way which seems 'hooded' and a large beak. It lives in Indo-China and Sumatra.*

*The **Smooth-Billed Ani** lives in the region of Euro-Asia. It likes to be with other birds of its kind. At hatching time, up to six females may share the same nest.*

*The **Livingstone Tauraco**, with a straight crest on its head, is also from Africa.*

BUFO LACTEUS
EAGLE OWL
PULSATRIX PERSPILLIATA
SPECTACLED OWL
AEGOLUS FUNEREUS
TAWNY OWL
GLAUCIDIUM PASSERINUM
PYGMY OWL
TYPO ALBA
BARN OWL
SPEOTYO CURICULARIA
LITTLE OWL

BARN OWL

Face shape: heart

Length: 30-40 cm

Wingspan: 90-95 cm

Habitat: old buildings, trees

The Barn Owl is a nocturnal bird of prey which can be found almost everywhere, although not in great numbers. Thanks to its special eyesight, it can find its way and capture prey in complete darkness.

Its ears are long clefts hidden at either side of its face, behind the feathers arranged in the shape of a heart. The Barn Owl can detect the slightest sound of a mouse or a vole.

Like many other owls, the exceptional hearing of the Barn Owl is made even better by its facial disc reflecting sound back to its ears. This also enables the owl to detect more easily the direction from which the sound comes.

The Barn Owl eats rodents and other birds such as doves, and also small mammals.

The Barn Owl does not build a nest, but in April and May and sometimes in November or December, the female lays her white eggs in clefts of buildings and trees, or in nests abandoned by other birds.

Thanks to the particular softness of its feathers, the flight of the Barn Owl is silent.

TENGMALM'S OWL

Face shape: round

Length: 25 cm

Habitat: conifer forests

Although the Tengmalm's Owl (also called the Boreal Owl) is quite rare in Europe, it can still be found in woodland in the region of the Alps. It is often confused with the Common Barn Owl, but it is smaller.

Tengmalm's Owl comes from the region of the Arctic Circle and likes conifer forests. Here it captures mammals, birds and insects which constitute its diet. It hunts by day as well as by night.

For nesting, it uses natural cavities in trees or nests left free by woodpeckers. The female lays her eggs from April to June, like the Common Barn Owl and the Tawny Owl.

The number of individuals sometimes increases following migration to the north, especially in winter – and this is also true of the Common Barn Owl, especially in central-southern zones of Europe.

Did you know...

*Nocturnal **birds of prey** have eyes which are more rounded than in other birds and more 'plate-like'. The range of sight in these birds is less, but they can see for long distances.*

*As in all Strigiformes, the eyes of the **Eagle Owl** are forward-looking and at the front of the head – whereas in all other birds the eyes are at the sides. The Eagle Owl cannot see out of the angle of its eye, to look to the side or over its 'shoulder'. So it has to move its head by continual and rapid movements of the neck. It can turn its head 180° from one side to the other.*

***Barn Owls** and **Eagle Owls** are able to see a dead mouse in light from 10 to 100 times weaker than that needed by the normal human eye, and a living mouse in almost complete darkness.*

*In ancient China, the **owl** was considered fierce and a bringer of death. For the North American Indians, however, the Eagle Owl had the power to help and to protect people during the night-time hours.*

*The **Eagle Owl** can find its way by hearing as well as by sight, and by tactile bristles at either side of its beak.*

*The **Pygmy Owl** is only found in a few Alpine areas in Europe. The **Little Owl** and the **Spectacled Owl** are both found in North and South America.*

*In Greek myths, the **owl** was the bird of Athene, the goddess of wisdom, and symbolized the powers of darkness. For Aztecs, the owl was a symbol of the god of the after-life, guardians of the Earth and the forces of nature.*

MARVELLOUS SPATULETAIL

Habitat: forests

Length: 16 cm

Flight: rapid taking off, or almost quite still in the air

Food: nectar

The Marvellous Spatuletail belongs to the Trochilidae family, commonly called hummingbirds, and much admired because of their splendid colouring.

These skilled fliers are generally small, with a roundish head, large eyes and a straight or curved beak, which is sometimes longer than its body.

The hummingbird, like the Marvellous Spatuletail, can be found in the high plains of northern Peru. The unusual appearance of the male makes its dance of love even more charming when he flies around the female showing off the rainbow shafts of light at the ends of his tail feathers. As with other birds in the same group, the male tends to have the more colourful plumage. One exception to this is a hummingbird with violet ears, in which the plumage of the two sexes is not all that different to each other.

ALPINE SWIFT

Habitat: rocks

Length: 20 cm

Speed of flight: up to 200 km/h

Food: insects

In mainland Europe, the Alpine Swift is not too rare, especially in the mountain zones of the Alps and on small and large islands, where it can often be seen flying around steep rocks, before diving into the sea.

The Alpine Swift likes to be with other birds. It feeds mainly on the insects which it catches as it flies, streaking down like lightning.

At nesting-time, Alpine Swift colonies use the clefts in rocks or holes in ruined buildings to lay white-coloured eggs at the end of May. The nest is in the shape of a crown roughly made of straw, feathers and different vegetable matter cemented together with the bird's secretions of saliva.

Did you know...

*The order of **Apodiformes** comprises 61 species, scattered around almost every part of the world. Among them are swifts and hummingbirds. Although these birds may seem to be very different, there are many common characteristics, such as very short legs and the way in which the feathers are arranged.*

*The flight of the **hummingbird** is so perfect that it can fly in any direction, within any space, or stay quite still in the air. With the structure of their shoulder joints together with the highly-developed pectoral (breast) muscles, hummingbirds can flap their wings up to a frequency of 70 flaps per second. In general, hummingbirds are very distinctive, especially during the mating season.*

They lead a solitary life and once they have established a hunting territory, they strongly defend it.

LODDIGESIA MIRABILIS
MARVELLOUS SPATULETAIL
CALYPTE ANNA
ANNA'S HUMMINGBIRD
OREOTROCHILUS ESTELLA
ANDEAN HILLSTAR
HEMIPROCNE LONGIPENNIS
GREY-RUMPED TREE SWIFT
TROCHILUS POLYTMUS
RED-BILLED STREAMERTAIL
ENISFERA ENISFERA
SWORD-BILLED
HUMMINGBIRD
APUS MELBA
ALPINE SWIFT
EUTOXERES AQUILA
WHITE-TIPPED SICKLE BILL

Did you know...

All ***toucans*** *have an over-large beak, jaws which are jagged at the edges and a long tongue. The* ***Toco Toucan*** *(in photograph), which measures 60 cm long, is among the largest.*

The family of Indicatoridae is made up of small birds with short beaks and legs and pointed wings. ***The Black-Throated Honey Guide*** *is found in tropical and southern Africa.*

The ***Collared Puff Bird*** *with its huge orange beak is fairly squat and emits low, weak sounds.*

The ***Red-Necked Aracari*** *cries in a sharp, awkward way, but it is very colourful, like the* ***Plate-Billed Mountain Toucan****.*

The ***Paradise Jacamar****, with it. melodious song, has colourin rather like that of th hummingbird.*

The ***Northern Flicke*** *is most common in Nort America. It is particularl colourful and a glutton fo*

ants, which it catches by extending its long, moist tongue into the ants' nest.

With its beautiful plumage, the ***Green Woodpecker*** *looks particularly splendid as it rises up in flight.*

The ***Lesser Spotted Woodpecker*** *from North America is about 50 cm long and catches insects under the bark of old trees. The killing of these birds has led to their near extinction.*

The ***Guiaman Piculet*** *belongs to the family of small woodpeckers and lives in Asia.*

The ***Red-Headed Woodpecker*** *is easily recognized by its bright red hood. It lives in Central America.*

The ***Black Woodpecker*** *lays its eggs in the cavities of conifer trees.*

The ***Greater Spotted Woodpecker*** *lives in woodland, forests and sometimes parks. Its repeated pecking in the bark of tree trunks stuns the insects and the grubs on which it feeds, but it is believed that this is also a way of communication between the sexes.*

BLACKCAP

Number of eggs: 4-6, each about 1.5 g

Weight of adult: 15-25 g

Length: 14 cm

Wingspan: 23 cm

The Blackcap belongs to the order of Passeriformes and the family of Muscicapidae which has over 1500 species with particular similarities. They all have a rather dull brown or grey plumage, sometimes with black, green or red. But, although the plumage of these birds may not be very attractive, their melodious song puts them among the best songbirds in all the world. The variation of tone and the unending modulation of their song repertoire are extraordinary; each species has an exclusive song and many species are able to recognize the song of their own kind.

The Blackcap is found mostly in Europe. The male has the top of its head clearly defined in black. As with all Muscicapidae, the Blackcap carries out a very private life in dense woodland or among thick hedges. It does not like to be seen and prefers the quiet, hiding itself among the greenery. At mating time, it builds its nest among honeysuckle or tufts of evergreen trees. Some species of blackcap stay in one place, others prefer to migrate at the first signs of cold weather, when the insects on which it feeds become scarce, returning in spring time.

SYLVIA ATRICAPILLA
BLACKCAP

Did you know...

*The **Wryneck** (see page 118) belongs to a family which is mostly non-migratory. However, this bird migrates from the Arctic to spend winter in Africa and southern Asia.*

*The **Brown Thrasher**, together with the other thirty species of Mimidae (mimics), lives in America. It has a remarkable ability to imitate the songs of other birds.*

*The **Lyrebird** gets its name from its spectacular tail in the form of a lyre. It lives in Australia. It is quite large and is an excellent mimic of other birds. However, its own song includes phrases which belong only to its species.*

*The **Cactus Wren** is the symbol of Arizona. It is a very small bird which lives in desert regions. When it becomes too hot, it opens the beak to lower its internal temperature, breathing very quickly and keeping its wings away from its body.*

*The **Blue Shrike** uses its beak to skewer its prey.*

*The **Asian Fairy Bluebird** is one of the most beautiful birds in tropical Asia. It has a plumage of intense blue which makes it very noticeable in the jungle, where it lives and where it eats ripe fruit.*

*The **Skylark** is common in Europe. Its sweet song heralds the beginning of spring.*

*The **Pied Wagtail**, common in Europe, belongs to the family of Motacillidae, birds distinguished by their long tails and which move with an elegance, as if they are on tip-toes. They eat insects which they find in open ground, sheltered by a stone or a bush, where it makes its nest.*

The high number of species belonging to the order of Passeriformes includes the Common Sparrow, shown in the photograph. Each species has preferences regarding habitat and diet – and so, to avoid competition for the same places and the same food, they are scattered throughout almost every part of the world, except for the Antarctic.

CAMPYLORHYNEUS BRUNNEICAPULLUS
CACTUS WREN
TOXOSTOMA RUFUM
BROWN THRASHER
IRENA PUELLA
ASIAN FAIRY BLUEBIRD
MENURA NOVAEHOLLANDIAE
LYREBIRD
LEPTOPERUS IADAGASCARNUS
BLUE SHRIKE
MOTACILLA ALBA
PIED WAGTAIL
ALAUDA ARUENSIS
SKYLARK

MAGPIE

Lenght: 46 cm, of which 23 cm is the tail

Number of eggs: 5-9

Habitat: hills and wood plains, parks and marine areas

Food: insects, seeds and eggs

The Magpie has a slender body with black plumage with green-bronze splashes. Its abdomen and the tops of its wings are white.

Magpies live in small family units, sometimes with jackdaws and ravens.

Always ready for combat, the Magpie never hesitates to face predators. It flies with ease, but on the ground it walks heavily, preferring to stay still. It builds its dome-shaped nest with grass and stalks cemented together with mud and lined with dry grass. In April, the female lays between 5–9 white-blue or greenish eggs, sometimes speckled with brown, and for 18 days the female and the male take turns to sit on them. The Magpie eats insects, grain, nuts and the eggs of other birds. It is widespread throughout Europe, Asia, North America and North Africa. The Magpie is protected in central Europe.

JACKDAW

Length: 65-70 cm

Wingspan: 125-140 cm

Number of eggs: 4-8

Food: carrion and small fish, but also game birds

The Jackdaw (in photograph) is the largest member of the Corvidae family and is also among the most intelligent of birds. It has a massive head with a black beak, as large as 10 cm, strong legs and claws and a tail 25 cm long. The feathers are shiny black, with metallic blue and purple splashes. The legs and eyes are also black. Living in North America, northern Africa and throughout Europe, they are also found in the Alps and the Apennines and in the mountain areas of Sardinia and the islands of Tuscany.

The Jackdaw is distrustful and usually lives in the wild and far away from human beings, going to populated areas only when it is cold or hungry.

Jackdaws live in pairs or small groups, whilst the young prefer to be alone. They argue fiercely with companions at mating time and with most other species of birds. Together, male and female Jackdaws prepare large nests ready for blue-green eggs speckled with brown to be laid.

Did you know...

*At daybreak throughout Europe and Asia, flocks of **Jackdaws** abandon their night-time homes to go to the fields in search of insects and larvae, grain, molluscs and birds' eggs. They often live together with ravens, and sometimes join with them in the search for food.*

*To protect itself against shortage of food, **Clark's Nutcracker** hides seeds, pines and nuts ready for the autumn. When the snow falls and covers its food store, this bird can still pick out the exact place where it is. It lives in the mountainous regions of North America.*

*The **Turquoise Jay**, 32 cm long, is most common in the mountainous regions of the Pacific coastline of the American continent. In Sri Lanka lives a similar bird, the Blue Magpie.*

*The **raven** is often regarded as a bird bringing bad luck, mostly because of its forbidding appearance and harsh voice.*

*Among the jays which spend more time on the ground, is the **Persian Ground Jay** which escapes from pursuers by running. It lives in the desert regions of Central Asia.*

*The **Jay** is often called the 'woodland guard' because at the arrival of a stranger, it alerts the other birds with its cries, so that they can flee from danger. It feeds on nuts, berries, fruit and the larvae of insects, but when rearing its young, it will also steal eggs from the nests of other birds to feed itself and its chicks.*

The Jay is over 30 cm long, with purple-red plumage. It lives in all the temperate regions of Europe and Asia.

NUCIFRAGA COLUMBIANA
CLARK'S NUTCRACKER
CYANOLYCA TURCOSA
TURQUOISE JAY
CISSA ORNATA
CEYLON BLUE MAGPIE
PICA PICA
MAGPIE
CORVUS MONEDULA
JACKDAW
PODOCES PANDERI
PERSIAN GROUND JAY
CORVUS CORAX
RAVEN
GARRULUS GLANDARIUS
JAY

MAMMALS

Common characteristics

The main thing which distinguishes mammals from other animals is the ability of the female to carry her unborn young and then to feed them with her own milk from her mammary glands (teats). Another important feature is hair on the skin in most species.

Evolution, distribution and habitat

Mammals developed from reptiles - from therapsids, huge lizards which disappeared more than 200 million years ago. Today, mammals live successfully in many different surroundings – deserts, tropical forests, polar zones, oceans – due more than anything else to their ability to regulate their body temperature. Mammals are found in all continents and at all latitudes.

The long and complex development of mammals has altered their external appearance. The skin of a mammal may be covered by hair, fur, spines (as with the porcupine) bony shields and there are also some species with skins which are bare.

The African Elephant *(Loxodonta Africana)* is the largest mammal in the world. Its size and its weight (average 7 tonnes) is surpassed only by some species of Cetacea, such as the whale. One of the smallest mammal is the Common Shrew *(Sorex Araneus)* shown above, which weighs only a few grammes.

Mammals include exceptional runners, jumpers, climbers, swimmers and fliers, each one perfectly suited to various surroundings.

Behaviour

Mammals have a well-developed brain, with a good memory which can receive and select information and learn from it. This has given them an advantage over the rest of the Animal Kingdom, enabling them to adapt most easily to every type of environment.

Food

Mammals can be herbivore, carnivore or omnivore (able to eat almost any type of food). The teeth are modified according to the type of food which each mammal eats – well-developed canine teeth for carnivores, molars for herbivores and incisors for rodents.

So many different kinds

Together with birds, mammals represent the most highly developed class of vertebrates. Although they make up only a very small proportion of the Animal Kingdom (5%) mammals have become the masters of the Earth. The development of a nervous system, and especially the **brain**, has enabled them to increase their intelligence, and, as a result, their ability to survive and to rise above other species.

The first mammals were tiny shrews. These appeared on Earth around 200 million years ago at the start of the Mesozoic Era. Unlike their reptile ancestors, mammals were able to regulate their body temperature and to keep this at between 36° and 41°, whatever the external temperature. This phenomena is called homeothermy, or thermal regulation. The female mammal also has **mammary glands** which secrete milk to feed her young.

All mammals – except for the Monotremata such as the duck-billed platypus and the spiny anteater, which lay eggs – are viviparous, meaning that the young develop inside the body of the mother.

The body of a mammal is almost always covered with **hair** which keeps in the warmth and protects the body from the external elements of heat and the cold, water and wind. Some mammals, such as the Porcupine and the Hedgehog have bodies covered with spines; others, such as the Pangolin have scales; other skins are bare – that is, without hair, such as the whale. The skin of a mammal always has **sweat glands** to get rid of perspiration and dead skin cells and **sebaceous glands** to lubricate and protect the skin. Mammals are also distinguished by their exceptionally well-developed senses, which are modified for the life which a mammal leads. One example of this is the acute hearing of bats, the only mammals which can really fly and which find their way by hearing the echo produced by their own cries. Due to their ability to adapt, mammals can live on the land, in the water or fly in the air. There are mammals of all shapes and all sizes.

PHYLUM	SUB-PHYLUM	CLASS	ORDER	SUB-ORDER	SPECIES
CHORDATA	Vertebrates	Mammals	Monotremata		Duck-Billed Platypus
			Marsupialia		kangaroo
			Insectivora		Mole, Fieldmouse
			Chiroptera		bat
			Edentata		sloth
			Pholidota		Pangolin
			Lagomorpha		rabbit
			Rodentia		squirrel, beaver
			Carnivora	Fissipedia	wolf, bear
				Pinnipedia	seal
			Cetacei	Odontoceti	dolphin
				Mysticetes	whale
			Proboscidata		elephant
			Sirenidia		Dugong, Manatee
			Hyracoidea		Hyrax
			Artiodatyl		pig, camel
			Perissodatyl		horse
			Primates	Proscimii	lemur
				Antropoidea	Baboon, Gorilla

DUCK-BILLED PLATYPUS

Reproduction:	oviparous
Number of eggs:	2
Length at birth:	2.5 cm
Length of adult:	30-45 cm
Length of tail:	10-15 cm

The Duck-Billed Platypus belongs to the order of Monotremata, strange animals with some characteristics of birds, some of mammals. The first example of the Duck-Billed Platypus was brought to Europe by a ship which had crossed the Indian Ocean at the end of the eighteenth century. Before then, it was thought that this extraordinary animal was only a fantasy of sailors becalmed in the seas of Asia. As the years passed, other examples were brought to Europe from Australia and finally, in 1884, a study carried out in Australia established that these animals, although mammals, laid eggs. The body of the Duck-Billed Platypus is 400 cm long, covered with thick fur, bristles and fine down. The tail, from 10–15 cm long, is similar to that of a beaver. The nose ends in a horny-type beak, large and flat like that of a duck, with teeth in the young, substituted by horny plates in adults. The front and back legs have webbed feet which enable the Duck-Billed Platypus to swim in the rivers and lakes of Australia and Tasmania where it lives. Underwater it hides its eyes and ears between folds in the skin, finding its way by using its sense of touch, which is particularly well-developed in the beak. At night it goes hunting for earthworms, small crustaceans and molluscs which it eats in great quantities; during the day it shelters in its lair which it builds in banks sloping down to the water. The lair is in the form of a tunnel which opens out into a round space. When she is ready to lay her eggs, the female lines this space with wet leaves. Then she lays two eggs and rolls herself up around them, enclosing them completely. Ten days later, the young ones are born. Even though they are only 2.5 cm long, they immediately suck the milk which flows from the skin of their mother, although she does not have proper mammary glands. The young leave the nest after about four months, when their bodies are completely developed and they can go in search of food on their own. These animals have an average life span of about 15 years. The male Duck-Billed Platypus has a hard spur connected to a poisonous gland near to each foot. It is among the few 'poisonous' mammals; its weapon is deadly for other animals, but 'only' causes severe pain to any human victim, and a swelling at the point where the spur pierced the skin.

ORNITHORHYNCHUS ANATINUS
DUCK-BILLED PLATYPUS

TACHYGLOSSUS ACULEATUS
SPINY ANTEATER

SPINY ANTEATER

Reproduction: oviparo

Number of eggs: 1

Length at birth: 12 mm

Length of adult: 35-53 cm

Length of tail: 9 cm

The Spiny Anteater, or Echidna also belongs to the order of Monotremata. It has a body similar to a Porcupine, about 50 cm long, and completely covered with thick spines. The long nose forms a sort of hard, tube-like beak, with two nostrils at the very end. The mouth, which has a thin opening, has no teeth but the palate is covered with a hard substance which enables the Spiny Anteater to chew its food. Its moist, threadlike tongue can project 20 centimetres outside its mouth and its thick, short legs have sharp claws which are ideal for breaking the nests of termites and ants.

There are two species of anteater but the best-known is the Long-Nosed Echidna (illustrated below) from Australia and New Guinea. This animal lives alone and in a variety of surroundings. It settles down without any problems, and although it only has poor vision, it can feel the slightest vibration of the ground. When it senses danger, it rapidly digs a tunnel and is completely buried within a few minutes. It only needs about nine minutes to disappear completely into its underground shelter. Unlike the Duck-Billed Platypus, it does not use this sort of tunnel as a permanent home, but prefers the lair of another animal. If the ground is too hard, and it does not have time to dig, it rolls itself into a ball, like a Hedgehog, anchoring itself to the ground with its claws.

When it needs to, it can also escape very quickly by lifting itself up on its hind legs and running along, hiding itself among the bushes. The Spiny Anteater mainly eats termites and ants which it catches with its long tongue, but in captivity it will also eat minced meat, milk, pieces of bread and almost anything it can get into its mouth. The female lays her egg in a stomach pouch which forms only during the time she carries her young. After about ten days after the egg is laid, the young Spiny Anteater is born, feeding on the milk which flows along the hairs in the stomach pouch of the mother. Here the baby remains for about eight weeks, by which time the young Spiny Anteater begins to sprout spines and reaches a length of 10 cm. At this point, it is hidden by the mother in a safe place, where she goes regularly to feed it.

At one year old, the Spiny Anteater reaches the size of an adult and is ready to mate.

The Spiny Anteater is very long-lived with an average life span of fifty years.

Did you know...

*The **Long-Nosed Bandicoot** is a marsupial of average size, from 21 to 43 cm in length, which lives in Australia, Tasmania and New Guinea. It is similar to a kangaroo, and as large as a badger. The hind legs are longer than the front legs, the nose is very long, the ears are pointed and it has a bristly skin. It feeds on insects and small animals, as well as bulbs and roots.*

*The **Spotted-Tail Quoll** is sometimes called the tiger cat. It lives in woodland regions and in the mountains along the southern coasts of Tasmania, Australia and New Guinea. It moves as lightly as a cat, and goes hunting for reptiles, birds and rabbits. Sometimes it will even attack small kangaroos.*

*When the young **marsupial** is born the mother guides it into a stomac pouch called a 'marsupium'. Here, th baby completes its development.*

*The **Marsupial Wolf** is also calle the 'Tasmanian Tiger' because of th stripes on its back. It is regarded as th largest predator of all marsupials. I looks very much like a dog, with*

TASMANIAN DEVIL

At birth: blind, without fur, 12 mm long	
Incubation: in marsupium	
Number of young: 2-4	
Length of adult: about 50 cm	
Length of tail: 23-30 cm	
Weight of adult: 4.5-9 kg	

ˉhe Tasmanian Devil is one of the ıargest marsupials, belonging to the amily of Dasyuridae – animals which are very active and bad-tempered. In he past it was widespread throughout Australia, but now it survives only in ˉasmania. It is a strong carnivore which looks rather like a small bear. Its avourite places are remote, rocky places where it can live a secluded xistence. During the day it seeks shelter in caves or cavities in rocks, or in shallow lairs which it digs among the roots of trees. ʇ comes out at night to go hunting for reptiles, small mammals and fish. It dives fearlessly into water and also swims ınderwater to shelter among the tufts of aquatic plants.

ˉhe Tasmanian Devil is a poor climber, but it will still go up the trunks of trees to capture a sleeping bird. But it is very quick .t running. The name 'devil' is because it is always ready for a fight. But when a Tasmanian Devil is adopted as a baby and ›rought up with care, it becomes docile and affectionate towards its owner. It reproduces in April and the young are born owards the end of May. They soon find the marsupium where they stay attached to the mammary glands of the mother, eeding on her milk for about 15 weeks.

OPOSSUM

At birth: blind, without fur	
Incubation: in marsupium	
Number of young: 10-20	
Length of adult: 60 cm	
Length of tail: 30 cm	
Weight: 6-7 kg	

ˉhe Opossum is a marsupial of the Didelphidae family, all of which live in the outh and in part of the north of the American continent. It has a pointed nose, ırge, bare ears, a thick coat and a tail covered with scales. During the day it stays .t the bottom of its lair or in the hole of a tree, coming out only after sunset to ;o in search of food.

ˉhe Opossum is an omnivore, feeding on fruit, roots, birds, small mammals, hrimps and eggs. At the least sound, it will either flee and hide under a tree, or how its teeth and bite. But if the danger is great, it will fall down on its side and pretend to be dead. A female can give birth o up to 20 young at any one time. But, as she only has 13 mammary glands in her marsupium, the number of new-born vhich survive is limited.

ody longer than 1 m and a tail 50 cm ɔng. There is only one species which ves in Tasmania and this was once hought to be extinct. But unconfirmed ightings have raised hopes that other xamples have survived.

ˉhe ***Numbat*** *comes from Australia. ʇ is easily recognized by its zebra-like oat and sharply-pointed nose. It lives ımong the eucalyptus trees and in woodland, where it uses its long, thin tongue to catch ants and termites to eat. The Numbat moves its tongue at surprising speed, getting it into every corner of a hole.*

The ***Yellow-Footed Antechinus*** *(marsupial mouse) lives only in Australia. It is a small mammal, rather like a rat, nocturnal and very active and fierce. It climbs easily on the branches of trees and along rocky walls.*

The ***Marsupial Mole****, about 18 cm long, lives in the sandy regions of Australia, and in its appearance and habits is like European moles. Like them, the Marsupial Mole digs underground tunnels (but only a few centimetres deep) and feeds on insects and worms.*

GIANT KANGAROO

At birth: blind, without fur

Incubation: in marsupium

Number of young: 1 or 2 at the most

Weight of adult: 40-70 kg

Length of body: 110-160 cm

Length of tail: 80-110 cm

There are ten species of kangaroo, all belonging to the Macropodidae family. The Giant Kangaroo is found in Australia, Tasmania and New Guinea. It is practically a two-legged animal. From the moment it can move about, it uses only its well-developed hind legs. Its front legs are much shorter and the kangaroo uses these to grasp and to hold. Its long, muscular tail is used for sitting and to keep the body balanced as the kangaroo runs.

The Giant Kangaroo is sometimes called the Grey Kangaroo and it lives in Australia and Tasmania. It eats grass and can withstand drought, going up to two or three months without drinking. It is the largest species of kangaroo and can continue growing throughout its life. It excels in running on its two legs, with a speed which can reach an average of 45 km/h and up to 80 km/h over short distances. It can also clear a length of up to 10 m in one jump. After carrying her young for 30-40 days, the female gives birth to a young kangaroo only as big as a bean, and which finds the marsupium of the mother immediately. There it remains for about eight months, feeding on her milk.

KOALA

At birth: blind, without fur

Incubation: in marsupium (6 months)

Number of young: 1

Weight at birth: 5-6 g

Weight of adult: 14-15 kg

The Koala lives only in some small Australian forests o eucalyptus trees and is not present in any other part of th world.

At the beginning of the 1900s, millions of koalas populate the whole of Australia, but they were hunted to such a extent that the numbers of these charming, shy animal were drastically reduced within just a few years. The Koal is tree-dwelling and descends to the ground only to lick th ground to get the mineral salts which it needs or to mov from one tree to another. It is an expert climber and th sharp claws at the end of its paws enable it to scale th highest, smoothest trunks.

It likes to be nocturnal and spends the day with its arm comfortably around a branch.

The Koala lives in groups of around 20 individuals whic communicate with each other by loud cries. Its one sourc of food is the hardest, toughest leaves of a special type o eucalyptus. The rhythm of reproduction of these animals i very slow and this adds to the problems of safeguardin them. However, for many years, the Koala has bee protected by the Australian Government, who have also re introduced the Koala back into the regions where it onc lived.

Did you know...

The ***Common Wombat*** *lives in Australia and Tasmania. It is a herbivore marsupial, which looks rather like a rodent, with a squat body covered by bristly fur. It is peaceful and timid, with nocturnal habits. During the day it stays hidden in its lair which it digs in the ground with tunnels up to 30 m long.*

The ***Sugar Glider*** *is a small marsupial similar to a squirrel, which lives in Australia. There are three species similar to a flying squirrel, each with a membrane at either side of the body. This enables the Sugar Glider to glide, rather lik using a parachute. It can jump lik this from one tree to another without losing height.*

The ***Tree Kangaroo*** *is a smal kangaroo which lives among th branches of trees, where it can mov about easily. In this animal, the tw pair of limbs are almost the sam*

ength. There are seven similar known species, living in the green forests of Australia and New Guinea.

The **Yellow-Footed Rock Wallaby** belongs to the same family as the kangaroo. Its tail is almost as long as its body, up to 70 cm, and it acts a counterweight as the wallaby jumps and climbs in the rocky regions of Australia. It is a nocturnal animal, which shelters during the day in caves situated in remote places.

The **Spotted Cuscus** is a marsupial with the joints of the claws well developed and perfect for gripping on trees. It has a thick fur, spotted in the male, and a long tail which it can also use for gripping. The Spotted Cuscus is a fine climber and spends its whole life in the trees of the Australian forests, without ever coming down to the ground. It moves slowly and only to hunt for food. For the rest of the time it stays still, sitting on a branch.

TENREC ECAUDATUS
TAILLESS TENREC
PARAECHINUS AETHIOPICUS
DESERT HEDGEHOG
POTAMOGALE VELOX
OTTER SHREW
SOREX ARANEUS
COMMON SHREW
ELEPHANTULUS
ELEPHANT SHREW
BLARINA BREVICAUDA
SHORT-TAILED SHREW
CONDYLURA CRISTATA
STAR-NOSED MOLE
TALPA EUROPAEA
COMMON MOLE

TENREC

Number of young: 12-15

Length of body: 27-39 cm

Length of tail: 1-1.6 cm

The Tenrec is a strange insectivore which looks like a hedgehog. There are about 20 known species of the same type, all living either in Madagascar or the Comoro Islands. The Common Tenrec is a mammal which gives birth to the largest number of young at any one time, up to fifteen at once. Shortly after the birth, the young are already able to run about and after only four weeks, they can look after themselves. The Common Tenrec prefers warm areas, with dry, sandy ground, in which it can dig tunnels easily. Many live in forests and woodland. It is strong, and, like the hedgehog has a coat of prickles – but it cannot roll itself up into a ball. It is mostly active at night. During the day it shelters in the hollows of tree trunks or among the roots. It digs a tunnel 1–2 metres long.

At the beginning of winter, the Tenrec retires into its underground lair and goes into hibernation, its breathing and its heart-rate slowing down. But before hibernation, the Tenrec eats as much as it can and accumulates reserves of fat. During spring and summer, it takes up its nocturnal hunting in search of insects, molluscs, various invertebrates and small reptiles.

COMMON SHREW

Number of young: 5-10

Length of body: 6-8 cm

Length of tail: 3-5 cm

Weight: 6-16 g

The Common Shrew lives mostly in hiding. In Europe there exist many species, all very similar and all belonging to the family of Soricidae. Among these is the Pygmy Shrew, the smallest of all mammals and just 3.5 cm in length.

The Common Shrew is a small animal with a long, sharp nose. It lives mostly in fields and damp areas; generally it hides under stones and bushes, in tunnels which it either digs itself or have already been dug by moles or mice. It is sometimes possible to find them up to about 1900 m in depth.

The main occupation of a Common Shrew is eating all that it can catch and carry in its sharp little teeth. Because it is such a small, warm-blooded animal, it has good reserves of energy. It is the most hungry of all predatory animals. Without food, it can be dying of hunger within just a few hours.

Insects are the most important prey for the Common Shrew, which is why it is so useful to farmers.

Did you know...

Insectivores *have a special type of teeth to chew their diet of insects and small animals.*

*The **Desert Hedgehog** lives in the dry desert regions of Africa. By day it shelters in the warmth of its lair dug in the sandy ground. It can withstand up to ten weeks of total fasting.*

*The **Elephant Shrew** gets its name from its long nose. It lives in Africa. It has very long hind legs which enable it to move by jumping. It is nocturnal; in the evening, it goes hunting for locusts and grass-hoppers, which it likes more than anything else.*

*The **Otter Shrew** likes the water and lives in the slow-flowing currents of the tropical African forests. It is a hungry hunter which searches for crustaceans, insects and fish, digging in the mud with its nose.*

*The **Mole** is common in Europe, 11–17 cm in length. It uses its broad front feet as tools for digging, living underground in long tunnels dug from soft earth. It has acute senses of hearing, smell and touch to discover its prey – earthworms, insects, mice, amphibians and reptiles.*

*The saliva of the **Short-Tailed Shrew**, which lives in North America, contains a poison which acts on the nervous systems of the animals it bites; it can cause severe pain even to humans.*

*The **Star-Nosed Mole** is an insectivore from the eastern regions of North America. Its nose ends in a crown of 22 hairless appendages, which the animal moves about during its search for food or to move food during a meal.*

FLYING FOX

The Flying Fox, or Fruit Bat, is a member of the order Chiroptera, or bats. It is a true giant among bats and native to Indonesia, India and Sri Lanka, where it eats the ripe fruits which abound in forest trees. Unlike other species of bat which live alone, the Flying Fox likes company. Flying Foxes gather in groups of one hundred individuals. By day they look for a place to sleep all together, and this is always a large tree generally without leaves. At twilight, they rise up in ranks to reach the trees which are rich in fruit. They are guided by their own echo-sounding system and by a highly developed sense of smell, which enables them to sense odours, even those present in tiny quantities.

The Flying Fox is also able to withstand strong winds, flying up to the top of trees or above water surfaces.

Number of young: 1

Weight of adult: 650-920 g

Wingspan: 120-170 cm

VAMPIRE BAT

Number of young: 1

Length: 6.5-10 cm

Weight: 25-90 g

Although it feeds only on blood, the Vampire Bat is not really like the popular image of this strange animal. In fact, it is a bat very much like the others. There are three species of Vampire Bat, all living in the tropical regions of South America. One of these is an expert in sucking the blood of chickens and other large birds, whilst the other two prefer the blood of large mammals and will also attack birds and cattle.

Vampire Bats pass the day hidden in the holes of trees or in caves and go in search of their victims when night falls. To find prey, they fly very near the ground, at a height of 1–3 m. When a Vampire Bat has identified a sleeping animal, it lands on the ground and draws near to capture it, moving in a semi-erect position, and jumping over obstacles like a frog. As soon as it nears its victim, it jumps on it and bites it on the neck or on the shoulder. The saliva of the Vampire Bat contains anti-clotting substances, and so the animal bleeds to death, even after the Vampire Bat has gone. Only rarely do Vampire Bats attack humans. It is considered to be a very dangerous animal, not only because it sucks out the blood of its victim, but also because its bite carries disease, including rabies, which is widespread in the areas where the Vampire Bat lives.

Did you know...

__Bats__ are the only mammals really able to fly, because their front limbs have been transformed into wings. Bats have a membrane which extends from the 'finger' of the front claws to either end of its tail. To find its way around, most __bats__ have a system based on the perception of sonar or echo. They make high-pitched squeaks at ultrasound – soundwaves not heard by the human ear. These soundwaves rebound off objects which they meet (such as walls, trees or prey) and the echo is then recaptured by the bat. Using this sonar system, it moves about safely and can hunt even in the dark.

The __Greater Bulldog Bat__ feeds on the fish which it catches with a particular technique. It flies just above the surface of the water, and using the same sort of echo system as all bats, when it becomes aware of the presence of a fish, it grabs it suddenly with the claws on its foot. It lives in Mexico, northern Argentina and in the Antilles.

The __Small Mouse-Tailed Bat__ lives in the dry places of Egypt, in the Middle East and in India. It uses its thin tail as an organ of touch.

This is almost as long as its body, which makes this bat easily recognized when it takes flight by squeezing into small holes.

The **Lesser False Vampire Bat** lives in South-East Asia. It can be recognized by its large ears and a long sort of bump on its nose. It lives in caves and is a carnivore.

The **Greater Horseshoe Bat** is quite common in Europe. It gets its strange name from its nose which is shaped like a horseshoe. It comes out of its hiding place late at night to go looking for insects, the only food it eats.
At the first sign of winter, the Greater Horseshoe Bat goes into hibernation and falls into a deep sleep. During hibernation, its heart-beat and breathing become slower. It wakes up from its hibernation in the spring.

The **Mexican Free-Tail Bat** belongs to a family of expert fliers which can also move quickly on the ground, thanks to their strong legs and wide feet.
The Mexican Free-Tail Bat lives in flocks of thousands of individuals, inside caves, old buildings and other hiding places, from where they give off a strong musky smell.

PANGOLIN

Number of young: 1

Length at birth: 20-30 cm for all species

Length of adult: 75-150 cm according to species

'Pangolin' is the common name for mammals belonging to the order Pholidota. It comes from the Malayan ***peng goling*** which means 'animal which wraps itself into a ball'.
In fact, Pangolins can wrap their bodies completely around themselves. The body is not covered with hair, but with hard scales, one overlapping the other like the tiles on a roof.
There are seven species of Pangolin which are scattered around the hottest and most humid regions of Africa and Asia. Its home is the forest, where it can live on the ground or in the trees. Generally, the largest species live on the ground.
The Giant Pangolin prefers to live in the rocky zones of the tropical regions of Africa, where it digs a deep lair under a stone. During the day it rests, going out in search of food at night. This is when it raids nearby savannahs to use the long talons at the end of its claws to tear apart the nests of termites, its favourite food. It can swallow a quantity equal to 2 litres. Once satisfied, it returns to its lair before dawn. The smallest species, such as the Tree Pangolin, prefer to live in trees, which they climb slowly, grasping the branches with their tails.
Pangolins have no teeth and catch termites and ants with their tongues, each one similar to a long sticky tape. In the largest Pangolins, the tongue can be 40 cm long and, when it is not being used, it is rolled up and withdrawn into a chest cavity.

Did you know...

Edentata *are mammals which have few or no teeth at all. This order comprises three families to which belong anteaters, sloths and armadillos, all living in South-Central America.*

The ***Silky Anteater****, with a length of 40 cm and a thick tail, is the smallest in the group of anteaters. It spends its whole life up in the trees.*

The ***Giant Anteater*** *can reach 2 m in length, including the tail and 35 kg in weight. Its very long head is filled almost entirely by its long tongue, about 60 m long, which it can extend and retract very rapidly.*

The ***sloth*** *is an animal with a thick coat and very long limbs. It moves from one tree to another in tropical forests, with movements which seem to be very slow. It is strictly herbivore.*

The ***Two-Toed Sloth*** *spends the whole of its life hanging from branches. Like other species of sloths, it lives in symbiosis with a microscopic algae which settles itself in the skin of the sloth and gives it a greenish colour. This enables it to blend in perfectly with the leafy branches of trees.*

The ***Giant Armadillo*** *weighs around 60 kg and is longer than 1.5 m. It is a nocturnal animal and during the day it stays at the bottom of its lair which it digs in the ground.*

The ***Nine-Banded Armadillo*** *is the most common and widespread of around 15 species of armadillo.*

The ***Pink Fairy Armadillo*** *is the smallest armadillo with a maximum length of only 15 cm. Its back is covered with bony plates but its sides and stomach are covered with a thick coat of white fur. It lives underground, in a network of tunnels.*

The ***Collared Anteater****, from the same family as anteaters, spends its life in the branches of trees, searching for ants and termites. It has a long tail covered with scales.*

Bradypus tridactylus
PALE-THROATED THREE-TOED SLOTH
Cyclopes didactylus
SILKY ANTEATER
Tamandua tetradactyla
COLLARED ANTEATER
Choleopus didactylus
TWO-TOED SLOTH
Priodontes giganteus
GIANT ARMADILLO
Dasypus novemcinctus
NINE-BANDED ARMADILLO
Chlamyphorus truncatus
PINK FAIRY ARMADILLO
Myrmecophaga tridactyla
GIANT ANTEATER

LEPUS TIMIDUS
MOUNTAIN HARE
summer coat
winter coat
LEPUS CALIFORNICUS
BLACK-TAILED JACK RAB
SYLVILAGUS FLORIDANUS
EASTERN COTTONTAIL
ORYCTOLAGUS CUICULUS
EUROPEAN WILD RABBIT
LEPUS EUROPAEUS
BROWN HARE
OCHOTONA ALPINA
ALPINE PIKA

JACK RABBIT

Teeth: 4 incisors in the jaw; no canine teeth

Rate of reproduction: 4 or more litters per year

Length of adult: 45 cm

Length of ears: 20 cm

The Jack Rabbit belongs to the order of Lagomorpha. It is distinguished by the length of its ears, the extreme rapidity of its movements and its hind legs which are longer than the front. Also, unlike other rabbits, it gives birth to young which are already covered with fur, have open eyes, and which eat grass from their first day. The Jack Rabbit is common in dry regions covered with undergrowth in the south-east United States to Mexico. It can be most easily recognized by its ears with their large auricles (external part of the ear). As the Jack Rabbit developed, these changed in size and shape to enable the animal to regulate its body temperature in dry climates. The large surface of the auricle disperses excess heat more easily, and so avoids perspiration, which would lose precious water from the body. The Jack Rabbit can live without drinking, getting the water it needs from the fleshy blades of cactus. The Jack Rabbit is active by night, when the temperature is at its lowest. It eats grass, leaves of plants rich in sap and also cultivated plants, causing damage to crops. When the sun rises, it goes back into its lair, which is never very deep.

ALPINE PIKA

Teeth: 4 incisors in the jaw; no canine teeth

Rate of reproduction: 2-3 litters per year

Length: 20-26 cm

Weight: 100-130 g

The Alpine Pika is a small mammal which, like the Jack Rabbit, belongs to the order of Lagomorpha. It has small ears but no tail. There are 14 species, two of which live in North America and the rest in central and north-east Asia. The Alpine Pika also lives at great heights and can be found in the Himalayas at over 5000 m, the highest altitude at which a mammal can survive. It is also sometimes called the 'whistling hare' because of the long whistles it gives as a signal to its companions. It lives in colonies or in large family groups.

The Alpine Pika builds its lair in woodland beneath roots of trees or digs tunnels in places sheltered by the mountainside. It does not hibernate. Instead, during the summer, it gathers large quantities of grass dried by the Sun and then takes this dried grass to store in its lair, ready to eat during the winter.

The Alpine Pika goes out within a radius of around 200 m from its lair to gather grass, twigs and pieces of bark. It does not fear the cold and goes out into the open even when the temperature descends to below -20°C.

Did you know...

__Lagomorphs__ are distinguished from rodents because they have four incisor teeth instead of two, which grow continually.

The __Mountain Hare__ changes the colour of its coat according to the season. In summer it has a red-brown or grey-brownish coat, whilst in winter its coat changes and becomes completely white, to blend in better with the snowy landscape.

The __Eastern Cottontail__ is a North American rabbit which lives in humid zones. Unlike other species, it does not fear swampy ground, nor water, where it is a fine swimmer.

The __European Wild Rabbit__ is widespread through most of Europe. There are also several species in Asia, Africa and America. It is a sociable animal which lives in small colonies which comprise numerous families. It spends half its life underground, where it digs an intricate series of deep tunnels as its lair. The female gives birth to 5–12 young up to six times a year.

The __Brown Hare__ is also widespread in Europe, but it has also settled in America, New Zealand, Australia and the Far East.

GOPHER

Teeth: 2 large incisors, non canine teeth

Number of young: 1-3

length of body: 13-20 cm

Length of tail: 6-10 cm

Weight of adult: 300-400 g

The gopher is a rodent of the family of Geomyidae. There are only a few species, the most common of which is the Eastern Pocket Gopher of North America.

It has two external pouches on its cheeks which are fur-lined and which open instead of lips. These cheek pouches are used to carry both food and the material to line its nest. Thanks to special muscles, a gopher can turn the pouches inside out to empty them of everything inside. Then the cheek pouches go back to normal and the opening is closed by the skin of the face.

A gopher spends its life mostly in its lair, built in an intricate network of tunnels and holes. The male only leaves the lair to go in search of a female. He leaves her directly after mating.

Gophers are burrowing animals with long, strong claws. They are vegetarian and gnaw roots and bulbs.

SQUIRREL

Teeth: 2 large incisors, no canine teeth

Number of young: 3-7

Length of body: 20-28 cm

Length of tail: 14-24 cm

Weight of adult: 230-480 g

The Common Squirrel, or Red Squirrel, lives among the branches of trees in European woodland. It changes its coat in spring and in autumn; during the winter, its tawny coat becomes grey. It is active only during the day, jumping between the branches, where it builds its nest, called a drey. It uses its tail as a counterweight and a rudder when jumping. When it sleeps, the squirrel also uses its tail as a 'cover' and at mating time, as a signal, moving it in a certain way. As it moves, it is guided by hairs at the side of the body which it uses as organs of touch, and with which it can identify obstacles. The squirrel is a sociable animal, living in small groups made up of different couples, each couple with their own territory. Its food is varied – seeds, berries, buds, mushrooms and toadstools, snails, insects and birds' eggs. It hides its food away in small holes.

Did you know...

*The order of **Rodentia** is the largest of mammals and distinguished by their teeth which continue growing throughout their lives. These teeth are kept short by gnawing.*

*The **Flying Squirrel** from North America is a nocturnal animal. The wide folds in its skin, between its front and back legs at either side of its body, enable it to glide in the air for quite long distances.*

*The **Black Squirrel** is widespread in Java, Bali and Sumatra. It is a large animal, up to 50 cm long. It can be easily tamed and can live up to 16 years.*

*The **Desert Kangaroo Rat** is a typical inhabitant of arid surroundings and is nocturnal. It moves by making long jumps with its well-developed hind legs. It uses its tail as a rudder.*

*The **Marmot** is quite common in the Alps and is active only by day. It is a sociable mammal, which lives in large colonies. In autumn the whole community goes into deep tunnels which they have dug under the rocks. Here the Marmot remains in hibernation for the whole of the winter, its heart-beat and breathing slowing down.*

*The **Mountain Beaver** is different to the common beaver because of its shorter tail and long, strong claws with which it digs intricate networks of shallow tunnels. It lives in the mountainous regions of the Pacific coasts of North America.*

*The **Eastern Grey Squirrel** is widespread among oak trees and the woodlands of North America. It is very agile and can reach a speed of 90 km/h over short distances.*

RATUFA BICOLOR
BLACK SQUIRREL
GEOMYS BURSARIUS
PLAINS POCKET GOPHER
GLAUCOMYS VOLANS ASSAPAN
FLYING SQUIRREL
DIPODOMYS DESERTI
DESERT KANGAROO RAT
APLODONTIA RUFA
MOUNTAIN BEAVER
MARMOTA MARMOTA
ALPINE MARMOT
SCIURUS CAROLINENSIS
EASTERN GREY SQUIRREL
CITELLUS SOUSLICUS
SOUSLIK
SCIURUS VULGARIS
RED SQUIRREL

Did you know...

*The **Jerboa** is a small rodent of the African and Asian deserts. It is a nocturnal animal and its strong, well-developed hind legs enable it to jump like a kangaroo.*

*The **Northern Birch Mouse** is only 5–7 cm long, which makes it the smallest mouse in central Europe. Its hibernation is the longest of all mammals, lasting at least eight months.*

*The **Mole Rat** lives mainly in the region of the eastern European steppes and in western Asia. Its eyes have become shrunken due to the animal living totally in darkness, deep underground.*

*The **Lemming** lives in the Tundra of the Artic regions. It is active throughout the year and has many young. It lives in deep lairs which it digs for itself.*

*The **Fat Dormouse** lives in the woodlands of Europe. Its long hibernation lasts from October to April. Before going into hibernation, it gorges itself on seeds and fruit.*

*The **Harvest Mouse** is widespread throughout Europe and eastern Asia. It is about 5.5–7.5 cm long.*

*The **Striped Grass Mouse** from Africa gets its name because its coat is striped with brown and black. It builds a*

ball-shaped nest in the undergrowth.

The **Common Dormouse** is a rodent which can be found in the woodland and countryside areas of Europe. It lives in trees and builds its nest in bushes and among the branches of nut trees, eating the nuts.

The **Capybara** grows to a length of about 1.5 metres. It lives on the banks of the large rivers of South America.

In the **African Porcupine**, a great deal of its fur has been transformed into spines. It can be found in southern Europe, as well as Africa.

The **Tuco Tuco** is a South American rodent. It gets its name from the strange guttural sound that it makes.

In the wild, the **Chinchilla** lives in the highest regions of the Andes mountain chain. It is hunted for its valuable fur, and so it is a protected species.

The **Brazilian Guinea Pig** lives in the grassy expanses of South America.

The **Mara** lives in the pampas and the dry steppes of South America. It can survive without drinking, getting the water it needs from vegetable matter.

The **Guinea Pig** is descended from a wild species reared by the Incas of Peru.

The **Cane Rat** lives in Africa and grows to a length of 60 cm with a weight of 7 kg.

Did you know...

*The **Corsac Fox** lives in Mongolia and Mancuria in Asia. It lives in small groups, eating rabbits, marmots and birds.*

*The **Arctic Fox** lives in Arctic regions, sometimes on ice floes. Its numbers have become drastically reduced, because of its valuable fur. It feeds on rodents, fish, birds and sometimes insects, as well as the remains of the prey of the Polar Bear.*

*The **coyote** is a distant relative of the wolf. It lives in Alaska, in Canada and throughout North America. Because it is very adaptable, it can often survive on the outskirts of big cities. It is an animal with a great sense of fun and sometimes it can be seen playing with small animals without doing them any harm.*

***The Swift Fox** from North America also belongs to the same family as the Red Fox, as well as the **Pale Fox** which lives in Africa and in parts of*

WOLF

Number of young: 4-7

Length: 1-1.50 m

Weight: 25-50 kg

Life span: 16 years

The Common Wolf is a carnivore which is widespread throughout Asia and in North America, but its numbers have been reduced and there are few examples in Europe.

It is one of the most well-known members of the family Canidae, as well as being one of the most intelligent mammals. It is able to adapt to different conditions both in climate and in the land and types of vegetation. All this has enabled the wolf to survive in many regions of the world – in the steppes, scrubland, forests and sometimes in farming areas, where, if it is hungry, it will kill sheep and chickens for food.

The wolf avoids human beings, not because it fears them, but because it prefers to live away from inhabited areas. Only when it is gripped by hunger will it venture towards urban areas in search of food.

The wolf is a sociable animal, which means it likes company. It lives in families or in small packs of two or three pairs with their young. An adult male wolf leads the group as pack leader, and he mates only with the most dominant female. The other members of the group are under the control of the leader.

RED FOX

Number of young: 3-8

Length: 60-70 cm

Weight: 6-10 kg

Life span: about 12 years

The Red Fox belongs to the family Vulpes. It is quite a common animal, living in the woods of Asia, North America and Europe. It is famous for its cunning and its craftiness, and, together with the wolf, is one of the most easily recognizable European carnivores.

The Red Fox can be small to medium size, with a long nose, big ears and bushy tail from 35–40 cm long and it continues to survive despite being the object of so much hunting. It prefers the quiet life, especially in woodland and in the country. It likes to stay in one place, and once it has chosen a home, it stays there and digs a lair in the form of a network of shallow tunnels. But sometimes, the Red Fox does not dig its own lair, but takes possession of a home belonging to another animal, such as a badger or a rabbit, after chasing away the rightful owner.

It lives in small family groups and goes out mostly at night-time. Its diet is made up of small rodents, fish, birds and reptiles. It has earned its dubious fame for theft and plundering because of its liking for poultry and small cattle.

Asia Minor.

Foxes have a 'language' which is very distinctive and communicate to each other by different sounds – loud cries during fights, whimpers, yelps, whines and growls between the mother and her cubs, cries of alarm and howls when they sense danger.

When two ***wolves*** *fight for the leadership of a pack, the one who loses lays down at the feet of the winner to avoid being torn to pieces. The winner, as a sign of victory, marks his territory with urine, and marks out his hunting territory in the same way.*

The ***Jackal*** *(in photograph) is a slender, agile canine. The* ***Golden Jackal*** *lives in northern Africa and all of southern Asia. It lives in all surroundings, from humid forests to arid zones.*

Díd you know...

All members of the Ursidae family have massive heads with small ears and a rather clumsy-looking large body with strong, solid legs and strong claws.

The ***Spectacled Bear*** *is the only bear which lives in the Southern Hemisphere in the forests of the Andes, South America. Here, this bear can reach altitudes of over 3000 m, but it will also descend down to the valleys as well as the moors and pasture areas on the valley bed. It prefers a vegetable diet and often makes a bed among the highest branches of a tree, where it spends the night.*

The ***Sun Bear*** *spends most of its time in trees, where it makes a bed of broken branches. It eats fruit, leaves, small animals and especially ants, which are its favourite food. It destroys young palm trees by eating away the centre.*

The ***Sloth Bear*** *is different in many ways to other bears in the Ursidae family. Its coat is long and hairy and its lips are fleshy and extremely mobile. Its bottom lip sticks out, making it look rather like a monkey. It is found on the island of Sri Lanka and in southern*

GRIZZLY BEAR

Number of cubs: 1-2

Length of body: 1.50-2.50 m

Weight of adult: 100-350 kg and over

Life span: 30-40 years

The Grizzly Bear is one of the most well-known mammals, found across a vast territory which extends from Europe to northern Asia and from Central to North America. Together with the Polar Bear, it is the largest carnivore in the world, but its size varies according to the region where it lives.

The Grizzly Bear is a solitary animal which lives only with its mate. It is a wanderer, not staying in one place for very long. Its diet is varied and includes vegetables, insects, fish, rodents and sometimes also larger prey such as reindeer and elks. Although it may look good-natured, the Grizzly Bear can change its mood very quickly, but only rarely is it a danger to humans.

When the weather begins to get cold, the Grizzly Bear takes shelter under a large rock or in a cave to spend the winter. It does not go into a proper hibernation, but passes the time mostly sleeping and using up the reserves of fat which it has put on during the summer. The female takes particular care in preparing the winter home ready for her cubs which are born towards the end of December or the beginning of January. She rears them with dedication and does not leave them for almost two years.

POLAR BEAR

Number of cubs: 1 or 2

Length of body: 2 m for the female; 2.5 m for the male

Weight of adult: 320-410 kg; sometimes up to 1 tonne

The habitat of the Polar Bear (in photograph) are the seas of the Arctic Circle covered with layers of floating ice. It has a slender neck and head, small ears and a round body. Its feet have short claws and the 'toes' are joined together with a membrane, which makes it easy for the Polar Bear to swim. The soles of the feet are also covered with fur to prevent it from slipping on the ice. The Polar Bear's coat is creamy-white, enabling it to blend in with its snowy surroundings, as well as being waterproof. The Polar Bear catches fish, sea birds and seals. Despite its massive shape, it can move lightly on the snow and, when it sees prey, it draws near slowly, taking its prey by surprise and killing it with its paw.

Mating takes place only in April. Only at this time the male lives with the female. At all other times it lives alone. In autumn when the female is expecting her cubs, she digs a lair out of the ice and snow ready for her babies when they are born towards the middle of December. The female, and in some cases also the male, pass the winter in one lair in the snow.

India, living in forests and among the trees in the savannahs and the plains. In summer the Sloth Bear goes in search of bee-hives, which it will destroy in order to eat the honey, the larvae and even the bees. It also sucks out termites from their nests.

The ***Asian Black Bear*** *is sometimes called the Collared Bear, because of the long hairs which cover the front and back of its neck and shoulders in a distinctive 'Y' shaped pattern on its black coat. It lives in the forests of the Asiatic mountains at altitudes of between 1400–3300 m. An expert climber, the Asian Black Bear spends most of its time in the trees. During winter it goes lower down into the valleys where it can find food more easily. It eats berries, fruits, buds and leaves.*

The ***American Black Bear*** *is one of the most well-known animals of North America. It is a peaceful animal, eating mainly fruit and berries. It is this animal which inspired a toy-maker in New York to make a 'teddy bear' named after the President of the United States at the time, Theodore (or 'Teddy') Roosevelt. From then on, toy bears have always been called 'teddy bears'.*

GIANT PANDA

Length of body: up to 1.5 m

Weight of adult: about 100 kg,the female; 120–140 kg, the male

Number of young: 1

The Giant Panda belongs to the family of Ailuridae and lives only in the mountains of western China. It looks rather like a bear, with a massive head, small, round ears and its black and white coat, thick and bristly. For a long time it has been the symbol of the International Association of the Directors of Zoos and from 1961 the emblem of the World Wildlife Fund (WWF).

Although the Giant Panda is regarded as a symbol of animal species in danger, it is at less risk than other animals. Because it lives within such a small, precise area, it can be well-protected by the measures which the Chinese government have put into place to prevent its extinction.

The diet of the Giant Panda consists of vegetable matter, especially bamboo shoots. Sometimes it will also eat rodents and other small animals. It lives mainly on the ground, climbing only rarely – although if it feels threatened, it can climb trees easily. The female rears her one cub in the hole of a tree or an underground lair.

The Giant Panda does not fear the cold and can live at altitudes of up to 3500 m.

COATI

Weight of adult: 3-6 kg and more in the male

Length of body: 75-135 cm (including tail)

Length of tail: 30-70 cm

Number of young: 2-6

The Coati is a small carnivore mammal belonging to the family Procyonidae – raccoons and their close relatives. It can be recognized by its long nose, shaped like the small proboscis of an insect, which it inserts in every hole, every nook and cranny, in its continuous search of something to eat. Its claws have long, strong talons which enable it to climb trees as well as burrowing in the ground. The Coati is an omnivore and eats invertebrates, fruits, birds and small rodents.

The Coati lives in groups of various sizes from 4 to 25 individuals, both male and female. At about two years old, the male leaves its group to lead a solitary life.

In the mating season, each adult male re-joins a group, chasing away all the other males at the time and marking out his territory with its body odours. A short time after the birth of the young, the male leaves the female and takes up his solitary life once again.

As well as being a very active animal, the Coati takes great care over its grooming, combing its coat with its talons and its teeth. It alternates its hunting and grooming for brief moments of rest lying down in the sunshine, whilst the young ones in the group play in the area.

Did you know...

*The **Olingo** is a member of the Procyonidae family which lives in Peru and Bolivia as far as the tropical forests of Mexico. It is an animal with a slender body covered with long, soft grey-brown hair. It is nocturnal and lives alone or in pairs, spending most of its time at the top of trees and mainly eating fruit.*

*The **Red Panda** is an Ailuridae. It is a small raccoon-like mammal which lives in the bamboo forests in the Himalayas at altitudes of between 1800 m and 4000 m. It is a nocturnal animal. It is very elegant and has a thick, soft coat and tail. It lives a solitary life, feeding on the buds of bamboo, roots, pods and fruit.*

*The **Ringtail**, often called the Ringtail Cat, can be recognized by its long, thick tail with ringed markings in black fur.*

It lives both in the rocky plains of North America and in the thick woodland and humid Central America. Strictly nocturnal, it is very quick and agile as it moves, going in search of small mammals, birds and eggs as well as reptiles and fruit.

*The **Crab-Eating Raccoon** lives in the tropical jungles of South America. It has a larger and thinner body than the Northern Raccoon, its fur is shorter and bristly and its tail is not so bushy. As well as living on the ground, it is also an excellent swimmer.*

*The **Kinkajou** is a graceful carnivore with a thick and very soft coat of a most beautiful golden colour. It also has a long prehensile tail by which it can hang or settle among the branches of trees in the tropical forests of South America where it lives.*

*The **Northern Raccoon** lives in the forests of North America and is recognized most of all by its strange habit of washing all its food in water – but not to make sure the food is clean, as many people think. In the wild, the Northern Raccoon does not wash anything and hunts aquatic animals, such as crustaceans, fish, amphibians, which it catches by putting an open claw into water and searching the depths. In captivity, not being able to catch aquatic prey, it gives vent to its instinct by washing whatever objects or food it can get hold of.*

HYAENA HYAENA
STRIPED HYENA

PROTELES CRISTATUS
AARDWOLF

CROCUTA CROCUTA
SPOTTED HYENA

Did you know...

*The similarity which the **Aardwolf** has with the Striped Hyena makes it look far more fierce than it actually is, especially when the long, stiff hairs on its back and neck stand on end to make it look even bigger. Generally, it will eat dead animals and birds, but when these are scarce, it may attack cattle. The Aardwolf can be found in South Africa, Tanzania and southern Egypt.*

***Hyenas** are distinguished by having four toes on each foot. They are important animals, because they are scavengers, clearing up the remains of dead animals which other carnivores have left behind. The hyena's strong jaws and sharp teeth can even shatter bone, so they are able to clean carcasses of every last shred of flesh – although a hungry hyena will also attack and kill animals for food.*

*The **Spotted Hyena** (see photograph, right) is a formidable predator. It hunts by day as well as night, and making its strange-sounding call rather like a peal of*

STRIPED HYENA

Weight: from 30 to 45 kg

Length of body: about 1 m

Number of young: 2-4

The Striped Hyena (in photograph) belongs to the family Hyaena, animals which have an appearance similar to the civet and the mongoose. Compared with its two cousins, the Aardwolf and the Spotted Hyena, the Striped Hyena is smaller in size, and it can be found over a wider area. The Spotted Hyena lives in Africa and Asia. Its colouring varies from region to region with shades of yellowish brown to sand. On the back, sides and legs it has black stripes. From its neck to the tail, it has a thick mane of hair which it can stand up straight as it likes. When the Striped Hyena is angry, it becomes a very impressive animal. At twilight and by night, it lives either alone or with a partner, more rarely in small packs. Whilst the Asiatic Striped Hyena feeds on herbivore animals which it captures during the night, the African species faces competition from other predators, such as the Spotted Hyena, and so prefers to eat the remains of carcasses, small prey and sometimes cattle.

Due to its rather timid nature and few bursts of aggression, the Striped Hyena poses no danger to humans. During the day, it lives mostly in a lair which it digs with its sharp claws and which is shaded by the courses of rivers, or in a natural lair such as the network of roots which spread out at the bottom of trees in the Savannah.

AARDWOLF

Weight of adult: 20-32 kg

Length of body: 55-80 cm

Number of young: 2-4

The Aardwolf is the only member of the family of Protelidae and is scattered from the Sudan to South Africa.

Its body structure is typical of hyenas, with a long nose, large, pointed ears and a coat which can vary in colour with black stripes, and a very bushy tail.

It is a mysterious animal, strictly nocturnal and never going out during the day. It spends the daylight hours hidden in the cleft of a rock or inside a lair which has been abandoned by another animal. However, the Aardwolf can also dig a lair for itself, but only in very soft earth.

Despite its appearance it is not a great predator; it does not have many teeth and its diet mainly consists of insects, preferably termites, which the Aardwolf catches in large quantities with its long, sticky tongue. When insects are scarce, the Aardwolf contents itself with small birds or lizards.

laughter, which is why this hyena is often called the 'Laughing Hyena'. Sometimes, it lives alone, sometimes in a pack which will go out together ready to trail and to attack living prey. The Spotted Hyena can easily catch a gnu or a zebra, because it can reach a running speed of 65 km/h. The Spotted Hyena is the strongest of the family of hyenas; its height to its shoulders is about 90 cm, and its weight averages 85 kg. Unlike the Brown and the Striped Hyena, the Spotted Hyena has a short tail and very small eyes. It lives south of the Tropic of Cancer in the desert zones and the African savannah, often going up into the mountains as well.

The Spotted Hyena is often the object of occult and superstition among many African tribes. Many people believe that the bile of this animal is both poisonous and able to bring about unexpected death.

LION

Weight at birth: 1200-1500 g

Weight of adult: 150-250 kg for males; 120-180 kg, females

Life span: 15-25 years

Number of cubs: 2-4; 5-6 in captivity

The lion is always regarded as king of the African savannah. The male, unlike all members of the cat family, has a thick mane which surrounds his head and protects the neck from the claws of an opponent when it fights.

Of all members of the family Felidae, the lion is the one which likes company the most. It lives in groups of 6–30 individuals made up mostly of female lions and their young, with one or two adult males to protect the group and to ensure reproduction. Each group occupies a definite territory which covers a vast area of around 250 sq km. Its famous roar seems to serve the lion purely as a signal of its presence and to establish territorial limits.

The male lion is very lazy, mostly 'robbing' prey from other carnivores or prey caught by a lioness, who is always more active and ready to attack than he is. The female lions are also excellent mothers, forming a very united community together to feed and to rear the young. When the cubs are about six months old, the females take them hunting. At the age of about three years, when they are able to mate, the young male lions leave their families in search of a group where they can be the leader. Whilst the female lions remain in a group for life, the males go away, substituted every two or three years by new arrivals which are stronger and more vigorous.

Did you know...

*The **Snow Leopard** can be found among the isolated valleys of the high mountains of central Asia, up to altitudes of 4000 m. It is a medium size, solitary animal, about 1 m long and weighing from 25 to 50 kg and with many features of familiar cat behaviour, purring and eating the remains of food stuck between its claws.*

*The **Cheetah** is the fastest animal on earth and can reach a speed of 90–98 km/h. It has long legs, a slender body and a long tail. It lives in the African and Asian savannahs and goes hunting by day. With its sharp eyes, the Cheetah can pick out an antelope at a considerable distance. It chases its prey in a short and extremely fast sprint of 300–400 m.*

*The **Puma** is a large solitary feline living in North and South America, from Alaska to the extreme end of Patagonia. It fears neither the heat nor the cold and finds itself perfectly at ease either among rocks or among the trees. It mostly eats deer but its diet can be very varied.*

*The **Clouded Leopard** is one of the most mysterious felines, because it is difficult to observe in its natural surroundings. It is a tree-dwelling animal and lives in the thickest parts of the Asian tropical jungle, leading a very secluded existence. It can be distinguished from other members of the family by its upper canine teeth which are particularly long and sharp and which remind people of the prehistoric sabre-toothed tiger.*

*The **Leopard** is a most beautiful predator, elegant and agile. The Black Panther, as well as being wrongly regarded as more fierce, is not truly 'black'. Instead, it is no different to the ordinary male leopard except for the black colour of its coat.*

The leopard lives in Africa and the hot regions of Asia, and adapts well to the plains in mountain regions. It is a carnivore which is easily pleased, eating any animal from a mouse to a porcupine, or a gazelle to a zebra.

TIGER

Weight at birth: 800-1600 g

Weight of adult: 120-280 kg

Life-span: 20 years

Number of young: usually 2

If the lion is the king of the savannah, then the tiger can be regarded as the ruler of the Asian jungle. It is an extremely agile animal which can clear 5 or 6 m in one jump, leap over obstacles 2 m high and swim with great ability. It is even stronger than its cousin, the lion, and its solitary way of life has made it well-respected in the Animal Kingdom. The ancient Chinese worshipped the tiger as a creature with supernatural powers. But, despite all this, the tiger has had to overcome continued persecution and now some species are at risk of total extinction. In order to live, the tiger needs three things – an abundance of large wild life (young elephants, goats, wild boar), water and places where it can rest (natural caves or thick vegetation).

The cutting down of forests and expansion of farmland has dramatically reduced the natural home of the tiger. At one time it was present in almost the whole of Asia. Now it survives in some safari parks and in the mangrove forests in the delta of the River Ganges.

Each tiger has its own personal territory where it comes and goes and remains for years. Only the females consider this territory as a place for hunting, moving away solely to find a male to mate with. When they give birth, the female goes in search of shelter. At about eight weeks after the birth, the young begin to follow the mother hunting and they stay with her until they are two or three years old.

Did you know...

*The **Serval** has very strong limbs, ideal for its life in the savannah and among the tall grasses. It has exceptional hearing and can detect the presence of rodents in their network of underground tunnels. Its diet consists of rodents, rabbits, young antelopes and birds.*

*The **Desert Lynx** likes hot, dry surroundings. It can be found mostly in open areas which have a lot of undergrowth and especially in the open desert regions of India and Pakistan. It is a solitary beast of the night. During the day, it shelters in the lair of a fox or a porcupine.*

*The **Bobcat** is quite common in the eastern United States. It particularly likes areas dotted with bushes and trees – although it lives in all surroundings, from the forest to the swampy expanses of Florida. It is an excellent hunter, and will kill rabbits and rodents.*

*The **Ocelot** is a well-known carnivore from South America. It is easily recognized by its beautiful coat, and cat-like ears. The Ocelot has been*

LYNX

Weight at birth: about 200 g

Weight of adult: 11-42 kg

Length of body: 70-110 cm

Number of young: 1-4

The Lynx is a very distinctive feline (member of the cat family) with its short body and long limbs. Its tall, pointed ears have tufts of hair at the top and these enable the Lynx to find the source of sounds more easily. Its feet are well-developed and covered with thick hair to protect it from the cold, the ice and the snow.

The coat of the Lynx can vary in colour, from grey to red, thinly spotted or with no spots at all. The Lynx can be found throughout northern Europe and most of North America, from the flat plains to the sea, and in mountain areas at altitudes of up to 3400 m. But the Lynx is mainly a solitary forest animal and its hunting territory can extend up to 20 sq km in any area – more, if wildlife is scarce. Most of all, the Lynx likes to catch deer, chamois and rabbits. However, it will not ignore smaller prey, which it kills and then swallows in one gulp.

The Lynx is very wary of humans, so it is very difficult for people to observe it for any length of time. It makes its lair in a well-protected place, in small caves or hollow trees. Both male and female leave their territory only to mate and then they return.

JAGUAR

Weight at birth: 700-980 g

Weigh of adult: 45-110 kg

Length of adult: 110-180 cm

Life-span: 22 years

Number of young: 1-4

The Jaguar is the largest feline in America. At one time it was even present in the south-west of the United States. It is still common in the province of Chaco in Argentina and in the Mato Grosso, Brazil. Elsewhere, it is becoming increasingly rare. For many years, the Jaguar has been killed for its precious fur and with the excuse that it attacks cattle, such as horses and sheep.

As well as the savannahs and rainforests, the Jaguar will also live in woodland areas where there is a plentiful supply of water and regions with tall grasses which give it shelter and with running water flowing through it. Its feeding routine is very varied, and it eats a great quantity of small animals – sheep, rodents, monkeys and sloths.

Although it looks rather like the leopard, the Jaguar is less agile and a poor climber, so it prefers to hunt on the ground. But it is an excellent hunter of fish, which it likes very much. It will sit on a rock patting and stroking the water, and, when a fish comes near, the Jaguar will stun it with a vicious blow of its paw at just the right time.

Like all large felines, the Jaguar is mostly nocturnal. It lives alone within its own territory and a male and female remain together only during the mating season.

known from the times of the ancient Aztecs. It has a peaceful nature and rarely attacks humans.

*The **African Golden Cat** is a very rare animal which lives in the tropical forests of western Africa. It is larger than the domestic cat and gets its name from the golden colour of its magnificent coat.*

It hunts monkeys, squirrels, porcupines and birds.

*The **Fishing Cat** is quite widely scattered throughout tropical Asia, where it lives in the humid forests among the mangroves, along the banks of flowing waters and among reeds.*

It is a fine climber, but prefers to remain on the ground. It eats molluscs and shrimps, as well as the fish which it catches – hence its name.

***Pallas's Cat** is a small feline which is broad and squat. It has a long tail which is ringed with black.*

It lives in the rocky regions of central Asia and can be found at altitudes of up to 4000 m.

It can withstand the extremely cold climates of Siberia and Mongolia.

WALRUS

Number of young: 1

Length: 3-4.5 m

Length of body: 1.50-2.50 m

Food: invertebrates

Home: Arctic zones of Europe, Asia and North America

Life-span: 12 years in captivity

The Walrus is the giant of the Arctic ice floes, a member of the sub-order of Pinnipeds (aquatic mammals with fins). It is an impressive-looking animal with its long tusks – two long, upper teeth which hang down, and which are present both in the male and female. In the female, however, these tusks are shorter and not as strong. These great teeth are weapons which males sometimes use in fierce battles during the mating season, when two rivals fight for the same female. These tusks also enable the Walrus to search among the mud of the sea bed for the bivalve molluscs which hide there and which the Walrus likes to eat. These are made up mostly of clams which it swallows whole, including the shell. The shell is broken up in its stomach by 'stones' which effectively mince up its food.

Peaceful by nature, the Walrus lives in colonies of various size, sometimes up to a hundred individuals. It spends most of its time dozing lazily on the stony banks. On the ground, the Walrus moves slowly and not without effort, but it can swim at speeds of up to 24 km/h. To find food, the Walrus can go underwater to a depth of 50 m, remaining submerged for about 10 minutes. When the sea is calm, the Walrus can sleep in the water, keeping afloat thanks to two pockets filled with air under its throat, which it inflates like balloons. Strict laws have been put in place to protect the Walrus, because it is hunted for its tusks, its fat and its skin.

CALIFORNIA SEA LION

Weight of male: up to 1 tonne

Length of male: 3.5 m

Feeds on: fish and cephalopod molluscs

Home: Pacific (from Mexico to California)

Life-span: 23 years

The California Sea Lion (in photograph below) is often mistaken for a seal. The main difference is that its rear legs are joined together to make a sort of fin, although it does still have two feet. Its coat is red-brown when dry, black when wet. It can climb rocks and coral reefs, diving from considerable heights. In water, the California Sea Lion is a champion swimmer, achieving speeds of 17–40 km/h and swimming down to depths of up to 75 m. It can also jump up to 2 m high outside the water. On sand, it can run more quickly than a man, although its style is not so elegant as when it is swimming.

As mating-time approaches, the males wait in groups for the arrival of the females, and, after some fierce battles, one conquers a 'harem'. Each female has just one baby, after it develops inside her body for one year. There are many kinds of sea lion, but the California Sea Lion, which lives along the coast of the Pacific Ocean of the United States and Mexico, is the most well-known. Today it is a protected species, but in the nineteenth century it was extensively hunted for its fat. It is the Californian Sea Lion which has astounded spectators by its incredible talent for play. Its remarkable sense of balance makes it unique; it is the only animal that will play naturally with a ball or similar object.

ODOBENTUS ROSMARUS DIVERGENS
PACIFIC WALRUS

ZALOPHUS CALIFORNIANUS
CALIFORNIA SEA LION

ARCTOCEPHALUS AUSTRALIS
SOUTH AMERICAN FUR SEAL

CALLORHINUS URSINUS
NORTHERN FUR SEAL

EUMETOPIAS JUBATUS
STELLER SEA LION

ODOBENTUS ROSMARUS ROSMARUS
ARCTIC WALRUS

Did you know...

Pinnipeds *belong to an order of carnivores which have many features in common with other carnivores – in particular, the shape of the skull and their teeth. For example, the long-shaped head of the sea lion is rather like that of a dog. Under the skin, a Pinniped has a considerable layer of fat to insulate the body in very cold waters, and to conserve energy.*

Sea lions are members of the family ***Otariidae*** *together with the* ***Eared Seal*** *and the* ***Fur Seal****, indicated by the fact that these animals all have large outer ears. This distinguishes them from other Pinnipeds which, in aquatic animals, are generally provided with some sticking-out structure, like pointed outer ears, which contrasts with the hydrodynamic shape of the body.*

The male and female ***Steller Sea Lion*** *live in the Bering Sea on the northern Pacific coast. At the time of reproduction they remain lazily in large groups along the rocky coastline.*

In the nineteenth century, the ***Northern Fur Seal*** *was almost completely exterminated by fur-trappers. This reduced numbers considerably; but, fortunately, the American Government have undertaken strict measures to protect the species and numbers have now risen.*

The adult ***Southern Fur Seal*** *which lives in the Antarctic, southern Africa and South America is brown-coloured with shades of pale yellow. However, the young seals have a darker back, which is almost black. Like other Pinnipeds, its body length is from 2–2.5 m.*

Did you know...

*The family of **Phocidae** is the largest family of Pinnipeds, and comprise animals with a hydrodynamic shape to the body, a very short neck and nostrils which become round during breathing. The short front legs have become flat and plate-like and the rear legs are joined in such a way that the soles of the feet face one another. Many species make a howl similar to that of a dog.*

*There are at least six million **Ringed Seals** – so many that there has been a census of those which have been studied – but there could be more. Even though they live only in Arctic waters, the Ringed Seal is one of the most common seals. The pups are born from the middle of March to the middle of April.*

*The **Hooded Seal** gets its name from the nasal cavity in adult males. When it is excited, the male closes its nostrils and this cavity swells with air until it becomes a 'crest' or 'hood' which extends from the point of its nose to*

MEDITERRANEAN MONK SEAL

Weight: 250-400 kg

Length: 2 m, sometimes 3 m

Home: Sardinia, Dalmatian Coast (rarely)

The Mediterranean Monk Seal, with its grey-brown, thickish coat, is the only seal which spends its life in warm seas. Because of the development of tourism in the regions where it lives, it is in danger of extinction. It spends long periods of time in the sea, coming on to the beaches of uninhabited zones only at mating time. In the relatively brief time spent on land, its coat becomes covered with green algae which disappears only when the Monk Seal returns to shallow waters or basks in the sun. Today there are quite a few Monk Seals which live in the open seas. Those surviving spend a great deal of time near the coast, and more rarely in places which remain isolated. Also, the places and the surroundings where the Monk Seal reproduces have changed. Many have left the sandy beaches and begun to prefer grottos, and this has certainly helped them to survive. The Monk Seal eats mostly mackerel, but, because of intense fishing and pollution, shoals of these fish are much less numerous today than ten years ago. There is no doubt that the reduction in numbers of these seals is the result of the reduction in their prey.

ELEPHANT SEAL

Weight of male: 3500 kg

Length: 6.5 m

Feeds on: fish

Home: islands in the Southern Hemisphere and along the North American Pacific coast

Life-span: 18 years

The Elephant Seal is the most impressive of Pinnipeds. Only the White Rhinocerous and the elephant weigh more than the male Elephant Seal. Its adaptation to its aquatic life has not only profoundly changed its shape, it has also given it a perfect ease in water, whereas it can only move with incredible effort on land. But, despite the heaviness of its body, the Elephant Seal can move fast enough when it needs to.

The Elephant Seal spends a large part of the year living in groups and finding food. Its teeth, numerous but small, do not enable it to catch large prey, but it can go without food for up to 100 days. In the mating season, Elephant Seals gather together in their favourite places for reproduction – small, secluded islands where, after carrying their young for one year, the females give birth, one seal pup for each couple. Each male will mate with more than one female. So, at mating time, he gathers females around, fighting off rival males to conquer them.

At the end of the summer the Elephant Seals go back to the salty swamps where they moult and the hairs of the previous winter falls off and rots, giving off a horrible smell. There are two different sub-species of Elephant Seal – one lives in the Southern Hemisphere, the other along the eastern coasts of North America, on the eastern side of the Pacific Ocean.

behind its eyes for about 20 cm.

*The **Leopard Seal** is the only one in the order of Pinnipeds where the female is larger than the male. This seal is very greedy for penguins, hunting for them as it dives into the sea in an ambush style of attack. The oesophagus (food pipe) of this seal is very elastic, so that it can take in enormous quantities of food.*

*In the past **seals** were very important for some populations – for example, the Inuits or Eskimos, because they provided these people with meat, fat and bone, and so helped them to survive.*

*The **Harbor Seal**, with its shiny, dark coat covered with spots are known to follow fishing boats feeding on the scraps thrown overboard. Occasionally Harbor Seals will haul themselves out of the water on to the dock or even on to a boat. When hauled out, they often lie, with their heads and hind flippers elevated, in what is often referred to as the 'banana-like' position.*

Even though they are wary of humans they have a reputation of being relatively easy to tame.

DUGONG

Weight: 150-200 kg

Length: up to 3 m

Feeds on: algae and marine plants

The Dugong belongs to the family of Sirenidae. It is a close relation of the sea cow, from which the Dugong can be distinguished by the tail fin divided into two parts called lobes.
This gentle animal is a member of a group of mammals which has survived despite a threat to its survival by humans. In the past, the Dugong was much more common and could be found as far south as Japan. Unfortunately, despite the laws made to protect them, secret hunting is still carried out in some countries, because the meat of the Dugong is very delicate, its fat is valuable and its teeth are valuable ivory.
The adult Dugong weighs about 150–200 kg and can grow to a length of up to 3 m. It has a large, round head, connected to the trunk by a very short neck. Its eyes are small and without eyelids, but with a membrane which keeps out dirt. The nostrils are in the shape of a half moon, and they are mobile, able to open and close as the Dugong wants. Its mouth is very large but it has few teeth, with horny plates to chew its food. Its two upper incisors have been transformed into tusks which can reach a length of 20–25 cm, overlapping the lower gums by 5–7 cm.
The Dugong is strictly herbivore, eating marine plants or algae. This is why it always stays near the coast, in water no deeper than 20 m. After a meal, it stays on the sea bed, coming up every so often to the surface to breathe. It breathes every 2–4 minutes, but it can stay immersed up to a quarter of an hour. Its top speed is about 18 km/h.
Mostly silent, this peaceful animal makes whistles and grunts only when it is particularly excited. But at mating time, the males become restless and engage in violent fights. The female Dugong carries her young for 11–12 months. The young are born under water and brought up to the surface immediately to breathe. They take milk from the female for about a year. The oldest Dugongs can live up to 70 years. Dugongs mostly stay in one place, living in pairs or in small family groups, but in the past they also lived in larger groups.
The adult Dugong has practically no enemies except humans, but the young, less than 1.5 m long, can be attacked by sharks and so they never go out alone without an adult. Dugongs live in the tropical waters of Africa, Asia, Australia and in the Indian Ocean. Although it has been known from ancient times, the Dugong has only been studied and described since the nineteenth century.

DUGONG DUGON
DUGONG

BALAENOPTERA PHYSALUS
FIN WHALE

WHALE

Weight: 60 tonnes

Length: up to 25 m

Feeds on: plankton

Whales belong to the order of Cetacea, a group of mammals which, at a certain time during their evolution, quite suddenly adapted to enable them to return to the sea. For a long time it was thought that whales might be 'fish which can spurt' but although they do have dorsal and pectoral fins and enormous tails, they are warm-blooded like all other mammals. Also, they breathe with lungs and the female carries her young inside her body. The easiest way to distinguish the whale from fish is to look at the tail. The tail of a mammal in the Cetacea order is horizontal and moves from the top to the bottom – whereas the tail of a fish is upright and moves from side to side. Some species of whale are long and slender, others short and squat; some have enormous dorsal fins, others have no fins at all. Some whales have teeth, others fangs, some with plates arranged like combs attached to the jaw and some which have rigid threads which filter the water and trap food inside. Female whales give birth under water, near to the surface, one baby a time which generally hides at the tail. To begin with, the baby whale is barely able to be pushed to the surface by its mother or an 'assistant whale' to take its first breath.

The Blue Whale is the largest of whales. This well-known whale with its long, hydrodynamic body, dark grey or silver or blue-black can grow up to 26 m. Its head is coloured the same on both sides. The lower lip, the oral cavity and some flanks on the right side are white, whilst those on the left are completely grey. This feature probably dates back to the time when the Blue Whale developed the habit of swimming on its side whilst eating. Another species is the Fin Whale, sometimes called the Finback Whale, so-called because of the ridge on its back and up to 80 or more grooves along its throat and chest to its tummy button.

Communication by whistles, and the intervals between one and another, from 2 to 5, each 10 to 20 seconds, are characteristic of the Fin Whale. Although other whales can remain under water for longer, the Fin Whale can stay immersed only from 5–15 minutes and can reach a depth of 230 m. It can swim rapidly up to 30 km/h and sometimes it will jump completely out of the water.

These jumps and spectacular dives are typical of all whales, but the reason for them is still a mystery; they could be a ritual for courting, a way of signalling something, a strategy to confuse fish or to scare away parasites, a demonstration of strength, or simply for fun.

Whales live in seas throughout the world but they are most common in the temperate waters of the Northern Hemisphere – it is quite common to see whales swimming in the Mediterranean. Some communities migrate south in winter and towards the north in summer, although their moving about is not so predictable as with other Cetacea. At one time they were more numerous, but today many species are still endangered by hunting.

TURSIOPS TRUNCATUS
BOTTLE-NOSED DOLPHIN

DOLPHIN

Weight: 200-370 kg

Length: up to 4 m

Food: fish

With 32 species, the family of dolphins is the largest and most varied of Cetacea. Among these, the examples with prominent beaks (upper jaws) are grouped under the heading of 'dolphins'. Those which can be seen most easily in European seas are the Bottle-Nosed Dolphin and the Stenidae or Long-Snouted Dolphin. The Bottle-Nosed Dolphin, grey-blue with a white stomach, seems always to be smiling and happy, due to the structure of its mouth and the round wrinkles which surround it. Weighing about 200–370 kg and because it is more or less always hungry, it spends a lot of its time hunting in the water and eating from 11–33 kg of fish every day. It is a friendly mammal and lives in groups of 10–20 individuals. These are the creatures we see jumping, swimming in circles, diving and coming to the surface, without, it seems, ever stopping, when we are lucky enough to see them at a distance.

The particular structure of the dolphin's skin gives very little resistance to water, which means that it can swim at incredible speed, by thrusting its nose out in front and propelling itself along by the vertical movements of its tail. It rides the waves on the 'wash' produced by whales as if 'surfing' with its body, just stroking the surface of the water with its tail and jumping out, reaching heights of up to several metres. A dolphin immerses itself for 3 or 4 minutes and it can swim together with sharks and sea turtles or other Cetacea. Members of the group can help each other and even play with fishermen, but only particular male individuals venture out towards any bathers.

Dolphins are found in all 'closed' seas – the Black Sea, the Red Sea, the Mediterranean and the Gulf of California. It seems that some communities migrate, whilst those near the coasts stay in the same zone for years. The dolphin, which is particularly active and curious, is one of the noisiest animals in the ocean; members of the same group 'speak' continually with each other by way of a complex language of cries and whistles.

Thanks to its ability to make sounds and ultrasounds, and to 'read' the sounds of the waves which, as they rebound off objects, echo back to the dolphin and become picked up by the bones of the skull, the dolphin can get a good idea of its surroundings, even in the dark. Many zoologists believe that this animal, with its highly-developed brain, is one of the most intelligent, if not the most intelligent, after humans.

KILLER WHALE

Weight: 2.5-9 tonnes

Length: 7 m

Food: carnivores

Whales belong to the largest species in the family of Delphinidae. The long body, weighing around 2.5–9 tonnes, the enormous dorsal fin of the male – which in individuals can reach up to 1.8 m – and the easily recognized shining black skin, with white splashes above the eyes and its sides and grey behind the dorsal fin, all make whales such as the Killer Whale easy to identify. The main difference between the two sexes is size – the female is about 1 m shorter than the male, which measures on average 7.3 m. The pectoral fins have a typical rounded shape and grow with age.

It is possible to see individual Killer Whales, but a family community, called a 'school', a 'pod' or a 'herd' is more usual, with between 5 to 20 individuals. Two or more schools can unite temporarily to form larger groups, which can comprise up to 150 individuals. The members of a 'school' stay close all their life to a larger 'clan' developing their own means of communication. Whales communicate by high cries and ultrasounds which have the power to scare away all the other marine mammals, except for the largest Cetacea which prefer to keep their distance.

The whale is a carnivore predator with the most varied diet of all Cetacea. It will eat squid, fish, birds, sea turtles, seals and dolphins and even attack animals as large as the Blue Whale. Often a 'school' will work together during a hunt, ignoring all other potential predators. But however hungry they are, whales will not attack humans and aggression within a group is rare. Curious and charming, the whale can travel at a speed of 55 km/h and display all the habits typical of Cetacea – jumping, flapping its fins on the water and slowly emerging until just its head is dry, gradually going back underwater soon after. At times, it is possible to catch sight of a whale rubbing itself on the sand.

It is very difficult to keep a whale in captivity, because it needs to eat a minimum of 50 kg of fresh fish every day, although it can be tamed easily and is as intelligent as other members of its family. The whale lives in most seas but is most commonly seen in cold waters. It prefers deep oceans, but it can also be found within deep bays, in inland seas and in estuaries. It can make its way easily through floating ice packs in search of prey. The whale does not undertake long migrations but stays generally in one area.

ORCINUS ORCA
KILLER WHALE

ELEPHANT

LOXODONTA AFRICANA
AFRICAN ELEPHANT

The land-dwelling mammal which grows to the largest size is the elephant, a member of the order of Proboscidata. In addition to the two famous species, the Indian Elephant and the African Elephant, a third has recently been added, which is the African Forest Elephant. Although they may look similar, the Indian and the African Elephant have quite a few differences. Their common features include a huge head, small eyes and large ears. However, the most noticeable feature is the trunk, which is the result of the elephant's nose fusing with the upper lip. The trunk is very versatile and most useful to the elephant, both as an organ for grasping and for touch, as well as eating, carrying water into its mouth and giving itself a shower. The elephant has a fine sense of hearing and smell, but its eyesight is poor.

The growth of a baby elephant takes about 660 days. At one time, elephants in the wild were nomads, continuously wandering from place to place in order to find the large amounts of food that they need. Now, due to the occupation of vast territories by humans, they have been confined to smaller zones where they have become more stationary animals.

Weight:	5 tonnes (Indian Elephant) 7 tonnes (African Elephant)
Height:	3 m (Indian Elephant) 3-4 m (African Elephant)
Habitat:	jungle (Indian Elephant) Savannah (African Elephant)

The skeleton of an Indian Elephant has 19 pairs of ribs and 33 caudal (back) vertebrae, whilst the African Elephant has 21 pairs of ribs and 26 caudal vertebrae. In the Indian Elephant the skull is flat at the front, whereas it is very curved in the African Elephant. In the Indian Elephant the legs are shorter than the African Elephant, which can be up to 3.40 m and weigh up to 100 kg. The Indian Elephant is a typical inhabitant of the jungle, avoiding open spaces and the grassy savannah. In mountain areas, it can reach altitudes of up to 3600 m and it lives in herds of 15–30 individuals, always guided by an old female. They prefer to browse juicy grasses and tender plants which grow in the humid undergrowth.

The African Elephant is the largest land dwelling animal in the world. Only some Cetacea surpass it in weight and size. It weighs at least 7 tonnes, 2 tonnes more than the Indian Elephant, with a body length of about 7 m and a height of 3–4 m. The huge ears probably amplify the sounds by which it communicates within its dry surroundings, such as the savannah. It has a lively, somewhat restless nature, and sometimes its great strength and size makes a male elephant dangerous, especially when it is ready to mate.

In the past, there were many examples of the African Elephant. Now, after many years of being hunted and the spread of farmland, resulting in a reduction of its habitat, the population of the African Elephant has diminished. However, many examples survive in reserves in eastern Africa.

ELEPHAS MAXIMUS
INDIAN ELEPHANT

HIPPOPOTAMUS AMPHIBIUS

HIPPOPOTAMUS

Weight: 2-3 tonnes

Length: 4.5 m

Habitat: Africa

The Hippopotamus belongs to the order of Artiodactyls and is closely related to pigs and wild boar. Its name means 'river horse' and it is a colossal 2–3 tonnes in weight and 4.5 m in length. It lives only in Africa, south of the Sahara Desert. It is a funny-looking animal, especially when it moves about on land. Its heavy, cylindrical body has short, column-like legs which seem to support it by a miracle.

Although it is not a great swimmer, the Hippopotamus spends most of its time semi-submerged in water, where it seems to be completely at ease. It has colonized the island of Zanzibar, reaching this location by swimming there. It prefers water which is no deeper than 1 metre; it rolls in mud to protect itself from parasites – in fact, rolling itself in mud also enables the Hippopotamus to get rid of them without too much effort.

The Hippopotamus is very important to the environment of a river, because every time it moves, it 'ploughs' the slime and so renews the river bed. Its adaptation to life in the water and on land is mostly due to a valve in its nostrils so that these close completely when it goes underwater. Its tail is almost 50 cm long, whilst the head is enormous, like a horse which could have been drawn by an animal cartoonist. The eyes and the ears are small, but the mouth has an impressive opening, with powerful tusks which can measure up to 60 cm. Its skin is 4 cm thick and without hair. The reddish colouring of the Hippopotamus is due to a red oily substance secreted by glands on its skin and which protects the skin when the Hippopotamus cannot bathe for any length of time.

Towards eight years of age, at the end of the dry season, the Hippopotamus mates in water, and, also in water, the female gives birth after 7–8 months. It feeds on land during the cool hours of sunset, when it separates from its group and goes to graze always in the same place, leaving ridges as a result of its continuous passages. The female stays with her young to defend them from predators – crocodiles, lions and leopards. In case of danger, both the male and female Hippopotamus will gallop towards the water. But when their grazing extends on to cultivated areas, they become a real nuisance to farmers. Uncontrolled hunting and people taking over the territory of the Hippopotamus has resulted in their extinction in some regions,. Today, the Hippopotamus is confined to restricted zones such as national parks, but in time it is hoped that they will be present once again at the delta of the River Nile and in the Middle East.

WILD CAMEL

Weight: 500-600 kg

Food: herbivore

Life-span: 45 years

Habitat: Asiatic deserts

The Wild Camel was identified and distinguished from the domestic camel only at the end of the nineteenth century, in the Gobi Desert, where it still lives today. It has two humps on its back and these are deposits of fat, each hump surmounted by a crest of hair. Some zoos maintain that the Wild Camel is a primitive form of Dromedary, which has just one hump and this continues to be a source of argument. The Wild Camel is also called 'ship of the desert'. At 2 m tall at the shoulder and with a body length of 3 m, it has a tail 50 cm long which ends in a tuft. It can be recognized by a hard area of skin on the knees and some growth on the rear legs. It is a friendly animal and is active by day. Unlike the Dromedary which withstands very hot temperatures, the Wild Camel withstands very well the severe cold and windy climates of the vast Asiatic deserts thanks to its very thick coat which re-grows in winter and insulates the body. During any length of time in which it cannot find water, it can lose up to 30% of its body weight without any ill effect, due to the reserves of liquid which it accumulates in more than 50 water cells in its stomach. This liquid does not depend on the water drunk by the Wild Camel, but is extracted from the vegetable matter which it eats. However, when it comes to an oasis or a rare well in the desert, the Wild Camel can drink up to 140 litres of water.

VICUNA

Height from the withers: 80 cm

Length: 1.5 m

Food: herbivore

Life-span: 20 years

Habitat: Peru, Argentina, Chile

The Vicuna is the smallest South American Camelidae. It is a close relative of the Llama and the Guanaco, but it has adapted even better than them to life in the high mountains. Not only is it never seen below altitudes of about 3500 m, it is not afraid of climbing to over 5000 m. It is one of the few mammals capable of running and jumping in the rarefied atmosphere at such heights. Like all members of the family Camelidae, it lives in groups of 6–12 individuals, among which is one adult male, females with their young and females under one year of age. Although it is careful and rather fussy, it spends a good deal of the day searching for food, browsing among the semi-dry grass of the Puna, the name given to the prairies of the Andes Mountains. It is not afraid of severe cold and so it does migrate away from its territory.

Each group occupies a determined area and they only move around within this.

From the time of the ancient Incas, the Vicuna has been hunted for its fine, silky-soft wool. Despite this type of hunting not being very widespread, it has still brought this animal to the brink of extinction. Fortunately, strict measures have now been taken to protect the Vicuna and to save it.

Did you know...

*Recent studies have put forward a theory that the domestic llama and the Alpaca are descendents of the **Wild Guanaco** which, because it had been tamed and cross-bred by the ancient Incas, became modified so that two new types of animal developed. The Alpaca is often reared because it is a source of abundant precious wool.*

*The **Guanaco** often succeeds in easily out-distancing a rider on horseback who is chasing it. Many times, humans have tried to tame it, but without success and so it is ridden less than the Vicuna. The Guanaco is found more to the south of the Southern Hemisphere and prefers the low hills and the plains of Patagonia. It has long legs and a proud bearing; it lives in groups of various size, from 10 to 30 individuals, but, in the past, before the species became decimated by hunters, it was possible to see herds consisting of some hundreds of individuals in the Argentine pampas.*

*The **Vicuna** and the **Alpaca** can cross-breed, generating fertile individuals.*

The ***Dromedary*** *(in photograph) has long eyelashes to protect its eyes from the sand. It can withstand the heat to amazing levels and can go without drinking for up to five days. But, if the temperature is not excessive and it can graze on green grass, it can last 10–15 days without a drink. Now the Dromedary exists only in the domesticated state.*

The ***Llama*** *(not illustrated) lives in the Andes mountains. It plays a very important part in the survival of the South American Indians.*
They use the Llama as a beast of burden, as well as using its excrement for fuel. They also use its wool to make warm clothes, its hide for footwear and gloves and its fat for making candles.

MOOSE

Height at withers (shoulder): over 2 m

Weight: up to 800 kg

Food: herbivore

Life-span: 20 years

Habitat: humid forests, with lakes and expanses of mud

The Moose is the giant of the Cervidae family. It has an enormous head and an odd-shaped nose which gives it rather a curious appearance. It is a forest animal with a liking for humid areas with lots of lakes and expanses of mud which it can cross safely, thanks to the hooves which it can open wide. It is also a fine swimmer and does not hesitate to cross rivers and expanses of sea. It lives alone, but within small groups – although, in the depths of winter, it is possible to observe numerous Moose working together to dig among the snow to find a source of grass.

During the summer the Moose often grazes in ponds and small lakes, chewing the leaves of water-lilies and other aquatic plants. In winter, the Moose contents itself with worms and young woody shoots which it can reach, thanks to its remarkable height. No other member of the family of Cervidae has such impressive antlers as the male Moose, not only for their length, but also for their span. The female, on the other hand, has no antlers at all. The mating season is from September to the end of October. Mating is temporary and even the mother stays with her baby only until it can eat solid food. The Moose has few natural enemies, among which is the wolf which hunts in packs and so almost always succeeds in catching its prey.

REINDEER

Food: grass and especially lichen

Habitat: tundra

Habitat: Northern territories of Europe, Asia and America

The Reindeer which, in America is better known as the Caribou is a relation of goats and deer and belongs to the same family – but it is the only species in which the female has antlers the same as the male. The Reindeer has very long hooves and its two 'toes' can divide, enabling the animal to step safely, even on snow. As it walks, the Reindeer makes a strange noise which is like a rhythmic crackling. This sound is caused by the rubbing together of the two bones in each leg. The skin of its coat, very thick and warm, covers even the nostrils and the muzzle.

For the people who live in the same regions of the far north, the reindeer is a valuable animal. As well as using it as an animal for pulling sledges, the Lapps and the Inuits cure the skin, eat the meat, use the antlers to make tools and the milk of the female reindeer to make a traditional cheese. These nomad people have tamed reindeer, completing long seasonal migrations across the tundra.

Did you know...

The most recognizable feature of the family of Cervidae is that they have antlers made of bone. These are renewed every year, becoming more and more elaborate as they grow. The budding antlers of young reindeer are covered with a soft, velvety skin.

*The **Wapiti** is a huge American deer, related to the European Red Deer. In the language of the Shawnee Indians its name means 'white deer'. The Wapiti was once very numerous throughout North America, from Canada to the Mexican border, and tens of millions of individuals could be counted, until white settlers began killing them for their skins. As a result, they are still very rare and survive only in some protected areas. However, they can still be found in the Rocky Mountains, in northern Canada and in reserves in the United States.*

*The **Fallow Deer** and **Roe Deer** (not illustrated above) also belong to the family Cervidae. The Fallow Deer has a yellowish-red coat spotted with white in summer.*
The antlers of the Fallow Deer are flatter than those of other deer and paddle-shaped. The Roe Deer is very small and weighs only 25 kg.

*The **Sambar Deer** is considered to be the largest, the most beautiful and the most majestic of all Asiatic deer.*
*There are many species all varying in size, but all living in forests areas, either alone or in small groups. The most common is the **Indian Sambar** which may be found in India and throughout South-East Asia, and is recognized by a thick ruff of fur around its neck. It weighs up to 300 kg, whereas the **Malayan Sambar** is the smallest species, weighing only 50 kg. The Malayan Sambar can be found as far south as Japan and on islands nearby.*

GIRAFFE

Height at withers (shoulders): 3 m

Height at head: 6 m

Weight: 250 kg

The first giraffes appeared on Earth more than 25 million years ago. The fossilized skulls and bone fragments of giraffes which lived in prehistoric times have enabled experts to be quite certain that these were the ancestors of the giraffes we know today. The prehistoric giraffe did not have such a long neck, and their territory was far more vast, from the south of the Sahara Desert to southern Africa.

Like all mammals with spotted coats, the various species of giraffe display a great variation in colourings and patterns due to cross-breeding. The giraffe is the only mammal which cannot learn to swim, and so any expanses of water are a natural barrier.

Bearing this in mind, it is not surprising to find giraffe communities isolated one from the other, and often with even more different patterns and colourings.

The Chequered Giraffe is the most beautiful among the many species which exist. It is 3 m high at the withers, and, with its very long neck, measures up to 6 m from the top of its head. Add to this, its long legs and the withers so much higher than its pelvis, and this giraffe is unmistakeable. The head, with the long, pointed nose, has two horns rather like those of a camel. The tail is long and terminates in a tuft of hair. Its eyes are at a considerable height, which means that eyesight is the most important sense for the giraffe. A giraffe likes to eat the leaves of the Acacia tree, picking these directly from the branches at 2–6 m high, and by grasping them with its prehensile tongue.

In order to lower its head close enough to the ground to drink, the giraffe can stretch its rear legs very wide apart. Because of its height, the circulation of blood around the giraffe's body is like no other animal, with its heart and network of arteries and veins specially adapted so that the blood can reach its head when this is held high, and yet not rush to the head when this is lowered down.

The habitat of the giraffe is the savannah with high trees, which can be quite sparse. The male lives alone or in small groups. Females live in groups of females and young or groups of young.

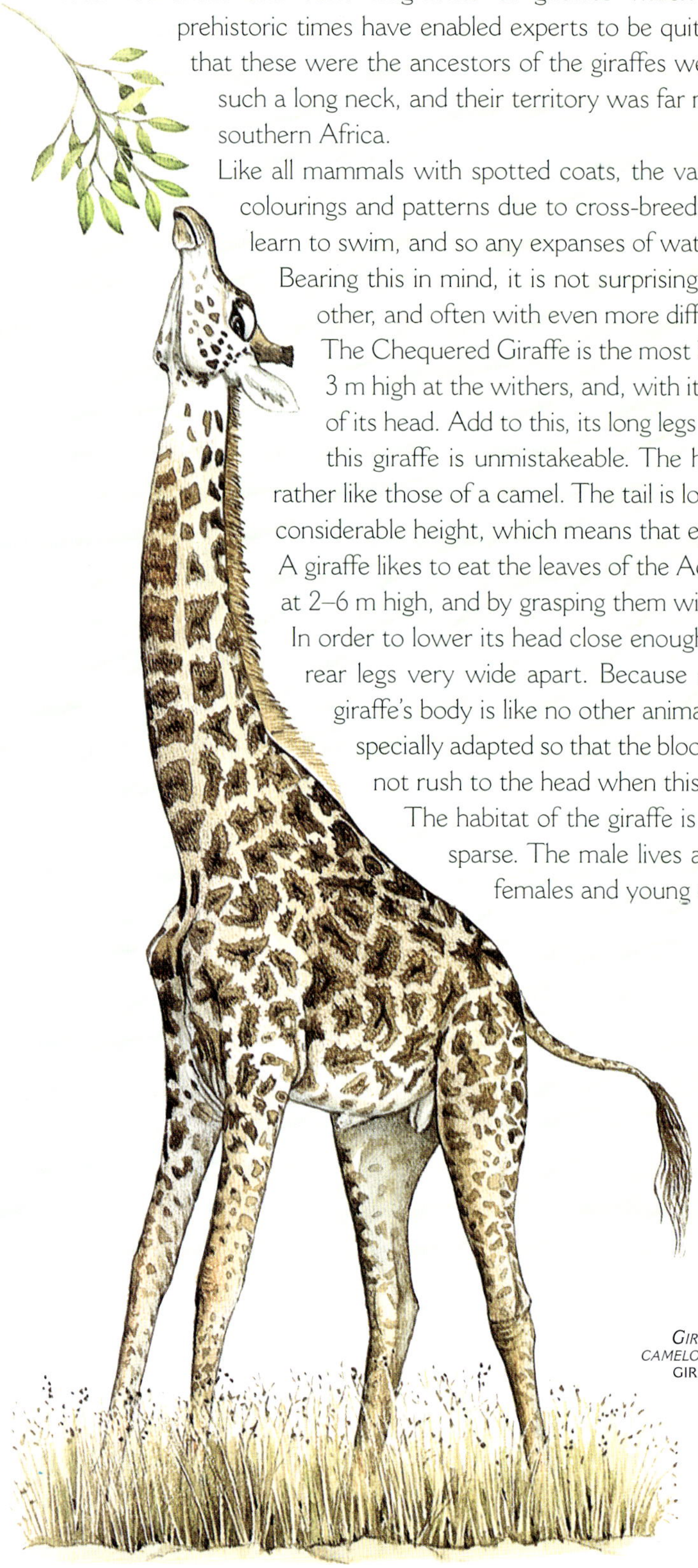

GIRAFFA CAMELOPARDALIS
GIRAFFE

ANTILOCAPRA AMERICANA
PRONGHORN ANTELOPE

PRONGHORN ANTELOPE

The Pronghorn Antelope is the only surviving species of a family which was scattered across northern America 20 million years ago and comprised many species. Before the arrival of the European settlers, who began to hunt them and to farm the land which had previously been their pastures, the Pronghorn Antelope was still abundant and their population was counted in millions. But by 1923, there were only 30,000. Today, strict measures taken to protect them have brought about an increase and the total population now totals around a half a million.

The male Pronghorn Antelope has a particular type of antler which provides a true link between actual horns and the antlers of Cervidae. Its antlers are divided into two parts and are made of a horny substance, which is renewed each year. Today, the Pronghorn Antelope may be found only in the western United States and, although it can live quite happily up to an altitude of 2000 m, it prefers the borders of the prairies, dry and semi-desert. It is an animal which likes company and which lives in groups of various sizes, although sometimes the older males prefer to live alone.

Weight: about 50-70 kg

Length: 1.40 m

Height at withers (shoulder): about 1 m

Average speed: 60 km/h

The Pronghorn Antelope is very timid yet very curious. In the past, settlers attracted their attention by waving handkerchieves tied at the tops of long sticks, and any unwary antelope usually fell into a man-made trap.

The Pronghorn Antelope is the fastest antelope in America. It can gallop even over long distances and reach speeds of up to 80 km/h, as well as covering a length of 6 m in one jump. It eats all vegetable matter which grows on the plain, but needs very little water and can withstand drought for some weeks without drinking.

The Pronghorn Antelope is ready to mate at an age from 15 to 24 months. The female carries her young inside her body from 230–240 days, giving birth to just one baby, rarely two or three,

During the mating season, the males are territorial and the females are isolated afterwards. When the mating season is over, they return to their groups.

YAK

Weight: around 1000 kg

Height at withers (shoulder): more than 2 m

Length of body: 3 m

Number of young: 1

The Wild Yak is a bovine (family of oxen and cows) which has adapted perfectly to life in the high mountains, where the air is thinnest and the most rarefied. It lives in the desert steppes of northern Tibet in groups of 20 to 300 animals, composed mainly of females and their young. The males live alone or grouped together, separated from the rest of the Yaks. Because of the severe climate in which it lives, the Yak has to undertake seasonal migrations, moving to the high plains of up to 5000 m altitude during the summer, when the grass is green, and coming down into the valleys in autumn in search of pastures which are less covered in snow. Thanks to its thick and long-haired coat, which moults at the beginning of June, these animals can withstand temperatures less than -40°C – just like the Musk Ox of the Arctic Tundra. The mating season begins in September and lasts about two months. During this time, the adult males mate with females, after having seen off their rivals. After about nine months, in June, the female gives birth to one baby, which she raises within a year.

The Tibetans and the inhabitants of the western regions of China use the Yak for domestic purposes, by taking the youngest one they can find from the wild. A tamed Yak will then give milk and male and female Yaks become used as working animal for pulling carts and wagons.

AMERICAN BISON

Weight: 450-1350 kg

Height at withers (shoulder): 1.60-1.80 m

Length: 2-3.50 m

Number of young: 1

Life-span: 33 years

The American Bison is a wild bovine, a close relative of the European Bison. It is the most powerful mammal living in the great, grassy plains of North America. Bison are herbivore, eating leaves, shoots and tubers during summer, lichen, moss and dry grass during winter.

During the mating season, males and females gather together in very large groups and the males fight furiously among themselves. The American Bison is a dangerous animal. It will attack at great speed, soon identifying its rival and being led towards its prey by an acute sense of smell and hearing. In the past, the American Bison was much valued by the North American Indians. These people followed the seasonal moves of the herds because they were their means of sustenance, but killing only the number of bison which they needed to survive.

The arrival of European settlers and the progress of the railways towards the west, brought about a massive disappearance. Only a few hundred American Bison survived and these were brought to reserves for reproduction. Today, some ten thousand American Bison live in protected areas.

Did you know...

__Bovines__ constitute a large family in the order of Artiodactyls and include all species of goat, sheep, chamois and oxen. Bovines are herbivore animals with horns and of varying size, from a tiny goat to a giant buffalo. They can be found all over the world, but there are none in their wild state in Australia and southern America.

The __African Buffalo__ is truly colossal, without fear of any other animal and considered by hunters the most dangerous prey in the whole of Africa. Despite its mythical strength and its aggression, at heart the African Buffalo likes the quiet life and wants to be left in peace. It lives in herds, each one in the charge of an adult female, in the savannah near to running water, because it needs to drink once or twice every day.

The __Gayal__ is a tamed species of wild cattle which is descended from another species, the Gaur. The Gayal can rarely be kept in cattle sheds, because in the wild it lives in groups in the jungle. If it ventures into nearby villages, it will return to the jungle each evening to

renew the supply of natural salt which its body needs.

The **Banteng** is a large-size bovine with a slender body, living in various regions of South-East Asia. It is considered to be the most beautiful of all wild bovines. The male is very different to the female. The male has a dark coat with showy white 'stockings' and pointed horns curving inward. The female is a beautiful red-brown colour and her antlers are less developed. The Banteng has a shy, timid nature and lives in groups, preferring the thickest parts of the jungle and gathering together in the forest clearings only towards the evening.

The **Anoa** is one of the most primitive of bovines and is found only on the island of Celebes. It is a small size buffalo, only 1.50 m long and with a weight of about 300 kg.
The Anoa leads a very secluded life, hidden in the thickest depths of the forests. Despite its small size, it is a very aggressive, dangerous animal, because of its pointed horns with which it is always ready to fight an enemy. Its diet includes grasses, ferns, saplings, palms, ginger and fallen fruit.

Did you know...

The ***Sable Antelope*** *has a shining black coat which contrasts to the white of its stomach. It lives in the tree-lined savannah of South-East Africa. It needs to drink a lot of water, unlike other antelopes which can withstand thirst very well.*

The ***Addax*** *is from the desert regions of North Africa. It is one of the antelopes which are the most wild and difficult to catch, because its large hooves enable it to run at great speed across the sandy ground.*

The ***Bontebok*** *is a very elegant African antelope. Only a small number survive in special reserves.*

The ***Lichenstein's Hartebeest*** *prefers the forests and the bushy zones of South-East Africa.*

The ***Gerenuk****, or* ***Giraffe Antelope****, is one of the most unusual of African bovines, mostly because it can stand erect on its rear legs with its body vertical and its long neck quite straight. It finds this position very useful when it is tearing leaves and twigs from low-growing trees.*

The ***White-Tailed Gnu*** *is a rare African animal, with quite a comical appearance. It has a thick beard and tufts of short, erect hair on its nose and between its eyes.*

The ***Arabian Oryx*** *is a large antelope recognized mostly by its very pale coat and perfectly suited to the severe conditions of the Arabian and*

African deserts. It is a wanderer, continuously going in search of food. It lives in small groups in the charge of an old female.

The ***Mountain Gazelle*** *was once common from Arabia to India. Today only a very few examples survive.*

The ***Dama Gazelle*** *lives mainly in the dry zones at the far south of the Sahara Desert and is one of the most beautiful gazelles. It trots and runs lifting its front legs high and keeping its head erect, which is why it can often be mistaken for a little horse from a distance.*

The ***Dibatag*** *or* ***Clarke's Gazelle*** *lives in the open territories of Somalia. Its legs and long, thin neck makes it similar in appearance to the Gerenuk. But it is smaller in size and its horns curve towards the front in the shape of a scythe.*

The ***Springbok*** *(in photograph) is the national emblem of South Africa. It gets its name because of the many jumps which it makes very quickly and in succession, reaching heights of 3 m.*

The ***Mongolian Gazelle*** *belongs to a family which can be found in central Asia. The horns are present only in the males. During the mating season, the males also develop a bulge on the throat, similar to the crop of a cockerel.*

ZEBRA

Weight: 300 kg

Height at withers (shoulder): up to 150 cm

Lives in: Africa

The Zebra belongs to the order of Perissodatyla, which comprises herbivore medium and large-size animals, distinguished by feet with an odd number of toes which are protected by a strong nail. In equines such as the horse and the zebra, only the third toe on each foot is developed as a hoof. Zebras are elegant animals living in the vast, open spaces, where they run rapidly at the trot and at the gallop, and live in groups. There are three different species of zebra – Grevy's Zebra, the Mountain Zebra and Burchell's Zebra. Grevy's Zebra is the largest, 140–160 cm at the withers. It lives alone in the scrublands of eastern Africa It has a large head, a very long nose, long ears which are large and round and its stripes are more narrow than the other species, except for its abdomen which is white. Its cry is similar to the braying of a donkey. Grevy's Zebras gather together in large mixed groups only at the time of their seasonal migration to find grasses to eat. At other times, the females usually keep apart from the males. Most often, these stay in groups of 6–30 and the males can be very aggressive towards each other, especially at mating time.

The Mountain Zebra lives in Namibia and some parts of South Africa. This is the smallest zebra, only 118–130 cm high at the withers. It has pointed ears and its stripes are widest on its thighs, more narrow and crosswise on its back. It is a fine climber, most suited to life in mountainous desert regions. Today, both the Mountain Zebra and Grevy's Zebra are endangered species, which means that they are threatened with extinction. Only a very few examples of the Mountain Zebra live in protected zones in South Africa.

There are higher numbers of Burchell's Zebra, which live in the grasslands across eastern and much of southern Africa – hence the reason why this Zebra is sometimes called the 'Zebra of the Steppes'. Burchell's Zebra is recognized by its shorter ears, the long tuft of hair at the end of its tail and the broad stripes across its back which gather into a central dark stripe. It lives in small family groups, with one male at the head and several females with their foals.

EQUUS QUAGGA
BURCHELL'S ZEBRA or
ZEBRA OF THE STEPPES

TAPIRUS INDICUS
MALAYAN TAPIR

TAPIR

Weight: up to 300 kg

Height at withers (shoulder): up to 120 cm

Number of young: 1

The Tapir belongs to the family Tapiridae, the least developed family of Perissodatyla. It is considered a 'living fossil' a survivor of a geological age many thousands of years ago, not only because Tapirs today are so much like those of 35–40 million years ago, but also because the areas where they live are very distant one from another. In America, there is Baird's Tapir (also called the Central American Tapir) which may be found from Mexico to Central America. From South America comes the Mountain Tapir which is smaller and lives at altitudes between 2000 m to over 4000 m in the Andes Mountains. Most widespread is the South American Lowland Tapir, from tropical America.

There are three Asian tapirs. The one most similar to the American tapirs is the Malayan Tapir. This is larger than its American relations, and is usually identified by its contrasting coat – black head, shoulders and legs, and white back and abdomen. Like all tapirs, it lives in the tropical and equatorial forest. To look at, it is something of a mixture between a pig, a horse and an elephant. Its strong, squat body moves in a zig-zag fashion when it walks slowly, whilst if it is fleeing, it goes into a fast gallop, hurling itself into thick vegetation.

The tapir has a long nose, mobile and retractable, like an elephant's trunk. But, unlike the elephant, the tapir does not use it for grasping and holding, but for touch and for smelling.

Tapirs prefer to live near flowing water and they are fine swimmers. The Malayan Tapir can also stay for long periods under water, walking on the sea bed. Another feature of tapirs, apart from the Mountain Tapir, is the very strong skin.

The food of the tapir is mainly leaves, shoots, small, tender twigs, but also fruit, grass and water plants. As it moves, the tapir always crosses the same paths and so 'builds' well-marked paths and tunnels through the thick vegetation. Tapirs live alone and they are not very friendly towards one another, except for the mating season. The female gives birth to just one baby at a time. When this is born, it is covered with stripes and little spots.

Did you know...

The family of Rhinocerotidae comprise animals of enormous size which usually have skin without hair and which are most easily recognized by a horn or horns on its muzzle. Its appearance is very heavy. The body is massive, supported by short legs terminating in three toes. The head is huge, with small eyes and wide, paddle-shaped outer ears.

The ***Sumatran Rhinoceros*** *is the only rhinoceros which is partly covered with hair. It is sometimes called the 'hairy rhinoceros' and seems to be descended from an extinct hairy rhinoceros found preserved in the ice of Siberia. It is the smallest species of rhinoceros with a height to the withers which reaches only 1.50 m and an average weight of 600–700 kg. There may be still more examples to be found in Borneo, Myanmar (previously Burma), Thailand, Sumatra and in the Malayan peninsular, but it is difficult to observe them because they hide in the depths of the forests.*

The horn of a ***rhinoceros*** *is made*

INDIAN RHINOCEROS

Weight: over 2000 kg

Length: 2-4 m

Height at withers (shoulder): 2 m

Life-span: about 47 years

Number of young: 1

The Indian Rhinoceros, also known as the Plated Rhinoceros is a survivor from the prehistoric era and is one of the largest mammals in the world. Its numbers have been reduced to about one thousand, because it was hunted for a long time for its one horn, which was thought to be a valuable tonic. The Indian Rhinoceros has a skin without any hair, covered by a sort of armour plating made of plates of very thick and rigid skin, jointed between each other by areas of thick, softer skin, which enable the plates to move. The Indian Rhinoceros usually lives alone or with one partner at the most and controls a vast territory. This is criss-crossed in all directions by paths battened down by the animal continually coming and going, and these passages sometimes forming types of tunnels through the vegetation. It lives among tall grasses in swampy zones, avoiding the thickest parts of the forest and it also likes to stay immersed in water. The Indian Rhinoceros is a herbivore, eating grass and young shoots which it takes to its mouth using its finger-like appendage on its upper lip. It is active by day and night, but when it senses the presence of a human being, it prefers to move about at night.

WHITE RHINOCEROS

Weight: 3 tonnes

Length of body: 3.50-5 m

Height at withers (shoulder): 1.60-2 m

Life-span: about 40 years

Number of young: 1

The African White Rhinoceros is the largest land-dwelling mammal, after the elephant. It lives in the African savannah. Its muzzle is long and flat and it has two horns, of which the front is the longer and more pointed. It has a highly developed sense of smell, whilst its eyesight and hearing are not so good. It is a sociable animal which likes company and lives in groups of up to 18 individuals, mostly females accompanied by their young. These stay with the mother up to three years old, and then go to find a mate to reproduce.
The day of the White Rhinoceros is influenced very much by weather conditions. If it is very hot, then it rests in shady areas, going out to browse in the evening. In the same way, it rests when there is rain and the temperature lowers. It spends a great deal of its time searching for grass to eat. Like all members of its family, it loves to stay immersed in mud. But, despite their huge size, rhinoceroses can move at a speed of about 45 km/h, even through thick undergrowth.

up keratin, a substance which is also found in its hair. Each horn rests on the hard ridge of the nasal bone. Sometimes the horns wear out, which is when they can often appear to be covered with hair.

The ***Black Rhinoceros*** *is the only species of rhinoceros which has a reasonable population. It is also the most widespread of the rhinoceros family, living in central-southern Africa and the largest populations are in Tanzania, Mozambique and South Africa.*
The habitat of the Black Rhinoceros is partly savannah, partly steppes and it can adapt to altitudes of up to 2000 m.
It has two horns, the first of which can reach a length of more than 50 cm. It is not very tall, about 1.50 m at the withers, whilst its weight can reach 2 tonnes. Its upper lip is particularly long and mobile and the Black Rhinoceros uses this as a sort of grasping finger and to carry leaves and the twigs of bushes and low-growing trees to its mouth.
A rhinoceros has poor sight, but well-developed hearing and a sense of smell as acute as that of a dog.

AYE-AYE

Length of body: 40 cm; with tail, 100 cm

Food: larvae, insects, dry and fresh fruit

Number of young: 1

Habitat: forest

Home: Madagascar

The Aye-Aye is a strange primate. It leads a nocturnal and private life in the remaining strips of forest in Madagascar, living in the holes of tree-trunks which it likes to keep tidy and comfortable for itself. Sometimes it makes a proper nest at the fork of a branch, with twigs and leaves. Since this mammal was first discovered, it has been endlessly discussed and argued about. At first, zoologists believed that the Aye-Aye was a species of squirrel, because its four scalpel-like incisor teeth are similar to those of a rodent. Also, its hands are very strange. The third finger is very thin and very long, compared to the other fingers, and it also has a solid claw which the Aye-Aye uses to extract larvae from the bark of trees. It eats mainly larvae and insects, which it finds simply by putting its head against a tree-trunk. As soon as it knows where its prey is, it quickly chews at the bark with its teeth and digs into the hole it has made with the claw of its third finger.

VERREAUX'S SIFAKA

Length: 95-150 cm including tail

Food: leaves, fruit and shoots

Number of young: 1

Habitat: forest

Home: Madagascar

Verreaux's Sifaka is a Prosimian, a member of a lower order of Primates. It lives only in Madagascar and two species are known – Verreaux's Sifaka and the Diademed Sifaka and each has developed differently according to the surroundings in which it lives - Verreaux's Sifaka lives in hot and humid zones and has a dark coat. The Diademed Sifaka lives in the dry, cold regions of Madagascar, and its coat is pale. It is a friendly animal which lives in groups of 6–10 individuals made up of two families. Like all daytime primates, the Sifaka is very fond of the Sun, and loses no opportunity to bask in the Sun on the branches of trees. It is only at these times that the Sifaka is still, usually chattering to itself. During the times when Sifakas are more active, they talk non-stop among themselves, making a sort of gurgling noise and short sounds. Thanks to a membrane which connects the front limbs to the bottom of its neck and at the trunk, the Sifaka can glide in the air from one branch to another. To gather up ripe fruit, it will also descend to the ground, where it moves by jumping on both feet and extending its long arms to keep its balance. Unlike other Prosimians, it rarely uses its hands to feed itself.

Did you know...

The term 'Primate' was coined by the famous naturalist Carolus Linnaeus as a sign of the Primates' superiority over all other animals. There are two orders of Primates - Prosimians, the lower class, and Anthropoid (man-like), the upper class.

***Prosimians** are the most ancient group of Primates. The transformation of legs and joints into flat claws have enabled them to perfect their grip and to become the dominant animals of all tree-dwellers. Almost all are nocturnal, with large eyes able to distinguish the minimum variations of intense daylight.*

*The **Weasel Lemur** is from Madagascar. The big toe of the hind leg has a long, curved nail which enables it to climb. It is covered by a woolly coat which is very thick and it has a short, pointed nose. Its body reaches a length of only 60 cm including the tail. Small and nervous, it is not too well-known because it lives by night.*

*Other lemurs are active by day. Among these is the **Black Lemur**, black in the male and red-brown in the female, and the **Ring-Tailed Lemur** with its tail ringed with black and white. These also live in Madagascar.*

*The **Lesser Mouse Lemur** and the **Fork-Marked Lemur** live by night and eat mainly insects. In the driest periods, which are also the most severe for them, they go into hibernation and survive by using up the fat which they have accumulated on their hind legs and in their tail.*

PHANER FURCIFER
FORK-MARKED LEMUR
LEPILEMUR MUSTELINUS
WEASEL LEMUR
PROPITHECUS
VERREAUXI
VERREAUX'S SIFAKA
LEMUR CATTA
RING-TAILED
LEMUR
MICROCEBUS
MURINUS
LESSER MOUSE
LEMUR
LEMUR MACACO
BLACK LEMUR
DAUBENTONIA
MADAGASCARIENSIS
AYE-AYE

Did you know...

*Various species of **macaque** monkeys belong to the family Cercopithecidae, including the **Pig-Tailed Macaque** with its rather forbidding appearance. This monkey is found mostly in south-east Asia and on the islands of Sumatra and Borneo.*

*The **Crab-Eating Macaque** lives in forests, near rivers and on the coasts of the Sonda Islands, Malaya, Myanmar (formerly Burma) and southern Indo-China.*

*The **Celebes Crested Macaque** (also known as the **Celebes Black Ape**) is recognized by a vertical tuft of hair on its head. It lives on the island of Celebes.*

*The **Drill** lives in New Guinea. It spends almost all its life on the ground.*

*The agile **Long-Tailed Macaque** is regarded as the most handsome of all macaques. It lives in southern India and on the island of Sri Lanka.*

***Baboons** and **Mandrills** are the largest macaques and these are distinguished by their powerful teeth with small tusks in place of normal canine teeth. Both live in the sub-Sahara area of Africa and in southern Arabia. Baboons and mandrills live in small communities – baboons in groups of males, females and young baboons, mandrills in a 'harem' of females*

with one or two males.

*The **Gelada** is found only in the mountains of Ethiopia and Eritrea.*

Monkeys belonging to the Cercopithecidae family are the most multi-coloured of all. They are sometimes called 'Old World Monkeys' because they come from Africa. There are about 100 species and almost all of them live in trees, within tribes of about ten individuals under the control of an elderly male.

*The **Grivet Guenon** is the species which is the most widely scattered.*

*Like the **Spot-Nosed Guenon**, the **Diana Monkey** lives in the rainforests of western Africa. This monkey is known for the bright colours of its coat. The male Diana Monkey has a white 'beard' and whiskers.*

*The **DeBrazza's Monkey** lives in Uganda, in the Congo and in Kenya.*

*The **Patas Monkey** is among the fastest-moving monkeys known. It lives in the dry, arid belt of land south of the Sahara Desert and spends its life mainly on the ground.*

*Like most other members of the Cercopithecidae family, the **Black Mangabey** has long limbs and a long tail and lives in the forests of equatorial Africa.*

CHIMPANZEE

Height: 94-170 cm

Weight of adult: 40-80 kg

Arm span: up to 2 m

Habitat: humid forest, woodland and plains

Home: Equatorial Africa

Of all animals on Earth, the Chimpanzee is the closest relative of human beings. It has an intelligence which is similar to ours. This enables them to use various objects very well and also to display their ability to work out the reason for their actions. For example – to catch ants, a Chimpanzee will dampen a little stick with saliva and poke it into the ant hill, so that the ants will stick to the moist surface. It will throw stones and sticks at enemies to make them go away, clean its skin with the peel of a banana, and collect water to quench its thirst by laying leaves out in the rain.

Although some Chimpanzee communities have adapted to live in the savannah, they generally live in forest areas. According to some zoologists, the species living in the savannah are more developed.

Unlike the gorilla, the Chimpanzee, an anthropoid primate, is tree-dwelling and descends to the ground only when necessary. However, it runs easily on its two feet for long distances. Its social life is very developed – the family comprises an adult male, two or three females and their young. Sometimes, other families join together to make quite a tribe. In general, Chimpanzees are vegetarian, but sometimes the adults catch medium-size mammals. Every evening, chimpanzees prepare on the branches a bed of leaves on which to spend the night.

GORILLA

Height: 1.80 m, male; 1.30 m, female

Weight: 200-270 kg, male; 80-100 kg, female

Food: leaves, seeds, fruit,
bamboo shoots and young bunches of bananas

Span of arms: 2-2.75 m

Habitat: forest

Home: Africa

Despite its proverbial strength, and being a true giant among primates, the gorilla is a mainly peaceful animal. It lives in the African jungle and lives quite a secluded life. Three species of gorilla are known – the Western Lowland Gorilla which lives in the coastal zones of the Gulf of Guinea and in the basin of Zaire and in Gabon, the Eastern Lowland Gorilla whose home is the rainforests of the region of the Congo and the Mountain Gorilla which lives among the highest mountains of central Africa up to altitudes of 4000 m, in forests of bamboo. Here, the topmost heights are always wrapped in mist. It may also live in zones bordered by high volcanoes.

Gorillas live in small groups, each one comprising one adult male, two or three females and their young. The male is the undisputed head of the group, who ensures the group's safety and defends their territory. The group does not move about very much, because gorillas can usually get all the food that they need from the jungle all year round. The gorilla is land-dwelling and only the young climb trees and play among the branches. As the male gets old, the hairs on its back become white, rather like older human beings who often become grey-haired.

Did you know...

*The **Gibbon** moves only by swinging along and dangling on its long arms, balancing in free flight for up to 15 – 20 m. This type of movement, swinging from one arm to another, is called 'brachiation' and the animal can reach a surprising speed in this way. Sometimes, however, a gibbon acrobat may fall to the ground – the skeleton of a gibbon often shows fractures to its arms. The front legs are shorter and are useful when the animal wants to rest on branches or to pick fruit.*

***Pan Paniscus** – the **Pygmy Chimpanzee** – also lives in Africa.*

*The name **Orang-Utan** comes from the Malayan **orang-utan** meaning 'man of the woods'. The most surprising feature of this primate is the fact that the different examples all have their own facial features, and, like human beings they can be recognized and distinguished one from the other by the expression on the face of each one. The males also*

have beards and whiskers, similar to the anthropoid primates to which they are related.

Unlike the Orang-Utang, which lives alone, the ***Siamang*** *lives in a family group. But, like the Orang-Utang, it has a throat sac which swells up as it cries and works like a sound-box. The cry of the Orang-Utang would be very soft if it were not for its throat sac to make it louder.*

The young female and male ***Mountain Gorilla*** *(in photograph) sleep alone at a certain height from the ground on large branches. The heavy adult males sleep on the ground, on a bed of ferns, grass and twigs.*

GLOSSARY

appendages – an appendage is usually a leg. It can, however, be anything which projects from the central part of the body of an animal, such as a leg, a wing or an antenna.

arachnida – a class of creatures which includes scorpions and spiders. An arachnid has a segmented body divided into two parts. The front part has four pairs of legs, but no antennae.

chelicerae – pincer-like appendages nearest the mouth of arachnids and some crustaceans. The creature uses them to bite its prey and to inject poison into its victim. Spiders also use their chelicerae for spinning silk. Crabs use them to grip their prey.

class – a broad category of animals which have some features in common. Many orders may make up one class.

crustacean – a class of mostly aquatic (water-dwelling) invertebrate animals with a horny or bony outer shell, a pair of appendages on each segment of the body and two pairs of antennae.

evolution – slow process of change, usually over many thousands or millions of years.

family – a group of animals which are closely related. There may be many genuses within a family, and many families within an order.

genus – one or more related species which form part of a family of animals.

habitat – a creature's natural home.

hermaphrodite – an animal with the organs of both sexes, male and female.

invertebrate – without a backbone.

metamorphosis – a complete transformation within the life-cycle of an animal, for example, a larva being transformed into a butterfly.

complete metamorphosis – usually comprises three stages of transformation, e.g. egg, larva, full-grown animal.

incomplete metamorphosis – comprises only two stages, with the egg hatching into an adult animal without going through the larva stage.

mollusc – a creature with a soft, compact body without an internal skeleton and which is sub-divided into a head and a trunk.

nocturnal – living by night and sleeping by day.

order – one section of a class, comprising animals which are quite closely related.

oviparous – egg-laying.

ovoviviparous – where fertilized eggs develop within the female body, but, unlike mammals, these are not nourished by her.

parasite – an animal which lives off another.

pedipalps – second pair of appendages in arachnids. Sometimes these are used for chewing, sometimes as huge, powerful pincers and for gripping, as organs of reproduction and for tasting.

phylum – major group of animals to which many classes may belong.

predator – a hunter of prey.

scavengers – animals which feed on the remains left by others.

sloughing – shedding a dead skin, which has usually become too tight for the animal.

species – a special kind of animal. One or more species can make a genus.

spermatophore – a capsule containing fertile sperm and which some male arachnids deposit on the ground ready for the female to lay her eggs.

symbiosis – where two different species exist together, each one helping the other.

vertebrate – with a backbone.

viviparous – where the young grow inside the body of the female, are nourished by her and born fully developed.